Methods of
ANALYTICAL DYNAMICS

Leonard Meirovitch

Department of Engineering Science and Mechanics
Virginia Polytechnic Institute and State University

DOVER PUBLICATIONS, INC.
Mineola, New York

Copyright

Copyright © 1970, 1998 by Leonard Meirovitch
All rights reserved.

Bibliographical Note

This Dover edition, first published in 2003, is an unabridged republication of the work originally published by McGraw-Hill Book Company, New York, in 1970.

Library of Congress Cataloging-in-Publication Data

Meirovitch, Leonard.
    Methods of analytical dynamics / Leonard Meirovitch.
       p. cm.
    Originally published: New York : McGraw-Hill, 1970, in series: Advanced engineering series.
    ISBN-13: 978-0-486-43239-7 (pbk.)
    ISBN-10: 0-486-43239-4 (pbk.)
    1. Dynamics. I. Title.

QA846.M45 2003
620.1'04—dc22

2003060379

Manufactured in the United States by Courier Corporation
43239404
www.doverpublications.com

To the memory of my dear mother,
ADELLE MEIROVITCH

Preface

For many years the emphasis in the study of mechanics has been on formalism and structure culminating with the developments of Lagrangian and Hamiltonian mechanics. Solutions to problems of dynamics have not kept pace with the formulations, primarily because of inadequate methods for treating nonlinear phenomena. Although there have been certain solutions, such as the classical Lagrange's solution of the restricted three-body problem, no general method for the direct integration of the differential equations of motion exists. The introduction of generalized coordinates by Lagrange took dynamics from a physical world to a more abstract analytical realm and at the same time opened the gates to a new approach to the integration problem, namely, the indirect approach based on coordinate transformations. The method consists of uncovering first integrals of motion by means of a coordinate transformation producing ignorable coordinates. Hamilton gave the idea new impetus and, replacing the set of n second-order Lagrangian equations by $2n$ first-order Hamiltonian equations, broadened the concept of transformations by regarding the generalized coordinates and momenta as independent canonical variables. The transformations, known as contact transformations, opened new possibilities for producing motion integrals by the indirect approach. It was left for Jacobi to follow through on Hamilton's idea and develop a workable transformation theory.

In the last century, the interest in solutions to nonlinear problems of mechanics led to substantial progress in this field. These developments have come to be known as nonlinear mechanics, and the methods of solutions can be divided into two broad classes, analytical and geometrical. The analytical approach was prompted by problems in celestial mechanics. This is the well-known perturbation theory, which consists in seeking an analytical solution valid in the neighborhood of a known solution. The solution is in the form of a power series in a small parameter, where the parameter reflects a perturbative effect, and it reduces to the known solution in the absence of the perturbation. The second approach is concerned with the qualitative behavior of the solution in the phase space, namely, the $2n$-dimensional space defined by Hamilton's canonical variables, and makes no attempt to obtain an explicit solution of the problem. The main interest lies in the stability characteristics of the solution in the neighborhood of an equilibrium point or a stationary motion.

Books on dynamics have traditionally tended to follow the development of the subject. As a result, many books are concerned primarily with formalism and structure, and others concentrate on methods of solutions, each treating the other aspect somewhat lightly. This book is written on the premise that a better understanding of the subject is achieved by presenting a balance of both aspects of analytical dynamics. The approach permits not only a broader but also a deeper view of the subject, and it is as natural as it is logical. The book contains material normally covered in graduate courses in dynamics and nonlinear mechanics, in addition to certain modern applications. The organization is such that the book can serve as a text on analytical dynamics taught either as an integrated course or in separate courses. If the latter route is chosen, increased benefits are bound to be derived by using this book. Quite recently the author taught the material on an integrated basis with remarkable improvement in student motivation and understanding, as opposed to the more traditional approach of teaching the material in separate courses. It is hoped that the book will eventually lead to a rearrangement of the graduate instruction in analytical dynamics. Dynamicists in industry with a strong interest in applications should regard the integrated approach as natural.

Perhaps the book can be best discussed by means of a brief review of its contents. Chapter 1 introduces the fundamental concepts of dynamics, following a historical survey of the development of mechanics. The basic ideas of Newtonian mechanics are presented by placing the emphasis on the physical aspects of dynamics. With the introduction of the generalized coordinates of Lagrange in Chap. 2, the emphasis shifts to the more abstract analytical mechanics. Here the variational principles are shown to be not only of esthetic value but also important tools in formulating problems of dynamics. Hamilton's principle is presented and then used to derive

Lagrange's equations of motion. With a view to the integration problem, certain conservation laws are discussed and Hamilton's canonical equations are introduced. Chapters 3 and 4 treat problems of rotational motion and rigid body dynamics, using both vector and matrix notation. Because of its importance, the subject of gyroscopic motion receives special consideration. Finally the problem of the general motion of a rigid body is formulated by means of Lagrange's equations in terms of quasi-coordinates. With Chap. 5 the attention turns from formulations to the behavior of dynamical systems. To this end, the geometric representation in the phase space is discussed, and such concepts as equilibrium and stability of motion are introduced. Further considerations are confined to the phase plane. Chapter 6 is concerned with the solution of sets of first-order ordinary differential equations, with emphasis on the stability characteristics of autonomous systems. General systems as well as canonical systems are discussed. The subject is extended in Chap. 7 to nonautonomous systems by giving special attention to systems with periodic coefficients. Chapters 6 and 7 are concerned primarily with the form and quality of the solution. By contrast, in Chap. 8 closed-form solutions by perturbation techniques are discussed. In Chap. 9 interest turns from direct to indirect integration methods. In this chapter, through a discussion of the transformation theory and the Hamilton-Jacobi method, an attempt is made to provide a deeper insight into the Hamiltonian mechanics. A perturbation method based on the Hamilton-Jacobi theory is also presented. The last three chapters are devoted primarily to applications in which methods of analytical dynamics can be illustrated. Chapter 10 is devoted entirely to gyroscopic motion. Various types of gyroscopic instruments, many of them used as navigational devices, are discussed. No book on analytical dynamics is complete without some of the classical problems of celestial mechanics. Indeed, in Chap. 11 the celebrated three-body problem receives special consideration. Also discussed are the libration of a rigid body under the effect of gravitational torques and the orbital changes caused by perturbing forces of a general nature. In spite of their vintage, these problems retain an amazing modern character. Chapter 12 presents modern applications from the area of spacecraft dynamics. Topics discussed are the orbital motion of satellites and the rotational motion of the satellites about their own mass center. Finally the subjects of variable-mass systems and rocket dynamics are introduced. Two appendixes present elements of dyadics and topology, respectively, for the benefit of those not familiar with the subjects.

If the book is to be used for separate courses in dynamics and nonlinear mechanics, Chaps. 1 to 4 can be used for the first and Chap. 5 to 8 for the second. Chapter 9 could be included in either course, whereas the applications of the last three chapters can be presented according to interest.

The author wishes to acknowledge the many valuable comments from colleagues and students. Special thanks are due to Peter W. Likins, Robert H. Tolson, and Kyle T. Alfriend for their careful review of the manuscript and useful suggestions. Thanks are also due to Ralph Pringle, Jr., William R. Wells, and Ronald L. Huston for their review of an early version of the manuscript, as well as to Robert A. Calico and John Bankovskis for their help with the proofreading and plotting of diagrams. Finally, the author expresses his appreciation to Joyce Helton and Rose Ann Butler for their expert typing of the manuscript.

LEONARD MEIROVITCH

Contents

Preface		vii
Chapter 1 Fundamentals of Newtonian Mechanics		1
1.1	Historical Survey of Mechanics	1
1.2	Newton's Laws	9
1.3	Impulse and Momentum	12
1.4	Moment of a Force and Angular Momentum	12
1.5	Work and Energy	14
1.6	Energy Diagrams	17
1.7	Systems of Particles	21
1.8	The Two-Body Central Force Problem	25
1.9	The Inverse Square Law. Orbits of Planets and Satellites	30
1.10	Scattering by a Repulsive Central Force	37
	Problems	40
	Suggested References	44
Chapter 2 Fundamentals of Analytical Mechanics		45
2.1	Degrees of Freedom. Generalized Coordinates	46
2.2	Systems with Constraints	48
2.3	The Stationary Value of a Function	53
2.4	The Stationary Value of a Definite Integral	55

xi

2.5	The Principle of Virtual Work	59
2.6	D'Alembert's Principle	65
2.7	Hamilton's Principle	66
2.8	Lagrange's Equations of Motion	72
2.9	Lagrange's Equations for Impulsive Forces	79
2.10	Conservation Laws	82
2.11	Routh's Method for the Ignoration of Coordinates	85
2.12	Rayleigh's Dissipation Function	88
2.13	Hamilton's Equations	91
	Problems	97
	Suggested References	100

Chapter 3 Motion Relative to Rotating Reference Frames 101

3.1	Transformation of Coordinates	102
3.2	Rotating Coordinate Systems	104
3.3	Expressions for the Motion in Terms of Moving Reference Frames	110
3.4	Motion Relative to the Rotating Earth	112
3.5	Motion of a Free Particle Relative to the Rotating Earth	114
3.6	Foucault's Pendulum	116
	Problems	119
	Suggested References	121

Chapter 4 Rigid Body Dynamics 122

4.1	Kinematics of a Rigid Body	123
4.2	The Linear and Angular Momentum of a Rigid Body	126
4.3	Translation Theorem for the Angular Momentum	130
4.4	The Kinetic Energy of a Rigid Body	132
4.5	Principal Axes	134
4.6	The Equations of Motion for a Rigid Body	137
4.7	Euler's Equations of Motion	138
4.8	Euler's Angles	140
4.9	Moment-Free Inertially Symmetric Body	143
4.10	General Case of a Moment-Free Body	147
4.11	Motion of a Symmetric Top	149
4.12	The Lagrangian Equations for Quasi-Coordinates	157
4.13	The Equations of Motion Referred to an Arbitrary System of Axes	160
4.14	The Rolling of a Coin	162
	Problems	164
	Suggested References	169

Chapter 5 Behavior of Dynamical Systems. Geometric Theory 170

5.1	Fundamental Concepts	171
5.2	Motion of Single-Degree-of-Freedom Autonomous Systems about Equilibrium Points	178

Contents   xiii

5.3 Conservative Systems. Motion in the Large   189
5.4 The Index of Poincaré   195
5.5 Limit Cycles of Poincaré   198
    Problems   206
    Suggested References   208

Chapter 6   Stability of Multi-Degree-of-Freedom Autonomous Systems   209

6.1 General Linear Systems   210
6.2 Linear Autonomous Systems   217
6.3 Stability of Linear Autonomous Systems. Routh-Hurwitz Criterion   222
6.4 The Variational Equations   225
6.5 Theorem on the First-Approximation Stability   226
6.6 Variation from Canonical Systems. Constant Coefficients   229
6.7 The Liapunov Direct Method   231
6.8 Geometric Interpretation of the Liapunov Direct Method   239
6.9 Stability of Canonical Systems   243
6.10 Stability in the Presence of Gyroscopic and Dissipative Forces   252
6.11 Construction of Liapunov Functions for Linear Autonomous Systems   258
    Problems   260
    Suggested References   262

Chapter 7   Nonautonomous Systems   263

7.1 Linear Systems with Periodic Coefficients. Floquet's Theory   264
7.2 Stability of Variational Equations with Periodic Coefficients   271
7.3 Orbital Stability   272
7.4 Variation from Canonical Systems. Periodic Coefficients   273
7.5 Second-Order Systems with Periodic Coefficients   277
7.6 Hill's Infinite Determinant   280
7.7 Mathieu's Equation   282
7.8 The Liapunov Direct Method   288
    Suggested References   292

Chapter 8   Analytical Solutions by Perturbation Techniques   293

8.1 The Fundamental Perturbation Technique   294
8.2 Secular Terms   297
8.3 Lindstedt's Method   299
8.4 The Krylov-Bogoliubov-Mitropolsky (KBM) Method   302
8.5 A Perturbation Technique Based on Hill's Determinants   309
8.6 Periodic Solutions of Nonautonomous Systems. Duffing's Equation   313
8.7 The Method of Averaging   322
    Problems   327
    Suggested References   328

Chapter 9 Transformation Theory. The Hamilton-Jacobi Equation — 329

9.1 The Principle of Least Action — 330
9.2 Contact Transformations — 334
9.3 Further Extensions of the Concept of Contact Transformations — 339
9.4 Integral Invariants — 346
9.5 The Lagrange and Poisson Brackets — 349
9.6 Infinitesimal Contact Transformations — 352
9.7 The Hamilton-Jacobi Equation — 355
9.8 Separable Systems — 361
9.9 Action and Angle Variables — 365
9.10 Perturbation Theory — 372
Problems — 378
Suggested References — 380

Chapter 10 The Gyroscope: Theory and Applications — 381

10.1 Oscillations of a Symmetric Gyroscope — 382
10.2 Effect of Gimbal Inertia on the Motion of a Free Gyroscope — 386
10.3 Effect of Rotor Shaft Flexibility on the Frequency of Oscillation of a Free Gyroscope — 389
10.4 The Gyrocompass — 393
10.5 The Gyropendulum. Schuler Tuning — 398
10.6 Rate and Integrating Gyroscopes — 403
Problems — 406
Suggested References — 407

Chapter 11 Problems in Celestial Mechanics — 408

11.1 Kepler's Equation. Orbit Determination — 409
11.2 The Many-Body Problem — 413
11.3 The Three-Body Problem — 416
11.4 The Restricted Three-Body Problem — 420
11.5 Stability of Motion Near the Lagrangian Points — 425
11.6 The Equations of Relative Motion. Disturbing Function — 428
11.7 Gravitational Potential and Torques for an Arbitrary Body — 430
11.8 Precession and Nutation of the Earth's Polar Axis — 438
11.9 Variation of the Orbital Elements — 442
11.10 The Resolution of the Disturbing Function — 447
Problems — 450
Suggested References — 451

Chapter 12 Problems in Spacecraft Dynamics — 452

12.1 Transfer Orbits. Changes in the Orbital Elements Due to a Small Impulse — 453

12.2	Perturbations of a Satellite Orbit in the Gravitational Field of an Oblate Planet	457
12.3	The Effect of Atmospheric Drag on Satellite Orbits	463
12.4	The Attitude Motion of Orbiting Satellites. General Considerations	466
12.5	The Attitude Stability of Earth-Pointing Satellites	470
12.6	The Attitude Stability of Spinning Symmetrical Satellites	475
12.7	Variable-Mass Systems	483
12.8	Rocket Dynamics	487
	Problems	491
	Suggested References	492

Appendix A Dyadics 494

Appendix B Elements of Topology and Modern Analysis 497

B.1	Sets and Functions	498
B.2	Metric Spaces	501
B.3	Topological Spaces	504
	Suggested References	506

Name Index 507

Subject Index 511

Methods of
ANALYTICAL DYNAMICS

chapter one

Fundamentals of Newtonian Mechanics

The oldest and most fundamental part of physics is concerned with the equilibrium and motion of material bodies. Problems of mechanics attracted the attention of such pioneers of physics as Aristotle and Archimedes. Although the field of classical mechanics is based on only a few fundamental ideas, it has inspired the creation of a large and elegant portion of mathematical analysis. The basic concepts of mechanics are *space*, *time*, and *mass* (or, alternately, *force*). These concepts are familiar from earlier encounters with the subject of dynamics, but we shall attempt to bring their meaning into sharper focus. To this end, a historical account of the development of mechanics will prove rewarding.

1.1 HISTORICAL SURVEY OF MECHANICS

The development of mechanics can be traced to the development of *geometry*, *kinematics*, and *dynamics*. By far the simplest and most widely used geometry is the *geometry of Euclid*. A Euclidean space is a space in which the length of a vector is defined by the Pythagorean theorem, according to which the length of the vector squared is equal to the sum of the vector components squared. The parallel and congruence postulates hold, so that Euclidean geometry is ideally suited for the study of the motion of a rigid body. The metrical structure of Euclidean space is homogeneous and isotropic, hence independent of the distribution of matter in the space. As a result, there is complete relativity of position or orientation in that space. A more general geometry, still consistent with the congruence concept, is the *geometry of Riemann*, which represents an extension of the two-dimensional geometry of curved surfaces developed by Gauss. By contrast with Euclidean geometry, in Riemannian geometry the space properties, as reflected by the metric coefficients, can vary from point to point. In such a geometry there is no longer relativity of position. Euclidean geometry was generally accepted as the framework of Newtonian mechanics. By means of a certain transformation, the concepts of Euclidean geometry could be extended to accommodate Einstein's special relativity theory, and it was not until the development of Einstein's general relativity, or gravitational, theory that Euclidean geometry was abandoned in favor of the more general Riemannian geometry.

Kinematics is concerned with the motion of material bodies, and, for this reason, it is sometimes referred to as the *geometry of motion*. But motion has meaning only when measured relative to a *system of reference*, which requires a well-defined system of coordinates and a time-measuring device. In Newtonian mechanics it is assumed that there exists an *absolute space* which is Euclidean and an *absolute time* whose flow is independent of the space. Since Euclidean space is homogeneous and isotropic, we must conclude that there is no preferred position or orientation, hence no preferred coordinate system. To measure the time at a certain point, an observer may choose any periodic phenomenon, such as the vibration of a tuning fork. With what is generally referred to as a clock, the question of synchronizing the clock with observers at other points remains. The local time can be extended to other observers by means of signals possessing infinite speed, provided such signals exist.

The study of dynamics considers the motion of material bodies under the influence of the surroundings. The question can be posed whether in the absence of interacting surroundings there exists a natural state of motion for a body. Aristotle explored the idea that there exists a frame of reference such that the natural state for any body is a state of rest with respect to that frame. Indeed, he believed that there was a natural place, namely, the center of the earth, toward which each body was striving and which it approached if no obstacles were encountered. Motion was regarded as starting from rest under the influence of "efficient causes" and determined by "final causes," which implies that once the motion was initiated, the body had a tendency to reach the natural place. Aristotle believed that wherever there was motion there must be a force, with the exception of bodies that moved themselves. This seems to imply that the law of motion he considered can be expressed in terms of modern concepts in the form *mass* × *velocity* = *force*. However, he was still puzzled by the observation that the motion of a body continued even when the source of motion was no longer in contact with the body. Although the notion of a natural place may have had a certain intuitive appeal in ancient Greece, when the earth was believed to be the center of things with the planets moving in circular orbits around it, this idea became questionable when Copernicus showed that the planets were really moving around the sun and not the earth, thus placing the sun at the center. Of course, later it became apparent that the sun did not occupy such a preferred position either, as it could be regarded as but one star among many in the universe. At this point we should recall that the idea of a natural position is in conflict with the concept of a Euclidean space. Aristotle's researches concerning the motion of bodies must be considered a complete failure as no unique frame of reference for velocity is possible.

The first step toward placing the study of dynamics on a truly scientific foundation was taken by Galileo. He set the study of the motion of bodies on the right track by developing the concept of acceleration. Galileo's investigation of falling objects led him to the observation that it is a fallacy to assume that wherever there is motion there must be a force. Indeed he concluded that force causes a change in velocity but no force is necessary to maintain a motion in which the magnitude and direction of the velocity does not change. This is basically the statement of Galileo's *law of inertia*. The measure of the tendency of a body to resist a change in its uniform motion is known as the *mass* or *inertia* of the body. Galileo also observed that while the acceleration is constant, the velocity varies with time for freely falling bodies, so that deviations from a state of rest or uniform motion must be attributed to the influence of other bodies. He recognized that the laws of motion are not affected by uniform motion, so that not only was there no natural position in space but there was no favored velocity of the reference frame either. Thus Galileo showed that there indeed exist preferred reference systems in which a body will move with uniform velocity or remain in a state of rest unless acted upon by external forces. Such a homogeneous and isotropic reference frame is referred to as an *inertial space* or *Galilean reference frame*. The inertial system is either at rest or translates with uniform velocity relative to a fixed space. If two systems move uniformly relative to each other, and if one of the systems is inertial, then the other system is also inertial. In fact there is an infinity of inertial frames translating uniformly relative to one another. The properties of space and time are the same and the laws of mechanics are identical in all these frames. This is the essence of *Galileo's principle of relativity*. It is impossible from observations of mechanical phenomena to detect a uniform motion relative to an inertial space. The relations between the motions expressed in terms of two inertial systems are known as *Galilean transformations*.

Newton expanded on the ideas of Galileo and toward the end of the seventeenth century formulated what have come to be known as *Newton's laws of motion*. In fact Newton's first law is simply Galileo's law of inertia. Recognizing that Galileo's results reflected the fact that the gravity force is constant in the vicinity of the earth's surface, Newton generalized these results by admitting variable forces as well. Moreover, Newton applied his laws to the motion of celestial bodies. In the process, he developed his *law of gravitation*, helped by his correct interpretation of Kepler's planetary laws. Newton's laws of motion assume the simplest form when referred to an inertial system, and, to this end, Newton introduced the notion of *absolute space* relative to which every motion should be measured. He proposed as an absolute space a coordinate system attached to the distant "fixed stars." Newton's fundamental equations are invariant under Galilean transformations, but this invariance does not hold for transformations involving accelerated reference frames. If we insist on treating mechanical phenomena in accelerated systems, we must introduce fictitious forces, such as centrifugal and Coriolis forces. These fictitious forces are strictly of a kinematical nature and appear when the motion is expressed in terms of rotating coordinate systems. According to Newton, the time is absolute and independent of space, which is another way of saying that the time is the same in any two inertial systems. Moreover, if there are forces exerted by one body upon another which depend only on the relative positions of the bodies, these forces are assumed to propagate instantaneously. Because the time is the same for all inertial frames, the concept of *simultaneity* presents no difficulty; i.e., if two events are observed to occur simultaneously by one observer traveling with one inertial frame, then the events occur simultaneously for all other inertial observers.

If the principle of relativity is to be valid for all fields of physics, including electrodynamics, then, among other things, the velocity of propagation of light waves, which according to Maxwell's theory of light are special electromagnetic waves, should be the same for all inertial observers. This turns out to be in conflict with the results predicted by observers whose motion with respect to one another are related by means of Galilean transformations. To show this, let us consider two inertial frames translating uniformly with respect to one another and assume that at the origin of one of the systems there is a source emitting light waves in the form of spherical waves centered at the origin and propagating with the speed c. Denoting a unit vector in the radial direction by $\mathbf{u}_r$, we can write the velocity of light as seen by an observer traveling with the light source in the form $c\mathbf{u}_r$. If a second inertial system is moving uniformly with velocity $\mathbf{v}$ with respect to the first, and if a Galilean transformation is used, we conclude that an observer on the second system will observe a light velocity $c\mathbf{u}_r - \mathbf{v}$, which is no longer equal in every direction. Hence for that observer the waves do not appear spherical. This, however, is in contradiction with the experimental evidence. Indeed the celebrated experiments performed by Michelson and Morley showed that the velocity of light is the same in all directions and does not depend on the relative motion of the observer and the source. Hence we must conclude that the Galilean transformation cannot be correct and should be replaced by a transformation preserving the constancy of the velocity of light in all systems. Such a transformation, known as the *Lorentz transformation*, is applicable to both mechanical and electromagnetic phenomena.

By the latter part of the nineteenth century the wave theory of light was set on a firm foundation by the researches of Faraday, Maxwell, and Hertz in electromagnetic field theory. According to this theory, light waves are simply electromagnetic waves propagating with constant velocity relative to an absolute space. But, in contrast with the equations of Newtonian mechanics, Maxwell's equations turned out not to be invariant under Galilean transformations. Indeed Maxwell insisted that the fundamental equations of electrodynamics were valid only in a unique privileged reference frame, known as the "luminiferous ether." By analogy with waves in gases and elastic solids, it was believed that the electromagnetic waves also needed a medium to propagate. The ether, imagined as an elastic medium permeating all transparent bodies, was assumed to be the carrier of all optical and electromagnetic phenomena. This elastic medium was supposed to provide an absolute reference frame for the electromagnetic phenomena in the same way that Newton's absolute space provided a reference frame for mechanical phenomena. As the ether was only a hypothesis, the need to produce conclusive proof of its existence remained. Because the earth was presumed to move relative to the ether at a certain velocity v, and since the speed of light c relative to the ether was supposed to be constant, it was anticipated that at least sometime during the year the speed of light relative to the earth should be different from c. If a ray of light is reflected from a mirror a distance L away from a light source, and if the source and the mirror are aligned with the direction of motion of the earth relative to the ether, then the time needed for the ray to return to the source has the value

$$T = \frac{L}{c+v} + \frac{L}{c-v} \approx \frac{2L}{c}\left[1 + \left(\frac{v}{c}\right)^2 + \cdots\right].$$

Hence, in order to demonstrate the motion of the earth relative to the ether, an apparatus capable of detecting quantities of order $(v/c)^2$ is necessary. In 1881 Michelson performed an experiment by means of an interferometer capable of detecting an effect much smaller than the anticipated one, but he could find no such effect. In 1887 Michelson and Morley repeated the experiment, using a more accurate apparatus, but the experiment again failed to detect the existence of the ether. However, it did demonstrate with a large degree of accuracy that the speed of light is the same in every direction, independent of the motion of the source. In an attempt to explain the experiment's failure to detect the motion of the earth relative to the ether, Fitzgerald advanced a hypothesis according to which a body contracts in the direction of motion. For a rod whose original length in the direction of motion is l_0 the length contracts to $l = l_0[1 - (v/c)^2]^{\frac{1}{2}}$ during motion. This *contraction hypothesis* was adopted by Lorentz, who generalized it by introducing a set of transformations rendering electromagnetic and optical phenomena independent of uniform motion of the system. In particular, he introduced a variable time, known as the *local time* because it differs from system to system. This amounted to a dilatation of the time scale. The difference between the Lorentzian and the Galilean transformations is of the order $(v/c)^2$. Although Lorentz realized that to account for the constancy of the light velocity a new kinematics, namely, the *Lorentzian kinematics*, was necessary, he did not question the validity of the classical principle of relativity, nor did he abandon the ether theory. In fact, the entire purpose of his transformations was to save the ether concept by providing an explanation for the failure of the experiments to detect uniform motion through the ether.

At about the same time, Poincaré also developed a set of transformations similar to the Lorentz transformation and achieving the same purpose, namely, rendering electromagnetic and optical phenomena independent of uniform motion of the reference frame. Both Lorentz and Poincaré realized that, as a result of these transformations, Maxwell's equations could be expressed in an infinite number of inertial systems, but Lorentz continued to believe that one of these systems represented the ether at rest. Poincaré, however, went one step further by recognizing that the mathematical equivalence of the inertial reference systems for the electromagnetic phenomena represented a new relativity principle. In fact, he proposed this principle as a general law of nature and suggested that the laws of mechanics be modified to conform to this law. However, he never understood the full physical implication of this relativity principle and regarded the transformation purely as a mathematical device. He did not take the important step of making the relativity principle independent of its derivation from the Maxwell equations.

Lorentz and Poincaré made a giant stride toward providing a new description of the physical world, basing that description on sound facts observed by means of reliable experiments rather than basing it on unproved hypotheses. However, they both failed to appreciate the far-reaching implications of their transformations. It remained for Einstein to demonstrate that the principle of relativity and the Lorentz transformation raised questions about the very fundamental concepts, such as the ether and absolute space, which were being assumed. Einstein proposed in 1905 to build new principles based on experimental evidence. He advanced two postulates:

Fundamentals of Newtonian Mechanics 7

1. The laws of nature (including the laws of mechanics and electrodynamics) are the same in all inertial frames.
2. The velocity of light has the same value for all inertial systems, independent of the velocity of the light source.

The two postulates form the basis of *Einstein's special theory of relativity*. Although the two postulates appear to be contradictory, Einstein showed that they can coexist if the concept of absolute time is discarded and time is added as a fourth coordinate to the three Euclidean spatial coordinates. Einstein did retain the Lorentz transformation, but it must be pointed out that he derived the corresponding equations from the general point of view of the principle of relativity. Later Minkowski concluded that the new *Einsteinian kinematics*, in which the space and time are inseparable, leads to a new geometrical structure consisting of a four-dimensional space formed by the ordinary space and time. This space-time world is referred to as *world space* or *Minkowski space*. It turns out that the Lorentz transformation is simply the orthogonal transformation of Minkowski space. The new relativity principle does not do away with Galilean inertial frames. On the other hand, by revising the time concept and relating the inertial reference frames by means of Lorentz transformations, Einstein succeeded in providing a common basis for the treatment of both mechanical and electromagnetic phenomena. Overwhelming experimental evidence corroborates the conclusions of special relativity.

In the relativistic mechanics of Einstein the concept of simultaneity requires further scrutiny. Suppose we wish to devise an experiment for measuring the speed of light. We consider two points A and B in a given inertial system, so that a light signal emitted at A at a certain time t_1 is received at B at time t_2. The light velocity is simply the ratio between the distance from A to B and the time interval $t_2 - t_1$. Whereas the measurement of the distance from A to B presents no problem, the measurement of the time interval is possible only if there are perfectly synchronized clocks at A and B. One of the best methods of synchronizing clocks is by means of light signals. But the speed of light is precisely the quantity we proposed to measure in the first place, so that we find ourselves in a vicious circle. The problem of synchronizing the clocks arises, of course, because the speed of the light signal is finite. But there is no known signal traveling faster than the light, so that the proposed experiment is the best we can produce. Moreover, unless the speed of the signal is infinite, any signal of finite speed is plagued by the same problem. The conclusion is that the concept of *absolute simultaneity* between two events occurring in two different places has no real meaning, as there is no experimental way of ascertaining simultaneity. It turns out, by virtue of the fact that the speed of light is finite although it is constant, that two events occurring simultaneously for an observer in one inertial system generally do not occur simultaneously for an observer in another inertial system. Einstein showed that it is possible to define a *relative simultaneity* for two events occurring at different points of a given inertial system.

In Newtonian mechanics *mass* is an inherent constant property of the body. It is independent of the motion relative to an inertial space or the flowing of time, and it represents a measure of the tendency of a body to preserve its uniform motion relative to a Galilean frame, where this tendency is called *inertia*. In relativistic mechanics the concept of constant mass must also be revised, as a result of the new kinematics. In particular, a *relativistic mass* depending on velocity obtains.

The laws of mechanics hold in their simplest classical form when referred to an inertial system. A noninertial observer traveling with a rotating reference frame will sense the so-called Coriolis and centrifugal forces. These forces, which are directly proportional to the mass, are kinematical and can be eliminated by referring to a Galilean system. But some very important forces in classical mechanics, namely, gravitational forces, are also proportional to the mass. By contrast, however, gravitational forces cannot be eliminated by a kinematical transformation which retains the Euclidean concept of space, as in special relativity. Moreover, in the presence of gravity the constancy of the velocity of light cannot be maintained. Einstein was equally discontent with the first postulate of special relativity restricting the class of acceptable reference systems to Galilean systems. In another stroke of genius, Einstein abandoned the world space of Minkowski in favor of a four-dimensional Riemannian space in which the gravitational forces disappear. The space possesses curvature in finite dimensions and is Minkowskian only in the small. The metric coefficients of Riemannian space are related to the gravitational mass at every point, so that the new geometrical world comprises *space*, *time*, and *matter*. The new theory, referred to as *Einstein's general theory of relativity* or *Einstein's gravitational theory*, was able to explain the anomalous motion of the perihelion of Mercury.

The nature of the dynamical problems with which we shall concern ourselves cannot justify abandoning the convenience of Euclidean space in order to remove gravitational forces. Moreover, we pointed out that the difference between the Galilean and Lorentzian transformations is of order $(v/c)^2$, where v is the velocity of the moving body and c the velocity of light. It follows that for dynamical problems involving velocities which are very small compared with the velocity of light or when we allow the velocity of light to become infinite, the Lorentzian transformations reduce to the Galilean transformations and the time becomes independent of the space. Hence the results of classical mechanics involving terrestrial phenomena and the results of celestial mechanics are correct to a high degree of approximation, which explains why Newtonian mechanics went unchallenged for two centuries.

It has become common to regard *classical mechanics* as the branch of physics concerned with the behavior of macroscopic systems, whereas the study of systems of atomic size is referred to as *quantum mechanics*. We shall be concerned only with the first, and in particular with nonrelativistic mechanics.

1.2 NEWTON'S LAWS

Newtonian mechanics is based on three laws stated for the first time by Sir Isaac Newton in his *Philosophiae Naturalis Principia Mathematica* (*The Mathematical Principles of Natural Sciences*), published in 1687. Some outstanding scientists preceded Newton in the field of mechanics, and he benefited considerably from their work, especially that of Galileo and Kepler. Although the first two laws were known to Galileo, Newton was the first to state the three laws clearly. The laws were formulated for single particles, and they postulate the existence of reference frames in which they are valid. These reference frames, called *inertial or Galilean systems of reference*, represent a class of preferred reference systems at rest or moving with uniform velocity with respect to one another. In Newtonian mechanics an inertial system is defined as one being at rest or moving with uniform velocity relative to an average position of the distant "fixed stars." For many practical purposes, however, it may be possible to assume an inertial system moving through space with the solar system, or perhaps one whose origin coincides with the center of the earth. We may even find it reasonable to regard a system of coordinates attached to the earth's surface as inertial, provided the accelerations resulting from the translation and rotation of the system are negligible compared with the acceleration of the body under consideration. Newton calls his laws of motion *axioms*. They can be stated as follows:

First Law *If there are no forces acting upon a particle, the particle will move in a straight line with constant velocity.*

A particle is an idealization of a material body whose dimensions are very small compared with the distance to other bodies and whose internal motion does not affect the motion of the body as a whole. Mathematically it is represented by a mass point having no extension in space. Denoting by **F** the *force vector* and by **v** the *velocity vector* measured relative to an inertial space, the first law can be stated mathematically:

$$\text{If } \mathbf{F} = \mathbf{0}, \text{ then } \mathbf{v} = \text{const.} \tag{1.1}$$

Second Law *A particle acted upon by a force moves so that the force vector is equal to the time rate of change of the linear momentum vector.*

The *linear momentum* is defined as the product of the *mass* and the *velocity* of the particle, $\mathbf{p} = m\mathbf{v}$, so that the second law can be written

$$\mathbf{F} = \frac{d\mathbf{p}}{dt} = \frac{d}{dt}(m\mathbf{v}). \tag{1.2}$$

The mass m of the particle is defined as a positive quantity whose value does not depend on time.

Third Law *When two particles exert forces upon one another, the forces lie along the line joining the particles and the corresponding force vectors are the negative of each other.*

This law is also known as the *law of action and reaction*. Denoting by $\mathbf{F}_{12}$ ($\mathbf{F}_{21}$) the force exerted by particle 2 (1) upon particle 1 (2), the law can be stated mathematically

$$\mathbf{F}_{12} = -\mathbf{F}_{21}, \tag{1.3}$$

where the vectors $\mathbf{F}_{12}$ and $\mathbf{F}_{21}$ are clearly collinear. It must be noted that the first law, which is Galileo's inertial law, is a special case of the second law. A notable exception to the third law are the electromagnetic forces between moving particles.

Perhaps the concepts can be further clarified by examining the motion of a particle of mass m along curve s, as illustrated in Fig. 1.1. Axes x, y, z represent an inertial space, and we shall refer to the motion with respect to such a system as *absolute*. The radius vector $\mathbf{r}_1$ ($\mathbf{r}_2$) denotes the absolute position of m at time t_1 (t_2). By definition, the *absolute velocity vector* is given by

FIGURE 1.1

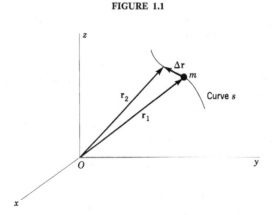

Fundamentals of Newtonian Mechanics 11

$$\mathbf{v} = \lim_{t_2 \to t_1} \frac{\mathbf{r}_2 - \mathbf{r}_1}{t_2 - t_1} = \lim_{\Delta t \to 0} \frac{\Delta \mathbf{r}}{\Delta t} = \frac{d\mathbf{r}}{dt}, \tag{1.4}$$

and it should be observed that **v** is a vector tangent to curve s at any instant t. In a similar way, we can write the expression for the *absolute acceleration vector* of the particle

$$\mathbf{a} = \lim_{\Delta t \to 0} \frac{\Delta \mathbf{v}}{\Delta t} = \frac{d\mathbf{v}}{dt} = \frac{d^2\mathbf{r}}{dt^2}. \tag{1.5}$$

Since the mass of the particle is assumed to be constant in time, Newton's second law reduces to

$$\mathbf{F} = m\mathbf{a} = m\frac{d^2\mathbf{r}}{dt^2}. \tag{1.6}$$

Denoting by F, T, and L the units of force, time, and length, respectively, it follows that the units of mass are FT^2L^{-1}.

Newton's law provides a complete formulation of the dynamical problem associated with the single free particle. The problem consists of the integration of a set of three simultaneous second-order differential equations. Assuming that the three components F_x, F_y, F_z of the force vector **F** are of class C_1† in a domain D of values of $x, y, z, \dot{x}, \dot{y}, \dot{z}, t$, where dots denote differentiations with respect to t, and that the initial values $x_0, y_0, z_0, \dot{x}_0, \dot{y}_0, \dot{z}_0$ at time $t = t_0$ are given, then Eq. (1.6) can be integrated to obtain the values of x, y, z at time t.

In addition to the three laws of motion, Newton formulated the *law of universal gravitation*, based on the three laws of planetary motion enunciated by Kepler, who, in turn, derived the laws from the observations of the motion of planets made by his teacher Tycho Brahe. The law of gravitation states that two particles of mass m_1 and m_2 attract each other with a force of magnitude

$$F(r) = \frac{Gm_1m_2}{r^2}, \tag{1.7}$$

whereas the direction of the force is along the line joining the two particles. In Eq. (1.7) r denotes the distance between the particles, and G is the universal gravitational constant. The law is commonly known as the *inverse square law*.

Together the law of gravitation and the laws of motion form the foundation of celestial mechanics.

For the most part, Newtonian mechanics predicts the motion of planets quite well. One notable exception is the anomalous behavior of the perihelion of Mercury. Many attempts were made to explain the discrepancy

---

† A function f of several variables is said to be of class C_k in a domain D of the variables if all the partial derivatives of f with respect to the variables through order k exist and are continuous in D.

between the observed motion of Mercury and the prediction based on Newton's law of gravitation but to no avail, until the discrepancy was finally explained by Einstein's general theory of relativity.

1.3 IMPULSE AND MOMENTUM

Let us multiply Eq. (1.2) through by dt and integrate with respect to time between the times t_1 and t_2 to obtain

$$\int_{t_1}^{t_2} \mathbf{F}\, dt = \int_{t_1}^{t_2} \frac{d\mathbf{p}}{dt}\, dt = \mathbf{p}_2 - \mathbf{p}_1 = (m\mathbf{v})_2 - (m\mathbf{v})_1. \tag{1.8}$$

The integral $\int_{t_1}^{t_2} \mathbf{F}\, dt$ is called the *linear impulse*, and $\mathbf{p} = m\mathbf{v}$ was defined as the *linear momentum*, so that the difference $\mathbf{p}_2 - \mathbf{p}_1 = (m\mathbf{v})_2 - (m\mathbf{v})_1 = \Delta \mathbf{p} = \Delta(m\mathbf{v})$ is simply the change in the linear momentum associated with the time interval between t_1 and t_2.

Hence, the linear impulse (or simply impulse) is equal to the incremental change in the linear momentum. We must never forget that the impulse as well as the momentum is a vector quantity.

In the absence of any force acting upon the particle, $\mathbf{F} = \mathbf{0}$, Eq. (1.8) yields

$$(m\mathbf{v})_2 = (m\mathbf{v})_1 = m\mathbf{v} = \text{const}, \tag{1.9}$$

which implies that the momentum at times t_1 and t_2, or any arbitrary time t, has the same value. Equation (1.9) is the mathematical statement of the theorem of the *conservation of linear momentum*.

1.4 MOMENT OF A FORCE AND ANGULAR MOMENTUM

Consider an inertial system x, y, z fixed in space and a particle of mass m at a distance $\mathbf{r}$ from the origin O (Fig. 1.2), and denote by $\dot{\mathbf{r}}$ the velocity of m relative to the inertial space, where the dot indicates differentiation with respect to time. The *moment of momentum*, or *angular momentum*, of m with respect to point O is defined as the cross product (vector product) of the radius vector $\mathbf{r}$ and the linear momentum $\mathbf{p}$. Denoting the angular momentum by $\mathbf{L}$, we have

$$\mathbf{L} = \mathbf{r} \times \mathbf{p} = \mathbf{r} \times m\dot{\mathbf{r}}. \tag{1.10}$$

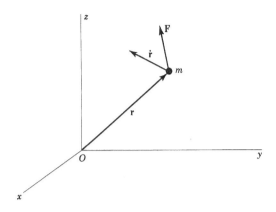

FIGURE 1.2

The rate of change of the angular momentum, for constant m, is

$$\dot{\mathbf{L}} = \dot{\mathbf{r}} \times m\dot{\mathbf{r}} + \mathbf{r} \times m\ddot{\mathbf{r}} = \mathbf{r} \times m\ddot{\mathbf{r}}. \qquad (1.11)$$

By definition, however, the moment of a force is

$$\mathbf{N} = \mathbf{r} \times \mathbf{F} = \mathbf{r} \times m\ddot{\mathbf{r}} = \dot{\mathbf{L}}. \qquad (1.12)$$

Hence, the moment of a force about a fixed point is equal to the time rate of change of the angular momentum about that point.

It should be noticed that the moment, or torque, $\mathbf{N}$ is a vector normal to the plane defined by $\mathbf{r}$ and $\mathbf{F}$, whereas the angular momentum $\mathbf{L}$ is a vector normal to the plane containing $\mathbf{r}$ and $\dot{\mathbf{r}}$.

In a manner analogous to that for the linear momentum, it is easy to show that

$$\int_{t_1}^{t_2} \mathbf{N} \, dt = \Delta \mathbf{L}, \qquad (1.13)$$

where the integral on the left side represents the *angular impulse* and $\Delta \mathbf{L}$ is the incremental change in the angular momentum associated with the time interval $t = t_2 - t_1$.

When the torque with respect to point O is zero, it follows from Eq. (1.12) that

$$\mathbf{L} = \text{const}, \qquad (1.14)$$

which is the statement of the theorem of the *conservation of angular momentum*.

1.5 WORK AND ENERGY

Let us consider a particle of mass m moving along curve s while being acted upon by a given force $\mathbf{F}$ (Fig. 1.3). By definition, the *increment of work* associated with the displacement of m from position $\mathbf{r}$ to position $\mathbf{r} + d\mathbf{r}$ is given by the dot product (scalar product)

$$\overline{dW} = \mathbf{F} \cdot d\mathbf{r}, \tag{1.15}$$

where the bar indicates that $\overline{dW}$ is not to be regarded as the true differential of a function W but simply as an infinitesimal expression.

By Newton's second law, however, we have $\mathbf{F} = m\ddot{\mathbf{r}}$, so that, substituting this equation into Eq. (1.15), we can write

$$\overline{dW} = \mathbf{F} \cdot d\mathbf{r} = m\ddot{\mathbf{r}} \cdot d\mathbf{r} = m\dot{\mathbf{r}} \cdot d\dot{\mathbf{r}} = d(\tfrac{1}{2} m\dot{\mathbf{r}} \cdot \dot{\mathbf{r}}) = dT. \tag{1.16}$$

By contrast with $\overline{dW}$, the quantity on the right side of Eq. (1.16) does represent the differential of a function, namely, the *kinetic energy* T given by the expression

$$T = \tfrac{1}{2} m\dot{\mathbf{r}} \cdot \dot{\mathbf{r}}, \tag{1.17}$$

in which $\dot{\mathbf{r}}$ is the particle velocity. If the particle moves from position $\mathbf{r}_1$ to position $\mathbf{r}_2$ under the force $\mathbf{F}$, the corresponding work is simply

$$\int_{\mathbf{r}_1}^{\mathbf{r}_2} \mathbf{F} \cdot d\mathbf{r} = \tfrac{1}{2} m\dot{\mathbf{r}}_2 \cdot \dot{\mathbf{r}}_2 - \tfrac{1}{2} m\dot{\mathbf{r}}_1 \cdot \dot{\mathbf{r}}_1 = T_2 - T_1, \tag{1.18}$$

where the subscripts 1 and 2 are used to denote quantities evaluated in the positions $\mathbf{r}_1$ and $\mathbf{r}_2$, respectively. Hence, Eq. (1.18) implies that the work

FIGURE 1.3

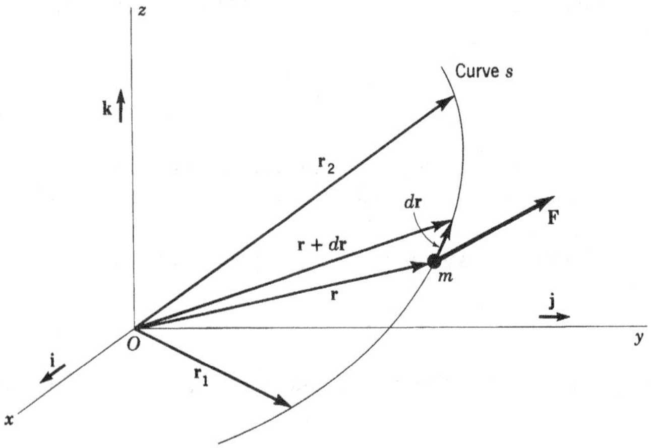

Fundamentals of Newtonian Mechanics

done in moving the particle from position $\mathbf{r}_1$ to $\mathbf{r}_2$ is responsible for an increase in the kinetic energy from T_1 to T_2.

In many physical problems the given force depends on the position alone, $\mathbf{F} = \mathbf{F}(\mathbf{r})$, and the quantity $\mathbf{F} \cdot d\mathbf{r}$ can be expressed in the form of a perfect differential

$$\mathbf{F}(\mathbf{r}) \cdot d\mathbf{r} = -dV(\mathbf{r}), \tag{1.19}$$

where the function $V(\mathbf{r})$ depends only on the position vector $\mathbf{r}$ and does not depend explicitly on the velocity $\dot{\mathbf{r}}$ or the time t. Such a force field is said to be *conservative*, and the function $V(\mathbf{r})$ is known as the *potential energy*. Assuming that the components of $\mathbf{F}$ belong to the class C_1 in a given domain of values of the variables x, y, z, it follows that V belongs to the class C_2. From Eq. (1.19) we can express the potential energy in the form of the integral

$$V(\mathbf{r}) = -\int_{\mathbf{r}_0}^{\mathbf{r}} \mathbf{F} \cdot d\mathbf{r} = \int_{\mathbf{r}}^{\mathbf{r}_0} \mathbf{F} \cdot d\mathbf{r}, \tag{1.20}$$

which does not depend on the path of integration but only on the initial and final positions. The initial position $\mathbf{r}_0$ serves as a reference position, and since in many cases we are concerned only with changes in the potential energy, this position is immaterial for the most part. It is clear that the integral over a closed path is zero,

$$\oint \mathbf{F} \cdot d\mathbf{r} = 0, \tag{1.21}$$

which is a different way of saying that the force field is conservative. From a mathematical point of view, we must conclude that forces for which the expression $\mathbf{F} \cdot d\mathbf{r}$ does not change sign over the entire path cannot be conservative because the integral would not vanish. Physically any dissipative forces, such as frictional forces, must be ruled out as being conservative.

Equation (1.21) involves a line integral and may not be very convenient for testing whether a force field is conservative or not. Using Stokes' theorem, Eq. (1.21) leads to

$$\boldsymbol{\nabla} \times \mathbf{F} = 0 \tag{1.22}$$

as the condition for a force field to be conservative, where $\boldsymbol{\nabla}$ is a differential operator called *del* or *nabla*. The operator can be written in terms of the cartesian components x, y, z in the form

$$\boldsymbol{\nabla} = \frac{\partial}{\partial x}\mathbf{i} + \frac{\partial}{\partial y}\mathbf{j} + \frac{\partial}{\partial z}\mathbf{k}, \tag{1.23}$$

where $\mathbf{i}$, $\mathbf{j}$, and $\mathbf{k}$ are the corresponding unit vectors. Equation (1.22) involves differentiation rather than integration and is simpler to use than

Eq. (1.21). Quite often, however, it is possible to decide whether a force field is conservative or not on physical grounds.

But the *curl* of **F** vanishes if and only if **F** is the *gradient* of some scalar function. Denoting this function by $-V$, we have

$$\mathbf{F} = -\nabla V, \tag{1.24}$$

where V is the same potential energy function as the one defined by Eq. (1.19). Equation (1.24) provides us with a way of deriving the conservative force components if the potential energy is known. These components are simply

$$F_x = -\frac{\partial V}{\partial x}, \quad F_y = -\frac{\partial V}{\partial y}, \quad F_z = -\frac{\partial V}{\partial z}. \tag{1.25}$$

It must be stated that we used the notation $-V$ for the scalar function in Eq. (1.24) in anticipation of the result, as the equation can be demonstrated in a different manner. To this end, we return to Eq. (1.19) and write it as

$$\mathbf{F} \cdot d\mathbf{r} = -dV = -\left(\frac{\partial V}{\partial x} dx + \frac{\partial V}{\partial y} dy + \frac{\partial V}{\partial z} dz\right) = -\nabla V \cdot d\mathbf{r}. \tag{1.26}$$

From Eq. (1.26) we conclude that the function V of Eq. (1.24) is indeed the same as the one of Eq. (1.19). Hence, the conservative force vector is the negative of the gradient of the potential energy.

Next let us consider the time derivative of T, which can be obtained from Eq. (1.16) in the form

$$\frac{dT}{dt} = \mathbf{F} \cdot \dot{\mathbf{r}}. \tag{1.27}$$

On the other hand, from Eq. (1.26), we have for a conservative force field

$$\frac{dV}{dt} = -\mathbf{F} \cdot \dot{\mathbf{r}}, \tag{1.28}$$

so that, adding Eqs. (1.27) and (1.28), we arrive at

$$\frac{d}{dt}(T + V) = 0, \tag{1.29}$$

which can be integrated to yield

$$T + V = E = \text{const}, \tag{1.30}$$

where the constant of integration E is referred to as the *total energy*. Equation (1.30) is the statement of the principle of the *conservation of energy*, which explains why the force field defined by Eq. (1.19) is called conservative.

Fundamentals of Newtonian Mechanics 17

When the increment of work $\overline{dW}$ can be expressed as a perfect differential, it is also customary to denote this differential by dW (without a bar), where $W = -V$ is referred to as the *work function* or the *potential function*. The advantage of using the negative of the potential energy instead of the work function is obvious in view of Eq. (1.30) with its physical implication.

If the force **F** depends not only on **r** but also on the time t, the possibility still exists that $\mathbf{F} \cdot d\mathbf{r}$ can be written in the form of a perfect differential. In this case the counterpart of Eq. (1.19) is

$$\mathbf{F}(\mathbf{r},t) \cdot d\mathbf{r} = -dV(\mathbf{r},t)|_{t=\text{const}}. \tag{1.31}$$

Hence, the vector **F** can be written again as the gradient of a scalar function. Such vector field is said to be *irrotational* or *lamellar*, and it reduces to a conservative field if **F** is independent of time. From Eq. (1.31) we conclude that

$$\frac{d}{dt}(T + V) = \frac{\partial V}{\partial t}, \tag{1.32}$$

so that an energy integral in the classical form, Eq. (1.30), does not exist for a lamellar field which is not conservative.

In general a force field consists of a force $\mathbf{F}_p$ which is derivable from a potential function $-V$ and a force $\mathbf{F}_{np}$ which is not, so that $\mathbf{F} = \mathbf{F}_p + \mathbf{F}_{np}$. In view of Eqs. (1.31) and (1.32), Eq. (1.27) leads to

$$\frac{d}{dt}(T + V) = \frac{\partial V}{\partial t} + \mathbf{F}_{np} \cdot \dot{\mathbf{r}}. \tag{1.33}$$

In the frequent case in which the potential energy V does not depend on time explicitly, the potential force can be identified as the conservative force $\mathbf{F}_c$ and the nonpotential force as the nonconservative force $\mathbf{F}_{nc}$. In this case Eq. (1.33) reduces to

$$\frac{d}{dt}(T + V) = \frac{dE}{dt} = \mathbf{F}_{nc} \cdot \dot{\mathbf{r}}, \tag{1.34}$$

which states that the rate of work done by the nonconservative force equals the rate of change of the total energy of the system.

1.6 ENERGY DIAGRAMS

A qualitative analysis of a dynamics problem can sometimes be carried out by means of energy diagrams. This appears to be the case with the harmonic

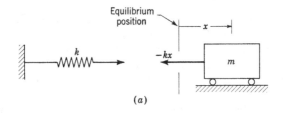

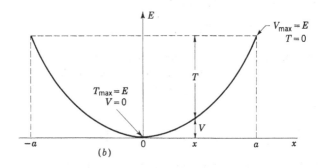

FIGURE 1.4

oscillator depicted in Fig. 1.4a. The oscillator consists of a mass m supported through a spring of stiffness k. The problem is one-dimensional, as the position of m is given by the displacement x from the equilibrium position, where x is a scalar. In this particular case, the equilibrium position corresponds physically to the position in which the spring is unstretched. It should be mentioned in passing that for a vertical spring with the mass hanging unobstructed, the equilibrium position differs from the unstretched spring position by an amount $\delta_{\text{static}} = mg/k$, where g is the acceleration due to gravity. Under the assumption of small displacements g is regarded as constant.

Denote by $\dot{x}$ the translational velocity of the mass when in position x, so that at this point the kinetic energy is

$$T = \tfrac{1}{2}m\dot{x}^2. \tag{1.35}$$

For a linear spring, there is a *restoring force* acting upon the mass which is proportional to the displacement. For any intermediate position ζ, $0 < \zeta < x$, the restoring force is

$$F = -k\zeta, \tag{1.36}$$

and, using as reference position the point $x = 0$, the potential energy corresponding to position x is

Fundamentals of Newtonian Mechanics

$$V = \int_x^0 F\, d\zeta = -k \int_x^0 \zeta\, d\zeta = -\tfrac{1}{2}k\zeta^2\big|_x^0 = \tfrac{1}{2}kx^2. \tag{1.37}$$

Since there are no nonconservative forces present, the total energy is a constant, having the value

$$E = T + V = \tfrac{1}{2}m\dot{x}^2 + \tfrac{1}{2}kx^2 = \text{const}, \tag{1.38}$$

which is equal to the total energy imparted to the system initially.

The energy diagram (Fig. 1.4b) consists of a plot of the various energies as functions of x. At the maximum displacement, $x = \pm a$, the potential energy attains its maximum value, and the kinetic energy is zero. It follows that

$$E = V_{\max} = \tfrac{1}{2}ka^2. \tag{1.39}$$

At the points $x = \pm a$ the velocity reduces to zero, and the mass is ready to reverse its motion. The velocity at any other point x is obtained by means of Eqs. (1.38) and (1.39)

$$\dot{x} = \pm \sqrt{\frac{2}{m}(E - V)} = \pm \sqrt{\frac{k}{m}(a^2 - x^2)}. \tag{1.40}$$

The sign in front of the radical merely indicates the direction of motion. Note that the value of a depends on the initial position and velocity.

Further insight can be gained by examining the equation of motion of the oscillator. Using Newton's second law, we obtain

$$m\ddot{x} = F = -kx, \tag{1.41}$$

and when we let

$$\omega^2 = \frac{k}{m}, \tag{1.42}$$

where ω is the natural frequency of the oscillator, Eq. (1.41) becomes

$$\ddot{x} + \omega^2 x = 0, \tag{1.43}$$

which is the well-known differential equation of a harmonic oscillator. The period of the oscillation is simply

$$P = \frac{2\pi}{\omega} = \frac{1}{f} = 2\pi\sqrt{\frac{m}{k}}, \tag{1.44}$$

where f is the natural frequency of the oscillator measured in cycles per second, in contrast with ω, which is measured in radians per second.

Another illustration of an energy diagram is furnished by a ball moving on a frictionless track in a vertical plane. The only force acting upon the ball is the gravity force

$$F = -mg, \tag{1.45}$$

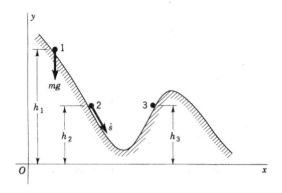

FIGURE 1.5

as shown in Fig. 1.5. The motion is confined to the neighborhood of the earth's surface, so that g is assumed to be constant.

Using as a reference position the point at which the height of the ball is zero, the potential energy at the height $y = h_1$ is

$$V_1 = -\int_{h_1}^{0} mg \, dy = mgh_1. \tag{1.46}$$

If the ball is released from rest at that point, then

$$E_1 = V_1 = mgh_1. \tag{1.47}$$

At any other height y we have

$$E_1 = T + V = \tfrac{1}{2}m\dot{s}^2 + mgy = \text{const}, \tag{1.48}$$

where $\dot{s}$ is the velocity of the ball parallel to the track.

If any other point to the right of point 1 is such that $y < h_1$, then $\dot{s} > 0$ at all times and the ball will continue to roll down the incline and never return. If, on the other hand, the ball is released from rest at point 2, then

$$E_2 = V_2 = mgh_2 \tag{1.49}$$

and at any other point y we have

$$E_2 = \tfrac{1}{2}m\dot{s}^2 + mgy. \tag{1.50}$$

Combining Eqs. (1.49) and (1.50), we obtain the velocity

$$\dot{s} = \pm\sqrt{2g(h_2 - y)}. \tag{1.51}$$

The motion is restricted to the region $y < h_2$, because at $y = h_2 = h_3$ the velocity reduces to zero and the ball reverses its motion. Hence, the motion consists of an oscillation between the points 2 and 3 with the maximum velocity reached at the lowest point on the track between these two points.

1.7 SYSTEMS OF PARTICLES

Newton's laws of motion were formulated for single particles. However, they can be extended without difficulty to systems of particles and bodies of finite dimensions. In extending the concepts of the preceding sections to systems of particles, we must distinguish between *external* and *internal forces*. The first are due to sources outside the system, and the latter are due to the interaction between the particles.

Let us consider a system of n particles of mass m_j ($j = 1, 2, \ldots, n$), as shown in Fig. 1.6. The position of the *center of mass* of the system is defined by

$$\mathbf{R} = \frac{1}{M} \sum_{j=1}^{n} m_j \mathbf{r}_j, \qquad (1.52)$$

where $M = \sum_{j=1}^{n} m_j$ is the total mass of the system. Physically the center of mass can be interpreted as a weighted average position of the system of particles. In the special case of a uniform gravitational field the center of mass coincides with the *center of gravity*.

FIGURE 1.6

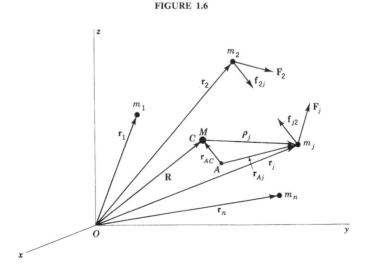

Let $\mathbf{F}_j$ be the external force acting upon particle j and let $\mathbf{f}_{jk}$ be the internal force representing the action of particle k upon j. Newton's second law applied to particle j yields

$$\mathbf{F}_j + \sum_{k=1}^{n} \delta_{jk}^* \mathbf{f}_{jk} = m_j \ddot{\mathbf{r}}_j, \qquad (1.53)$$

where δ_{jk}^* is a *complementary Kronecker delta* defined by

$$\delta_{jk}^* = 1 - \delta_{jk} = \begin{cases} 0 & \text{for } j = k, \\ 1 & \text{for } j \neq k, \end{cases} \qquad (1.54)$$

in which δ_{jk} is the ordinary Kronecker delta (see Sec. 3.1). The symbol δ_{jk}^*, in the manner used in Eq. (1.53), takes into account automatically the fact that there are no interacting forces between a particle and itself. Summing up over the entire system of particles, we obtain

$$\sum_{j=1}^{n} \mathbf{F}_j + \sum_{j=1}^{n} \sum_{k=1}^{n} \delta_{jk}^* \mathbf{f}_{jk} = \sum_{j=1}^{n} m_j \ddot{\mathbf{r}}_j. \qquad (1.55)$$

By virtue of Newton's third law, however,

$$\mathbf{f}_{jk} = -\mathbf{f}_{kj}, \qquad (1.56)$$

so that the double sum in Eq. (1.55) reduces to zero. Furthermore, letting $\mathbf{F}$ be the resultant of all the external forces acting upon the system,

$$\mathbf{F} = \sum_{j=1}^{n} \mathbf{F}_j, \qquad (1.57)$$

and recalling the definition (1.52) of the mass center, we obtain

$$\sum_{j=1}^{n} m_j \ddot{\mathbf{r}}_j = M\ddot{\mathbf{R}}. \qquad (1.58)$$

Introduction of Eqs. (1.56) through (1.58) into (1.55) yields

$$\mathbf{F} = M\ddot{\mathbf{R}} = \dot{\mathbf{P}}, \qquad (1.59)$$

where

$$\mathbf{P} = M\dot{\mathbf{R}} \qquad (1.60)$$

is the linear momentum vector of the system of particles. Equation (1.59) indicates that the motion of the center of mass of the system is the same as the motion of a fictitious body equal in mass to the mass of the system, concentrated at the center of mass, and acted upon by the resultant of all the external forces.

It appears that internal forces do not affect the motion of the center of mass. For example, the center of mass of an exploding projectile free of

Fundamentals of Newtonian Mechanics

external forces, such as air resistance and gravitational forces, will continue after the explosion to move along the original trajectory. This statement can be put on a more formal basis: In the absence of any external forces acting upon a system of particles, $\mathbf{F} = 0$, the linear momentum of the system is conserved, $\mathbf{P} = M\dot{\mathbf{R}} = \text{const}$. This is the theorem of the *conservation of linear momentum for a system of particles*, and it assumes that the total mass of the system does not change in time.

The angular momentum of the system of particles with respect to any moving point A is defined[1] as

$$\mathbf{L}_A = \sum_{j=1}^{n} \mathbf{L}_{Aj} = \sum_{j=1}^{n} \mathbf{r}_{Aj} \times m_j \dot{\mathbf{r}}_j. \tag{1.61}$$

Differentiation through Eq. (1.61) with respect to time yields

$$\dot{\mathbf{L}}_A = \sum_{j=1}^{n} \dot{\mathbf{r}}_{Aj} \times m_j \dot{\mathbf{r}}_j + \sum_{j=1}^{n} \mathbf{r}_{Aj} \times m_j \ddot{\mathbf{r}}_j, \tag{1.62}$$

where, of course, it is assumed that the masses m_j do not vary with time. But from Fig. 1.6 we have

$$\mathbf{r}_{Aj} = \mathbf{r}_{AC} + \boldsymbol{\rho}_j, \qquad \mathbf{r}_j = \mathbf{R} + \boldsymbol{\rho}_j, \tag{1.63}$$

where $\mathbf{r}_{AC}$ is the radius vector from point A to the center of mass C and $\boldsymbol{\rho}_j$ is the vector from C to particle j, so that introducing Eqs. (1.53) and (1.63) into (1.62) and recalling Eq. (1.56) and the fact that these forces are collinear, we obtain

$$\dot{\mathbf{L}}_A = \sum_{j=1}^{n} (\dot{\mathbf{r}}_{AC} + \dot{\boldsymbol{\rho}}_j) \times m_j (\dot{\mathbf{R}} + \dot{\boldsymbol{\rho}}_j) + \sum_{j=1}^{n} \mathbf{r}_{Aj} \times \mathbf{F}_j$$
$$= M\dot{\mathbf{r}}_{AC} \times \dot{\mathbf{R}} + \mathbf{N}_A, \tag{1.64}$$

because, from the definition of the center of mass, $\sum_{j=1}^{n} m_j \boldsymbol{\rho}_j = 0$. Moreover, $\mathbf{N}_A$ is recognized as the torque produced by the external forces with respect to point A.

In the event *point A coincides with the fixed origin O*, $\mathbf{r}_{AC} = \mathbf{R}$, Eq. (1.64) reduces to the simple form

$$\dot{\mathbf{L}}_0 = \mathbf{N}_0, \tag{1.65}$$

and when *point A coincides with the moving center of mass C*, $\mathbf{r}_{AC} = 0$, Eq. (1.64) becomes

$$\dot{\mathbf{L}}_C = \mathbf{N}_C. \tag{1.66}$$

---

[1] Another common definition is $\mathbf{L}_A = \sum \mathbf{r}_{Aj} \times m_j \dot{\mathbf{r}}_{Aj}$. Since $\dot{\mathbf{r}}_{Aj}$ is the velocity of m_j relative to point A rather than the absolute velocity $\dot{\mathbf{r}}_j$, the two definitions represent in general different angular momenta.

Hence, the time rate of change of the angular momentum with respect to the fixed origin O or with respect to the moving center of mass C is equal to the resultant of the external torques about O or C respectively.

From Eq. (1.65) we conclude that if the external torque about a fixed point O is zero, the angular momentum of the system of particles about O is conserved. Equation (1.66) leads to a similar conclusion concerning the angular momentum about the moving center of mass C. These two statements represent the *conservation of angular momentum for a system of particles* about a fixed point and the moving center of mass, respectively.

Next let us return to Eq. (1.61) and write the angular momentum about A in the form

$$\mathbf{L}_A = \sum_{j=1}^{n} \mathbf{r}_{Aj} \times m_j \dot{\mathbf{r}}_j$$

$$= \mathbf{r}_{AC} \times M\dot{\mathbf{R}} + \sum_{j=1}^{n} \boldsymbol{\rho}_j \times m_j \dot{\boldsymbol{\rho}}_j = \mathbf{r}_{AC} \times \mathbf{P} + \mathbf{L}'_C, \quad (1.67)$$

where
$$\mathbf{L}'_C = \sum_{j=1}^{n} \boldsymbol{\rho}_j \times m_j \dot{\boldsymbol{\rho}}_j \quad (1.68)$$

is the angular momentum of motion about the center of mass, which is sometimes referred to as *apparent angular momentum*. This is the angular momentum seen by an observer stationed at the center of mass and unaware of the translation of the system as a whole relative to the inertial space. In the special case in which point A coincides with C, the actual and apparent angular momenta are equal to each other, and Eq. (1.66) can be rewritten as

$$\dot{\mathbf{L}}'_C = \mathbf{N}_C. \quad (1.69)$$

Hence, if the nature of the problem necessitates that the moment equations be written with respect to a moving point, the advantage of choosing that point as the center of mass is quite evident. The moment equations with respect to the center of mass of the system retain the same simple form as for a fixed point. Although there are other special situations leading to this simple form, they are not sufficiently important to elaborate upon.

The kinetic energy of a system of particles deserves special attention. Of course, the kinetic energy has the simple expression

$$T = \frac{1}{2} \sum_{j=1}^{n} m_j \dot{\mathbf{r}}_j \cdot \dot{\mathbf{r}}_j, \quad (1.70)$$

but it will prove of interest to derive an expression in terms of the center of mass of the system. To accomplish this, introduce the second of Eqs. (1.63) into (1.70) and write

$$\begin{aligned}
T &= \frac{1}{2}\sum_{j=1}^{n} m_j(\dot{\mathbf{R}} + \dot{\boldsymbol{\rho}}_j) \cdot (\dot{\mathbf{R}} + \dot{\boldsymbol{\rho}}_j) \\
&= \tfrac{1}{2}\dot{\mathbf{R}} \cdot \dot{\mathbf{R}} \sum_{j=1}^{n} m_j + \dot{\mathbf{R}} \cdot \sum_{j=1}^{n} m_j \dot{\boldsymbol{\rho}}_j + \frac{1}{2}\sum_{j=1}^{n} m_j \dot{\boldsymbol{\rho}}_j \cdot \dot{\boldsymbol{\rho}}_j \\
&= \tfrac{1}{2} M \dot{R}^2 + \frac{1}{2}\sum_{j=1}^{n} m_j \dot{\rho}_j^{\,2}.
\end{aligned} \qquad (1.71)$$

Thus, the kinetic energy of a system of particles is equal to the kinetic energy obtained by regarding the entire mass of the system as concentrated at the center of mass plus the kinetic energy of motion about the center of mass.

The advantage of working with the center of mass will be amply demonstrated later.

1.8 THE TWO-BODY CENTRAL FORCE PROBLEM

A problem recurring frequently in mechanics is that of two particles which are free to move in space under the influence of forces exerted by the particles on each other along the line joining them. The motion of a planet around the sun or of an artificial satellite around the earth can be described to a large degree by the two-body theory. On a considerably smaller scale, the electron in an atom lends itself to such a treatment. It turns out that the problem of two interacting particles is reducible to an equivalent problem of a single free particle acted on by a central force.

Let us consider the two-body system of Fig. 1.7. Particles 1 and 2 are

FIGURE 1.7

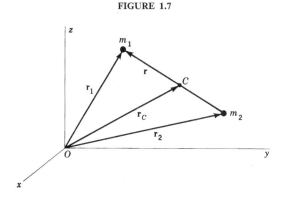

acted on by the mutual forces $\mathbf{f}_{12}$ and $\mathbf{f}_{21}$ along the line connecting the particles, as well as by the external forces $\mathbf{F}_1$ and $\mathbf{F}_2$, respectively. If C denotes the center of mass of the system and $\mathbf{r}$ the radius vector from m_2 to m_1, then the radius vectors from C to m_1 and m_2 are $[m_2/(m_1 + m_2)]\mathbf{r}$ and $-[m_1/(m_1 + m_2)]\mathbf{r}$, respectively, and it follows that the absolute positions of the particles are

$$\mathbf{r}_1 = \mathbf{r}_C + \frac{m_2}{m_1 + m_2} \mathbf{r}, \qquad \mathbf{r}_2 = \mathbf{r}_C - \frac{m_1}{m_1 + m_2} \mathbf{r}. \tag{1.72}$$

The equations of motion of the two particles are simply

$$\mathbf{F}_1 + \mathbf{f}_{12} = m_1 \ddot{\mathbf{r}}_1 = m_1 \ddot{\mathbf{r}}_C + \frac{m_1 m_2}{m_1 + m_2} \ddot{\mathbf{r}}$$

$$\mathbf{F}_2 + \mathbf{f}_{21} = m_2 \ddot{\mathbf{r}}_2 = m_2 \ddot{\mathbf{r}}_C - \frac{m_1 m_2}{m_1 + m_2} \ddot{\mathbf{r}}. \tag{1.73}$$

Next let us examine the case in which there are no external forces acting upon the system and the only forces present are the mutual forces. Adding the two equations of motion, Eqs. (1.73), and recalling that by Newton's third law $\mathbf{f}_{12} = -\mathbf{f}_{21}$, we obtain

$$\mathbf{F} = \mathbf{F}_1 + \mathbf{F}_2 = (m_1 + m_2)\ddot{\mathbf{r}}_C = \mathbf{0}, \qquad \ddot{\mathbf{r}}_C = \mathbf{0}, \tag{1.74}$$

or the center of mass either travels with uniform velocity or is stationary. Of course, we could have reached the same conclusion from Eq. (1.59). On the other hand, either of Eqs. (1.73) yields

$$\mathbf{f} = \mathbf{f}_{12} = -\mathbf{f}_{21} = \frac{m_1 m_2}{m_1 + m_2} \ddot{\mathbf{r}} = m\ddot{\mathbf{r}}, \tag{1.75}$$

which can be interpreted as describing the motion of a single body of equivalent mass $m = m_1 m_2/(m_1 + m_2)$. Because, by Eqs. (1.74), the center of mass C is unaccelerated, we can assume without loss of generality that it is at rest. Hence, the two-body problem can be treated as that of a single body of mass m moving with acceleration $\ddot{\mathbf{r}}$ relative to the fixed center of force C.

It turns out that many of the motion characteristics of a two-body system do not depend on the law of force but simply on the fact that the force is central. Denoting the magnitude of the force $\mathbf{f}$ by f and the unit vector in the radial direction by $\mathbf{u}_r$, the central force takes the form

$$\mathbf{f} = f\mathbf{u}_r. \tag{1.76}$$

Furthermore, the radius vector has the expression

$$\mathbf{r} = r\mathbf{u}_r, \tag{1.77}$$

where r is the radial distance.

Because the force $\mathbf{f}$ passes through the center C, the moment of force about that point is zero

$$\mathbf{N} = \mathbf{0}, \tag{1.78}$$

from which it follows that the *angular momentum about C is conserved*

$$\mathbf{L} = \text{const}, \tag{1.79}$$

implying that *both the magnitude and direction* of **L** *are constant*. Since **L** does not change its direction in an inertial space, *the motion of m must take place in a plane*, namely, the plane whose normal is parallel to **L**. This reduces the problem to a two-dimensional one.

The choice of coordinates is often of critical importance, as the right choice can expedite solution of the problem. In this particular case the use of polar coordinates appears warranted. The velocity vector in polar coordinates can be shown to have the expression

$$\mathbf{v} = \dot{r}\mathbf{u}_r + r\dot{\theta}\mathbf{u}_\theta, \tag{1.80}$$

where $\mathbf{u}_r$ and $\mathbf{u}_\theta$ are unit vectors in the radial and transverse directions. The direction of these unit vectors changes continuously, so that they cannot be regarded as constant vectors.

The angular momentum vector is

$$\begin{aligned}\mathbf{L} &= \mathbf{r} \times m\mathbf{v} = r\mathbf{u}_r \times m(\dot{r}\mathbf{u}_r + r\dot{\theta}\mathbf{u}_\theta) \\ &= mr^2\dot{\theta}\mathbf{u}_n = \text{const},\end{aligned} \tag{1.81}$$

where $\mathbf{u}_n$ is a constant unit vector normal to the plane of motion. But the magnitude of **L** is constant, so that Eq. (1.81) leads to

$$L = |\mathbf{L}| = mr^2\dot{\theta} = \text{const}. \tag{1.82}$$

Equation (1.82) has some notable geometrical interpretations. From Fig. 1.8 we conclude that an element of area in polar coordinates has the form

$$dA = \tfrac{1}{2}r^2\,d\theta. \tag{1.83}$$

FIGURE 1.8

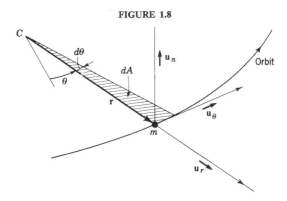

Dividing Eq. (1.83) through by dt, we obtain

$$\dot{A} = \tfrac{1}{2}r^2\dot{\theta} = \frac{L}{2m} = \text{const}, \qquad (1.84)$$

which is precisely the mathematical statement of *Kepler's second law* for planetary motion: *Every planet moves in such a way that its radius vector sweeps over equal areas in equal times.* It must be noted that this statement is true regardless of the form of the function f.

In general the central force field need not be conservative. However, if the force field is to be conservative, then the central force f must depend on the radial distance r alone. To show this, we use Eqs. (1.24) and (1.76) and obtain the radial and transverse components of ∇V in the form

$$\frac{\partial V}{\partial r} = -f, \qquad \frac{1}{r}\frac{\partial V}{\partial \theta} = 0. \qquad (1.85)$$

The second of Eqs. (1.85) indicates that f cannot depend on θ and must depend only on r, $f = f(r)$. On the other hand, the first of Eqs. (1.85) can be integrated to yield

$$V(r) = \int_r^{r_0} f(r)\, dr = V(r_0) - \int^r f(r)\, dr. \qquad (1.86)$$

Hence, the necessary and sufficient condition for the force f to be conservative is that it depend only on r.

At this point, let us turn our attention to the derivation and integration of the equations of motion. For this purpose we need expressions for the radial and transverse components of the acceleration, which can be obtained by differentiating Eq. (1.80). Since the unit vectors $\mathbf{u}_r$ and $\mathbf{u}_\theta$ are not constant, we must first obtain their time derivatives. Figure 1.9a shows the position of m after a time increment Δt, and Fig. 1.9b shows the unit vectors

FIGURE 1.9

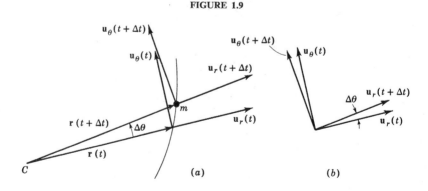

associated with the angle change corresponding to that time increment. From Fig. 1.9b we conclude that

$$\begin{aligned} \mathbf{u}_r(t + \Delta t) &= \mathbf{u}_r(t) + \Delta\theta \times 1 \times \mathbf{u}_\theta(t), \\ \mathbf{u}_\theta(t + \Delta t) &= \mathbf{u}_\theta(t) - \Delta\theta \times 1 \times \mathbf{u}_r(t), \end{aligned} \quad (1.87)$$

leading to

$$\begin{aligned} \dot{\mathbf{u}}_r &= \lim_{\Delta t \to 0} \frac{\mathbf{u}_r(t + \Delta t) - \mathbf{u}_r(t)}{\Delta t} = \lim_{\Delta t \to 0} \frac{\Delta\theta}{\Delta t} \mathbf{u}_\theta(t) = \dot{\theta}\mathbf{u}_\theta, \\ \dot{\mathbf{u}}_\theta &= \lim_{\Delta t \to 0} \frac{\mathbf{u}_\theta(t + \Delta t) - \mathbf{u}_\theta(t)}{\Delta t} = \lim_{\Delta t \to 0} -\frac{\Delta\theta}{\Delta t} \mathbf{u}_r(t) = -\dot{\theta}\mathbf{u}_r. \end{aligned} \quad (1.88)$$

With this in mind, a differentiation of Eq. (1.80) yields the acceleration

$$\mathbf{a} = \dot{\mathbf{v}} = (\ddot{r} - r\dot{\theta}^2)\mathbf{u}_r + (r\ddot{\theta} + 2\dot{r}\dot{\theta})\mathbf{u}_\theta, \quad (1.89)$$

so that the equations of motion corresponding to the radial and transverse directions are

$$\begin{aligned} m(\ddot{r} - r\dot{\theta}^2) &= f(r), \\ m(r\ddot{\theta} + 2\dot{r}\dot{\theta}) &= 0. \end{aligned} \quad (1.90)$$

The second of Eqs. (1.90) can be shown to lead to Eq. (1.82). On the other hand, introducing Eq. (1.82) into the first of Eqs. (1.90), we obtain

$$m\ddot{r} - \frac{L^2}{mr^3} = f(r). \quad (1.91)$$

A simple algebraic manipulation shows that $\ddot{r} = \frac{1}{2}\frac{d}{dr}(\dot{r}^2)$, so that Eq. (1.91) can be integrated with respect to r to obtain

$$\tfrac{1}{2}m\dot{r}^2 + \frac{L^2}{2mr^2} - \int^r f(r)\,dr = \text{const.} \quad (1.92)$$

The first two terms in Eq. (1.92) are recognized as the kinetic energy components, whereas the negative of the integral represents the potential energy, as can be seen from Eq. (1.86). Hence, Eq. (1.92) expresses the conservation of the total energy

$$\tfrac{1}{2}m\dot{r}^2 + \frac{L^2}{2mr^2} + V(r) = E = \text{const}, \quad (1.93)$$

which is to be expected.

From Eq. (1.93) we obtain the radial velocity component

$$\dot{r} = \pm \left[\frac{2}{m}\left(E - V - \frac{L^2}{2mr^2}\right)\right]^{\frac{1}{2}}, \quad (1.94)$$

so that the total velocity of the particle is

$$v = [\dot{r}^2 + (r\dot{\theta})^2]^{\frac{1}{2}} = \left[\frac{2}{m}(E - V)\right]^{\frac{1}{2}}, \quad (1.95)$$

which depends on the radial distance alone.

Eliminating the time from Eqs. (1.82) and (1.94), we arrive at

$$\frac{dr}{d\theta} = \frac{\dot{r}}{\dot{\theta}} = \frac{mr^2}{L}\left[\frac{2}{m}\left(E - V - \frac{L^2}{2mr^2}\right)\right]^{\frac{1}{2}}, \quad (1.96)$$

where the positive sign in Eq. (1.94) has been used. An integration of Eq. (1.96) yields the orbit equation

$$\theta = \theta_0 + \frac{L}{m}\int_{r_0}^{r}\frac{dr}{r^2[(2/m)(E - V - L^2/2mr^2)]^{\frac{1}{2}}}, \quad (1.97)$$

in which r_0 and θ_0 represent a reference position. It is customary to make the substitution $r = u^{-1}$, $dr = -u^{-2}\,du$, so that Eq. (1.97) becomes

$$\theta = \theta_0 - \frac{L}{(2m)^{\frac{1}{2}}}\int_{u_0}^{u}\frac{du}{(E - V - L^2u^2/2m)^{\frac{1}{2}}}. \quad (1.98)$$

Of particular interest is the case in which the force is proportional to a given power of the radial distance, namely,

$$f(r) = \alpha r^n, \quad (1.99)$$

for which the potential energy becomes

$$V(r) = -\int^{r} f(r)\,dr = -\frac{\alpha}{n+1}r^{n+1} \quad (1.100)$$

or, in terms of u,

$$V(u) = -\frac{\alpha}{n+1}u^{-(n+1)}. \quad (1.101)$$

The integral in Eq. (1.98) assumes the general form

$$\int (a + bu^2 + cu^{-(n+1)})^{-\frac{1}{2}}\,du, \quad (1.102)$$

and it can be shown that when $n = +1, -2, -3$ integral (1.102) yields trigonometric functions, whereas when $n = +5, +3, 0, -4, -5, -7$, elliptic functions are obtained.

1.9 THE INVERSE SQUARE LAW. ORBITS OF PLANETS AND SATELLITES

The treatment of the two-body system in Sec. 1.8 is perfectly general in the sense that it is valid for any central force $f(r)$. As special cases, we have shown that if $f(r)$ is proportional to cer in powers of the radial distance, then solutions in the form of trigonometric or elliptic functions should be anticipated. By far the most important of the central force laws is the inverse square law. We need only remember that Newton's gravitational law and Coulomb's law, describing the force of two electrically charged particles, are of this type. In the first case the particles attract each other, and in the second case they repel each other.

When finite bodies possess spherically symmetric mass distribution, or when the distance between the bodies is very large compared with their dimensions, finite bodies behave like particles (see Sec. 11.7). This enables us to use the two-body theory, in conjunction with Newton's gravitational law, to derive the equations describing the orbit of planets around the sun or the orbit of artificial satellites around a planet. The latter case, of course, affords a larger variety of orbits and represents a subject of current interest.

Let the motion in the central force field be governed by Newton's inverse square law [see Eq. (1.7)]

$$f(r) = -\frac{Gm_1 m_2}{r^2}, \tag{1.103}$$

where the minus sign indicates that this is an attractive force. Perhaps we shall find it profitable to derive the orbit equation in a different way than in Sec. 1.8. We recall the definition of the equivalent mass m and write Eqs. (1.90) in the form

$$\frac{m_1 m_2}{m_1 + m_2} (\ddot{r} - r\dot{\theta}^2) = -\frac{Gm_1 m_2}{r^2},$$

$$\frac{m_1 m_2}{m_1 + m_2} \frac{1}{r} \frac{d}{dt}(r^2 \dot{\theta}) = 0. \tag{1.104}$$

The second of Eqs. (1.104) can easily be verified as being equivalent to the second of Eqs. (1.90). We let

$$G(m_1 + m_2) = \mu, \tag{1.105}$$

and Eqs. (1.104) lead to

$$\ddot{r} - r\dot{\theta}^2 = -\frac{\mu}{r^2}, \qquad r^2 \dot{\theta} = k, \tag{1.106}$$

where k is a constant representing the angular momentum per unit mass.

To derive the orbit equation, we must first eliminate the time dependence from the first of Eqs. (1.106). But

$$\dot{r} = \frac{dr}{d\theta} \frac{d\theta}{dt} = \frac{k}{r^2} \frac{dr}{d\theta} = -k \frac{d}{d\theta}\left(\frac{1}{r}\right) = -k \frac{du}{d\theta}, \tag{1.107}$$

where the substitution $r = u^{-1}$ has been made. It also follows that

$$\ddot{r} = \frac{d\dot{r}}{d\theta}\frac{d\theta}{dt} = -\frac{k^2}{r^2}\frac{d^2u}{d\theta^2} = -k^2u^2\frac{d^2u}{d\theta^2}. \tag{1.108}$$

Introduction of the second of Eqs. (1.106) and Eq. (1.108) into the first of Eqs. (1.106) yields the differential equation

$$\frac{d^2u}{d\theta^2} + u = \frac{\mu}{k^2}, \tag{1.109}$$

which does not contain the time explicitly. Its solution is simply

$$u = \frac{\mu}{k^2} + C\cos(\theta - \theta_0), \tag{1.110}$$

where C and θ_0 are constants of integration, whose determination follows.

By analogy with Eq. (1.86), we define the potential energy per unit mass as

$$V(r) = -\mu \int_r^\infty \frac{d\zeta}{\zeta^2} = \frac{\mu}{\zeta}\bigg|_r^\infty = -\frac{\mu}{r}, \tag{1.111}$$

where ζ is a dummy variable of integration. Notice that for convenience the reference point for the potential energy was chosen at infinity. The kinetic energy per unit mass is simply

$$T = \tfrac{1}{2}v^2 = \tfrac{1}{2}(\dot{r}^2 + r^2\dot{\theta}^2), \tag{1.112}$$

so that the total energy per unit mass is

$$E = T + V = \tfrac{1}{2}(\dot{r}^2 + r^2\dot{\theta}^2) - \frac{\mu}{r}. \tag{1.113}$$

Use of the second of Eqs. (1.106) and Eqs. (1.107) and (1.110) leads to

$$E = \frac{k^2}{2}\left[\left(\frac{du}{d\theta}\right)^2 + u^2\right] - \mu u = \frac{k^2}{2}\left[C^2 - \left(\frac{\mu}{k^2}\right)^2\right], \tag{1.114}$$

from which it follows that

$$C^2 = \left(\frac{\mu}{k^2}\right)^2 e^2, \tag{1.115}$$

where

$$e^2 = 1 + \frac{2Ek^2}{\mu^2}. \tag{1.116}$$

Introducing Eqs. (1.115) and (1.116) into Eq. (1.110) and recalling the definition of u, we obtain

$$r = \frac{k^2}{\mu}\frac{1}{1 + e\cos(\theta - \theta_0)}, \tag{1.117}$$

which represents the equation of a *conic section*. The constant e is recognized as the *eccentricity* of the orbit. A point where $dr/d\theta = 0$ is called an *apsis*. For an open orbit, such as a parabola or a hyperbola, there is only one apsis, whereas for an ellipse there are two *apsides*. For orbits around any center of force the shorter apsis is called the *pericentron*, and the longer one is called the *apocentron*. For orbits around the sun these points are called the *perihelion* and *aphelion*, and for orbits around the earth they are called the *perigee* and *apogee*, respectively. If θ is measured from the pericentron, $\theta_0 = 0$. We shall assume that this is the case here; however, in Sec. 11.1 we relax this assumption and indeed examine the physical meaning of the angle θ_0.

The type of conic depends mathematically on the eccentricity and physically on the total energy. The various possibilities are:

1. Hyperbola: $e > 1$, $E > 0$
2. Parabola: $e = 1$, $E = 0$
3. Ellipse: $0 < e < 1$, $-\mu^2/2k^2 < E < 0$
4. Circle: $e = 0$, $E = -\mu^2/2k^2$

The various orbits are illustrated in Fig. 1.10. Denoting by r_p and r_a the pericentron and apocentron, respectively, we obtain

$$r_p = \frac{k^2}{\mu} \frac{1}{1 + e\cos 0°} = \frac{k^2}{\mu} \frac{1}{1 + e},$$
$$r_a = \frac{k^2}{\mu} \frac{1}{1 + e\cos 180°} = \frac{k^2}{\mu} \frac{1}{1 - e},$$
(1.118)

where θ is assumed to be measured from the pericentron.

FIGURE 1.10

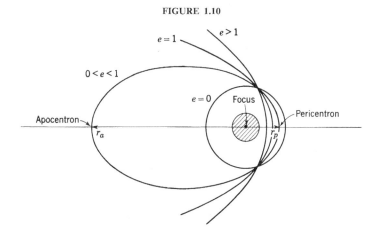

The behavior of the system can be presented in the form of an energy diagram with the radial distance r as the spatial variable. This is so because the problem is reducible to a second-order one. Indeed we can use the angular momentum integral, namely, the second of Eqs. (1.106), to eliminate $\dot{\theta}$ from the energy integral, Eq. (1.113), and obtain

$$E = \tfrac{1}{2}\dot{r}^2 + \frac{1}{2}\frac{k^2}{r^2} - \frac{\mu}{r}. \tag{1.119}$$

Coordinates such as θ, which can be eliminated from the problem formulation, are said to be *ignorable*. In Sec. 2.10 we shall provide the mathematical definition of ignorable coordinates and discuss the physical implication of a coordinate's being ignorable. It is customary to define an equivalent potential energy in the form

$$V'(r) = -\frac{\mu}{r} + \frac{k^2}{2r^2}, \tag{1.120}$$

which includes, in addition to the true gravitational potential energy $-\mu/r$, the term $k^2/2r^2$, which is called the *centrifugal potential* as it is due to the centrifugal force arising from the curvilinear motion. Of course, this has been done at the expense of the kinetic energy, which now includes only the term due to the radial motion. This procedure leads to an equivalent one-dimensional problem, with the axis of motion r being noninertial as it rotates in the orbital plane with the angular velocity $\dot{\theta} = k/r^2$. Introduction of Eq. (1.120) into (1.119) leads to the expression for the radial velocity

$$\dot{r} = \pm\sqrt{2(E - V')}, \tag{1.121}$$

which is similar in structure to Eq. (1.40).

A plot of V' versus r enables us to obtain the energy diagram depicted in Fig. 1.11, which, together with Fig. 1.10, completes the description of the motion. The two dashed curves in Fig. 1.11 represent the component $-\mu/r$, which is due to the true potential energy, and the component $k^2/2r^2$, which corresponds to the centrifugal potential energy. The solid curve, representing the equivalent potential energy V', is the sum of the two dashed curves.

Following is a brief discussion of some of the orbits.

1. *Hyperbola:* $E > 0$. The particle comes in from infinity at the right, reaches the pericentron r_p, where it reverses its motion, and returns to infinity. As there is no upper limit on r, this is an open orbit. The maximum radial velocity is reached at $r = r_0$, where the equivalent potential energy has a minimum and, hence, the equivalent kinetic energy a maximum. At infinity there is a residual velocity.

2. *Parabola:* $E = 0$. The statements made in connection with the hyperbola are valid for the parabola except that the velocities are

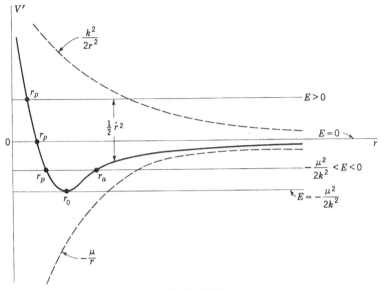

FIGURE 1.11

smaller and at infinity the velocity reduces to zero. The parabola is the open orbit with the lowest energy requirement.

In the case of an earth satellite launched at the perigee, $r_p = R + h$, where R is the radius of the earth and h the perigee height, the total energy can be written as

$$E = \tfrac{1}{2} r_p^2 \dot\theta^2 - \frac{\mu}{r_p} = 0, \qquad (1.122)$$

since the radial velocity is zero at that point. Denoting the tangential velocity there by $v_e = r_p \dot\theta$, it follows that

$$v_e = \sqrt{\frac{2\mu}{r_p}}, \qquad (1.123)$$

where v_e is known as the *escape velocity*; it is the minimum velocity for which an open orbit is achieved.

3. *Ellipse:* $-\mu^2/2k^2 < E < 0$. The motion is periodic, and the radial distance varies between r_p and r_a, corresponding to the intersection of the horizontal line E with the solid curve corresponding to V'.

At this point it may be of interest to calculate the period of oscillation defining the time interval between two successive passages through a given point on the orbit. Let us denote by a the *semimajor axis* of the ellipse so that

$$a = \tfrac{1}{2}(r_p + r_a) = \frac{k^2}{\mu}\frac{1}{1 - e^2} = -\frac{\mu}{2E}, \qquad (1.124)$$

where Eq. (1.116) has been used. The period is obtained by using Eq. (1.117), with $\theta_0 = 0$, and the second of Eqs. (1.106). Upon integrating and subsequently using Eqs. (1.116) and (1.124), we have

$$
\begin{aligned}
P &= \int_0^{2\pi} \frac{r^2\, d\theta}{k} = \frac{2k^3}{\mu^2}\int_0^{\pi} \frac{d\theta}{(1 + e\cos\theta)^2} \\
&= \frac{2k^3}{\mu^2}\frac{1}{1 - e^2}\left[\frac{-e\sin\theta}{1 + e\cos\theta} + \frac{2}{\sqrt{1 - e^2}}\tan^{-1}\frac{\sqrt{1 - e^2}\tan\tfrac{1}{2}\theta}{1 + e}\right]_0^{\pi} \\
&= \frac{2k^3}{\mu^2}\frac{\pi}{(1 - e^2)^{\frac{3}{2}}} = 2\pi\frac{a^{\frac{3}{2}}}{\mu^{\frac{1}{2}}}. \qquad (1.125)
\end{aligned}
$$

Equation (1.125) is the mathematical statement of *Kepler's third law: The squares of the periodic times of the planets are proportional to the cubes of the semiaxes major of the ellipses.* Actually the law as stated by Kepler implies that the proportionality constant is the same for all planets in the solar system, which is only approximately true. From Eq. (1.125), representing the period of a planet as obtained from a two-body system comprising the planet in question and the sun, and Eq. (1.105) we conclude that the proportionality constant is $2\pi/\sqrt{G(m_1 + m_2)}$. If m_1 denotes the mass of the planet and m_2 the mass of the sun, then, assuming that m_1 is much smaller than m_2, the constant can be approximated by $2\pi/\sqrt{Gm_2}$, which is the same for all planets.

The total velocity at any point on the ellipse can be calculated from

$$v^2 = \dot{r}^2 + r^2\dot{\theta}^2 = 2(E - V) = \mu\left(\frac{2}{r} - \frac{1}{a}\right), \qquad (1.126)$$

if the semimajor axis is known.

4. *Circle:* $E = -\mu^2/2k^2$. The constant radial distance corresponds to the point r_0 in Fig. 1.11. The circle can be regarded as the special case of an ellipse with zero eccentricity. In a circular orbit the gravitational and centrifugal forces balance each other, so that the radial velocity is zero and the circular velocity has only the transverse, tangent to the orbit component. The circular velocity is readily calculated by letting $r = a = r_0$ in Eq. (1.126) with the result

$$v_c = \sqrt{\frac{\mu}{r_0}}. \tag{1.127}$$

We notice that for a satellite launched tangentially from a circular orbit the escape velocity is $\sqrt{2}$ times larger than the circular velocity.

1.10 SCATTERING BY A REPULSIVE CENTRAL FORCE

Central forces can occur not only in celestial mechanics but also in atomic and nuclear physics. Although on the atomic scale we can expect the quantum effects to be large, it turns out that in many instances the classical approach leads to satisfactory results. This appears to be the case with the alpha particles from naturally radioactive material scattered by a gold or silver nucleus.

We shall be concerned with the one-body problem in which a particle of mass m is propelled into a repulsive central force field with the fixed center of force at F (Fig. 1.12). In problems of scattering the orbit details are not so much of interest as the initial and final elements. Even these are observed statistically, for the most part, in the form of the probability that scattering over a given angle will take place.

Let the particle be propelled into an inverse square force field of the Coulomb type with an initial velocity v_0 at a point relatively far from the

FIGURE 1.12

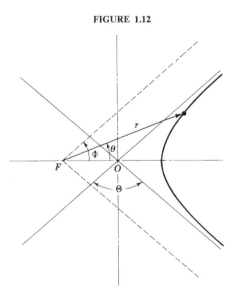

force center. The *impact parameter s* is defined as the distance of nearest approach to center F at which the particle would pass in the absence of the central force. Since the system is conservative, the total energy E is constant, and because the force field is central, the angular momentum L is conserved. Assuming that v_0 is the velocity at infinity, where the potential energy is zero, we have

$$E = T = \tfrac{1}{2}mv_0^2 = \text{const}, \qquad L = mv_0 s = \text{const}. \tag{1.128}$$

If the force field is described by $f(r) = \alpha r^{-2}$, where α is a positive constant, it follows from Eq. (1.101) that $V(u) = \alpha u$, in which u is the reciprocal of the radial distance r. Introducing this expression into Eq. (1.98) and integrating, we obtain

$$u = \frac{1}{r} = \frac{m\alpha}{L^2}(e\cos\theta - 1), \tag{1.129}$$

in which the eccentricity e has the form

$$e = \left(1 + \frac{2EL^2}{m\alpha^2}\right)^{\frac{1}{2}} = \left[1 + \left(\frac{2Es}{\alpha}\right)^2\right]^{\frac{1}{2}}. \tag{1.130}$$

Equation (1.129) represents a conic section, and since $e > 1$, the conic is a hyperbola. But, by definition, the radial distance r is a positive quantity, and so is its reciprocal u. It follows that θ is restricted to the values

$$\cos\theta > \frac{1}{e}. \tag{1.131}$$

Letting $r \to \infty$ in Eq. (1.129), we obtain the asymptotes

$$\Phi = \cos^{-1}\frac{1}{e}. \tag{1.132}$$

The hyperbola and the asymptotes are shown in Fig. 1.12. We notice that while for the attractive force field the center of force coincides with the interior focus, for the repulsive force field it coincides with the exterior one. The angle Θ is referred to as the *scattering angle*. Whereas in classical mechanics this angle is determined uniquely by the initial conditions E and s, in quantum mechanics it is given only in the form of a probability distribution.

It is customary to express the solution of the problem in terms of a *cross section σ for scattering in a given direction*. If there is a flux of N particles per unit area toward the force center F, and if n of these particles are scattered into the corresponding unit solid angle per unit time, then we can write the definition of σ in the form

$$\sigma = \frac{n}{N}. \tag{1.133}$$

Fundamentals of Newtonian Mechanics

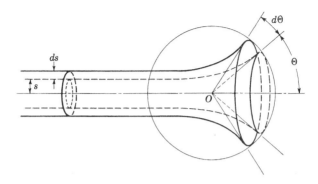

FIGURE 1.13

The total incident flux for a ring of width ds at a distance s from the symmetry axis (see Fig. 1.13) is given by

$$2\pi N s \, ds, \tag{1.134}$$

and if the particles are deflected into an annular region of angular width $d\Theta$, as shown in Fig. 1.13, the solid angle corresponding to this region has the value

$$d\Omega = 2\pi \sin \Theta \, d\Theta. \tag{1.135}$$

Since the number of particles scattered into the solid angle $d\Omega$ must be equal to the number of the incident particles corresponding to the ring of width ds, it follows that

$$2\pi N s \, ds = -2\pi N \sigma \sin \Theta \, d\Theta, \tag{1.136}$$

where the negative sign accounts for the fact that an increase in s brings about a decrease in Θ. Equation (1.136) yields

$$\sigma(\Theta) = -\frac{s \, ds}{\sin \Theta \, d\Theta}. \tag{1.137}$$

But from Eqs. (1.130) and (1.132) we have

$$\cot \frac{\Theta}{2} = \cot\left(\frac{\pi}{2} - \Phi\right) = \tan \Phi = \frac{2Es}{\alpha}, \tag{1.138}$$

so that, introducing Eq. (1.138) into (1.137), we obtain

$$\sigma(\Theta) = \frac{1}{4}\left(\frac{\alpha}{2E}\right)^2 \csc^4 \frac{\Theta}{2}, \tag{1.139}$$

which is the well-known *Rutherford scattering formula*. Equation (1.139) has been verified over a large range of angles, thus providing strong evidence for the nuclear model of the atom.

The *total scattering cross section* is obtained by integrating Eq. (1.137) over the complete solid angle about O

$$\sigma_t = \int_{4\pi} \sigma(\Omega)\, d\Omega = 2\pi \int_0^\pi \sigma(\Theta) \sin\Theta\, d\Theta. \qquad (1.140)$$

Introducing Eq. (1.139) into (1.140) and integrating, we conclude that for this particular scattering process the total cross section diverges. The reason is that the Coulomb potential decreases slowly with increasing distance from the force center, so that its effect extends to infinity. This is another way of saying that all particles in an incident beam of infinite cross section will be scattered to a certain extent.

In the preceding discussion we have regarded the scattering problem as a one-body problem with a fixed scattering center. In reality we have a two-body system which is subject only to the mutual repulsive forces and in which the scattering particle also moves. We have shown in Sec. 1.8 that a two-body problem can be reduced to a one-body problem in which we are concerned with the motion of the center of mass and the motion of one particle relative to the other. As the equivalent one-body problem furnished the relative motion only, we must distinguish between motion in the fixed laboratory frame of reference and the relative motion between the two particles as viewed from the system's moving center of mass. Whereas the scattering angle in the laboratory coordinates, denoted by ϕ, is the angle between the final and incident directions of the scattered particle, in the equivalent one-body problem the angle Θ is the angle between the final and initial directions of the relative vector between the two particles. Of course, when the scattering particle is very large compared with the scattered one, so that the first can be considered at rest, the two angles are the same. This turns out to be so in the case of the alpha particle and the gold or silver nucleus. In this case the mass of the nucleus is so large compared with the mass of the scattered particle that the laboratory and the center of mass systems of reference can be taken as coincident.

PROBLEMS

1.1 A particle lies at rest on the top of a smooth sphere of radius R. If the particle begins to slide from rest under the action of gravity, find the point at which it leaves the sphere and its velocity at that point. The gravitational field is uniform and parallel to the vertical axis of the sphere.

Fundamentals of Newtonian Mechanics

1.2 Find the range R for a projectile fired onto the inclined plane as shown in Fig. 1.14. What is the maximum range for a given muzzle velocity u? The gravitational field is uniform.

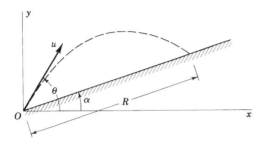

FIGURE 1.14

1.3 A particle m is thrown vertically upward with an initial velocity v. Assuming a resisting medium proportional to the velocity, where the proportionality factor is c, calculate the velocity with which the particle will strike the ground upon its return if there is a uniform gravitational field.

1.4 Two identical particles are dropped from rest in a gravitational field and resisting medium as in Prob. 1.3. If the second particle is released T sec after the first, plot the distance between the particles as a function of time. What is the limiting distance?

1.5 A man of mass m_1 stands in the stern of a boat of mass m_2, the bow of which touches the pier. If he decides to walk toward the pier, what will be his position relative to the pier when he reaches the bow? The boat has a length L and is perpendicular to the pier at all times. Neglect friction.

1.6 A horizontal platform in the form of a disk is supported only at its center. A man of mass m, originally at rest at a distance D from the center, begins to walk with uniform velocity v in the circumferential direction, stopping after an interval of time T. Plot the torque exerted on the support as a function of time.

1.7 Consider a force field in which the force is inversely proportional to the distance squared and directed toward a fixed point C. Use the results of Sec.

1.5 and derive expressions for the potential energy and for the cartesian components of the force.

1.8 Consider the system of Fig. 1.15 and plot the corresponding energy diagram. The mass is released from rest in the position $x = -L/4$, and there is no energy loss in the system.

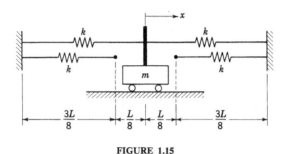

FIGURE 1.15

1.9 A particle moves under a central force inversely proportional to the cube of the distance from the force center. Determine the orbit equation and discuss the possible types of orbits.

1.10 Calculate the energy required to place a satellite in a circular orbit at an altitude h above the earth's surface.

1.11 Derive an expression relating the angular momentum of a system of particles about a moving point A to the angular momentum about the center of mass C.

1.12 A dumbbell is struck by an impulsive force $\hat{F}$, as shown in Fig. 1.16. Derive expressions for the resulting motion.

FIGURE 1.16

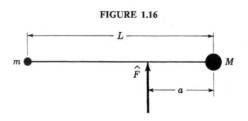

1.13 Consider a three-body system rotating with angular velocity ω about the common center of mass C. The three bodies lie in one line at all times, as shown in Fig. 1.17. If the mass m_0 is so small compared to $1 - m$ and m that it does not affect the motion of the latter two, find the equation for the distance ρ between m_0 and m corresponding to an equilibrium position of m_0 between the masses $1 - m$ and m.

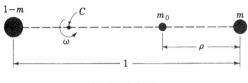

FIGURE 1.17

1.14 Two particles repel each other with forces inversely proportional to the square of the distance between the particles. Derive the orbit equation by integrating the differential equations (Sec. 1.8).

1.15 Consider an elliptic orbit and define the *mean angular motion* by $n = 2\pi/P$, where P is the period. The angle $M = nt$, where t is the time measured from the pericentron, is called the *mean anomaly*, and the angle w (see Fig. 1.18) is known as the *eccentric anomaly*. Derive an expression relating M and w.

FIGURE 1.18

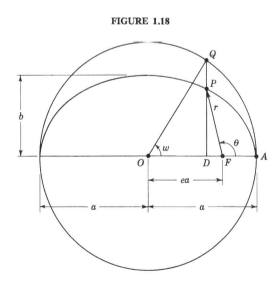

1.16 Consider the two-body problem in an inverse square repulsive central force field for the case in which the two particles are of comparable mass. Let the scattering particle be initially at rest and derive an expression relating the angle ϕ between the final and incident directions of the scattered particle relative to a fixed laboratory system of coordinates and the angle Θ between the final and initial directions of the relative vector between the two particles.

SUGGESTED REFERENCES

1. Corben, H. C., and P. Stehle: *Classical Mechanics*, John Wiley & Sons, Inc., New York, 1960.
2. Goldstein, H.: *Classical Mechanics*, Addison-Wesley Publishing Company, Inc., Reading, Mass., 1950.
3. Lanczos, C.: *The Variational Principles of Mechanics*, University of Toronto Press, Toronto, 1962.
4. Leech, J. W.: *Classical Mechanics*, John Wiley & Sons, Inc., New York, 1965.
5. Robertson, H. P.: in L. N. Ridenour (ed.), *Modern Physics for the Engineer*, McGraw-Hill Book Company, New York, 1954.

chapter two

Fundamentals of Analytical Mechanics

Newton's laws were formulated for single particles and can be extended to systems of particles. An approach to the problems of mechanics, referred to as *vectorial mechanics*, is based directly on Newton's laws and analyzes the motion by means of such concepts as force and momentum, both of which can be regarded as vector quantities. This approach considers the individual bodies and the forces acting upon them separately. It necessitates the calculation of constraint forces resulting from kinematical relations although these forces may be of no interest.

A different approach to the study of mechanics, attributed to Leibnitz and Lagrange and referred to as *analytical mechanics*, considers the system as a whole rather than its individual components. The problems of mechanics are formulated in terms of two fundamental quantities, the kinetic energy and work, both of which are scalar quantities. Whereas the kinematical relations must be properly accounted for, the constraint forces play no particular role and may be obtained only in an incidental way. Added versatility is acquired by introducing generalized coordinates; this represents a departure from the previous exclusive use of physical coordinates and is particularly significant in view of the fact that the mathematical formulations are rendered independent of any special system of coordinates. The special needs of analytical mechanics led to the development of the calculus of variations, for which reason analytical mechanics is often referred to as the *variational approach to mechanics*. The equations of motion are derived from a unified principle which, for a holonomic system, reduces to rendering the value of a definite scalar integral stationary. The higher mathematical abstraction of the analytical mechanics not only completely eliminates the dependence of the formulation on coordinate systems but also permits an efficient treatment of problems associated with multi-degree-of-freedom systems, as well as problems involving curvilinear coordinates or various types of constraints.

2.1 DEGREES OF FREEDOM. GENERALIZED COORDINATES

Let us consider a mechanical system consisting of N particles whose positions are given by the vectors $\mathbf{r}_i = \mathbf{r}_i(x_i, y_i, z_i)$, where x_i, y_i, z_i ($i = 1, 2, \ldots, N$) are the cartesian coordinates of the ith particle. The motion of the mechanical system is completely defined if the positions of all particles i are known functions of the time t, $x_i = x_i(t)$, $y_i = y_i(t)$, $z_i = z_i(t)$. This motion can be conveniently interpreted in terms of geometrical concepts by regarding the $3N$ coordinates x_i, y_i, z_i ($i = 1, 2, \ldots, N$) as the coordinates of a *representative point* P in a $3N$-dimensional space, known as the *configuration space*. The motion of the mechanical system can be visualized simply as the curve traced by the representative point P in the configuration space. The trace is referred to as the *C curve*. In our future discussions we shall make frequent use of the geometrical description of the motion.

Quite frequently we may find it more advantageous to express the motion of the system in terms of a different set of coordinates, say, $q_1, q_2, \ldots, q_n$, where $n = 3N$. The relation between the old and new coordinates can be given by the general form of *coordinate transformation*

$$x_1 = x_1(q_1, q_2, \ldots, q_n),$$
$$y_1 = y_1(q_1, q_2, \ldots, q_n),$$
$$\ldots \ldots \ldots \ldots \ldots \ldots$$
$$z_N = z_N(q_1, q_2, \ldots, q_n),$$
(2.1)

where the transformation is so chosen as to render the dynamical problem in terms of the new coordinates easier. As an illustration, we recall from the preceding chapter that the planar motion of a free particle can be conveniently expressed in terms of the polar coordinates r and θ. The transformation from the cartesian coordinates x and y to the polar coordinates $r = q_1$ and $\theta = q_2$ has the simple form

$$x = q_1 \cos q_2, \qquad y = q_1 \sin q_2. \tag{2.2}$$

Fundamentals of Analytical Mechanics 47

The solution for the motion of the equivalent particle corresponding to the two-body system was considerably simplified by the choice of polar rather than cartesian coordinates.

In many physical problems the mass particles are not free but are subject to kinematical conditions restricting their freedom of motion. In such cases the advantage of using a more general type of coordinate, rather than cartesian ones, becomes more apparent. As a simple example, let us consider a dumbbell, namely, two particles connected by a rigid rod. The motion of two free particles is completely defined by the six coordinates $x_1, y_1, \ldots, z_2$. In the case of the dumbbell these coordinates are not independent but related by

$$(x_1 - x_2)^2 + (y_1 - y_2)^2 + (z_1 - z_2)^2 = L^2 = \text{const}, \quad (2.3)$$

where L is the distance between the two particles. It is clear that in this case the problem can be regarded as solved if we know only five of the coordinates because the sixth coordinate can be determined by using Eq. (2.3). The motion of the dumbbell may be more conveniently defined by the three coordinates of one of the particles and two angles establishing the orientation of the rod.

In general, the motion of a system of N particles subject to c kinematical conditions can be described uniquely by n independent coordinates q_k ($k = 1, 2, \ldots, n$), where

$$n = 3N - c \quad (2.4)$$

is the number of *degrees of freedom* of the system. The number of degrees of freedom of a system coincides with the minimum number of independent coordinates necessary to describe the system uniquely. The n coordinates $q_1, q_2, \ldots, q_n$ are referred to as *generalized coordinates*. Constraints may be interpreted as restricting the motion to a *subspace* of a correspondingly smaller dimension.

At times it may not be possible, or advisable, to eliminate the excess coordinates, making it necessary to work with a larger number of coordinates than the number of degrees of freedom would require. In such cases, auxiliary conditions must be retained in a number equal to the number of coordinates exceeding the degrees of freedom of the system. In Sec. 2.2 we shall explore the question of constraint conditions in detail.

The generalized coordinates $q_1, q_2, \ldots, q_n$ may not always have physical meaning. Nor are they unique, which implies that there may be more than one set of coordinates capable of describing the system completely. They must, however, be finite, single-valued, and continuous and differentiable with respect to time. The concept of generalized coordinates provides us with the opportunity to enlarge our horizon by accepting as coordinates such quantities as the amplitudes of Fourier series expansions or certain functions of

2.2 SYSTEMS WITH CONSTRAINTS

It was pointed out in Sec. 2.1 that, with the exception of a system of free particles, the displacements defining the motion of a system of particles are not independent but subject to certain auxiliary conditions, such as Eq. (2.3). It turns out that Eq. (2.3) represents but one type of kinematical constraints and more general forms are possible.

To introduce the ideas, let us first consider a single particle whose motion is confined to a smooth surface. The equation of the surface is

$$f(x,y,z) = 0, \tag{2.5}$$

where f is a function of class C_2 (see Sec. 1.2 for definition). Since the coordinates x, y, z of the particle must satisfy the kinematical condition (2.5), it follows that the particle is not free. Because the particle velocity must be tangent to the surface at any point, the velocity components must satisfy the homogeneous linear relation

$$\frac{\partial f}{\partial x}\dot{x} + \frac{\partial f}{\partial y}\dot{y} + \frac{\partial f}{\partial z}\dot{z} = 0, \tag{2.6}$$

which is obtained by differentiating Eq. (2.5) with respect to time. We note that $\partial f/\partial x$, $\partial f/\partial y$, and $\partial f/\partial z$, which are functions of class C_1, are simply the direction numbers associated with the direction cosines between the normal to f and axes x, y, z. Equation (2.6) also implies that the displacements dx, dy, and dz must be such that the homogeneous differential expression

$$\frac{\partial f}{\partial x}dx + \frac{\partial f}{\partial y}dy + \frac{\partial f}{\partial z}dz = 0 \tag{2.7}$$

is satisfied. The differential expression (2.7) is said to be in a *Pfaffian form*. We shall encounter more general Pfaffian forms later. The class of displacements $d\mathbf{r}$ with components dx, dy, dz, which the particle may undergo in the time interval dt, constitutes a class referred to as *possible displacements*. The *actual displacement* that the particle undergoes is one of the totality of possible displacements.

To confine the motion to the surface (2.5), there must be, besides the given forces, an additional force acting on the particle, namely, the reaction of the surface. This force, called the *constraint force*, is assumed to be normal to the surface at any point, which implies a perfectly smooth surface. Reaction forces due to friction, such as those caused by sliding on rough surfaces, are not included in this formulation but are treated separately (see Sec. 2.12). Denoting the constraint force by $\mathbf{F}'$, we conclude that its components F_x', F_y', F_z' must be proportional to the direction numbers $\partial f/\partial x$, $\partial f/\partial y$, $\partial f/\partial z$, respectively. Hence

$$\frac{F_x'}{\partial f/\partial x} = \frac{F_y'}{\partial f/\partial y} = \frac{F_z'}{\partial f/\partial z}. \tag{2.8}$$

In view of Eq. (2.7) and recalling definition (1.15), we see that Eq. (2.8) is equivalent to

$$\overline{dW} = F_x'\,dx + F_y'\,dy + F_z'\,dz = \mathbf{F}' \cdot d\mathbf{r} = 0, \tag{2.9}$$

which simply states that *the work performed by the constraint force in any possible displacement is zero.*

Because the motion must satisfy one constraint equation, the system is reduced from three degrees of freedom to two. If there are two constraint equations,

$$f_1(x,y,z) = 0, \qquad f_2(x,y,z) = 0, \tag{2.10}$$

the motion is confined to the intersection of the two surfaces given by Eqs. (2.10), and a single-degree-of-freedom system obtains.

The situation is entirely different when the motion of the particle is confined to the surface having the equation

$$f(x,y,z,t) = 0, \tag{2.11}$$

where f is again of class C_2. In contrast with the preceding case, however, Eq. (2.11) indicates that f is a function not only of the spatial variables x, y, z but also of the time t, which implies that the surface is in motion. As a result, the velocity components must satisfy the relation

$$\frac{\partial f}{\partial x}\dot{x} + \frac{\partial f}{\partial y}\dot{y} + \frac{\partial f}{\partial z}\dot{z} + \frac{\partial f}{\partial t} = 0, \tag{2.12}$$

which is no longer homogeneous in the velocities $\dot{x}$, $\dot{y}$, and $\dot{z}$. Moreover, the partial derivatives in Eq. (2.12) are functions of the spatial variables x, y, z as well as the time t. In an analogous manner, the displacements dx, dy, and dz satisfy

$$\frac{\partial f}{\partial x}dx + \frac{\partial f}{\partial y}dy + \frac{\partial f}{\partial z}dz + \frac{\partial f}{\partial t}dt = 0. \tag{2.13}$$

On the other hand, the relation between the components of the constraint force $\mathbf{F}'$ and the partial derivatives $\partial f/\partial x$, $\partial f/\partial y$, and $\partial f/\partial z$ still retains the form (2.8). Hence, we must conclude that *when the constraint is time-dependent, the work performed by the constraint force in any possible displacement is no longer zero.*

At this point we introduce another class of displacements, namely, the *virtual displacements* δx, δy, and δz. These are not true displacements but infinitesimal changes in the coordinates, consistent with the system constraints, and imagined to take place contemporaneously. This is another way of saying that the virtual displacements are merely the result of imagining the system in a slightly different position, a process which does not involve the time element. (For further discussion of virtual displacements, see Sec. 2.3.) As no time change is involved, the virtual displacements satisfy

$$\frac{\partial f}{\partial x}\delta x + \frac{\partial f}{\partial y}\delta y + \frac{\partial f}{\partial z}\delta z = 0, \qquad (2.14)$$

rather than Eq. (2.13), which is satisfied by the true displacements. Since the components of the constraint force $\mathbf{F}'$ are proportional to the partial derivatives in Eq. (2.14), we can use the analogy with Eq. (2.9) and conclude that *the work performed by the constraint force in any virtual displacement is zero*

$$\overline{\delta W} = F'_x\,\delta x + F'_y\,\delta y + F'_z\,\delta z = 0. \qquad (2.15)$$

It is also clear that in the case of a moving constraint the virtual displacements are not possible displacements, as shown in Fig. 2.1.

Equation (2.7) can be regarded as representing the perfect differential $df(x,y,z)$, and Eq. (2.13) similarly represents the perfect differential $df(x,y,z,t)$. Hence, if the constraint equations are given in the form of expressions such as (2.6) and (2.12) relating the velocity components $\dot{x}$, $\dot{y}$, and $\dot{z}$ (as is often the case in a practical situation), and if these expressions are integrable, then they are, in fact, equivalent to constraints of the form (2.5) and (2.11). A more

FIGURE 2.1

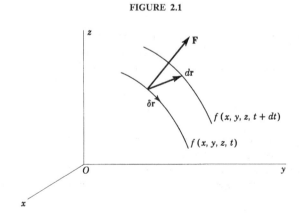

general equation of constraint is represented by the Pfaffian form

$$a_x \, dx + a_y \, dy + a_z \, dz + a_0 \, dt = 0, \tag{2.16}$$

where a_x, a_y, a_z, and a_0 are functions of class C_1 depending on x, y, z, and t. By contrast with the preceding cases, however, Eq. (2.16) is not integrable, so that it is not possible to reduce it to an expression of the type (2.11). Whereas the possible displacements satisfy Eq. (2.16), the virtual displacements satisfy

$$a_x \, \delta x + a_y \, \delta y + a_z \, \delta z = 0. \tag{2.17}$$

As a result of the constraint, a constraint force $\mathbf{F}'$ is produced. The constraint force and the given force together determine the actual motion, namely, one of the possible motions satisfying Eq. (2.16). But the constraint force is such that it performs no work in any virtual displacement. It follows that the components of $\mathbf{F}'$ are proportional to the coefficients a_x, a_y, and a_z. Denoting the proportionality constant by λ, we have

$$F'_x = \lambda a_x, \qquad F'_y = \lambda a_y, \qquad F'_z = \lambda a_z. \tag{2.18}$$

We are now in a position to classify the various types of constraints. If the coefficient a_0 in Eq. (2.16) is zero, the system is said to be *catastatic*, in which case the classes of virtual and possible displacements are the same; otherwise the system is called *acatastatic*. If the Pfaffian form (2.16) is integrable, so that the constraint can be reduced to an equation of the type (2.11), the system is *holonomic*. Equations (2.7) and (2.13) are examples of integrable Pfaffian forms, as they represent perfect differentials which can be integrated to yield Eqs. (2.5) and (2.11) (within an additive constant), respectively. If the Pfaffian form (2.16) admits no integrating factor, the system is *nonholonomic*. Finally, if a holonomic constraint does not contain the time explicitly and hence is of the form (2.5), the system is said to be *scleronomic*. If, on the other hand, the holonomic constraint is of the type (2.11), the system is *rheonomic*. It is clear, in retrospect, that Eq. (2.3) represents a scleronomic constraint.

An entirely different class of constraints restricts the particle not to a surface but to the domain bounded by a surface. These are known as *inequality constraints* and are of the form

$$f(x,y,z) \geq 0 \tag{2.19}$$

if the bounding surface is stationary, and

$$f(x,y,z,t) \geq 0 \tag{2.20}$$

if the bounding surface is in motion. Both constraints are nonholonomic.

The subject of constraints can be generalized by considering a system defined by n coordinates u_k ($k = 1, 2, \ldots, n$) which are not independent but related by m constraint equations having the Pfaffian forms

$$\sum_{k=1}^{n} a_{jk}\, du_k + a_{j0}\, dt = 0, \qquad j = 1, 2, \ldots, m. \tag{2.21}$$

The coefficients a_{jk} ($k = 0, 1, 2, \ldots, n$) are known functions of class C_1 in the variables $u_1, u_2, \ldots, u_n$ and the time t. We shall assume that Eqs. (2.21) are independent, which implies that the matrix of the coefficients must be of rank m. If *all* the coefficients a_{j0} ($j = 1, 2, \ldots, m$) are zero, the system is catastatic, and if at least one of the coefficients a_{j0} is different from zero, the system is acatastatic.

Whereas the possible displacements du_k satisfy Eqs. (2.21), the virtual displacements δu_k ($k = 1, 2, \ldots, n$) satisfy

$$\sum_{k=1}^{n} a_{jk}\, \delta u_k = 0, \qquad j = 1, 2, \ldots, m, \tag{2.22}$$

and, by analogy with the conclusions concerning the single constraint equation, we can state that the work performed by the constraint forces through the virtual displacements δu_k is zero. These forces will be shown to have the form (see Sec. 2.5)

$$F_k' = \sum_{j=1}^{m} \lambda_j a_{jk}, \qquad k = 1, 2, \ldots, n, \tag{2.23}$$

where λ_j are undetermined multipliers related to the magnitude of the constraint forces.

When all the Pfaffians (2.21) are integrable, hence reducible to the perfect differentials

$$df_j = 0, \qquad j = 1, 2, \ldots, m, \tag{2.24}$$

the system is holonomic, and if, in addition, the time t does not appear explicitly in any of the integrals f_j, the system is scleronomic.

The number of degrees of freedom of the system is $n' = n - m$. However, when the system is nonholonomic, it is not possible to solve the m constraint equations for the corresponding coordinates u_k, so that we are forced to work with a number of coordinates exceeding the degrees of freedom of the system. This is permissible provided the surplus number of coordinates matches the number of constraint equations. Although in the case of holonomic systems it may be possible to solve for the excess coordinates, thus eliminating them, this is not always necessary or desirable. If surplus coordinates are used, the corresponding constraint equations must be retained.

Since for holonomic constraints we can eliminate, at least in principle, the surplus coordinates, we shall refer to unconstrained systems as holonomic.

2.3 THE STATIONARY VALUE OF A FUNCTION

The problem of finding the *extremum* of a given function, i.e., the *maximum* or *minimum* of that function, has always excited the interest of mathematicians. Problems of dynamics are less complicated, in the sense that it is often sufficient to find merely the *stationary value* of the function. A function is said to have a stationary value at a certain point if the rate of change in every direction at that point is zero. In problems associated with the stability of motion of a dynamical system in the neighborhood of a stationary point we find it necessary to examine the nature of that point and determine whether it is a maximum, a minimum, or a saddle point. Such problems will be discussed in detail in Chaps. 5, 6, and 12. For a function depending only on two variables, stationary points can be easily visualized in a three-dimensional space in terms of a relief map in which the maximum corresponds to the top of a hill, etc. The same ideas can be extended to functions of n variables of the type

$$f = f(u_1, u_2, \ldots, u_n), \qquad (2.25)$$

which can be interpreted as a surface in an $(n + 1)$-dimensional space consisting of the n-dimensional configuration space and f. It must be mentioned that on the boundaries of the configuration space it is possible to have an extremum without actually having a stationary value.

An entire branch of mathematics, known as the *calculus of variations*, has been developed as a generalization of the elementary theory of maxima and minima. The method of finding a stationary point is to examine the immediate neighborhood of that point by introducing *variations* in the coordinates u_i ($i = 1, 2, \ldots, n$) at a given time. These small variations, called *virtual displacements*, are imagined infinitesimal changes in the coordinates which must be *consistent with the constraints of the system* but are otherwise *arbitrary*. The concept of virtual displacements was first introduced in Sec. 2.2. Lagrange introduced the special symbol δ to emphasize the virtual character of the instantaneous variations, as opposed to the symbol d designating actual differentials of displacements taking place in the time interval dt during which the forces and constraints may change. The virtual displacements obey the rules of differential calculus, so that the variation in the function f corresponding to the virtual displacements δu_i ($i = 1, 2, \ldots, n$) is

$$\delta f = \frac{\partial f}{\partial u_1} \delta u_1 + \frac{\partial f}{\partial u_2} \delta u_2 + \cdots + \frac{\partial f}{\partial u_n} \delta u_n = \sum_{i=1}^{n} \frac{\partial f}{\partial u_i} \delta u_i. \qquad (2.26)$$

The function f has a stationary value at points for which $\delta f = 0$. Due to the arbitrary nature of the virtual displacements, the conditions for f to have a stationary value are

$$\frac{\partial f}{\partial u_i} = 0, \quad i = 1, 2, \ldots, n, \tag{2.27}$$

provided that the system is not subject to any constraints. A variation problem involving no constraints is referred to as a *free variation problem*.

Next let us consider the case in which the variables u_i are not independent but subject to m *nonholonomic constraints* of the type

$$\sum_{k=1}^{n} a_{jk}\, du_k + a_{j0}\, dt = 0, \quad j = 1, 2, \ldots, m, \tag{2.28}$$

where the coefficients a_{jk} ($k = 0, 1, 2, \ldots, n$) are functions of the coordinates u_i. We recall from Sec. 2.2 that the matrix of the coefficients must be of rank m. To obtain stationary values of f for this case, we can use the elegant *method of undetermined multipliers* devised by Lagrange. Because the coordinates u_i are not independent, the variations δu_i are not arbitrary but must obey the relations

$$\sum_{k=1}^{n} a_{jk}\, \delta u_k = 0, \quad j = 1, 2, \ldots, m. \tag{2.29}$$

Thus we cannot conclude that the conditions for a stationary value of f are $\partial f/\partial u_i = 0$ ($i = 1, 2, \ldots, n$). It is possible, however, to solve Eqs. (2.29) for m values of δu_i ($i = n - m + 1, n - m + 2, \ldots, n$), introduce these values into Eq. (2.26), and treat the remaining δu_i ($i = 1, 2, \ldots, n - m$) as independent. The method of Lagrange multipliers provides a more refined approach. According to this method, we multiply each of the Eqs. (2.29) by an undetermined multiplier λ_j, correspondingly, and add all resulting expressions to Eq. (2.26). This operation implies adding zero to δf, but since the components of the added sum are not zero, the condition for a stationary value of f can be written as

$$\sum_{i=1}^{n} \left(\frac{\partial f}{\partial u_i} + \sum_{j=1}^{m} \lambda_j a_{ji} \right) \delta u_i = 0. \tag{2.30}$$

The virtual displacements δu_i are still not independent, but the m values λ_j can be chosen so that

$$\frac{\partial f}{\partial u_i} + \sum_{j=1}^{m} \lambda_j a_{ji} = 0, \quad i = n - m + 1, n - m + 2, \ldots, n, \tag{2.31}$$

whereas the remaining virtual displacements δu_i ($i = 1, 2, \ldots, n - m$) can be regarded as independent and chosen arbitrarily. This leads to

Fundamentals of Analytical Mechanics

$$\frac{\partial f}{\partial u_i} + \sum_{j=1}^{m} \lambda_j a_{ji} = 0, \qquad i = 1, 2, \ldots, n - m. \tag{2.32}$$

We conclude that f has a stationary value if the coefficient of each virtual displacement δu_i ($i = 1, 2, \ldots, n$) in Eq. (2.30) vanishes. This procedure enables us to treat all the virtual displacements as independent provided that the variation δf is augmented as shown in Eq. (2.30). Now we have $n + m$ equations, (2.31), (2.32), and (2.28), and $n + m$ unknowns, n values of u_i, and m values of λ_j.

The method for treating the case in which there are m *holonomic* constraints

$$g_j(u_1, u_2, \ldots, u_n, t) = 0, \qquad j = 1, 2, \ldots, m \tag{2.33}$$

follows immediately by writing

$$\delta g_j = \sum_{k=1}^{n} \frac{\partial g_j}{\partial u_k} \delta u_k = 0, \qquad j = 1, 2, \ldots, m, \tag{2.34}$$

which is always possible. Hence, we simply replace a_{ji} by $\partial g_j / \partial u_i$ in Eqs. (2.31) and (2.32).

The variation problem subject to the constraint equations (2.33) can be regarded as a free variation problem if the function to be rendered stationary is augmented so that it takes the form

$$\bar{f} = f + \sum_{j=1}^{m} \lambda_j g_j, \tag{2.35}$$

leading to n stationarity conditions. These conditions must be supplemented by the m constraint equations (2.33), yielding a total of $n + m$ equations in the variables u_k ($k = 1, 2, \ldots, n$) and the parameters λ_j ($j = 1, 2, \ldots, m$).

The advantage of using the method of Lagrange multipliers is that we need not choose which variables to eliminate and which to retain, thus preserving the symmetry of the formulation.

2.4 THE STATIONARY VALUE OF A DEFINITE INTEGRAL

A large number of problems in mechanics can be reduced to the problem of finding the stationary value of a definite integral. The well-known brachistochrone problem was first formulated and solved by Bernoulli, leading to the foundation of the calculus of variations. The problem consists of finding a curve in the vertical plane, beginning at point a and ending at point b, along which a particle descends under gravity in the minimum amount of time. The problem can be generalized by considering the determination of a function $y(x)$ which renders the definite integral

$$I = \int_{x_1}^{x_2} F(x,y,y')\, dx \tag{2.36}$$

an extremum, or at least stationary, where the function y satisfies the end conditions

$$y(x_1) = y_1, \qquad y(x_2) = y_2. \tag{2.37}$$

Primes indicate differentiations with respect to the independent variable x. The function F is twice continuously differentiable with respect to its three arguments x, y, and y'.

To find the stationary value of a definite integral we use a different approach than for ordinary functions. For this purpose we form the one-parameter family of varied functions $\bar{y}(x)$, which differ from the original function $y(x)$ by an arbitrary virtual change $\delta y(x)$ (see Fig. 2.2)

$$\bar{y}(x) = y(x) + \delta y(x) = y(x) + \epsilon \eta(x), \tag{2.38}$$

where the parameter ϵ of the family is a quantity sufficiently small for all the varied functions $\bar{y}$ to lie in an arbitrarily small neighborhood of y. The arbitrary new function $\eta(x)$ satisfies the same differentiability conditions as $y(x)$, and, in addition, it satisfies the end conditions

$$\eta(x_1) = \eta(x_2) = 0. \tag{2.39}$$

We note that in this particular type of path variation only the dependent function y is varied, whereas $\delta x = 0$. In a more general process we can also vary x (see Sec. 9.1).

Next let us replace y by $\bar{y}$ and y' by $\bar{y}'$ in Eq. (2.36) and obtain

$$I(\epsilon) = \int_{x_1}^{x_2} F(x,\bar{y},\bar{y}')\, dx, \tag{2.40}$$

FIGURE 2.2

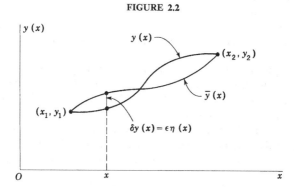

Fundamentals of Analytical Mechanics

which may be regarded as a function of ϵ for any given function η. When $\epsilon = 0$, integral (2.40) reduces to (2.36). Because y is the actual function rendering the integral stationary, it follows that the integral $I(\epsilon)$ has a stationary value at $\epsilon = 0$ with respect to all values of ϵ in a sufficiently small neighborhood of 0. As the variable of integration is x rather than ϵ, the procedure reduces to an ordinary stationary-value problem, so that we must solve

$$\left.\frac{dI(\epsilon)}{d\epsilon}\right|_{\epsilon=0} = 0. \tag{2.41}$$

The above must be true regardless of the function η. The derivative of the integral (2.40) with respect to the parameter ϵ is

$$\frac{dI(\epsilon)}{d\epsilon} = \int_{x_1}^{x_2} \left(\frac{\partial F}{\partial \bar{y}}\frac{\partial \bar{y}}{\partial \epsilon} + \frac{\partial F}{\partial \bar{y}'}\frac{\partial \bar{y}'}{\partial \epsilon}\right) dx = \int_{x_1}^{x_2} \left(\frac{\partial F}{\partial \bar{y}}\eta + \frac{\partial F}{\partial \bar{y}'}\eta'\right) dx. \tag{2.42}$$

Next we let $\epsilon = 0$, integrate by parts, and set the result equal to zero

$$\left.\frac{dI(\epsilon)}{d\epsilon}\right|_{\epsilon=0} = \int_{x_1}^{x_2}\left(\frac{\partial F}{\partial y}\eta + \frac{\partial F}{\partial y'}\eta'\right)dx = \int_{x_1}^{x_2}\left[\frac{\partial F}{\partial y} - \frac{d}{dx}\left(\frac{\partial F}{\partial y'}\right)\right]\eta\, dx = 0, \tag{2.43}$$

where conditions (2.39) have been taken into account. Because η may be any arbitrary function, it follows that y must satisfy the differential equation

$$\frac{\partial F}{\partial y} - \frac{d}{dx}\left(\frac{\partial F}{\partial y'}\right) = 0, \qquad x_1 < x < x_2, \tag{2.44}$$

for expression (2.43) to vanish. Equation (2.44), known as the *Euler-Lagrange differential equation*, represents the necessary condition for the definite integral I to have a stationary value provided conditions (2.37) are satisfied.

The same problem can be treated by a direct variational approach. To this end, the derivative of the variation can be written

$$\frac{d}{dx}\delta y = \frac{d}{dx}(\epsilon\eta) = \epsilon\eta'. \tag{2.45}$$

On the other hand,

$$\delta\frac{dy}{dx} = \bar{y}' - y' = (y' + \epsilon\eta') - y' = \epsilon\eta' = \frac{d}{dx}\delta y, \tag{2.46}$$

so that the derivative of the variation is equal to the variation of the derivative. Next we define the first variation of the integral I as

$$\delta I = \epsilon \left.\frac{dI(\epsilon)}{d\epsilon}\right|_{\epsilon=0} = \epsilon\int_{x_1}^{x_2}\left(\frac{\partial F}{\partial y}\eta + \frac{\partial F}{\partial y'}\eta'\right)dx$$

$$= \int_{x_1}^{x_2}\left(\frac{\partial F}{\partial y}\delta y + \frac{\partial F}{\partial y'}\delta y'\right)dx. \tag{2.47}$$

Integrating Eq. (2.47) by parts and using Eq. (2.46), we obtain

$$\delta I = \int_{x_1}^{x_2} \left[\frac{\partial F}{\partial y} - \frac{d}{dx}\left(\frac{\partial F}{\partial y'}\right)\right] \delta y \, dx, \quad (2.48)$$

where the end conditions

$$\delta y(x_1) = \delta y(x_2) = 0 \quad (2.49)$$

have been considered. Hence, as expected, the Euler-Lagrange differential equation is again obtained as the condition for the variation δI to vanish.

The method can be extended to multi-degree-of-freedom systems unconstrained or constrained. Holonomic as well as nonholonomic constraints can be treated by a procedure similar to the one in Sec. 2.3. All these topics will be discussed in this chapter.

Example 2.1

Determine the curve $y(x)$ passing through the points (x_1, y_1) and (x_2, y_2) (see Fig. 2.3) which, when rotated about the x axis, yields a surface of revolution with the minimum surface area.

The area of the surface of revolution thus obtained is given by the definite integral

$$I = 2\pi \int_{x_1, y_1}^{x_2, y_2} y \, ds = 2\pi \int_{x_1}^{x_2} y(1 + y'^2)^{\frac{1}{2}} \, dx. \quad (a)$$

Hence, the problem reduces to that of minimizing the definite integral I subject to the conditions

$$y(x_1) = y_1, \qquad y(x_2) = y_2. \quad (b)$$

FIGURE 2.3

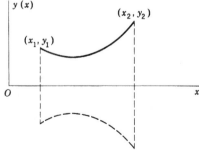

Fundamentals of Analytical Mechanics

This can be accomplished by solving the Euler-Lagrange equation (2.44), in which

$$F = y(1 + y'^2)^{\frac{1}{2}}. \tag{c}$$

The constant 2π is immaterial.

Introducing Eq. (c) into Eq. (2.44), we can write

$$\frac{d}{dx}\left[\frac{yy'}{(1 + y'^2)^{\frac{1}{2}}}\right] - (1 + y'^2)^{\frac{1}{2}} = 0, \tag{d}$$

which reduces to

$$yy'' - y'^2 - 1 = 0. \tag{e}$$

Next we introduce the substitution $y' = p$, $y'' = p\, dp/dy$, which enables us to separate the variables as follows

$$\frac{p\, dp}{1 + p^2} = \frac{dy}{y}, \tag{f}$$

yielding

$$y = A(1 + p^2)^{\frac{1}{2}} = A(1 + y'^2)^{\frac{1}{2}}. \tag{g}$$

Equation (g) leads to

$$\frac{dy}{dx} = \left(\frac{y^2 - A^2}{A^2}\right)^{\frac{1}{2}}, \tag{h}$$

which can be integrated to obtain

$$y = A \cosh \frac{x + B}{A}. \tag{i}$$

Equation (i) represents a catenary. The question remains whether it is possible to use boundary conditions (b) to evaluate the constants. This involves the solution of a transcendental equation, and the solution may not always exist.

2.5 THE PRINCIPLE OF VIRTUAL WORK

The principle of virtual work is basically a statement of the static equilibrium of a mechanical system. It is the first variational principle in mechanics, and it can be employed to facilitate the transition from Newtonian to Lagrangian mechanics.

Let us concern ourselves with a system of N particles, each acted on by a set of forces with resultant $\mathbf{R}_i$ ($i = 1, 2, \ldots, N$). For a system in equilibrium the total force on each particle vanishes, $\mathbf{R}_i = \mathbf{0}$, and the same can be said about the dot product $\mathbf{R}_i \cdot \delta\mathbf{r}_i$, representing the virtual work done by the

ith particle over the corresponding virtual displacement $\delta \mathbf{r}_i$. It follows that the virtual work of the entire system vanishes

$$\overline{\delta W} = \sum_{i=1}^{N} \mathbf{R}_i \cdot \delta \mathbf{r}_i = 0. \qquad (2.50)$$

The above result is quite trivial, as nothing new has been gained. However, if the system is subject to constraints, this approach has interesting implications. Indeed, in this case we can distinguish between the *applied*, or *impressed, forces* $\mathbf{F}_i$ and the constraint forces $\mathbf{F}'_i$, so that

$$\mathbf{R}_i = \mathbf{F}_i + \mathbf{F}'_i, \qquad i = 1, 2, \ldots, N. \qquad (2.51)$$

Introducing Eq. (2.51) into (2.50), we obtain simply

$$\overline{\delta W} = \sum_{i=1}^{N} \mathbf{F}_i \cdot \delta \mathbf{r}_i + \sum_{i=1}^{N} \mathbf{F}'_i \cdot \delta \mathbf{r}_i. \qquad (2.52)$$

By the definition of the virtual displacements (Sec. 2.2), however, the work of the constraint forces through virtual displacements compatible with the system constraints is zero

$$\sum_{i=1}^{N} \mathbf{F}'_i \cdot \delta \mathbf{r}_i = 0. \qquad (2.53)$$

The type of virtual displacements for which Eq. (2.53) holds is referred to as *reversible* in the geometric sense. Confining ourselves to such virtual displacements, we arrive at

$$\sum_{i=1}^{N} \mathbf{F}_i \cdot \delta \mathbf{r}_i = 0 \qquad (2.54)$$

or *the work done by the applied forces in infinitesimal reversible virtual displacements compatible with the system constraints is zero*. This is the statement of the *principle of virtual work*.

Equation (2.54) represents a substantially new result. Because for systems with constraints the virtual displacements $\delta \mathbf{r}_i$ are not all independent, Eq. (2.54) cannot be interpreted to imply that $\mathbf{F}_i = \mathbf{0}$ ($i = 1, 2, \ldots, N$). If the problem is described by a system of generalized coordinates, Eq. (2.54) assumes the form

$$\sum_{k=1}^{n} Q_k \, \delta q_k = 0, \qquad (2.55)$$

Fundamentals of Analytical Mechanics 61

where Q_k $(k = 1, 2, \ldots, n)$ are known as *generalized forces*. We shall return to this concept in Sec. 2.8. Since now the number of generalized coordinates coincides with the number of degrees of freedom of the system, the virtual generalized displacements δq_k are all independent, hence arbitrary, so that $Q_k = 0$ $(k = 1, 2, \ldots, n)$.

The principle of virtual work can be used to calculate the position of static equilibrium of a system.

As an interesting illustration let us consider a conservative system and write the virtual work expression

$$\overline{\delta W} = \sum_{i=1}^{N} \mathbf{F}_i \cdot \delta \mathbf{r}_i = -\delta V = -\sum_{i=1}^{N} \left(\frac{\partial V}{\partial x_i} \delta x_i + \frac{\partial V}{\partial y_i} \delta y_i + \frac{\partial V}{\partial z_i} \delta z_i \right) = 0. \quad (2.56)$$

For no constraints acting on the system, the virtual displacements δx_i, δy_i, and δz_i $(i = 1, 2, \ldots, N)$ are all independent, so that the conditions for static equilibrium are

$$F_{x_i} = -\frac{\partial V}{\partial x_i} = 0, \qquad F_{y_i} = -\frac{\partial V}{\partial y_i} = 0, \qquad F_{z_i} = -\frac{\partial V}{\partial z_i} = 0, \quad (2.57)$$

or, as expected, all the components of the applied forces must be zero. However, in terms of V, conditions (2.57) are precisely the ones rendering the potential energy stationary.

If the system is subject to the holonomic constraints

$$g_j(x_1, y_1, z_1, x_2, \ldots, z_N) = 0, \qquad j = 1, 2, \ldots, m, \quad (2.58)$$

we can use the method of Lagrange multipliers (Sec. 2.3) and seek to render stationary the augmented potential energy

$$\overline{V} = V - \sum_{j=1}^{m} \lambda_j g_j, \quad (2.59)$$

leading to the equilibrium conditions

$$\frac{\partial V}{\partial x_i} - \sum_{j=1}^{m} \lambda_j \frac{\partial g_j}{\partial x_i} = 0,$$

$$\frac{\partial V}{\partial y_i} - \sum_{j=1}^{m} \lambda_j \frac{\partial g_j}{\partial y_i} = 0, \qquad i = 1, 2, \ldots, N. \quad (2.60)$$

$$\frac{\partial V}{\partial z_i} - \sum_{j=1}^{m} \lambda_j \frac{\partial g_j}{\partial z_i} = 0,$$

Equations (2.60) imply that the equilibrium of the system is ensured by the presence of the constraint forces

$$F'_{x_i} = \sum_{j=1}^{m} \lambda_j \frac{\partial g_j}{\partial x_i},$$

$$F'_{y_i} = \sum_{j=1}^{m} \lambda_j \frac{\partial g_j}{\partial y_i}, \quad i = 1, 2, \ldots, N, \quad (2.61)$$

$$F'_{z_i} = \sum_{j=1}^{m} \lambda_j \frac{\partial g_j}{\partial z_i},$$

which verifies the statement that the multipliers λ_j ($j = 1, 2, \ldots, m$) are related to the magnitudes of the constraint forces resulting from the kinematical conditions (2.58) (see Sec. 2.2). The same approach can be extended to problems of dynamical equilibrium of systems with holonomic and nonholonomic constraints.

Example 2.2

The system in Fig. 2.4 consists of a mass m suspended through a massless link of length L and a spring of stiffness k, as shown. When the spring is unstretched, the link is in the horizontal position. Use the principle of virtual work to calculate the angle θ corresponding to the equilibrium position of the system.

Let us first denote the position of the ends of the link in the equilibrium position by

$$x = L(1 - \cos \theta), \quad y = L \sin \theta, \quad (a)$$

FIGURE 2.4

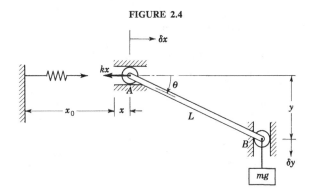

where x is the elongation of the spring and y the lowering of m. The virtual work can be written in the form

$$\overline{\delta W} = -kx\,\delta x + mg\,\delta y = 0, \qquad (b)$$

where the virtual displacements δx and δy must be equal to

$$\delta x = L \sin\theta\,\delta\theta, \qquad \delta y = L\cos\theta\,\delta\theta, \qquad (c)$$

by virtue of Eqs. (a). Introducing the first of Eqs. (a) and both Eqs. (c) into (b), we obtain

$$\overline{\delta W} = -kL(1 - \cos\theta)L\sin\theta\,\delta\theta + mgL\cos\theta\,\delta\theta = 0, \qquad (d)$$

which gives us the transcendental equation

$$(1 - \cos\theta)\tan\theta = \frac{mg}{kL}, \qquad (e)$$

the solution of which yields the value of θ in the equilibrium position.

It may prove of interest to solve the problem by means of Lagrange multipliers. With this in mind, let us write the constraint equations (a) in the form

$$f_1 = x - L(1 - \cos\theta) = 0, \qquad f_2 = y - L\sin\theta = 0, \qquad (f)$$

yielding

$$\delta f_1 = \delta x - L\sin\theta\,\delta\theta = 0, \qquad \delta f_2 = \delta y - L\cos\theta\,\delta\theta = 0. \qquad (g)$$

Multiplying the first of Eqs. (g) by λ_1 and the second by λ_2 and adding to Eq. (b), we have

$$-kx\,\delta x + mg\,\delta y + \lambda_1(\delta x - L\sin\theta\,\delta\theta) + \lambda_2(\delta y - L\cos\theta\,\delta\theta) = 0. \qquad (h)$$

Substituting x from the first of Eqs. (a) into the above and setting the coefficients of δx, δy, and $\delta\theta$ to zero, we obtain the equations

$$\begin{aligned} -kL(1 - \cos\theta) + \lambda_1 &= 0, \\ mg + \lambda_2 &= 0, \\ -\lambda_1 L\sin\theta - \lambda_2 L\cos\theta &= 0, \end{aligned} \qquad (i)$$

which yield the multipliers

$$\lambda_1 = mg\cot\theta, \qquad \lambda_2 = -mg \qquad (j)$$

and, of course, the same equation, (e), for θ.

The parameter λ_1 can be identified as the horizontal reaction at point B, whereas λ_2 is the vertical reaction at point A. They are the forces resulting from the kinematical constraints imposed on the system.

Example 2.3

A uniform cable of mass per unit length μ is suspended at two points (x_1, y_1) and (x_2, y_2) (see Fig. 2.3). It is required to find the function $y(x)$ corresponding to the equilibrium position of the cable if the length of the cable is L.

For an unconstrained system the position of static equilibrium corresponds to a minimum for the potential energy, which in this particular case is given by the definite integral

$$V = \mu g \int_{x_1, y_1}^{x_2, y_2} y \, ds = \mu g \int_{x_1}^{x_2} y(1 + y'^2)^{\frac{1}{2}} \, dx, \qquad (a)$$

where y is subject to the boundary conditions

$$y(x_1) = y_1, \qquad y(x_2) = y_2. \qquad (b)$$

The system at hand, however, is constrained by the equation

$$L = \int_{x_1, y_1}^{x_2, y_2} ds = \int_{x_1}^{x_2} (1 + y'^2)^{\frac{1}{2}} \, dx = \text{const.} \qquad (c)$$

The problem can be approached by using the method of Lagrange multipliers and seeking the minimum value of the augmented definite integral

$$\overline{V} = \int_{x_1}^{x_2} (y - \lambda)(1 + y'^2)^{\frac{1}{2}} \, dx, \qquad (d)$$

where λ is the undetermined multiplier. The constant μg is irrelevant and has been ignored.

Letting

$$F = (y - \lambda)(1 + y'^2)^{\frac{1}{2}} \qquad (e)$$

and using the Euler-Lagrange differential equation (2.44), it can be shown that y must satisfy the equation

$$(y - \lambda)y'' - y'^2 - 1 = 0. \qquad (f)$$

Following the pattern of Example 2.1 and making the same substitution as there, it can be shown that the solution of Eq. (f) is

$$y - \lambda = A \cosh \frac{x + B}{A}, \qquad (g)$$

which again is a catenary. The constants λ, A, and B are evaluated (if at all possible) by means of Eqs. (b) and (c).

2.6 D'ALEMBERT'S PRINCIPLE

The principle of virtual work is concerned with the static equilibrium of systems and by itself is not suitable for use in problems of dynamics. Its usefulness can be extended to dynamics, however, by means of a principle attributed to d'Alembert. Although there is a difference of opinion about the form of the principle, there is general agreement about its far-reaching consequences. It turns out that the interest lies primarily in these consequences.

Newton's second law for a particle of mass m_i can be written in the form

$$\mathbf{F}_i + \mathbf{F}'_i - \dot{\mathbf{p}}_i = 0, \tag{2.62}$$

where $\mathbf{F}_i$ and $\mathbf{F}'_i$ are the applied and constraint forces, respectively, and $\mathbf{p}_i = m_i \dot{\mathbf{r}}_i$ represents the linear momentum vector. Equation (2.62) is often referred to as *d'Alembert's principle*. To interpret Eq. (2.62), we may look upon the negative of the momentum rate of change, $-\dot{\mathbf{p}}_i$, as an *inertia force*. In this manner, Eq. (2.62) can be regarded as the statement of dynamic equilibrium of the particle in question.

D'Alembert's principle enables us to treat dynamical problems as if they were statical. As a consequence of that principle, namely, Eq. (2.62), we can extend the principle of virtual work to problems of dynamics by writing for the ith particle

$$(\mathbf{F}_i + \mathbf{F}'_i - \dot{\mathbf{p}}_i) \cdot \delta \mathbf{r}_i = 0, \tag{2.63}$$

Considering a system of N particles and assuming that the virtual displacements $\delta \mathbf{r}_i$ are reversible, so that Eq. (2.53) holds, we can write for the system of particles

$$\sum_{i=1}^{N} (\mathbf{F}_i - \dot{\mathbf{p}}_i) \cdot \delta \mathbf{r}_i = 0. \tag{2.64}$$

We shall refer to Eq. (2.64), which embodies both the principle of virtual work of statics and d'Alembert's principle, as the *generalized principle of d'Alembert*. The sum of the impressed force and the inertia force, $\mathbf{F}_i - \dot{\mathbf{p}}_i$, is sometimes called the *effective force*. In view of this definition, the generalized principle of d'Alembert can be stated as follows: *The total virtual work performed by the effective forces through infinitesimal virtual displacements, compatible with the system constraints, is zero*. The constraint forces resulting from the kinematical conditions can be calculated by the same method as in Sec. 2.5.

D'Alembert's principle, Eq. (2.64), represents the most general formulation of the problems of dynamics, and all the various principles of mechanics, including Hamilton's principle, are derived from it. Its main drawback is that the virtual work of the inertia forces, if we can justify using this name, is not reducible to a single scalar function. As a result, the use of curvilinear coordinates is precluded. In this respect Hamilton's principle proves superior.

As a special application of d'Alembert's principle, we can derive the principle of the conservation of energy. Let us consider a scleronomic system for which the virtual displacements $\delta \mathbf{r}_i$ can be chosen to coincide with the actual displacements $d\mathbf{r}_i$. Then, writing $d\mathbf{r}_i = \dot{\mathbf{r}}_i\, dt$ and recalling the definition of $\mathbf{p}_i$, we can rewrite Eq. (2.64) in the form

$$\sum_{i=1}^{N} \mathbf{F}_i \cdot d\mathbf{r}_i - \sum_{i=1}^{N} \frac{d}{dt}(m_i \dot{\mathbf{r}}_i) \cdot \dot{\mathbf{r}}_i\, dt = 0. \qquad (2.65)$$

For a scleronomic system the first sum in Eq. (2.65) is simply $-dV$, where V is the potential energy of the system, and the second sum yields

$$\sum_{i=1}^{N} \frac{d}{dt}(m_i \dot{\mathbf{r}}_i) \cdot \dot{\mathbf{r}}_i\, dt = \frac{d}{dt}\left(\sum_{i=1}^{N} \tfrac{1}{2} m_i \dot{\mathbf{r}}_i \cdot \dot{\mathbf{r}}_i\right) dt = dT, \qquad (2.66)$$

where T is the kinetic energy of the system. Hence, Eq. (2.65) is equivalent to

$$d(T + V) = 0, \qquad (2.67)$$

from which it follows that

$$T + V = E = \text{const}, \qquad (2.68)$$

or the total energy E of the system is conserved. The conclusion that emerges from this simple example is that *the law of conservation of energy is restricted to systems for which not only the potential energy but also the kinematical constraints are independent of time.*

2.7 HAMILTON'S PRINCIPLE

D'Alembert's principle provides a complete formulation of the problems of mechanics, and it occupies a central place in that field. Unfortunately, it is not very convenient for deriving the equations of motion, as the problems are formulated in terms of position coordinates, which, in contrast with generalized coordinates, may not all be independent. A different formulation, based on d'Alembert's principle, avoids this difficulty. The formulation in question is the *Hamilton principle*, which is perhaps the most famous variational principle of mechanics. Hamilton's principle considers the motion

Fundamentals of Analytical Mechanics

of the entire system between two times t_1 and t_2. It is an integral principle, and it reduces the problems of dynamics to the investigation of a scalar definite integral. This formulation has the remarkable advantage of being invariant to the coordinate system used to express the integrand.

Let us consider a system of N particles. From Sec. 2.6, we can write d'Alembert's principle, Eq. (2.64), in the form

$$\sum_{i=1}^{N} (m_i \ddot{\mathbf{r}}_i - \mathbf{F}_i) \cdot \delta \mathbf{r}_i = 0, \qquad (2.69)$$

where the virtual displacements $\delta \mathbf{r}_i$ are compatible with the system constraints. First

$$\sum_{i=1}^{N} \mathbf{F}_i \cdot \delta \mathbf{r}_i = \overline{\delta W} \qquad (2.70)$$

is recognized as the virtual work performed by the applied forces. Next we consider the following

$$\frac{d}{dt}(\dot{\mathbf{r}}_i \cdot \delta \mathbf{r}_i) = \ddot{\mathbf{r}}_i \cdot \delta \mathbf{r}_i + \dot{\mathbf{r}}_i \cdot \delta \dot{\mathbf{r}}_i = \ddot{\mathbf{r}}_i \cdot \delta \mathbf{r}_i + \delta(\tfrac{1}{2}\dot{\mathbf{r}}_i \cdot \dot{\mathbf{r}}_i),$$

or
$$\ddot{\mathbf{r}}_i \cdot \delta \mathbf{r}_i = \frac{d}{dt}(\dot{\mathbf{r}}_i \cdot \delta \mathbf{r}_i) - \delta(\tfrac{1}{2}\dot{\mathbf{r}}_i \cdot \dot{\mathbf{r}}_i). \qquad (2.71)$$

Multiplying Eq. (2.71) by m_i and summing over the entire system of particles, we obtain

$$\sum_{i=1}^{N} m_i \ddot{\mathbf{r}}_i \cdot \delta \mathbf{r}_i = \sum_{i=1}^{N} m_i \frac{d}{dt}(\dot{\mathbf{r}}_i \cdot \delta \mathbf{r}_i) - \delta T, \qquad (2.72)$$

where T represents the kinetic energy of the system of particles. Introducing Eqs. (2.70) and (2.72) into (2.69), we obtain

$$\delta T + \overline{\delta W} = \sum_{i=1}^{N} m_i \frac{d}{dt}(\dot{\mathbf{r}}_i \cdot \delta \mathbf{r}_i). \qquad (2.73)$$

We indicated in Sec. 2.1 that the instantaneous configuration of a system is given by the values of n generalized coordinates defining a representative point in the n-dimensional configuration space. The system configuration changes with time tracing a path, called the *true path* (also called *Newtonian* or *dynamical path*), in the configuration space. A slightly different path, known as the *varied path*, is obtained if at any instant we allow virtual changes $\delta \mathbf{r}_i$ in position which, by definition, involve no change in time, $\delta t = 0$. From

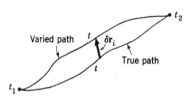

FIGURE 2.5

all the varied paths, however, we consider only those which at the two end points t_1 and t_2 coincide with the true path (see Fig. 2.5)

$$\delta \mathbf{r}_i(t_1) = \delta \mathbf{r}_i(t_2) = \mathbf{0}. \tag{2.74}$$

We must point out that in general the varied path is not a possible path if the system is nonholonomic. Multiplying Eq. (2.73) by dt and integrating between the times t_1 and t_2, we obtain

$$\int_{t_1}^{t_2} (\delta T + \overline{\delta W}) \, dt = \int_{t_1}^{t_2} \sum_{i=1}^{N} m_i \frac{d}{dt} (\dot{\mathbf{r}}_i \cdot \delta \mathbf{r}_i) \, dt = \sum_{i=1}^{N} m_i \dot{\mathbf{r}}_i \cdot \delta \mathbf{r}_i \bigg|_{t_1}^{t_2}, \tag{2.75}$$

which, in view of conditions (2.74), reduces to

$$\int_{t_1}^{t_2} (\delta T + \overline{\delta W}) \, dt = 0. \tag{2.76}$$

We shall refer to Eq. (2.76) as the *extended Hamilton's principle*.

In the case in which the virtual work can be expressed as the variation of a single scalar function, we have

$$\overline{\delta W} = \delta W = -\delta V, \tag{2.77}$$

and, introducing the *Lagrangian L* in the form

$$L = T - V, \tag{2.78}$$

Eq. (2.76) reduces to

$$\int_{t_1}^{t_2} \delta L \, dt = 0. \tag{2.79}$$

If the system is holonomic, Eq. (2.79) may be written

$$\delta I = \delta \int_{t_1}^{t_2} L \, dt = 0, \tag{2.80}$$

which is the mathematical statement of *Hamilton's principle* in its most familiar version. The principle can be stated as follows: *The actual path in the configuration space renders the value of the definite integral* $I = \int_{t_1}^{t_2} L\, dt$ *stationary with respect to all arbitrary variations of the path between two instants* t_1 *and* t_2 *provided that the path variations vanish at these two end points.* In many cases the stationary value is a minimum for the actual path. Hence for holonomic systems both Eqs. (2.79) and (2.80) are valid. On the other hand, for nonholonomic systems Hamilton's principle assumes the form (2.79) because the transition from Eq. (2.79) to (2.80) cannot be made if the varied path is not in general a possible path. (For a discussion of this subject see Ref. 5, sec. 3.7.[1]) We note that Eq. (2.77) does not require that the system be conservative but only lamellar (see Sec. 1.5), as V can depend on time. However, the most frequently encountered case for which Eq. (2.77) holds is that in which the system is conservative.

Although we have formulated Hamilton's principle for a system of particles whose motion is defined in terms of generalized coordinates depending on time alone, the principle is equally valid for continuous systems whose motion is defined by coordinates which are functions not only of time but of spatial variables as well. Example 2.5 shows how to derive the boundary-value problem of a continuous system by means of Hamilton's principle.

Hamilton's principle is an example of a variational principle which reduces the problems of dynamics to the investigation of a scalar integral independent of the coordinates used to describe the Lagrangian. The condition rendering the integral I stationary leads to all the equations of motion. Hamilton's principle provides a formulation rather than a solution of the problems of mechanics, and it belongs to a broader class of principles known as *principles of least action* (see Sec. 9.1).

Example 2.4

Consider the system of Fig. 2.4 and derive the equation of motion by means of Hamilton's principle.

Because the system is holonomic, we can use Eq. (2.80) to derive the equation of motion. This is a single-degree-of-freedom system, and to describe its motion we choose θ as the generalized coordinate. Using Eqs. (a) of Example 2.2, we can write the Lagrangian

$$L = T - V = \tfrac{1}{2}m\dot{y}^2 - \tfrac{1}{2}kx^2 + mgy$$
$$= \tfrac{1}{2}mL^2\left[\dot{\theta}^2 \cos^2\theta - \frac{k}{m}(1 - \cos\theta)^2 + \frac{2g}{L}\sin\theta\right]. \qquad (a)$$

---

[1] Numbered references will be found among the Suggested References at the end of each chapter.

Introducing Eq. (a) into Hamilton's principle, Eq. (2.80), we obtain

$$\delta I = \delta \int_{t_1}^{t_2} L \, dt = \int_{t_1}^{t_2} \left(\frac{\partial L}{\partial \theta} \delta\theta + \frac{\partial L}{\partial \dot\theta} \delta\dot\theta \right) dt$$

$$= -mL^2 \int_{t_1}^{t_2} \left\{ \left[\dot\theta^2 \sin\theta \cos\theta + \frac{k}{m}(1 - \cos\theta)\sin\theta - \frac{g}{L}\cos\theta \right] \delta\theta \right.$$

$$\left. - \dot\theta \cos^2\theta \, \delta\dot\theta \right\} dt = 0. \quad (b)$$

The second term in the integrand contains $\delta\dot\theta$, making it incompatible with the remaining terms, which are all multiplying $\delta\theta$. But $\delta\dot\theta = d(\delta\theta)/dt$, so that after an integration by parts of the term containing $\delta\dot\theta$ we arrive at

$$\int_{t_1}^{t_2} \left[\frac{d}{dt}(\dot\theta \cos^2\theta) + \dot\theta^2 \sin\theta \cos\theta + \frac{k}{m}(1 - \cos\theta)\sin\theta - \frac{g}{L}\cos\theta \right] \delta\theta \, dt$$

$$- \dot\theta \cos^2\theta \, \delta\theta\Big|_{t_1}^{t_2} = 0, \quad (c)$$

where the constant $-mL^2$ has been ignored.

Invoking the requirement that the variation $\delta\theta$ vanish at the two instants t_1 and t_2, the second term in Eq. (c) reduces to zero. Moreover, $\delta\theta$ is arbitrary in the time interval between t_1 and t_2, so that the only way for the integral to be zero is for the coefficient of $\delta\theta$ to vanish for any time t. Hence, we must set

$$\frac{d}{dt}(\dot\theta \cos^2\theta) + \dot\theta^2 \sin\theta \cos\theta + \frac{k}{m}(1 - \cos\theta)\sin\theta - \frac{g}{L}\cos\theta = 0, \quad (d)$$

which is the desired equation of motion.

We note that by letting $\dot\theta = \ddot\theta = 0$ we obtain the same equation for the system equilibrium position as the one derived in Example 2.2.

Example 2.5

Hamilton's principle can also be used to formulate boundary-value problems associated with continuous elastic systems. As an illustration, let us consider the vibrating string of Fig. 2.6. The mass per unit length of

FIGURE 2.6

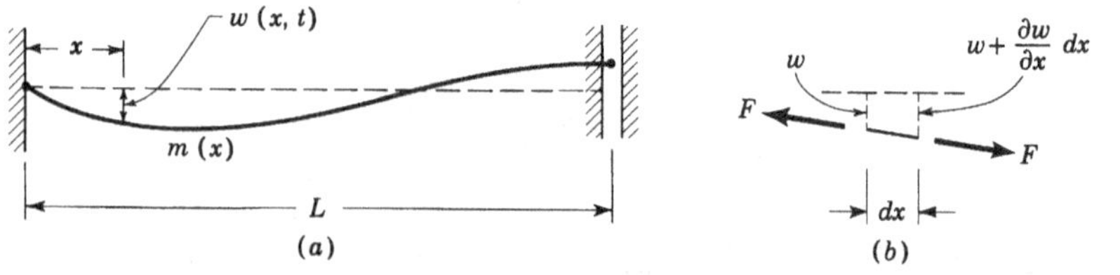

Fundamentals of Analytical Mechanics

string at any point x is designated by $m(x)$, and the transverse displacement at that point is denoted by $w(x,t)$. The string is subject to the tension F, which, for small transverse displacements, can be assumed to remain constant. At the end $x = 0$ the displacement is zero, and at the end $x = L$ the transverse component of force acting upon the string vanishes.

The kinetic energy of the string is simply

$$T(t) = \frac{1}{2} \int_0^L m(x) \left[\frac{\partial w(x,t)}{\partial t} \right]^2 dx. \qquad (a)$$

The potential energy $V(t)$ is proportional to the increase in the length of the string compared to the string at rest, with the factor of proportionality being equal to the tension F. For the element of length shown in Fig. 2.6b the increase in length is

$$\left\{ \left[w(x,t) + \frac{\partial w(x,t)}{\partial x} dx - w(x,t) \right]^2 + (dx)^2 \right\}^{\frac{1}{2}} - dx$$

$$= \left\{ 1 + \left[\frac{\partial w(x,t)}{\partial x} \right]^2 \right\}^{\frac{1}{2}} dx - dx \approx \frac{1}{2} \left[\frac{\partial w(x,t)}{\partial x} \right]^2 dx. \qquad (b)$$

It follows that the potential energy of the entire string has the value

$$V(t) = \frac{1}{2} \int_0^L F \left[\frac{\partial w(x,t)}{\partial x} \right]^2 dx. \qquad (c)$$

The system is holonomic, so that, according to Hamilton's principle, the solution w must render the integral

$$I = \int_{t_1}^{t_2} (T - V) \, dt = \frac{1}{2} \int_{t_1}^{t_2} \int_0^L \left[m \left(\frac{\partial w}{\partial t} \right)^2 - F \left(\frac{\partial w}{\partial x} \right)^2 \right] dx \, dt \qquad (d)$$

stationary, subject to the conditions

$$\delta w(x,t) = 0 \quad \text{at } t = t_1, t_2. \qquad (e)$$

To calculate δI, let us concentrate on the first term of the integrand in (d), integrate by parts with respect to t, consider conditions (e), and obtain

$$\delta \int_{t_1}^{t_2} \tfrac{1}{2} m \left(\frac{\partial w}{\partial t} \right)^2 dt = \int_{t_1}^{t_2} m \frac{\partial w}{\partial t} \delta \frac{\partial w}{\partial t} dt = \int_{t_1}^{t_2} m \frac{\partial w}{\partial t} \frac{\partial}{\partial t} (\delta w) \, dt$$

$$= \left(m \frac{\partial w}{\partial t} \right) \delta w \Big|_{t_1}^{t_2} - \int_{t_1}^{t_2} \frac{\partial}{\partial t} \left(m \frac{\partial w}{\partial t} \right) \delta w \, dt$$

$$= - \int_{t_1}^{t_2} m \frac{\partial^2 w}{\partial t^2} \delta w \, dt. \qquad (f)$$

The second term in the integrand in (d), when integrated with respect to x, yields

$$\delta \int_0^L \tfrac{1}{2} F \left(\frac{\partial w}{\partial x}\right)^2 dx = \int_0^L F \frac{\partial w}{\partial x} \delta \frac{\partial w}{\partial x} dx = \int_0^L F \frac{\partial w}{\partial x} \frac{\partial}{\partial x} (\delta w) \, dx$$

$$= \left(F \frac{\partial w}{\partial x}\right) \delta w \bigg|_0^L - \int_0^L F \frac{\partial^2 w}{\partial x^2} \delta w \, dx. \qquad (g)$$

Introducing Eqs. (f) and (g) into the mathematical statement of Hamilton's principle, namely,

$$\delta \int_{t_1}^{t_2} (T - V) \, dt = 0, \qquad (h)$$

we obtain

$$\int_{t_1}^{t_2} \left[\int_0^L \left(F \frac{\partial^2 w}{\partial x^2} - m \frac{\partial^2 w}{\partial t^2} \right) \delta w \, dx - \left(F \frac{\partial w}{\partial x} \right) \delta w \bigg|_0^L \right] dt = 0. \qquad (i)$$

But Eq. (i) must be satisfied for those w which vanish at $x = 0$ and whose slope is such that $F \, \partial w / \partial x$ vanishes at $x = L$. Since otherwise δw is arbitrary over the domain $0 < x < L$, Eq. (i) is satisfied if

$$F \frac{\partial^2 w}{\partial x^2} - m \frac{\partial^2 w}{\partial t^2} = 0 \qquad (j)$$

throughout the domain. Furthermore, if we write

$$\left(F \frac{\partial w}{\partial x} \right) \delta w \bigg|_0^L = 0, \qquad (k)$$

we take into account that δw is zero at the end $x = 0$ and that $F \, \partial w / \partial x$ is zero at the end $x = L$, where the latter implies that the transverse component of force vanishes at $x = L$. Equation (j) represents the differential equation and Eq. (k) the corresponding boundary conditions of the desired boundary-value problem. For an extended coverage of boundary-value problems associated with vibrating elastic systems see Ref. 4, chap. 5.

2.8 LAGRANGE'S EQUATIONS OF MOTION

We indicated in Sec. 2.7 that the generalized d'Alembert's principle, as given by Eq. (2.64), is not convenient for the derivation of the equations of motion. Indeed it is more advantageous to express Eq. (2.64) in terms of a set of generalized coordinates q_i $(i = 1, 2, \ldots, n)$, where for a holonomic system these coordinates are independent, so that the virtual displacements δq_i are independent and arbitrary. Under these circumstances the coefficients of δq_i $(i = 1, 2, \ldots, n)$ can be set equal to zero separately, thus obtaining a set of

Fundamentals of Analytical Mechanics

differential equations in terms of generalized coordinates, known as *Lagrange's equations of motion*. Instead of using d'Alembert's principle, the same equations can be derived by using Hamilton's principle, which is simply an integrated form of d'Alembert's principle. The Lagrange equations of motion retain many of the advantages of Hamilton's principle, in the sense that the equations can be derived from a single function, namely, the Lagrangian, if the system is holonomic. The differential equations thus obtained have the same general appearance as the Euler-Lagrange differential equation (2.44), and this is no coincidence, as the name indicates. Moreover, the formulation can be extended to nonholonomic systems.

Let us concentrate first on the case of a holonomic system and express the dependent variables $\mathbf{r}_i$ in terms of n generalized coordinates q_k and time t in the form

$$\mathbf{r}_i = \mathbf{r}_i(q_1, q_2, \ldots, q_n, t), \qquad i = 1, 2, \ldots, N, \qquad (2.81)$$

where n is the number of degrees of freedom of the system. The velocity vectors can be written in terms of the generalized coordinates and velocities by differentiating Eqs. (2.81)

$$\dot{\mathbf{r}}_i = \frac{d\mathbf{r}_i}{dt} = \frac{\partial \mathbf{r}_i}{\partial q_1}\dot{q}_1 + \frac{\partial \mathbf{r}_i}{\partial q_2}\dot{q}_2 + \cdots + \frac{\partial \mathbf{r}_i}{\partial q_n}\dot{q}_n + \frac{\partial \mathbf{r}_i}{\partial t}$$
$$= \sum_{k=1}^{n} \frac{\partial \mathbf{r}_i}{\partial q_k}\dot{q}_k + \frac{\partial \mathbf{r}_i}{\partial t}, \qquad i = 1, 2, \ldots, N. \qquad (2.82)$$

Equations (2.82) enable us to write the kinetic energy as follows:

$$T = \frac{1}{2}\sum_{i=1}^{N} m_i \dot{\mathbf{r}}_i \cdot \dot{\mathbf{r}}_i$$
$$= \frac{1}{2}\sum_{i=1}^{N} m_i \left(\sum_{r=1}^{n}\sum_{s=1}^{n} \frac{\partial \mathbf{r}_i}{\partial q_r} \cdot \frac{\partial \mathbf{r}_i}{\partial q_s} \dot{q}_r \dot{q}_s + 2 \frac{\partial \mathbf{r}_i}{\partial t} \cdot \sum_{r=1}^{n} \frac{\partial \mathbf{r}_i}{\partial q_r} \dot{q}_r + \frac{\partial \mathbf{r}_i}{\partial t} \cdot \frac{\partial \mathbf{r}_i}{\partial t} \right). \qquad (2.83)$$

Introducing the coefficients

$$\alpha_{rs} = \sum_{i=1}^{N} m_i \frac{\partial \mathbf{r}_i}{\partial q_r} \cdot \frac{\partial \mathbf{r}_i}{\partial q_s}, \qquad \beta_r = \sum_{i=1}^{N} m_i \frac{\partial \mathbf{r}_i}{\partial t} \cdot \frac{\partial \mathbf{r}_i}{\partial q_r},$$
$$\gamma = \frac{1}{2}\sum_{i=1}^{N} m_i \frac{\partial \mathbf{r}_i}{\partial t} \cdot \frac{\partial \mathbf{r}_i}{\partial t}, \qquad (2.84)$$

we can write Eq. (2.83) in the form

$$T = T_2 + T_1 + T_0, \qquad (2.85)$$

where

$$T_2 = \frac{1}{2}\sum_{r=1}^{n}\sum_{s=1}^{n} \alpha_{rs}\dot{q}_r\dot{q}_s \qquad (2.86)$$

is a homogeneous quadratic function in the generalized velocities,

$$T_1 = \sum_{r=1}^{n} \beta_r \dot{q}_r \tag{2.87}$$

is a linear homogeneous function in the generalized velocities, and

$$T_0 = \gamma \tag{2.88}$$

is a nonnegative function of the generalized coordinates and time. Noting that the coefficients α_{rs}, β_r, and γ depend on the generalized coordinates and time, it follows that the functional dependence of T is

$$T = T(q_1, q_2, \ldots, q_n, \dot{q}_1, \dot{q}_2, \ldots, \dot{q}_n, t). \tag{2.89}$$

For a holonomic system the Hamilton principle, Eq. (2.80), becomes

$$\delta I = \delta \int_{t_1}^{t_2} L(q_1, q_2, \ldots, q_n, \dot{q}_1, \dot{q}_2, \ldots, \dot{q}_n, t)\, dt = 0, \tag{2.90}$$

and, following the pattern of Sec. 2.4, it can be easily shown that I has a stationary value if the generalized coordinates q_k satisfy the following equations

$$\frac{d}{dt}\left(\frac{\partial L}{\partial \dot{q}_k}\right) - \frac{\partial L}{\partial q_k} = 0, \quad k = 1, 2, \ldots, n. \tag{2.91}$$

Equations (2.91) represent a set of n simultaneous differential equations, known as *Lagrange's equations*, which describe the motion of a holonomic system in a lamellar field.

In the case of a system for which the virtual work cannot be expressed as the variation of a single scalar function the extended Hamilton principle, Eq. (2.76), must replace Eq. (2.80). Whereas the kinetic energy can be treated in the same manner as the one in which we have just treated the Lagrangian, for the virtual work a different approach is necessary. Let us consider the case in which there are p forces acting upon the system so that the virtual work is

$$\overline{\delta W} = \sum_{j=1}^{p} \mathbf{F}_j \cdot \delta \mathbf{r}_j, \tag{2.92}$$

where the virtual displacement $\delta \mathbf{r}_j$ can be obtained from Eq. (2.81) in the form

$$\delta \mathbf{r}_j = \sum_{k=1}^{n} \frac{\partial \mathbf{r}_j}{\partial q_k} \delta q_k. \tag{2.93}$$

Introducing Eq. (2.93) into (2.92), we obtain

Fundamentals of Analytical Mechanics

$$\overline{\delta W} = \sum_{j=1}^{p} \mathbf{F}_j \cdot \sum_{k=1}^{n} \frac{\partial \mathbf{r}_j}{\partial q_k} \delta q_k = \sum_{k=1}^{n} \left(\sum_{j=1}^{p} \mathbf{F}_j \cdot \frac{\partial \mathbf{r}_j}{\partial q_k} \right) \delta q_k. \quad (2.94)$$

The virtual work, however, can be regarded as the product of n *generalized forces* Q_k acting over the virtual generalized displacements δq_k

$$\overline{\delta W} = \sum_{k=1}^{n} Q_k \, \delta q_k, \quad (2.95)$$

so that, comparing Eqs. (2.94) and (2.95), we conclude that the generalized forces have the form

$$Q_k = \sum_{j=1}^{p} \mathbf{F}_j \cdot \frac{\partial \mathbf{r}_j}{\partial q_k}, \qquad k = 1, 2, \ldots, n. \quad (2.96)$$

The generalized forces Q_k take the place of the single forces acting on each particle, and they form the components of an n-dimensional vector associated with the configuration space. Introducing Eq. (2.95) into the extended Hamilton principle, Eq. (2.76), we obtain

$$\int_{t_1}^{t_2} \delta T \, dt + \int_{t_1}^{t_2} \sum_{k=1}^{n} Q_k \, \delta q_k \, dt$$
$$= -\int_{t_1}^{t_2} \sum_{k=1}^{n} \left[\frac{d}{dt} \left(\frac{\partial T}{\partial \dot{q}_k} \right) - \frac{\partial T}{\partial q_k} - Q_k \right] \delta q_k \, dt = 0, \quad (2.97)$$

and, since the virtual generalized displacements are arbitrary, we must have

$$\frac{d}{dt} \left(\frac{\partial T}{\partial \dot{q}_k} \right) - \frac{\partial T}{\partial q_k} = Q_k; \qquad k = 1, 2, \ldots, n, \quad (2.98)$$

which are Lagrange's equations for the case in question.

If the system is acted upon by some forces which are derivable from a potential function $-V$ and some forces which are not, we can write the virtual work in the form

$$\overline{\delta W} = \overline{\delta W_p} + \overline{\delta W_{np}} = -\delta V + \sum_{k=1}^{n} Q_{knp} \, \delta q_k, \quad (2.99)$$

where Q_{knp} designates the generalized forces not derivable from a potential function. Introduction of Eq. (2.99) into (2.76) yields

$$\int_{t_1}^{t_2} \delta(T - V) \, dt + \int_{t_1}^{t_2} \sum_{k=1}^{n} Q_{knp} \, \delta q_k \, dt = 0, \quad (2.100)$$

and when we recall the definition of the Lagrangian, Eq. (2.100) leads to

$$\frac{d}{dt}\left(\frac{\partial L}{\partial \dot{q}_k}\right) - \frac{\partial L}{\partial q_k} = Q_k, \qquad k = 1, 2, \ldots, n, \qquad (2.101)$$

where, assuming that all the forces derivable from a potential function are accounted for in the Lagrangian, we denote the nonpotential forces simply by Q_k. When V does not depend explicitly on time, Q_k coincides with the nonconservative force associated with q_k.

In the above derivations it was tacitly assumed that the coordinates q_k were independent, by the definition of the generalized coordinates. This fact was used in the last step in the derivation of Lagrange's equations from Hamilton's principle, namely, when the arbitrariness of the virtual displacements δq_k was invoked. As in Sec. 2.3 (in fact the remainder of this section parallels very closely the developments of that section), it may be desirable at times to work with a surplus of coordinates. This is certainly the case when the coordinates q_k are related by nonholonomic constraints of the type

$$\sum_{k=1}^{n} a_{lk}\, dq_k + a_{l0}\, dt = 0, \qquad l = 1, 2, \ldots, m, \qquad (2.102)$$

where a_{lk} ($k = 0, 1, 2, \ldots, n$) are functions of the coordinates q_k. We note in passing that this is not the most general type of nonholonomic constraint, as it does not cover constraints in the form of inequalities. It follows from Eqs. (2.102) that the virtual displacements δq_k are no longer independent (hence not arbitrary) but related by

$$\sum_{k=1}^{n} a_{lk}\, \delta q_k = 0, \qquad l = 1, 2, \ldots, m. \qquad (2.103)$$

The interest lies in the exploration of the method of Lagrange multipliers to derive Lagrange's equations of motion from Hamilton's principle. To this end, we multiply in sequence each of Eqs. (2.103) by λ_l ($l = 1, 2, \ldots, m$) and add all the resulting equations to Eq. (2.100) to obtain

$$-\int_{t_1}^{t_2} \sum_{k=1}^{n} \left[\frac{d}{dt}\left(\frac{\partial L}{\partial \dot{q}_k}\right) - \frac{\partial L}{\partial q_k} - Q_k - \sum_{l=1}^{m} \lambda_l a_{lk}\right] \delta q_k\, dt = 0. \qquad (2.104)$$

While the virtual displacements δq_k are still not independent, we can choose the multipliers λ_l ($l = 1, 2, \ldots, m$) so as to render the bracketed coefficients of δq_k ($k = n - m + 1, \ldots, n$) equal to zero. The remaining δq_k, being independent, can be chosen arbitrarily, which leads to the conclusion that the coefficients of δq_k ($k = 1, 2, \ldots, n - m$) are zero. It follows that

Fundamentals of Analytical Mechanics

$$\frac{d}{dt}\left(\frac{\partial L}{\partial \dot{q}_k}\right) - \frac{\partial L}{\partial q_k} = Q_k + \sum_{l=1}^{m} \lambda_l a_{lk}, \qquad k = 1, 2, \ldots, n, \quad (2.105)$$

which are the desired Lagrange equations. The conclusion is that we can regard all the coordinates q_k as independent provided we modify Lagrange's equations (2.101) by adding on the right side equivalent forces

$$Q_k' = \sum_{l=1}^{m} \lambda_l a_{lk}, \qquad k = 1, 2, \ldots, n, \quad (2.106)$$

which are recognized as the constraint forces. There are $n + m$ equations, Eqs. (2.102) and (2.105), and $n + m$ unknowns, n coordinates q_k and m multipliers λ_l. Hence, this procedure should enable us to solve not only for the coordinates q_k but also for the constraint forces Q_k'.

The case of holonomic constraints of the type

$$g_l(q_1, q_2, \ldots, q_n, t) = 0, \qquad l = 1, 2, \ldots, m, \quad (2.107)$$

can be treated in an obvious fashion by substituting

$$a_{lk} = \frac{\partial g_l}{\partial q_k} \quad (2.108)$$

in Eqs. (2.102), (2.105), and (2.106).

The simplest case (and fortunately the most common) is the one in which Eqs. (2.81) do not contain the time explicitly, the force field is conservative, and the coordinates are independent. Under these conditions the kinetic energy reduces to the homogeneous quadratic form

$$T = T_2 = \frac{1}{2} \sum_{r=1}^{n} \sum_{s=1}^{n} \alpha_{rs} \dot{q}_r \dot{q}_s, \quad (2.109)$$

where the coefficients α_{rs} depend on the generalized coordinates alone. In this case the Lagrange equations assume the simple form (2.98), in which the forces Q_k are conservative. A system possessing all these properties is said to be a *natural system*. We shall refer to systems for which the kinetic energy is of the form (2.85) as *nonnatural*. Although Eqs. (2.81) involve the time explicitly, the kinetic energy may be independent of time. This turns out to be the case when Eqs. (2.81) represent a transformation of coordinates involving rotation. Whereas the term T_0 behaves like an apparent potential energy, giving rise to the so-called centrifugal forces, the term T_1 produces forces of the Coriolis type. Since the latter are related to gyroscopic phenomena associated with spinning bodies, the terms in T which are linear in the velocities are referred to as *gyroscopic terms*. We shall return to some of these ideas in Secs. 2.10 and 2.11.

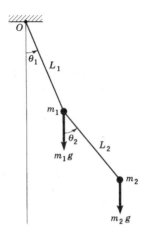

FIGURE 2.7

Example 2.6

Derive Lagrange's equations of motion for the double pendulum shown in Fig. 2.7. The bobs m_1 and m_2 are connected by inextensible strings.

Because the system is conservative and the constraints of the system are accounted for by choosing θ_1 and θ_2 as generalized coordinates, we shall use Eqs. (2.91) to derive the corresponding Lagrange equations. The kinetic and potential energies of the system are

$$T = \tfrac{1}{2}m_1(L_1\dot\theta_1)^2 + \tfrac{1}{2}m_2$$
$$\times [(L_1\dot\theta_1 \cos\theta_1 + L_2\dot\theta_2 \cos\theta_2)^2 + (L_1\dot\theta_1 \sin\theta_1 + L_2\dot\theta_2 \sin\theta_2)^2]$$
$$= \tfrac{1}{2}m_1(L_1\dot\theta_1)^2 + \tfrac{1}{2}m_2[(L_1\dot\theta_1)^2 + (L_2\dot\theta_2)^2 + 2L_1L_2\dot\theta_1\dot\theta_2 \cos(\theta_2 - \theta_1)], \qquad (a)$$

$$V = m_1 g L_1(1 - \cos\theta_1) + m_2 g[L_1(1 - \cos\theta_1) + L_2(1 - \cos\theta_2)], \qquad (b)$$

so that the Lagrangian has the form

$$L = T - V = \tfrac{1}{2}(m_1 + m_2)(L_1\dot\theta_1)^2 + \tfrac{1}{2}m_2(L_2\dot\theta_2)^2 + m_2 L_1 L_2 \dot\theta_1 \dot\theta_2 \cos(\theta_2 - \theta_1)$$
$$- (m_1 + m_2)L_1 g(1 - \cos\theta_1) - m_2 L_2 g(1 - \cos\theta_2). \qquad (c)$$

Introducing Eq. (c) into Eqs. (2.91) with $q_1 = \theta_1$ and $q_2 = \theta_2$, we obtain the equations of motion

$$\frac{d}{dt}\left(\frac{\partial L}{\partial \dot\theta_1}\right) - \frac{\partial L}{\partial \theta_1} = (m_1 + m_2)L_1^2 \ddot\theta_1 + m_2 L_1 L_2 \ddot\theta_2 \cos(\theta_2 - \theta_1)$$
$$- m_2 L_1 L_2 \dot\theta_2^2 \sin(\theta_2 - \theta_1) + (m_1 + m_2)L_1 g \sin\theta_1 = 0, \qquad (d)$$

$$\frac{d}{dt}\left(\frac{\partial L}{\partial \dot\theta_2}\right) - \frac{\partial L}{\partial \theta_2} = m_2 L_2^2 \ddot\theta_2 + m_2 L_1 L_2 \ddot\theta_1 \cos(\theta_2 - \theta_1)$$
$$+ m_2 L_1 L_2 \dot\theta_1^2 \sin(\theta_2 - \theta_1) + m_2 L_2 g \sin\theta_2 = 0.$$

Example 2.7

Derive the equation of motion for the system of Fig. 2.4 by using the method of Lagrange multipliers.

As in Example 2.2, we can work with three coordinates although this is only a single-degree-of-freedom system. The kinetic and potential energies of the system are

$$T = \tfrac{1}{2}m\dot{y}^2, \qquad (a)$$

$$V = \tfrac{1}{2}kx^2 - mgy, \qquad (b)$$

and, in addition, we have the constraint equations

$$g_1 = x - L(1 - \cos\theta) = 0, \qquad (c)$$

$$g_2 = y - L\sin\theta = 0. \qquad (d)$$

Following the procedure outlined in this section, we derive the coefficients

$$a_{1x} = \frac{\partial g_1}{\partial x} = 1, \quad a_{1y} = \frac{\partial g_1}{\partial y} = 0, \quad a_{1\theta} = \frac{\partial g_1}{\partial \theta} = -L\sin\theta,$$
$$a_{2x} = \frac{\partial g_2}{\partial x} = 0, \quad a_{2y} = \frac{\partial g_2}{\partial y} = 1, \quad a_{2\theta} = \frac{\partial g_2}{\partial \theta} = -L\cos\theta, \qquad (e)$$

and, with $q_1 = x$, $q_2 = y$, and $q_3 = \theta$, Eqs. (2.105) yield

$$kx - \lambda_1 = 0,$$
$$m\ddot{y} - mg - \lambda_2 = 0, \qquad (f)$$
$$\lambda_1 L \sin\theta + \lambda_2 L \cos\theta = 0.$$

The two Lagrange multipliers have the values

$$\lambda_1 = kx = kL(1 - \cos\theta) = kL - kL\left[1 - \left(\frac{y}{L}\right)^2\right]^{\frac{1}{2}}, \qquad (g)$$

$$\lambda_2 = -\lambda_1 \tan\theta = ky\left\{1 - \left[1 - \left(\frac{y}{L}\right)^2\right]^{-\frac{1}{2}}\right\}, \qquad (h)$$

and the equation of motion becomes

$$m\ddot{y} - mg - ky\left\{1 - \left[1 - \left(\frac{y}{L}\right)^2\right]^{-\frac{1}{2}}\right\} = 0, \qquad (i)$$

which can be shown to be the same as the equation of motion obtained in Example 2.4 in terms of θ.

2.9 LAGRANGE'S EQUATIONS FOR IMPULSIVE FORCES

When bodies collide, large forces develop over small intervals of time. Such forces, called *impulsive forces*, are defined by means of a limiting process in which the forces tend to infinity as the time interval during which they act tends to zero in such a way that the integral

$$\lim_{\epsilon \to 0} \int_{t}^{t+\epsilon} P(\tau)\, d\tau = \hat{P}(t) \tag{2.110}$$

exists, where τ is a dummy variable of integration. The quantity $\hat{P}(t)$ is called the *impulse* and has units FT.

Thanks to the relatively short duration of the impact, displacements may be assumed to remain unchanged throughout the impact whereas velocities change instantaneously. This is so because, for finite changes in velocities, displacements require a finite time to develop. Since energy is dissipated during impacts, systems subjected to impulsive forces are nonconservative.

Lagrange's equations of motion can be modified to allow for the treatment of impulsive forces. To this end, we replace t by τ in Eqs. (2.98), multiply the result by $d\tau$, and integrate over the duration of the impact to obtain

$$\int_{t}^{t+\epsilon} \frac{d}{d\tau}\left(\frac{\partial T}{\partial \dot{q}_k}\right) d\tau - \int_{t}^{t+\epsilon} \frac{\partial T}{\partial q_k}\, d\tau = \int_{t}^{t+\epsilon} Q_k\, d\tau. \tag{2.111}$$

Letting $\epsilon \to 0$, the first term becomes in the limit

$$\lim_{\epsilon \to 0} \int_{t}^{t+\epsilon} \frac{d}{d\tau}\left(\frac{\partial T}{\partial \dot{q}_k}\right) d\tau = \lim_{\epsilon \to 0} \left.\frac{\partial T}{\partial \dot{q}_k}\right|_{t}^{t+\epsilon} = \Delta p_k, \tag{2.112}$$

where

$$p_k = \frac{\partial T}{\partial \dot{q}_k} \tag{2.113}$$

is known as the *generalized momentum*. Equations (2.111) and (2.112) can be combined into

$$\Delta p_k = \lim_{\epsilon \to 0} \int_{t}^{t+\epsilon} \left(\frac{\partial T}{\partial q_k} + Q_k\right) d\tau. \tag{2.114}$$

But $\partial T/\partial q_k$ remains finite independently of the impulse duration, whereas Q_k tends to infinity as $\epsilon \to 0$ such that the integral

$$\lim_{\epsilon \to 0} \int_{t}^{t+\epsilon} Q_k\, d\tau = \hat{Q}_k \tag{2.115}$$

exists, where $\hat{Q}_k$ denotes the *generalized impulse* associated with the coordinate q_k. Introducing (2.115) into (2.114), we obtain Lagrange's equations for impulsive forces

$$\Delta p_k = \hat{Q}_k, \quad k = 1, 2, \ldots, n, \tag{2.116}$$

stating that the incremental change in the generalized momentum is equal to the generalized impulse. The value of the generalized impulse can be obtained from the expression for virtual work written in terms of impulsive forces

$$\sum_{k=1}^{n} \hat{Q}_k \, \delta q_k = \sum_{j=1}^{p} \hat{\mathbf{F}}_j \cdot \delta \mathbf{r}_j. \qquad (2.117)$$

In a manner similar to that of Sec. 2.8 we obtain

$$\hat{Q}_k = \sum_{j=1}^{p} \hat{\mathbf{F}}_j \cdot \frac{\partial \mathbf{r}_j}{\partial q_k}. \qquad (2.118)$$

The formulation above assumes an unconstrained system. (For the subject of impulsive constraints, see Ref. 5, chap. 14.)

Example 2.8

Consider the system of Example 2.6, in which the masses m_1 and m_2 are connected by means of massless rigid links of length L_1 and L_2, respectively. Derive Lagrange's equations of motion for the case in which an impulsive force $\hat{P}$ strikes horizontally at a distance D from the support when the links are at rest in the vertical position. Consider the case in which $L_1 < D < L_1 + L_2$.

The kinetic energy was calculated in Example 2.6. Its value, corresponding to the vertical position, is

$$T = \tfrac{1}{2} m_1 (L_1 \dot{\theta}_1)^2 + \tfrac{1}{2} m_2 (L_1 \dot{\theta}_1 + L_2 \dot{\theta}_2)^2. \qquad (a)$$

Denoting by $\hat{\Theta}_1$ and $\hat{\Theta}_2$ the impulsive forces associated with the generalized coordinates θ_1 and θ_2, we can write the virtual work expression

$$\hat{P} \, \delta[L_1 \theta_1 + (D - L_1)\theta_2] = \hat{P}[L_1 \, \delta\theta_1 + (D - L_1) \, \delta\theta_2] = \hat{\Theta}_1 \, \delta\theta_1 + \hat{\Theta}_2 \, \delta\theta_2, \quad (b)$$

from which we conclude that

$$\hat{\Theta}_1 = \hat{P} L_1, \qquad \hat{\Theta}_2 = \hat{P}(D - L_1), \qquad (c)$$

and we note that $\hat{\Theta}_1$ and $\hat{\Theta}_2$ are impulsive moments rather than forces.

Because the momentum before the application of $\hat{P}$ is zero, we have

$$\Delta \frac{\partial T}{\partial \dot{\theta}_i} = \frac{\partial T}{\partial \dot{\theta}_i}, \qquad i = 1, 2, \qquad (d)$$

so that, using Eqs. (2.116) in conjunction with Eqs. (a) and (c), we obtain

$$\begin{aligned} m_1 L_1^2 \dot{\theta}_1 + m_2 L_1 (L_1 \dot{\theta}_1 + L_2 \dot{\theta}_2) &= \hat{P} L_1, \\ m_2 L_2 (L_1 \dot{\theta}_1 + L_2 \dot{\theta}_2) &= \hat{P}(D - L_1), \end{aligned} \qquad (e)$$

which are the desired equations. Equations (e) represent the state of the system just after the impulse and can be regarded as the initial conditions for the subsequent force-free motion.

2.10 CONSERVATION LAWS

The Lagrange equations of motion constitute a set of n second-order differential equations. A complete solution of the system of equations requires $2n$ constants of integration. These constants may consist either of the initial values of the n coordinates q_i and n velocities $\dot{q}_i$ or of the values of the n coordinates at two different times. Problems which can be solved completely either in terms of known elementary functions or indefinite integrals are called *problems soluble by quadratures*. Complete solutions of the equations of motion of a dynamical system are for the most part not possible. Sometimes, however, the system exhibits certain peculiarities which enable us to derive a surprisingly large amount of information about its behavior without actually obtaining a complete solution of the equations. This is the case when the system admits the so-called *first integrals* of the motion. These integrals contain derivatives of the variables of one order lower than the order of the differential equations, and their values remain constant in time. We encountered such integrals in Chap. 1 in the form of the conserved momentum and energy. These types of integral may not always be immediately apparent, but a proper choice of coordinates can uncover them. In the sequel we shall expand on these ideas.

There exist systems for which a certain coordinate, say q_s, is absent from the Lagrangian L although its time derivative $\dot{q}_s$ does appear in L. If such a system is holonomic, the corresponding Lagrange equation, obtained from Eqs. (2.91), reduces to the form

$$\frac{d}{dt}\left(\frac{\partial L}{\partial \dot{q}_s}\right) = \frac{\partial L}{\partial q_s} = 0. \tag{2.119}$$

Let us denote the *conjugate generalized momentum* by

$$p_s = \frac{\partial L}{\partial \dot{q}_s} \tag{2.120}$$

and note that definitions (2.113) and (2.120) are equivalent because the potential energy V does not depend on the generalized velocities. A coordinate which does not appear in the Lagrangian is called *ignorable* or *cyclic*. Integrating Eq. (2.119) with respect to time, we obtain

$$p_s = \text{const.} \tag{2.121}$$

Hence, *the generalized momentum associated with an ignorable coordinate is conserved*. Equation (2.121) constitutes a first integral of the motion corresponding to the law of *conservation of momentum*.

As a simple illustration, let us recall the motion of a two-body system in the central force field governed by Newton's inverse square law (Sec. 1.8).

The Lagrangian per unit mass, expressed in terms of planar rectangular coordinates, assumes the form

$$L = T - V = \tfrac{1}{2}(\dot{x}^2 + \dot{y}^2) + \frac{\mu}{(x^2 + y^2)^{\frac{1}{2}}}, \qquad (2.122)$$

which gives no indication that an ignorable coordinate actually exists. But the same Lagrangian when expressed in terms of polar coordinates becomes

$$L = \tfrac{1}{2}(\dot{r}^2 + r^2\dot{\theta}^2) + \frac{\mu}{r}, \qquad (2.123)$$

from which it is immediately evident that θ is an ignorable coordinate and the conjugate momentum p_θ is conserved.

The term ignorable can be traced to the fact that the degree of freedom corresponding to q_s can be ignored from formulation of the problem. This can be accomplished by a procedure known as *Routh's method for the ignoration of coordinates*, discussed in Sec. 2.11. The ignorable coordinates are often angular coordinates, a fact to which the term cyclic is attributed.

Next let us consider the case in which the Lagrangian does not depend explicitly on time

$$\frac{\partial L}{\partial t} = 0. \qquad (2.124)$$

Recalling Lagrange's equations (2.91), we can write the total time derivative of L as

$$\begin{aligned}\frac{dL}{dt} &= \sum_{k=1}^{n} \frac{\partial L}{\partial q_k} \dot{q}_k + \sum_{k=1}^{n} \frac{\partial L}{\partial \dot{q}_k} \ddot{q}_k \\ &= \sum_{k=1}^{n} \left[\frac{d}{dt}\left(\frac{\partial L}{\partial \dot{q}_k}\right)\dot{q}_k + \frac{\partial L}{\partial \dot{q}_k}\frac{d\dot{q}_k}{dt} \right] = \sum_{k=1}^{n} \frac{d}{dt}\left(\frac{\partial L}{\partial \dot{q}_k}\dot{q}_k\right).\end{aligned} \qquad (2.125)$$

Equation (2.125) can be rewritten in the form

$$\frac{d}{dt}\left(\sum_{k=1}^{n} \frac{\partial L}{\partial \dot{q}_k} \dot{q}_k - L\right) = 0, \qquad (2.126)$$

and an integration with respect to time yields

$$\sum_{k=1}^{n} \frac{\partial L}{\partial \dot{q}_k} \dot{q}_k - L = h = \text{const}, \qquad (2.127)$$

which constitutes another type of integral of motion, known as *Jacobi's integral*.

In the general case, in which the kinetic energy is given by Eq. (2.85), the Lagrangian has the expression

$$L = T_2 + T_1 + T_0 - V = \frac{1}{2}\sum_{r=1}^{n}\sum_{s=1}^{n}\alpha_{rs}\dot{q}_r\dot{q}_s + \sum_{r=1}^{n}\beta_r\dot{q}_r + \gamma - V. \quad (2.128)$$

This time, however, the coefficients α_{rs}, β_r, and γ depend on the generalized coordinates alone and not on time. Introducing Eq. (2.128) into (2.127) and using Euler's theorem on homogeneous functions, we obtain the Jacobi integral

$$h = \sum_{k=1}^{n}\frac{\partial L}{\partial \dot{q}_k}\dot{q}_k - L = 2T_2 + T_1 - (T_2 + T_1 + T_0 - V) = T_2 - T_0 + V.$$
$$(2.129)$$

In the special case in which the kinetic energy is a homogeneous quadratic function of the generalized velocities of the form

$$T = T_2 = \frac{1}{2}\sum_{i=1}^{n}\sum_{j=1}^{n}\alpha_{ij}\dot{q}_i\dot{q}_j, \quad (2.130)$$

in which case the system is referred to as a natural system (see Sec. 2.8), the Jacobi integral reduces to

$$h = 2T - (T - V) = T + V = E = \text{const}, \quad (2.131)$$

which is the mathematical statement of the *law of conservation of energy*. Hence, *in a natural system for which the Lagrangian does not depend explicitly on time the total energy of the system E is conserved.*

It should be noted that the conservation law expressed by Eq. (2.127) is perceptibly more general than the law of conservation of energy in the form (2.131), which, in fact, it contains as a special case. The Jacobi integral finds its applications in problems involving rotating systems of coordinates.

Example 2.9

Consider the spherical pendulum of Fig. 2.8 and check for any first integrals of the motion. Use as generalized coordinates the angles θ and ϕ as shown.

The kinetic and potential energies of the spherical pendulum are

$$T = \tfrac{1}{2}m[(L\dot{\theta})^2 + (\dot{\phi}L\sin\theta)^2], \quad (a)$$
$$V = mgL(1 - \cos\theta), \quad (b)$$

so that the Lagrangian has the form

$$L = T - V = \tfrac{1}{2}mL^2[\dot{\theta}^2 + (\dot{\phi}\sin\theta)^2] - mgL(1 - \cos\theta). \quad (c)$$

Fundamentals of Analytical Mechanics

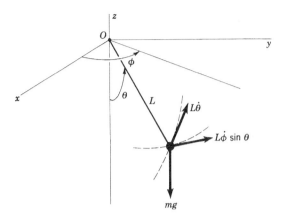

FIGURE 2.8

An examination of Eq. (*c*) shows that ϕ does not appear in the Lagrangian, so that ϕ must be a cyclic coordinate. Indeed we have the immediate integral

$$p_\phi = \frac{\partial L}{\partial \dot{\phi}} = mL^2 \dot{\phi} \sin^2 \theta = \text{const}, \qquad (d)$$

which merely states that the angular momentum about axis z is conserved, as there is no torque about that axis.

Since the Lagrangian does not depend explicitly on time, according to Eq. (2.127) we must have the Jacobi integral

$$\begin{aligned}
h &= \frac{\partial L}{\partial \dot{\theta}} \dot{\theta} + \frac{\partial L}{\partial \dot{\phi}} \dot{\phi} - L \\
&= mL^2[\dot{\theta}^2 + (\dot{\phi} \sin \theta)^2] - \tfrac{1}{2} mL^2[\dot{\theta}^2 + (\dot{\phi} \sin \theta)^2] + mgL(1 - \cos \theta) \\
&= \tfrac{1}{2} mL^2[\dot{\theta}^2 + (\dot{\phi} \sin \theta)^2] + mgL(1 - \cos \theta) \\
&= \text{const}. \qquad (e)
\end{aligned}$$

But the spherical pendulum is a natural system because the kinetic energy is a homogeneous quadratic function in the velocities $\dot{\theta}$ and $\dot{\phi}$. Indeed Eq. (*e*) is recognized as stating the fact that the total energy of the system is conserved

$$h = T + V = E = \text{const}. \qquad (f)$$

2.11 ROUTH'S METHOD FOR THE IGNORATION OF COORDINATES

We shall be concerned with a conservative holonomic system for which the Lagrangian does not involve the time explicitly. Furthermore, we shall assume that the m coordinates q_s ($s = n - m + 1, \ldots, n$) are cyclic, so that the corresponding Lagrange equations yield the momentum integrals

$$\frac{\partial L}{\partial \dot{q}_s} = p_s = \beta_s = \text{const}, \quad s = n - m + 1, \ldots, n, \quad (2.132)$$

where the constants β_s are determined by the initial conditions. Because Eqs. (2.132) are linear functions in the velocities $\dot{q}_k$ ($k = 1, 2, \ldots, n$), we can solve for the velocities $\dot{q}_s$ ($s = n - m + 1, \ldots, n$) in terms of the remaining velocities $\dot{q}_k$ ($k = 1, 2, \ldots, n - m$) and the constants β_s ($s = n - m + 1, \ldots, n$) and write

$$\dot{q}_s = \dot{q}_s(q_1, q_2, \ldots, q_{n-m}, \dot{q}_1, \dot{q}_2, \ldots, \dot{q}_{n-m}, \beta_{n-m+1}, \ldots, \beta_n). \quad (2.133)$$

Next let us construct the function

$$R = L - \sum_{s=n-m+1}^{n} \frac{\partial L}{\partial \dot{q}_s} \dot{q}_s = L - \sum_{s=n-m+1}^{n} \beta_s \dot{q}_s, \quad (2.134)$$

where the function

$$R = R(q_1, q_2, \ldots, q_{n-m}, \dot{q}_1, \dot{q}_2, \ldots, \dot{q}_{n-m}, \beta_{n-m+1}, \ldots, \beta_n) \quad (2.135)$$

is called the *Routhian function*, which contains the constants β_s instead of the associated velocities $\dot{q}_s$. The Routhian can be used as a Lagrangian function to write the equations of motion for the coordinates q_k ($k = 1, 2, \ldots, n - m$), thereby ignoring the cyclic coordinates. To show this, let us form

$$\begin{aligned}
\delta R &= \delta L - \sum_{s=n-m+1}^{n} \delta \beta_s \dot{q}_s - \sum_{s=n-m+1}^{n} \beta_s \, \delta \dot{q}_s \\
&= \sum_{k=1}^{n-m} \frac{\partial L}{\partial q_k} \delta q_k + \sum_{k=1}^{n} \frac{\partial L}{\partial \dot{q}_k} \delta \dot{q}_k - \sum_{s=n-m+1}^{n} \dot{q}_s \, \delta \beta_s - \sum_{s=n-m+1}^{n} \frac{\partial L}{\partial \dot{q}_s} \delta \dot{q}_s \\
&= \sum_{k=1}^{n-m} \frac{\partial L}{\partial q_k} \delta q_k + \sum_{k=1}^{n-m} \frac{\partial L}{\partial \dot{q}_k} \delta \dot{q}_k - \sum_{s=n-m+1}^{n} \dot{q}_s \, \delta \beta_s \quad (2.136)
\end{aligned}$$

on the one hand, and

$$\delta R = \sum_{k=1}^{n-m} \frac{\partial R}{\partial q_k} \delta q_k + \sum_{k=1}^{n-m} \frac{\partial R}{\partial \dot{q}_k} \delta \dot{q}_k + \sum_{s=n-m+1}^{n} \frac{\partial R}{\partial \beta_s} \delta \beta_s \quad (2.137)$$

on the other hand. We note that for a complete variation even the conserved momenta were varied. Comparing Eqs. (2.136) and (2.137), we conclude that

$$\frac{\partial L}{\partial q_k} = \frac{\partial R}{\partial q_k}, \quad \frac{\partial L}{\partial \dot{q}_k} = \frac{\partial R}{\partial \dot{q}_k}, \quad k = 1, 2, \ldots, n-m \quad (2.138)$$

and, in addition,

$$-\dot{q}_s = \frac{\partial R}{\partial \beta_s}, \quad s = n-m+1, \ldots, n. \quad (2.139)$$

Hence, substituting Eqs. (2.138) into the first $n-m$ of Eqs. (2.91), the corresponding equations of motion become

$$\frac{d}{dt}\left(\frac{\partial R}{\partial \dot{q}_k}\right) - \frac{\partial R}{\partial q_k} = 0, \quad k = 1, 2, \ldots, n-m, \quad (2.140)$$

which can be regarded as a new Lagrangian set of equations of motion for a system with only $n-m$ degrees of freedom. If a solution of Eqs. (2.140) can be found, we can obtain the cyclic coordinates by integrating Eqs. (2.139) and writing

$$q_s = -\int \frac{\partial R}{\partial \beta_s}\, dt, \quad s = n-m+1, \ldots, n. \quad (2.141)$$

Next let us assume that the original system is a natural one. However, when Eqs. (2.132) are solved for the cyclic velocities $\dot{q}_s$ in terms of the remaining velocities and the constants β_s, as described at the beginning of this section, and the results are introduced into Eq. (2.134), we conclude that the Routhian R contains not only quadratic but also linear terms in the velocities $\dot{q}_k$ ($k = 1, 2, \ldots, n-m$) as well as terms not involving the velocities at all. It follows that the new system is *nonnatural*. Linear terms in the velocities in the kinetic energy expression are called *gyroscopic*, and systems possessing such linear terms are referred to as *gyroscopic systems* because they are characteristic of systems involving gyroscopic motion. On the other hand, terms in the kinetic energy expression not involving the velocities are generally associated with centrifugal forces.

Example 2.10

Consider the spherical pendulum of Example 2.9 and use the method of ignoration of coordinates to reduce the degrees of freedom of the system to one.

Corresponding to the cyclic coordinate ϕ we have the momentum integral

$$\frac{\partial L}{\partial \dot{\phi}} = p_\phi = mL^2 \dot{\phi} \sin^2 \theta = \beta_\phi = \text{const.} \quad (a)$$

As a result, we can eliminate $\dot{\phi}$ by introducing the Routhian

$$R = L - \frac{\partial L}{\partial \dot{\phi}}\dot{\phi} = L - \beta_\phi \dot{\phi}$$

$$= \tfrac{1}{2}mL^2\left[\dot{\theta}^2 + \frac{\sin^2\theta\, \beta_\phi^2}{(mL^2 \sin^2\theta)^2}\right] - mgL(1-\cos\theta) - \frac{\beta_\phi^2}{mL^2 \sin^2\theta}$$

$$= \tfrac{1}{2}mL^2\left(\dot{\theta}^2 - \frac{\beta_\phi^2}{m^2L^4 \sin^2\theta}\right) - mgL(1-\cos\theta), \qquad (b)$$

which is a function of θ and $\dot{\theta}$ alone.

Next we use Eqs. (2.140) and derive the equation of motion

$$mL^2\ddot{\theta} - \frac{\beta_\phi^2 \cos\theta}{mL^2 \sin^3\theta} + mgL \sin\theta = 0, \qquad (c)$$

which can be identified as the moment equation about point O (see Fig. 2.8). The last term in Eq. (c) is recognized as being due to gravity. On the other hand, the second term results from the rotational motion of the mass about axis z, giving rise to what is commonly known as the centrifugal force, having the value

$$mL\dot{\phi}^2 \sin\theta = \frac{\beta_\phi^2}{mL^3 \sin^3\theta}. \qquad (d)$$

Equation (c) can be interpreted as representing a single-degree-of-freedom system with the kinetic energy $\tfrac{1}{2}mL^2\dot{\theta}^2$ and the equivalent potential energy $\beta_\phi^2/(2mL^2 \sin^2\theta) + mgL(1-\cos\theta)$. However, the first term in the equivalent potential energy is really the kinetic energy term T_0 due to the centrifugal force.

Note that for $\dot{\phi} = 0$ Eq. (c) reduces to the equation for a simple pendulum.

2.12 RAYLEIGH'S DISSIPATION FUNCTION

The generalized forces in Lagrange's equations (2.101) include any nonconservative forces which are not derivable from a potential function. Of these a certain class of forces must receive special consideration, namely, forces *proportional to the velocity* of a given particle and *resisting the motion*, as they act in a direction opposite to that of the velocity. They are *dissipative forces* since the system loses energy by their action. It turns out that it is possible to devise a new single function in terms of the generalized coordinates and velocities to account for these forces in Lagrange's equations of motion. To do so, let us denote the dissipative force components acting on particle i in the form

$$F_{x_i} = -c_{x_i}\dot{x}_i, \qquad F_{y_i} = -c_{y_i}\dot{y}_i, \qquad F_{z_i} = -c_{z_i}\dot{z}_i, \qquad i = 1, 2, \ldots, N, \tag{2.142}$$

where c_{x_i}, c_{y_i}, and c_{z_i} are functions depending on the coordinates only and not on the velocities. The virtual work performed by these forces through arbitrary virtual displacements is

$$\sum_{i=1}^{N} \mathbf{F}_i \cdot \delta \mathbf{r}_i = \sum_{i=1}^{N} (F_{x_i} \delta x_i + F_{y_i} \delta y_i + F_{z_i} \delta z_i)$$

$$= -\sum_{i=1}^{N} (c_{x_i}\dot{x}_i \delta x_i + c_{y_i}\dot{y}_i \delta y_i + c_{z_i}\dot{z}_i \delta z_i). \tag{2.143}$$

But, from Eqs. (2.82) and (2.93), we can write

$$\delta \mathbf{r}_i = \sum_{k=1}^{n} \frac{\partial \mathbf{r}_i}{\partial q_k} \delta q_k = \sum_{k=1}^{n} \frac{\partial \dot{\mathbf{r}}_i}{\partial \dot{q}_k} \delta q_k, \tag{2.144}$$

and, introducing the components δx_i, δy_i, and δz_i into Eq. (2.143), we obtain

$$\sum_{i=1}^{N} \mathbf{F}_i \cdot \delta \mathbf{r}_i = -\sum_{i=1}^{N} (c_{x_i}\dot{x}_i \delta x_i + c_{y_i}\dot{y}_i \delta y_i + c_{z_i}\dot{z}_i \delta z_i)$$

$$= -\sum_{k=1}^{n} \left[\sum_{i=1}^{N} \left(c_{x_i}\dot{x}_i \frac{\partial \dot{x}_i}{\partial \dot{q}_k} + c_{y_i}\dot{y}_i \frac{\partial \dot{y}_i}{\partial \dot{q}_k} + c_{z_i}\dot{z}_i \frac{\partial \dot{z}_i}{\partial \dot{q}_k} \right) \right] \delta q_k$$

$$= -\sum_{k=1}^{n} \left[\sum_{i=1}^{N} \frac{1}{2} \frac{\partial}{\partial \dot{q}_k} (c_{x_i}\dot{x}_i^2 + c_{y_i}\dot{y}_i^2 + c_{z_i}\dot{z}_i^2) \right] \delta q_k$$

$$= -\sum_{k=1}^{n} \frac{\partial F}{\partial \dot{q}_k} \delta q_k, \tag{2.145}$$

where

$$F = \sum_{i=1}^{N} \tfrac{1}{2}(c_{x_i}\dot{x}_i^2 + c_{y_i}\dot{y}_i^2 + c_{z_i}\dot{z}_i^2) \tag{2.146}$$

is known as *Rayleigh's dissipation function*.

If the only nonconservative forces present are of the dissipative type, the virtual work expression reduces to

$$\overline{\delta W}_{nc} = \sum_{k=1}^{n} Q_k \delta q_k = -\sum_{k=1}^{n} \frac{\partial F}{\partial \dot{q}_k} \delta q_k, \tag{2.147}$$

from which we conclude that the dissipative generalized forces can be derived from Rayleigh's dissipation function according to

$$Q_k = -\frac{\partial F}{\partial \dot{q}_k}, \quad k = 1, 2, \ldots, n. \tag{2.148}$$

Introducing Eqs. (2.148) into Eqs. (2.101), we obtain Lagrange's equations for a system with dissipative forces proportional to the velocity

$$\frac{d}{dt}\left(\frac{\partial L}{\partial \dot{q}_k}\right) - \frac{\partial L}{\partial q_k} + \frac{\partial F}{\partial \dot{q}_k} = 0, \quad k = 1, 2, \ldots, n, \tag{2.149}$$

so that all the equations of motion can be derived from two scalar functions, the Lagrangian and Rayleigh's dissipation function.

Equations (2.142) imply that the motion of one particle does not produce any dissipative force on another. This is an unnecessary restriction, and for the general case Rayleigh's dissipation function must be altered to account for any possible coupling effects. This can be done by introducing the additional terms required in expressions (2.142) and retracing the steps leading to Eq. (2.146). Following this route, it can be shown that in terms of the generalized coordinates the more general form of dissipation function is given by

$$F = \frac{1}{2} \sum_{r=1}^{n} \sum_{s=1}^{n} c_{rs} \dot{q}_r \dot{q}_s. \tag{2.150}$$

Example 2.11

Consider the two-degree-of-freedom system shown in Fig. 2.9 and derive the Lagrange equations of motion associated with that system. The force in the spring of stiffness k_i ($i = 1, 2$) is proportional to the difference in the displacements of the terminal points, and the force in the dashpot with the damping coefficient c_i ($i = 1, 2$) is proportional to the difference in the velocities of the terminal points.

The kinetic energy of the system is simply

$$T = \tfrac{1}{2}(m_1 \dot{q}_1^2 + m_2 \dot{q}_2^2), \tag{a}$$

FIGURE 2.9

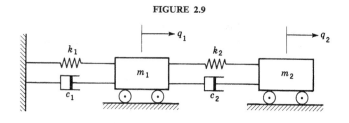

whereas the potential energy has the expression

$$V = \tfrac{1}{2}[k_1 q_1{}^2 + k_2(q_2 - q_1)^2]$$
$$= \tfrac{1}{2}[(k_1 + k_2)q_1{}^2 - 2k_2 q_1 q_2 + k_2 q_2{}^2]. \quad (b)$$

The potential energy due to the spring of stiffness k_2 is easily explained by the fact that the extension of that spring is $q_2 - q_1$ and the force in the spring is proportional to that extension. In an entirely analogous manner, the dissipation function has the form

$$F = \tfrac{1}{2}[c_1 \dot{q}_1{}^2 + c_2(\dot{q}_2 - \dot{q}_1)^2]$$
$$= \tfrac{1}{2}[(c_1 + c_2)\dot{q}_1{}^2 - 2c_2 \dot{q}_1 \dot{q}_2 + c_2 \dot{q}_2{}^2]. \quad (c)$$

We note that the cross-product terms in both F and V are the ones reflecting the effect of the motion of one mass relative to another.

Introducing Eqs. (a) to (c) into Eqs. (2.149), we obtain the desired Lagrange equations

$$m_1 \ddot{q}_1 + (c_1 + c_2)\dot{q}_1 - c_2 \dot{q}_2 + (k_1 + k_2)q_1 - k_2 q_2 = 0,$$
$$m_2 \ddot{q}_2 - c_2 \dot{q}_1 + c_2 \dot{q}_2 - k_2 q_1 + k_2 q_2 = 0. \quad (d)$$

Equations (d) can be written in matrix form as

$$[m]\{\ddot{q}\} + [c]\{\dot{q}\} + [k]\{q\} = 0, \quad (e)$$

where $[m]$ is a diagonal matrix and $[c]$ and $[k]$ are symmetric matrices, with the coupling terms as the off-diagonal elements. To solve these equations we must find a linear transformation which reduces the matrices $[m]$, $[c]$, and $[k]$ to a diagonal form simultaneously. For details of this problem see Ref. 4, sec. 9.2.

2.13 HAMILTON'S EQUATIONS

Lagrange's equations in any of their forms, Eqs. (2.91), (2.98), (2.101), or (2.105), consist of a set of simultaneous second-order differential equations, generally nonlinear in the coordinates and velocities. As they are second-order equations, the accelerations must appear at the most to the first power. We can express the accelerations as explicit functions of the coordinates, velocities, and time. To show this, we recall that the general form of the kinetic energy, obtained for a nonnatural system, can be written

$$T = T_2 + T_1 + T_0 = \frac{1}{2}\sum_{i=1}^{n}\sum_{j=1}^{n} \alpha_{ij}\dot{q}_i\dot{q}_j + \sum_{i=1}^{n} \beta_i \dot{q}_i + \gamma, \quad (2.151)$$

where α_{ij}, β_i, and γ represent functions of the coordinates and time, so that the notation T_2, T_1, and T_0 indicates the power to which the velocities appear in the corresponding expressions. Introducing Eq. (2.151) into (2.98), we obtain equations of the type

$$\sum_{j=1}^{n} \alpha_{ij}\ddot{q}_j = f_i(q_1,q_2,\ldots,q_n,\dot{q}_1,\dot{q}_2,\ldots,\dot{q}_n,t) = f_i(q,\dot{q},t), \quad (2.152)$$

where α_{ij} are the coefficients of the positive quadratic form T_2 and the functions f_i depend only on the coordinates, velocities, and time but not on accelerations. For simplicity we denoted by q and $\dot{q}$ the n-dimensional vectors with components q_i and $\dot{q}_i$, respectively.

Equations (2.152) can be written in the matrix form

$$[\alpha]\{\ddot{q}\} = \{f\}, \quad (2.153)$$

where $[\alpha]$ is a positive definite matrix. Hence it must have an inverse. It follows that Eq. (2.153) can be premultiplied by $[\alpha]^{-1}$ to obtain

$$\{\ddot{q}\} = [\alpha]^{-1}\{f\} = \{g\}, \quad (2.154)$$

yielding *explicit expressions for the accelerations* in terms of the vectors q and $\dot{q}$ as well as the time t.

At times it is desirable to formulate the problem in terms of first-order differential equations instead of second-order ones. To this end, we make the substitution

$$\begin{array}{ll} q_i = x_i, & \dot{q}_i = x_{i+n}, \\ \dot{x}_i = X_i, & g_i = X_{i+n}, \end{array} \quad i = 1, 2, \ldots, n, \quad (2.155)$$

so that we can replace Eqs. (2.154) by the $2n$ first-order equations

$$\{\dot{x}\} = \{X\}, \quad (2.156)$$

where $\{x\}$ and $\{X\}$ are column matrices representing the $2n$-dimensional vectors **x** and **X**, in which **X** depends on **x** and t. Now the configuration of the system is given by a $2n$-dimensional vector consisting of n coordinates and n velocities. Of the $2n$ equations the first n are purely kinematical, whereas the remaining n equations result from the dynamical laws governing the motion, as reflected by Eqs. (2.154).

A different procedure for the replacement of n second-order equations by $2n$ first-order ones consists of formulating the problem in terms of $2n$ Hamilton's equations instead of n Lagrange's equations. In this case the auxiliary coordinates are the momenta rather than the velocities. We recall that the definition of the generalized momentum is

$$p_k = \frac{\partial L}{\partial \dot{q}_k}, \quad k = 1, 2, \ldots, n, \quad (2.157)$$

where the p_k's are easily shown to be linear functions of the generalized velocities $\dot{q}_k$. Conversely, the generalized velocities can be expressed as linear functions of the generalized momenta. The transformation from the set of variables $(q,\dot{q},t)$ to the set (q,p,t) can be brought about by a procedure particularly suitable for this purpose, known as *Legendre's dual transformation*.

Let us consider a function of n variables u_i

$$F = F(u_1, u_2, \ldots, u_n) \tag{2.158}$$

and introduce a new set of n variables v_i related to the first by the transformation

$$v_i = \frac{\partial F}{\partial u_i}, \qquad i = 1, 2, \ldots, n. \tag{2.159}$$

To ensure that the new variables v_i are independent, the *Hessian matrix*, namely, the matrix with the elements consisting of the second partial derivatives of F, must be nonsingular. This enables us to solve Eqs. (2.159) for the variables u_i in terms of v_i.

Next we consider a new function G defined by

$$G = \sum_{i=1}^{n} u_i v_i - F, \tag{2.160}$$

so that, expressing the variables u_i in terms of v_i, the function G will depend only on the variables v_i

$$G = G(v_1, v_2, \ldots, v_n). \tag{2.161}$$

Because the sets u_i and v_i are independent, we can assign them infinitesimal arbitrary variations producing corresponding variations in G and F. Writing the variation in G from Eq. (2.160), we obtain

$$\delta G = \sum_{i=1}^{n} (u_i \, \delta v_i + v_i \, \delta u_i) - \sum_{i=1}^{n} \frac{\partial F}{\partial u_i} \delta u_i$$
$$= \sum_{i=1}^{n} \left[u_i \, \delta v_i + \left(v_i - \frac{\partial F}{\partial u_i} \right) \delta u_i \right], \tag{2.162}$$

and from Eq. (2.161) we have simply

$$\delta G = \sum_{i=1}^{n} \frac{\partial G}{\partial v_i} \delta v_i. \tag{2.163}$$

In view of Eqs. (2.159), however, the coefficients of δu_i in Eq. (2.162) are zero, and a comparison of that equation and Eq. (2.163) yields

$$u_i = \frac{\partial G}{\partial v_i}, \qquad i = 1, 2, \ldots, n, \tag{2.164}$$

which together with Eqs. (2.159) forms the dual transformation.

Following this pattern, let us introduce the *Hamiltonian function* defined by

$$H = \sum_{k=1}^{n} \frac{\partial L}{\partial \dot{q}_k} \dot{q}_k - L = \sum_{k=1}^{n} p_k \dot{q}_k - L \qquad (2.165)$$

and substitute for the generalized velocities $\dot{q}_k$ the corresponding generalized momenta p_k, so that the Hamiltonian becomes

$$H = H(q_1, q_2, \ldots, q_n, p_1, p_2, \ldots, p_n, t) = H(q, p, t), \qquad (2.166)$$

where q and p are n-dimensional vectors.

Next we write the variation of H from both Eqs. (2.165) and (2.166) and recall Eqs. (2.157), to obtain

$$\begin{aligned}
\delta H &= \sum_{k=1}^{n} \left(\dot{q}_k \, \delta p_k + p_k \, \delta \dot{q}_k - \frac{\partial L}{\partial q_k} \delta q_k - \frac{\partial L}{\partial \dot{q}_k} \delta \dot{q}_k \right) \\
&= \sum_{k=1}^{n} \left(\dot{q}_k \, \delta p_k - \frac{\partial L}{\partial q_k} \delta q_k \right) = \sum_{k=1}^{n} \left(\frac{\partial H}{\partial q_k} \delta q_k + \frac{\partial H}{\partial p_k} \delta p_k \right),
\end{aligned} \qquad (2.167)$$

from which it follows that

$$\dot{q}_k = \frac{\partial H}{\partial p_k}, \qquad -\frac{\partial L}{\partial q_k} = \frac{\partial H}{\partial q_k}, \qquad k = 1, 2, \ldots, n. \qquad (2.168)$$

Equations (2.168) are strictly a result of the Legendre transformation. To complete the transformation we invoke the laws of dynamics. For a *holonomic system in a lamellar field*, we obtain from Lagrange's equations (2.91)

$$\dot{p}_k = \frac{d}{dt}\left(\frac{\partial L}{\partial \dot{q}_k}\right) = \frac{\partial L}{\partial q_k}, \qquad k = 1, 2, \ldots, n, \qquad (2.169)$$

so that, introducing Eqs. (2.169) into the second half of (2.168), we obtain

$$\dot{q}_k = \frac{\partial H}{\partial p_k}, \qquad \dot{p}_k = -\frac{\partial H}{\partial q_k}, \qquad k = 1, 2, \ldots, n, \qquad (2.170)$$

which represent a set of $2n$ first-order differential equations known as the *Hamilton canonical equations*.

Next let us differentiate Eq. (2.166) with respect to time and write

$$\frac{dH}{dt} = \sum_{k=1}^{n} \left(\frac{\partial H}{\partial q_k} \dot{q}_k + \frac{\partial H}{\partial p_k} \dot{p}_k \right) + \frac{\partial H}{\partial t}. \qquad (2.171)$$

In view of Eqs. (2.170) we conclude from the above that

$$\frac{dH}{dt} = \frac{\partial H}{\partial t}. \qquad (2.172)$$

Moreover, when the Hamiltonian depends explicitly on time, so does the Lagrangian. Hence, differentiating Eq. (2.165) with respect to time and considering Eqs. (2.157) and (2.169), we obtain the relation

$$\frac{\partial H}{\partial t} = -\frac{\partial L}{\partial t}. \tag{2.173}$$

The Hamiltonian function fully describes the motion, as all the differential equations of motion can be derived from it. The obvious advantage of Hamilton's equations over Lagrange's equations is that in Hamilton's equations time derivatives of the variables appear only on the left side of the equations. Other advantages will be discussed in Sec. 5.1 in connection with the geometric representation of the solution.

It is clear that the first half of n equations in (2.170) is strictly a result of the Legendre transformation and the definition of the Hamiltonian, whereas the latter half reflects the laws of dynamics governing the motion. Equations (2.170) can be written in the form $\{\dot{x}\} = \{X\}$, in which $q_i = x_i$, $p_i = x_{i+n}$, and X_i and X_{i+n} ($i = 1, 2, \ldots, n$) are functions of q_i, p_i, and t.

If the system is holonomic but subject to *forces not derivable from a potential function*, Eqs. (2.169) are no longer valid and we see from Eqs. (2.101) that they must be replaced by

$$\dot{p}_k = \frac{d}{dt}\left(\frac{\partial L}{\partial \dot{q}_k}\right) = \frac{\partial L}{\partial q_k} + Q_k, \quad k = 1, 2, \ldots, n, \tag{2.174}$$

where Q_k represents the forces not derivable from a potential function. Hence, Hamilton's equations assume the modified form

$$\dot{q}_k = \frac{\partial H}{\partial p_k}, \quad \dot{p}_k = -\frac{\partial H}{\partial q_k} + Q_k, \quad k = 1, 2, \ldots, n. \tag{2.175}$$

In the event the *system is subject to m nonholonomic constraints*

$$\sum_{k=1}^{n} a_{lk}\dot{q}_k + a_{l0} = 0, \quad l = 1, 2, \ldots, m, \tag{2.176}$$

the method of Lagrange's multipliers leads to Hamilton's equations

$$\dot{q}_k = \frac{\partial H}{\partial p_k}, \quad \dot{p}_k = -\frac{\partial H}{\partial q_k} + \sum_{l=1}^{m} \lambda_l a_{lk}, \quad k = 1, 2, \ldots, n, \tag{2.177}$$

which must be considered in conjunction with the constraint equations (2.176).

Finally if the *nonconservative forces are derivable from Rayleigh's dissipation function* according to Eqs. (2.148), Hamilton's equations become

$$\dot{q}_k = \frac{\partial H}{\partial p_k}, \quad \dot{p}_k = -\frac{\partial H}{\partial q_k} - \frac{\partial F}{\partial \dot{q}_k}, \quad k = 1, 2, \ldots, n, \tag{2.178}$$

where the dissipation function F is given by Eq. (2.150).

In the case of a holonomic conservative system, in which the Hamiltonian does not depend explicitly on time, we can conclude from Eq. (2.172) that

$$\frac{dH}{dt} = 0. \tag{2.179}$$

It follows that, under these circumstances,

$$H = h = \text{const}, \tag{2.180}$$

which is recognized as the Jacobi integral, Eq. (2.127). Recalling Eq. (2.129), we conclude that for a nonnatural system the Hamiltonian has the form

$$H = T_2 - T_0 + V, \tag{2.181}$$

which does not contain linear terms in the velocities. On the other hand, for a natural system we see from Eq. (2.131) that the Hamiltonian reduces to the total energy of the system

$$H = T + V = E. \tag{2.182}$$

Example 2.12

Derive the Hamilton equations for the spherical pendulum of Example 2.9.

The Lagrangian for the spherical pendulum of Example 2.9 is

$$L = T - V = \tfrac{1}{2}mL^2[\dot{\theta}^2 + (\dot{\phi}\sin\theta)^2] - mgL(1 - \cos\theta), \tag{a}$$

from which it follows that the generalized momenta are

$$p_\theta = \frac{\partial L}{\partial \dot{\theta}} = mL^2\dot{\theta} \tag{b}$$

and

$$p_\phi = \frac{\partial L}{\partial \dot{\phi}} = mL^2\dot{\phi}\sin^2\theta. \tag{c}$$

Introducing Eqs. (a) to (c) into Eq. (2.165), we obtain the Hamiltonian

$$H = p_\theta\dot{\theta} + p_\phi\dot{\phi} - L = \frac{1}{2mL^2}\left(p_\theta^2 + \frac{p_\phi^2}{\sin^2\theta}\right) + mgL(1 - \cos\theta). \tag{d}$$

Using Eqs. (2.170), Hamilton's equations become

$$\dot{\theta} = \frac{\partial H}{\partial p_\theta} = \frac{p_\theta}{mL^2},$$

$$\dot{\phi} = \frac{\partial H}{\partial p_\phi} = \frac{p_\phi}{mL^2\sin^2\theta},$$

$$\dot{p}_\theta = -\frac{\partial H}{\partial \theta} = \frac{p_\phi^2\cos\theta}{mL^2\sin^3\theta} - mgL\sin\theta, \tag{e}$$

$$\dot{p}_\phi = -\frac{\partial H}{\partial \phi} = 0,$$

Fundamentals of Analytical Mechanics

which can be written in the form $\{\dot{x}\} = \{X\}$, where the four-dimensional vector $\mathbf{X}$ depends on the coordinates θ and ϕ and the momenta p_θ and p_ϕ. Because ϕ does not appear explicitly in H, $p_\phi = $ const and the coordinate ϕ is cyclic. Thus the set can be reduced to two first-order equations instead of four. Moreover, the Hamiltonian does not depend explicitly on time, so that it constitutes another first integral, namely, the Jacobi integral. Since this is a natural system, the implication is that the total energy is conserved, which we have already established in Example 2.9.

PROBLEMS

2.1 Consider three masses m_1, m_2, and m_3 suspended on a massless string, as shown in Fig. 2.10. The string is subjected to a constant tension T which remains unchanged during the motion of the masses. Use the method of virtual work to obtain the equilibrium configuration of the system assuming small deflections. Plot the configuration for $m_1 = m_2 = 0.3M$ and $m_3 = 0.4M$.

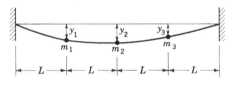

FIGURE 2.10

2.2 A bead of mass m is free to slide along a circular hoop of radius R. Find the positions of equilibrium if the hoop is rotating about a vertical diametrical axis with constant angular velocity ω.

2.3 Two masses sliding on smooth inclined planes, as shown in Fig. 2.11, are connected to a massless pulley consisting of two wheels rigidly attached to one

FIGURE 2.11

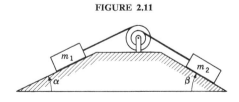

another. The diameter of one wheel is twice as large as the diameter of the other wheel. Use d'Alembert's principle to derive an expression for the acceleration of mass m_2.

2.4 Let the system of Fig. 2.10 rotate with a uniform angular velocity ω about an axis through the supports and use Hamilton's principle to derive the equations of motion.

2.5 Derive Lagrange's equations of motion of an elastic spherical pendulum (see Fig. 2.12). Assume that the bob is suspended through a spring of stiffness k. The unstretched length of the pendulum is L.

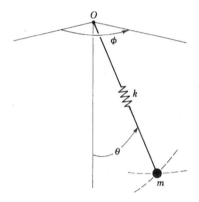

FIGURE 2.12

2.6 Solve Prob. 1.1 by the method of Lagrange multipliers.

2.7 Derive Lagrange's equations of motion for the system shown in Fig. 2.13. The massless links L_1 and L_2 are rigid, and the spring of stiffness k is con-

FIGURE 2.13

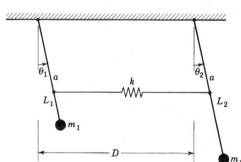

nected at a distance a from the supports. The spring is unstretched when the links are vertical.

2.8 Solve Prob. 2.7 under the assumption that the spring stiffness is infinite. Formulate the problem in terms of the two coordinates θ_1 and θ_2 by means of the method of Lagrange multipliers.

2.9 Derive the equation of motion for the bead of Prob. 2.2.

2.10 The motion of the double pendulum of Fig. 2.14 takes place in the vertical plane. Assume that the massless links L_1 and L_2 are rigid and derive the equation of motion for the system for the case in which the link L_1 rotates with constant angular velocity ω.

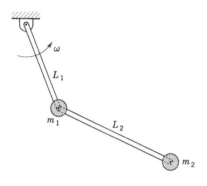

FIGURE 2.14

2.11 Solve Prob. 1.12 by means of Lagrange's equations for impulsive forces.

2.12 Consider the spherical pendulum of Prob. 2.5 and check whether there are any first integrals of the motion. If such integrals exist, explain their physical significance.

2.13 Consider the system of Prob. 2.2 and check whether there is any first integral of the motion. What can be said of the total energy $E = T + V$ of the system?

2.14 For the system of Prob. 2.5 use the Routh method for the ignoration of coordinates and reduce the degree of freedom of the system. Derive the equations of motion.

2.15 Derive Hamilton's equations of motion for the spherical pendulum of Prob. 2.5.

SUGGESTED REFERENCES

1. Goldstein, H.: *Classical Mechanics*, Addison-Wesley Publishing Company, Inc., Reading, Mass., 1950.
2. Lanczos, C.: *The Variational Principle of Mechanics*, University of Toronto Press, Toronto, 1962.
3. Leech, J. W.: *Classical Mechanics*, John Wiley & Sons, Inc., New York, 1965.
4. Meirovitch, L.: *Analytical Methods in Vibrations*, The Macmillan Company, New York, 1967.
5. Pars, L. A.: *A Treatise on Analytical Dynamics*, William Heinemann, Ltd., London, 1965.
6. Whittaker, E. T.: *Analytical Dynamics*, Cambridge University Press, London, 1937.

chapter three

Motion Relative to Rotating Reference Frames

Newton's laws imply the use of preferred reference frames, called inertial systems, as the acceleration assumes a simple form only when the motion is referred to such systems. Nevertheless, on various occasions it is more convenient to use noninertial systems, in particular rotating coordinate systems. This is the case when the interest lies in examining the motion of particles relative to spinning bodies or the motion of spinning rigid bodies relative to a fixed reference system. An example of the former is the motion of a particle relative to the rotating earth and of the latter the motion of gyroscopes or related devices with respect to an inertial space.

In this chapter the transformation between two sets of coordinates where one set is rotated with respect to another is first discussed. This leads to the development of expressions for the motion relative to rotating reference frames. Several applications of Newton's second law to the motion of a mass particle relative to the rotating earth are presented. Of particular significance is Foucault's pendulum, a device used to demonstrate the rotation of the earth.

3.1 TRANSFORMATION OF COORDINATES

Any vector **r** can be resolved in one or more systems of coordinates. In many problems of dynamics relations between the components of **r** in various coordinate systems prove extremely useful. To derive such relations, let us consider Fig. 3.1 and write the vector **r** in terms of components along two rectangular sets of axes, x_i ($i = 1, 2, 3$) and ξ_i ($i = 1, 2, 3$). The unit vectors along these axes are denoted by **i, j, k** and **i′, j′, k′**, respectively, so that

$$\mathbf{r} = x_1 \mathbf{i} + x_2 \mathbf{j} + x_3 \mathbf{k} = \xi_1 \mathbf{i}' + \xi_2 \mathbf{j}' + \xi_3 \mathbf{k}'. \tag{3.1}$$

Since these are two ways of expressing the same vector, the components x_i and ξ_i must evidently be related. The relation between ξ_1 and the components x_i can be obtained by writing the scalar product of **r** and **i′** with the result

$$\begin{aligned}
\xi_1 &= (\mathbf{i}' \cdot \mathbf{i})x_1 + (\mathbf{i}' \cdot \mathbf{j})x_2 + (\mathbf{i}' \cdot \mathbf{k})x_3 \\
&= x_1 \cos(\xi_1, x_1) + x_2 \cos(\xi_1, x_2) + x_3 \cos(\xi_1, x_3) \\
&= l_{11} x_1 + l_{12} x_2 + l_{13} x_3,
\end{aligned} \tag{3.2}$$

where $l_{1j} = \cos(\xi_1, x_j)$ ($j = 1, 2, 3$) are the direction cosines between axis ξ_1 and axes x_j. Similarly, we can express ξ_2 and ξ_3 in terms of the x_j components so that Eq. (3.2) can be generalized to

$$\xi_i = l_{i1} x_1 + l_{i2} x_2 + l_{i3} x_3 = \sum_{j=1}^{3} l_{ij} x_j, \quad i = 1, 2, 3. \tag{3.3}$$

Next let us define the 3×1 column matrices $\{x\} = \{x_j\}$ and $\{\xi\} = \{\xi_i\}$ representing the vector **r** in terms of the corresponding components, as well

FIGURE 3.1

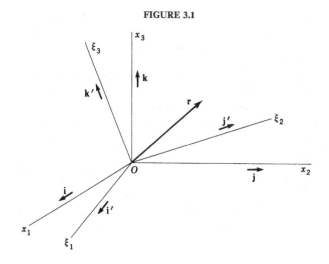

as the 3 × 3 square matrix $[l] = [l_{ij}]$ of the direction cosines, and write Eq. (3.3) in the matrix form

$$\{\xi\} = [l]\{x\}, \tag{3.4}$$

where the matrix $[l]$ may be regarded as an operator transforming the vector $\{x\}$ into the vector $\{\xi\}$. Equation (3.4) represents a *coordinate transformation* between two cartesian sets of axes. As such, it is not the most general type of coordinate transformation; a more general one would be the transformation between cartesian coordinates and generalized coordinates given by Eq. (2.81). In the case of the *linear transformation*

$$\{y\} = [a]\{x\}, \tag{3.5}$$

in which $[a]$ is a square matrix of coefficients a_{ij}, we can obtain the vector $\{x\}$ by premultiplying both sides of Eq. (3.5) by the reciprocal $[a]^{-1}$ of $[a]$

$$\{x\} = [a]^{-1}\{y\}, \tag{3.6}$$

provided that the matrix $[a]$ is not singular. In the special case of the transformation matrix $[l]$, however, the coefficients l_{ij} are not all independent. To show this, we can write the scalar products of **r** and **i**, **j**, and **k** in sequence and obtain the relation

$$x_r = \sum_{s=1}^{3} l_{sr}\xi_s, \qquad r = 1, 2, 3, \tag{3.7}$$

which assumes the matrix form

$$\{x\} = [l]^T\{\xi\}, \tag{3.8}$$

where $[l]^T$, defined by $[l_{ij}]^T = [l_{ji}]$ ($i, j = 1, 2, 3$), denotes the transpose of the matrix $[l]$. Introducing Eq. (3.4) into (3.8), we arrive at

$$[l]^T[l] = [1] = [\delta_{ij}], \tag{3.9}$$

where $[1]$ is the *identity matrix* or *unit matrix*, namely, a matrix with all its elements equal to the *Kronecker delta* defined by

$$\delta_{ij} = \begin{cases} 1 & i = j, \\ 0 & i \neq j. \end{cases} \tag{3.10}$$

From Eq. (3.9), we can easily conclude that the matrix of the direction cosines l_{ij} satisfies the relation

$$[l]^{-1} = [l]^T, \tag{3.11}$$

which implies that the reciprocal, or inverse, of $[l]$ is equal to the transpose of $[l]$. A transformation satisfying relation (3.11) is called an *orthonormal transformation.* Equation (3.9) can be written in index notation as

$$\sum_{k=1}^{3} l_{ki} l_{kj} = \delta_{ij}, \qquad i,j = 1, 2, 3, \tag{3.12}$$

which expresses the fact that axes x_i on the one hand and axes ξ_j on the other form *orthogonal sets of axes* and also that the length of the vector **r** is the same regardless of the set of axes in which it is resolved. The latter statement can be written

$$\mathbf{r} \cdot \mathbf{r} = \{x\}^T \{x\} = \{\xi\}^T \{\xi\}, \tag{3.13}$$

which can also be used to derive Eq. (3.9).

It may prove of interest to calculate the value of the determinant $|l|$ of the matrix $[l]$. From matrix algebra we have the relation

$$|[a][b]| = |a|\,|b|, \tag{3.14}$$

or the determinant of a product of two matrices is equal to the product of the determinants of the two matrices. But the determinant of the identity matrix is equal to 1, and since the determinant of a matrix is equal to the determinant of the transposed matrix, it follows from Eqs. (3.9) and (3.14) that

$$|l|^2 = 1. \tag{3.15}$$

Equation (3.15) indicates that $|l|$ may assume the value $+1$ or -1. For the type of transformations we shall be concerned with the value is $+1$ (see Sec. 4.1).

The matrix $[l]$ can be regarded as being the result of three successive rotations leading from system $\{x\}$ to system $\{\xi\}$, as we are going to see in the next section.

3.2 ROTATING COORDINATE SYSTEMS

In many dynamical problems involving spinning bodies it is convenient to express the motion in terms of components along rotating frames of reference, which, by definition, are noninertial frames. If this motion is to be related to the inertial space again, we must produce expressions relating the components of the rotating and the fixed systems of axes. In doing so, we recognize that one of the coordinate systems of Sec. 3.1 may be regarded as inertial and the other one as rotating in space, in which case the associated direction cosines are implicit functions of time. In the sequel we shall develop explicit expressions for the direction cosines between an inertial set of axes, say x_i, and a rotating one, denoted by ξ_i. The latter is obtained from the former by means of three successive rotations θ_1, θ_2, and θ_3 about axes x_1, y_2, and z_3, resulting in the systems y_i, z_i, and ξ_i, respectively.

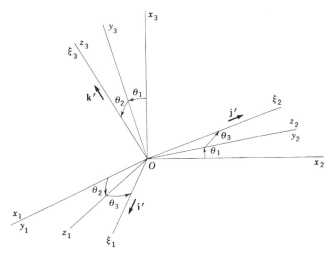

FIGURE 3.2

From Fig. 3.2 we conclude that the relation between the systems of coordinates x_i and y_i is as follows:

$$\begin{aligned} y_1 &= x_1, \\ y_2 &= x_2 \cos \theta_1 + x_3 \sin \theta_1, \\ y_3 &= -x_2 \sin \theta_1 + x_3 \cos \theta_1, \end{aligned} \qquad (3.16)$$

which can be written in matrix form as

$$\begin{Bmatrix} y_1 \\ y_2 \\ y_3 \end{Bmatrix} = \begin{bmatrix} 1 & 0 & 0 \\ 0 & \cos \theta_1 & \sin \theta_1 \\ 0 & -\sin \theta_1 & \cos \theta_1 \end{bmatrix} \begin{Bmatrix} x_1 \\ x_2 \\ x_3 \end{Bmatrix} \qquad (3.17)$$

or in more compact notation as

$$\{y\} = [R_1(\theta_1)]\{x\}. \qquad (3.18)$$

The *rotation matrix* $[R_1(\theta_1)]$, denoting the square matrix of the coefficients in Eq. (3.17), represents the rotation of a system of axes originally coincident with axes x_i by an angle θ_1 about axis x_1. In a similar fashion, we can write

$$\{z\} = \begin{bmatrix} \cos \theta_2 & 0 & -\sin \theta_2 \\ 0 & 1 & 0 \\ \sin \theta_2 & 0 & \cos \theta_2 \end{bmatrix} \{y\} = [R_2(\theta_2)]\{y\} \qquad (3.19)$$

and
$$\{\xi\} = \begin{bmatrix} \cos\theta_3 & \sin\theta_3 & 0 \\ -\sin\theta_3 & \cos\theta_3 & 0 \\ 0 & 0 & 1 \end{bmatrix} \{z\} = [R_3(\theta_3)]\{z\}. \tag{3.20}$$

Combining Eqs. (3.18) to (3.20), we obtain

$$\{\xi\} = [R_3(\theta_3)][R_2(\theta_2)][R_1(\theta_1)]\{x\} = [l]\{x\}, \tag{3.21}$$

where the matrix of direction cosines has the form

$$[l] = \begin{bmatrix} c\theta_2\, c\theta_3 & c\theta_1\, s\theta_3 + s\theta_1\, s\theta_2\, c\theta_3 & s\theta_1\, s\theta_3 - c\theta_1\, s\theta_2\, c\theta_3 \\ -c\theta_2\, s\theta_3 & c\theta_1\, c\theta_3 - s\theta_1\, s\theta_2\, s\theta_3 & s\theta_1\, c\theta_3 + c\theta_1\, s\theta_2\, s\theta_3 \\ s\theta_2 & -s\theta_1\, c\theta_2 & c\theta_1\, c\theta_2 \end{bmatrix}, \tag{3.22}$$

in which the abbreviations $s\theta_i = \sin\theta_i$ and $c\theta_i = \cos\theta_i$ are used to save space.

Next we wish to explore the possibility of expressing *finite rotations* as vector quantities. The natural thing to investigate is whether the preceding rotations can be represented as vectors $\boldsymbol{\theta}_i$ ($i = 1, 2, 3$) directed along the axes x_1, y_2, and z_3, respectively. For this purpose, let us check whether the rotations satisfy the addition rule for vectors, namely, that addition is a commutative process. This is equivalent to requiring that the order of the rotations be immaterial. The check can be made by using the rotation matrices developed above and going from system $\{x\}$ to system $\{z\}$ in two ways: (1) according to the sequence given by Eqs. (3.18) and (3.19) and (2) by a rotation θ_2 about axis x_2 followed by a rotation θ_1 about axis y_1. If the results obtained in the two ways are equal, it can be concluded that finite rotations are commutative and satisfy at least the addition rule for vectors. The two different sequences yield

$$\{z\} = [R_2(\theta_2)][R_1(\theta_1)]\{x\} = \begin{bmatrix} c\theta_2 & s\theta_1\, s\theta_2 & -c\theta_1\, s\theta_2 \\ 0 & c\theta_1 & s\theta_1 \\ s\theta_2 & -s\theta_1\, c\theta_2 & c\theta_1\, c\theta_2 \end{bmatrix} \{x\} \tag{3.23}$$

and

$$\{z\} = [R_1(\theta_1)][R_2(\theta_2)]\{x\} = \begin{bmatrix} c\theta_2 & 0 & -s\theta_2 \\ s\theta_1\, s\theta_2 & c\theta_1 & s\theta_1\, c\theta_2 \\ c\theta_1\, s\theta_2 & -s\theta_1 & c\theta_1\, c\theta_2 \end{bmatrix} \{x\}, \tag{3.24}$$

and it is not difficult to see that the addition rule for vectors is violated, as the two resulting vectors $\{z\}$ are not the same. This is not at all surprising since, in general, matrix products are not commutative

$$[R_2(\theta_2)][R_1(\theta_1)] \neq [R_1(\theta_1)][R_2(\theta_2)]. \tag{3.25}$$

Hence, *finite angles of rotation cannot be represented by vectors.*

Fortunately, however, considerable interest remains in representing *infinitesimal rotations*, rather than finite rotations, by vectors. Indeed, in the particular case in which the angles of rotation are sufficiently small to permit higher-order terms to be ignored Eqs. (3.23) and (3.24) both yield the same result. Hence, following this line of thought, we can show that infinitesimal rotations can be represented by vectors. As we are not so much interested in representing rotations by vectors as in representing rates of change of rotations, namely, angular velocities, by vectors, we shall find it to our advantage to change the approach slightly.

Let us consider a vector **r** fixed with respect to a set of moving axes ξ_i. Because the set of axes ξ_i rotates relative to an inertial space, say the set x_i, the vector **r** undergoes some change. We may recall that a change in the direction of a vector is sufficient to bring about a change in the vector, and our interest lies in calculating the rate of change of **r** due to the angular velocity of the reference frame ξ_i. We have the choice of expressing **r** and the rate of change of **r** in terms of components along the inertial system x_i or along the moving system ξ_i. This may sound like a paradox in view of the fact that the vector **r** is fixed relative to the system ξ_i. The fact remains, however, that, due to the rotation of the system ξ_i, the vector **r** changes continuously with respect to an inertial space and the vector representing the corresponding rate of change can be resolved into components along the set x_i or the set ξ_i. It turns out that in the study of spinning bodies it is frequently more useful to express the motion in terms of components along the moving system ξ_i. To accomplish this, let us assume that the angles of rotation $\Delta\theta_i$ are sufficiently small for the approximations $\sin \Delta\theta_i \approx \Delta\theta_i$ and $\cos \Delta\theta_i \approx 1$ ($i = 1, 2, 3$) to be justified. Then premultiplying both sides of Eq. (3.21) by $[l]^T$, we arrive at

$$\{x\} = \{\xi\} + \{\Delta\xi\} = [l]^T\{\xi\} = \begin{bmatrix} 1 & -\Delta\theta_3 & \Delta\theta_2 \\ \Delta\theta_3 & 1 & -\Delta\theta_1 \\ -\Delta\theta_2 & \Delta\theta_1 & 1 \end{bmatrix}\{\xi\}$$

$$= [1]\{\xi\} + \begin{bmatrix} 0 & -\Delta\theta_3 & \Delta\theta_2 \\ \Delta\theta_3 & 0 & -\Delta\theta_1 \\ -\Delta\theta_2 & \Delta\theta_1 & 0 \end{bmatrix}\{\xi\}, \quad (3.26)$$

where $[l]^T$ was obtained by transposing Eq. (3.22) and letting $\theta_i \to \Delta\theta_i$ ($i = 1, 2, 3$). Equation (3.26) leads to

$$\{\Delta\xi\} = [\Delta\theta]\{\xi\}, \qquad (3.27)$$

where
$$[\Delta\theta] = \begin{bmatrix} 0 & -\Delta\theta_3 & \Delta\theta_2 \\ \Delta\theta_3 & 0 & -\Delta\theta_1 \\ -\Delta\theta_2 & \Delta\theta_1 & 0 \end{bmatrix} \qquad (3.28)$$

is a *skew-symmetric*, or *antisymmetric*, matrix. Hence, a vector **r**, fixed with respect to a moving system of coordinates ξ_i, will undergo an incremental change $\Delta \mathbf{r}$ relative to an inertial space as a result of an incremental rotation with components $\Delta \theta_i$ ($i = 1, 2, 3$) of the moving system about axes ξ_i, respectively. In terms of components along the inertial system, the vector **r** + $\Delta \mathbf{r}$ is given by $\{x\}$, whereas in terms of components along the system ξ_i, it is expressed by $\{\xi\} + \{\Delta \xi\}$, where the column matrix $\{\Delta \xi\}$ represents the increment $\Delta \mathbf{r}$. All quantities in Eqs. (3.27) and (3.28) are in terms of components along the moving system ξ_i. Clearly, the results (3.27) and (3.28) do not depend on the order of the rotations if the rotations are small. Dividing Eqs. (3.27) and (3.28) by the associated time increment Δt and letting $\Delta t \to 0$, we obtain the time derivative

$$\{\dot{\xi}\} = \lim_{\Delta t \to 0} \left\{\frac{\Delta \xi}{\Delta t}\right\} = \lim_{\Delta t \to 0} \left[\frac{\Delta \theta}{\Delta t}\right]\{\xi\} = [\omega]\{\xi\}, \qquad (3.29)$$

where $[\omega]$ is the skew-symmetric matrix

$$[\omega] = \begin{bmatrix} 0 & -\omega_3 & \omega_2 \\ \omega_3 & 0 & -\omega_1 \\ -\omega_2 & \omega_1 & 0 \end{bmatrix}, \qquad (3.30)$$

in which ω_i ($i = 1, 2, 3$) are the angular velocity components of the moving system ξ_i relative to the inertial space x_i when expressed in terms of components along the system ξ_i.

It is easy to show that the vector counterpart of Eq. (3.29) is

$$\dot{\mathbf{r}} = \boldsymbol{\omega} \times \mathbf{r}, \qquad (3.31)$$

where
$$\mathbf{r} = \xi_1 \mathbf{i}' + \xi_2 \mathbf{j}' + \xi_3 \mathbf{k}' \qquad (3.32)$$
and
$$\boldsymbol{\omega} = \omega_1 \mathbf{i}' + \omega_2 \mathbf{j}' + \omega_3 \mathbf{k}'. \qquad (3.33)$$

Equation (3.31) can easily be interpreted physically by means of Fig. 3.3, in which **r** represents the position vector, **ω** the angular velocity vector which is coincident with the instantaneous axis of rotation, and $\dot{\mathbf{r}}$ a vector normal to both **r** and **ω** and in the direction shown.

As a special application of Eq. (3.31), we can obtain expressions for the time derivatives of the unit vectors $\mathbf{i}'$, $\mathbf{j}'$, and $\mathbf{k}'$, respectively, in the form

$$\frac{d\mathbf{i}'}{dt} = \boldsymbol{\omega} \times \mathbf{i}' = \omega_3 \mathbf{j}' - \omega_2 \mathbf{k}',$$

$$\frac{d\mathbf{j}'}{dt} = \boldsymbol{\omega} \times \mathbf{j}' = \omega_1 \mathbf{k}' - \omega_3 \mathbf{i}', \qquad (3.34)$$

$$\frac{d\mathbf{k}'}{dt} = \boldsymbol{\omega} \times \mathbf{k}' = \omega_2 \mathbf{i}' - \omega_1 \mathbf{j}'.$$

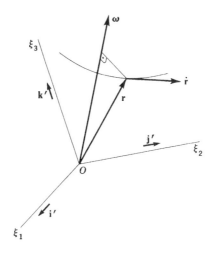

FIGURE 3.3

In fact, Eq. (3.31) can be interpreted as the time derivative of **r**, in which the magnitudes ξ_1, ξ_2, and ξ_3 are constant in time and the unit vectors have time derivatives according to Eqs. (3.34).

In the above discussion, although not specifically stated, it was implied that the elements of the skew-symmetric matrix $[\omega]$ form a vector $\boldsymbol{\omega}$ in all cartesian coordinate systems. For $\boldsymbol{\omega}$ to qualify as a vector, however, it must also transform like the components of a vector, which has yet to be shown. In the sequel we shall examine this question.

The relation between the components of the vector $\boldsymbol{\omega}$ and the matrix $[\omega]$ can be written in the form

$$\omega_i = \frac{1}{2} \sum_{j=1}^{3} \sum_{k=1}^{3} \epsilon_{ijk} \omega_{kj}, \tag{3.35}$$

in which ϵ_{ijk} is the standard *epsilon symbol*, defined to be equal to zero if any two of the three indices are equal, equal to $+1$ if the indices are in cyclic order, and equal to -1 if they are not. The inverse relation corresponding to Eq. (3.35) is

$$\omega_{nm} = \sum_{l=1}^{3} \epsilon_{mnl} \omega_l. \tag{3.36}$$

For $\boldsymbol{\omega}$ to qualify as a vector, it must be shown that for an orthonormal transformation defined by the matrix $[a]$, $[a]^T = [a]^{-1}$, relating the elements

ω_{nm} to the elements ω'_{kj} of the transformed matrix by

$$\omega'_{kj} = \sum_{m=1}^{3} \sum_{n=1}^{3} a_{kn}\omega_{nm}a_{jm}, \quad (3.37)$$

the components in the new coordinate system have the form

$$\omega'_i = \sum_{j=1}^{3} a_{ij}\omega_j. \quad (3.38)$$

It turns out that the coordinate transformation in question has the form

$$\omega'_i = |a| \sum_{j=1}^{3} a_{ij}\omega_j, \quad (3.39)$$

where $|a|$ is the determinant of the matrix $[a]$, rather than the form (3.38) (see Ref. 1, p. 130). Thus for $\boldsymbol{\omega}$ to qualify as a vector, $|a|$ must be equal to $+1$. This actually happens only in the case of an orthonormal transformation corresponding to a *proper rotation* defined by a transformation matrix with the determinant equal to $+1$, as opposed to the *improper rotation* with the determinant equal to -1. An orthonormal transformation corresponding to a proper rotation transforms a right-handed system into another right-handed system, as in the case described by Eq. (3.4). Quantities transforming according to Eq. (3.39) are called *pseudovectors* or *axial vectors*. Hence, the elements of any 3 × 3 skew-symmetric matrix form the components of a pseudovector. For all practical purposes, however, we need not make the distinction and can regard $\boldsymbol{\omega}$ as a vector.

3.3 EXPRESSIONS FOR THE MOTION IN TERMS OF MOVING REFERENCE FRAMES

In Sec. 3.2 we considered systems of coordinates rotating relative to an inertial system. In particular, we derived an expression for the rate of change of a vector fixed in the rotating system. Now we wish to obtain an expression for the time derivative of a vector whose components along the moving system vary with time. Refer to Fig. 3.4 and denote by **r** the position of point P relative to O when expressed in terms of the components x_i in an inertial space and by **r**′ when expressed in terms of the components ξ_i along the rotating system. Of course, the two represent the same vector

$$\mathbf{r} = \mathbf{r}' = \xi_1\mathbf{i}' + \xi_2\mathbf{j}' + \xi_3\mathbf{k}'. \quad (3.40)$$

We pointed out in Sec. 3.2 that it is often advantageous to express the motion

Motion Relative to Rotating Reference Frames

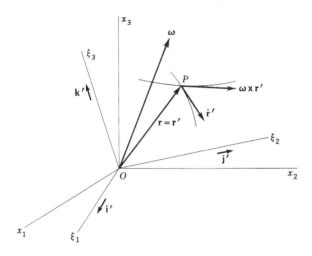

FIGURE 3.4

in terms of components along the rotating system. Differentiating (3.40) with respect to time, we obtain

$$\dot{\mathbf{r}} = \frac{d\xi_1}{dt}\mathbf{i}' + \frac{d\xi_2}{dt}\mathbf{j}' + \frac{d\xi_3}{dt}\mathbf{k}' + \xi_1\frac{d\mathbf{i}'}{dt} + \xi_2\frac{d\mathbf{j}'}{dt} + \xi_3\frac{d\mathbf{k}'}{dt}$$
$$= \dot{\mathbf{r}}' + \boldsymbol{\omega} \times \mathbf{r}', \tag{3.41}$$
where $\quad \dot{\mathbf{r}}' = \dot{\xi}_1\mathbf{i}' + \dot{\xi}_2\mathbf{j}' + \dot{\xi}_3\mathbf{k}' \tag{3.42}$

denotes the rate of change of $\mathbf{r}'$ relative to the system ξ_1, ξ_2, ξ_3 and $\boldsymbol{\omega} \times \mathbf{r}'$ denotes the rate of change of $\mathbf{r}'$ due to the rotational motion of the system ξ_1, ξ_2, ξ_3. Notice that the latter is precisely the expression derived in Sec. 3.2. In terms of velocities, $\dot{\mathbf{r}}'$ is the velocity of point P relative to the rotating system, and $\boldsymbol{\omega} \times \mathbf{r}'$ is the velocity of the coincident point (a point coinciding with P instantaneously).

Equation (3.41) represents the time derivative of the vector $\mathbf{r}$ in an inertial space when the vector is expressed in terms of a rotating frame of reference and *is valid for any vector*, such as the velocity or the angular momentum vector. Under these circumstances, the second derivative of $\mathbf{r}$ assumes the form

$$\ddot{\mathbf{r}} = \frac{d}{dt}(\dot{\mathbf{r}}') + \frac{d\boldsymbol{\omega}}{dt} \times \mathbf{r}' + \boldsymbol{\omega} \times \frac{d}{dt}(\mathbf{r}')$$
$$= \ddot{\mathbf{r}}' + \boldsymbol{\omega} \times \dot{\mathbf{r}}' + \dot{\boldsymbol{\omega}} \times \mathbf{r}' + (\boldsymbol{\omega} \times \boldsymbol{\omega}) \times \mathbf{r}' + \boldsymbol{\omega} \times \dot{\mathbf{r}}' + \boldsymbol{\omega} \times (\boldsymbol{\omega} \times \mathbf{r}')$$
$$= \ddot{\mathbf{r}}' + 2\boldsymbol{\omega} \times \dot{\mathbf{r}}' + \dot{\boldsymbol{\omega}} \times \mathbf{r}' + \boldsymbol{\omega} \times (\boldsymbol{\omega} \times \mathbf{r}'), \tag{3.43}$$
where $\quad \ddot{\mathbf{r}}' = \ddot{\xi}_1\mathbf{i}' + \ddot{\xi}_2\mathbf{j}' + \ddot{\xi}_3\mathbf{k}' \tag{3.44}$

is the second derivative of $\mathbf{r}'$ relative to the rotating system. In terms of accelerations, $\ddot{\mathbf{r}}$ is the acceleration of P in an inertial space, $\ddot{\mathbf{r}}'$ is the acceleration of P relative to the rotating frame, $2\boldsymbol{\omega} \times \dot{\mathbf{r}}'$ is known as the *Coriolis acceleration*, and $\dot{\boldsymbol{\omega}} \times \mathbf{r}' + \boldsymbol{\omega} \times (\boldsymbol{\omega} \times \mathbf{r}')$ is the acceleration of the coincident point. The term $\boldsymbol{\omega} \times (\boldsymbol{\omega} \times \mathbf{r}')$ is called the *centripetal acceleration* and is directed toward the instantaneous axis of rotation.

If the origin O translates with velocity $\mathbf{v}_0$ and acceleration $\mathbf{a}_0$ with respect to the inertial space, the absolute velocity and acceleration of point P are

$$\mathbf{v} = \mathbf{v}_0 + \dot{\mathbf{r}} \qquad (3.45)$$

and
$$\mathbf{a} = \mathbf{a}_0 + \ddot{\mathbf{r}}, \qquad (3.46)$$

respectively, where $\dot{\mathbf{r}}$ is given by Eq. (3.41) and $\ddot{\mathbf{r}}$ by Eq. (3.43).

3.4 MOTION RELATIVE TO THE ROTATING EARTH

Although Newton's second law has an extremely simple form when the motion is referred to an inertial system, it frequently is more convenient to refer the motion to a noninertial system. It is natural to assume that there are advantages in referring motion in the vicinity of a point on the earth's surface to a coordinate system rigidly attached to that surface. This indeed proves to be the case. Such a reference frame, however, is not inertial because the earth's center revolves around the sun and the earth rotates about its own axis. The expressions developed in Sec. 3.3 in connection with noninertial reference frames are extremely useful in treating these types of problems.

Although the center of the earth is moving around the sun, the acceleration is relatively small compared to the acceleration due to gravity or even the acceleration of a point on the earth's surface due to the earth's spin, where the point in question is reasonably far from the poles. Furthermore, for the present purpose, the earth's axis of rotation can be assumed to be fixed in space (for a discussion of this assumption, see Sec. 11.8). Hence, we can choose as an inertial system a rectangular set X, Y, Z with the origin at the earth's center C and axis Z aligned with the earth's axis of rotation. Axes X and Y are in the equatorial plane, with X axis pointing toward the vernal equinox (see Fig. 3.5). The earth rotates with an angular velocity Ω, which, for all practical purposes, can be assumed constant and equal to one rotation per day.

We shall be concerned with the motion of a particle in the neighborhood of the earth's surface, the earth being assumed to be a perfect sphere. To express the motion of the particle relative to the earth, we attach a coordinate

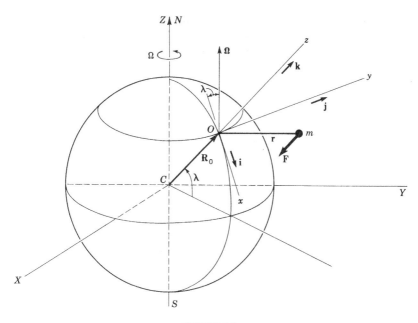

FIGURE 3.5

system x, y, z to the surface of the earth with the origin O situated at a given latitude λ; the longitude turns out to be immaterial. The x axis is tangent to the meridian circle pointing south, y is tangent to the parallel pointing east, and z, since it is directed toward the zenith, coincides with the local vertical. The corresponding unit vectors are $\mathbf{i}$, $\mathbf{j}$, and $\mathbf{k}$. The system x, y, z, being rigidly attached to the earth's surface, possesses the same angular velocity as the earth. From Fig. 3.5 we conclude that the angular velocity can be expressed in terms of components along the x, y, z axes in the following form

$$\mathbf{\Omega} = -(\Omega \cos \lambda)\mathbf{i} + (\Omega \sin \lambda)\mathbf{k} = \text{const.} \tag{3.47}$$

The position of the mass m relative to the system x, y, z is denoted by $\mathbf{r}$, and the radius vector from center C to the origin O is denoted by $\mathbf{R}_0$. Hence, using Eq. (3.46), the acceleration of m in an inertial space is

$$\begin{aligned}\mathbf{a} &= \mathbf{a}_0 + \ddot{\mathbf{r}}' + 2\boldsymbol{\omega} \times \dot{\mathbf{r}}' + \dot{\boldsymbol{\omega}}' \times \mathbf{r}' + \boldsymbol{\omega} \times (\boldsymbol{\omega} \times \mathbf{r}') \\ &= \mathbf{\Omega} \times (\mathbf{\Omega} \times \mathbf{R}_0) + \mathbf{a}' + 2\mathbf{\Omega} \times \mathbf{v}' + \mathbf{\Omega} \times (\mathbf{\Omega} \times \mathbf{r}),\end{aligned} \tag{3.48}$$

where $\mathbf{v}'$ and $\mathbf{a}'$ are, respectively, the velocity and acceleration of m relative to the x, y, z system.

Let the forces acting upon m be gravity and forces from yet unspecified sources, where the latter are combined into a resultant **F**. For motion in the vicinity of O, the effect of the earth's curvature may be neglected and the gravitational field assumed uniform. It follows that the gravity force vector is constant in magnitude, always parallel to the z axis, and pointing in the negative direction of the z axis. Under these circumstances, Newton's second law takes the form

$$\mathbf{F} - mg\mathbf{k} = m\mathbf{a} = m[\mathbf{\Omega} \times (\mathbf{\Omega} \times \mathbf{R}_0) + \mathbf{a}' + 2\mathbf{\Omega} \times \mathbf{v}' + \mathbf{\Omega} \times (\mathbf{\Omega} \times \mathbf{r})]. \quad (3.49)$$

The last term on the right side of (3.49) is generally very small, so that the differential equations of motion become

$$\frac{F_x}{m} = \ddot{x} - 2\Omega\dot{y}\sin\lambda - \Omega^2 R_0 \sin\lambda\cos\lambda,$$

$$\frac{F_y}{m} = \ddot{y} + 2\Omega(\dot{x}\sin\lambda + \dot{z}\cos\lambda), \quad (3.50)$$

$$-g + \frac{F_z}{m} = \ddot{z} - 2\Omega\dot{y}\cos\lambda - \Omega^2 R_0 \cos^2\lambda,$$

which can be solved for the relative motion x, y, z.

3.5 MOTION OF A FREE PARTICLE RELATIVE TO THE ROTATING EARTH

The problem of a particle moving freely relative to the rotating earth has interesting implications. For simplicity let us neglect air resistance and any other forces except gravity, $F_x = F_y = F_z = 0$. Because the rotation of the earth is very small, $\Omega = 7.29 \times 10^{-5}$ rad/sec, second-order terms in Ω lead to accelerations which are negligible compared with the acceleration due to gravity and will be ignored throughout. Under these assumptions, Eqs. (3.50) reduce to

$$\ddot{x} - 2\Omega\dot{y}\sin\lambda = 0,$$
$$\ddot{y} + 2\Omega(\dot{x}\sin\lambda + \dot{z}\cos\lambda) = 0, \quad (3.51)$$
$$\ddot{z} - 2\Omega\dot{y}\cos\lambda + g = 0.$$

Assuming that at $t = 0$ the relative position and velocity components are

$$\begin{aligned} x(0) &= 0, & y(0) &= 0, & z(0) &= h, \\ \dot{x}(0) &= u_0, & \dot{y}(0) &= v_0, & \dot{z}(0) &= w_0, \end{aligned} \quad (3.52)$$

Motion Relative to Rotating Reference Frames

we can integrate Eqs. (3.51) once and obtain

$$\dot{x} - 2\Omega y \sin \lambda = u_0,$$
$$\dot{y} + 2\Omega(x \sin \lambda + z \cos \lambda) = v_0 + 2\Omega h \cos \lambda, \quad (3.53)$$
$$\dot{z} - 2\Omega y \cos \lambda + gt = w_0.$$

Introducing the first and third of Eqs. (3.53) into the second of Eqs. (3.51) and neglecting terms in Ω^2, we obtain

$$\ddot{y} + 2\Omega(u_0 \sin \lambda + w_0 \cos \lambda) - 2\Omega gt \cos \lambda = 0, \quad (3.54)$$

which can be integrated twice with the result

$$y = v_0 t - \Omega t^2 (u_0 \sin \lambda + w_0 \cos \lambda) + \tfrac{1}{3}\Omega g t^3 \cos \lambda. \quad (3.55)$$

A substitution of Eq. (3.55) into the first and third of Eqs. (3.53) and integrations with respect to time yield

$$x = u_0 t + \Omega v_0 t^2 \sin \lambda,$$
$$z = h + w_0 t + \Omega v_0 t^2 \cos \lambda - \tfrac{1}{2} g t^2, \quad (3.56)$$

where again terms in Ω^2 have been ignored.

If a particle is dropped from rest, $u_0 = v_0 = w_0 = 0$, at a height h, it will land on the earth's surface after an interval of time

$$t = \sqrt{\frac{2h}{g}}. \quad (3.57)$$

The coordinates of the landing point are

$$x = z = 0, \quad y = \tfrac{2}{3}\Omega h \left(\frac{2h}{g}\right)^{\frac{1}{2}} \cos \lambda, \quad (3.58)$$

where the positive sign of y indicates that it lands at a point on the y axis *east of the origin*. This result may seem a little puzzling in view of the fact that the earth is rotating from west to east. This can be easily explained, however, since as the particle drops with a downward velocity gt, it is being deflected eastward by the Coriolis effect. Due to the nature of the assumptions, the above result is difficult to verify experimentally.

Another case of interest which can be attributed directly to the Coriolis effect but is easier to detect is the cyclone, created when a point of low pressure is surrounded by points of high pressure. The winds rushing toward the point of low pressure are deflected so as to generate a cyclone. In Fig. 3.6, which shows an idealized situation as it might occur in the Northern Hemisphere, the concentric circles represent isobars, with the pressure decreasing toward the center. The dashed arrows represent the direction the winds would have in the absence of the earth's rotation, whereas the solid arrows include the deflection produced by the Coriolis effect, as reflected by the

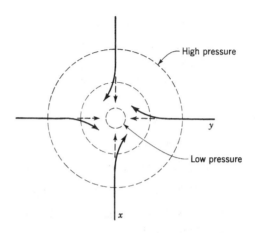

FIGURE 3.6

terms $v_0 t^2 \sin \lambda$ and $-u_0 t^2 \sin \lambda$ in the equations for x and y, respectively. This effect causes the winds to travel around the low-pressure point in a counterclockwise direction in the Northern Hemisphere and in a clockwise direction in the Southern Hemisphere. It also causes the water to spin in a likewise manner when draining out of a sink.

3.6 FOUCAULT'S PENDULUM

A simple pendulum can be used to demonstrate the earth's rotation, a device first suggested by Foucault. Consider a pendulum close to the earth's surface at a latitude λ. If the initial motion of the pendulum is such that in the absence of the earth's rotation it oscillates as a simple pendulum in a plane containing the local vertical, then to an observer on the surface of the rotating earth the plane of the pendulum will appear to rotate with an angular velocity $-\Omega \sin \lambda$ about an axis coinciding with the local vertical. This effect is strongest at the North (or South) Pole, where the earth can be envisioned as rotating beneath the pendulum; the effect disappears at the equator. The general equations of Sec. 3.4 can be used to demonstrate mathematically the rotation of the pendulum plane relative to the earth.

Let us consider a pendulum consisting of a bob of mass m supported by a relatively long wire of length L, where the mass of the wire is negligible. The wire is attached at a point $z = L$ on the local vertical axis in a manner that allows free rotation about the z axis. The motion of the bob must be small, so that, in effect, it takes place in the xy plane, with the result

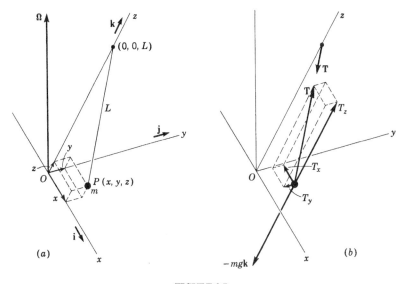

FIGURE 3.7

that $\dot{z}$ and $\ddot{z}$ can be ignored. Moreover, since the motion is small, second-order terms in the coordinates x and y and their time derivatives are negligible. Figure 3.7 shows the pendulum and the corresponding free-body diagram.

According to the free-body diagram, the force components due to the tension T in the wire are

$$F_x = -T_x = -\frac{x}{L}T,$$
$$F_y = -T_y = -\frac{y}{L}T, \qquad (3.59)$$
$$F_z = T_z = \frac{L-z}{L}T,$$

so that Eqs. (3.50) assume the form

$$-\frac{x}{mL}T = \ddot{x} - 2\Omega\dot{y}\sin\lambda - \Omega^2 R_0 \sin\lambda\cos\lambda,$$
$$-\frac{y}{mL}T = \ddot{y} + 2\Omega(\dot{x}\sin\lambda + \dot{z}\cos\lambda), \qquad (3.60)$$
$$-g + \frac{L-z}{mL}T = \ddot{z} - 2\Omega\dot{y}\cos\lambda - \Omega^2 R_0 \cos^2\lambda.$$

It turns out that terms in Ω^2 are again negligible (note that there are cases in which this is not true and each case must be decided independently). Furthermore, by the assumption that L is very large compared with the rise z and its time derivatives, the last of Eqs. (3.60) reduces to

$$\frac{T}{m} = g - 2\Omega \dot{y} \cos \lambda, \qquad (3.61)$$

which can be introduced into the first two of Eqs. (3.60) to obtain

$$\begin{aligned} \ddot{x} &= -\frac{x}{L}(g - 2\Omega \dot{y} \cos \lambda) + 2\Omega \dot{y} \sin \lambda \\ &= -\frac{x}{L}g + 2\Omega \dot{y} \sin \lambda, \\ \ddot{y} &= -\frac{y}{L}(g - 2\Omega \dot{y} \cos \lambda) - 2\Omega(\dot{x} \sin \lambda + \dot{z} \cos \lambda) \\ &= -\frac{y}{L}g - 2\Omega \dot{x} \sin \lambda, \end{aligned} \qquad (3.62)$$

where second-order terms in x and y and the term in $\dot{z}$ have been ignored. Introducing the notation $g/L = p^2$, where p is the natural frequency of the simple pendulum, Eqs. (3.62) reduce to

$$\begin{aligned} \ddot{x} - 2\Omega \dot{y} \sin \lambda + p^2 x &= 0, \\ \ddot{y} + 2\Omega \dot{x} \sin \lambda + p^2 y &= 0, \end{aligned} \qquad (3.63)$$

which represents a set of two simultaneous linear equations in x and y. The coupling of Eqs. (3.63) is provided by the velocities $\dot{x}$ and $\dot{y}$.

Equations (3.63) constitute a typical example of a set of simultaneous equations which can be successfully treated by means of a complex vector representation defined by

$$u = x + iy, \qquad i = \sqrt{-1}. \qquad (3.64)$$

With this in mind, we can multiply the second of Eqs. (3.63) by i and add the result to the first to obtain

$$\ddot{u} + (2i\Omega \sin \lambda)\dot{u} + p^2 u = 0. \qquad (3.65)$$

Let the solution of Eq. (3.65) have the form

$$u = Ae^{\alpha t}, \qquad (3.66)$$

where the exponent α must satisfy the characteristic equation

$$\alpha^2 + (2i\Omega \sin \lambda)\alpha + p^2 = 0. \qquad (3.67)$$

Equation (3.67) yields two roots

$$\left.\begin{array}{c}\alpha_1\\ \alpha_2\end{array}\right\} = -i\Omega \sin \lambda \pm \sqrt{(i\Omega \sin \lambda)^2 - p^2} \approx -i\Omega \sin \lambda \pm ip, \quad (3.68)$$

so that the general solution of Eq. (3.65) is

$$u = u_0 \exp(-i\Omega t \sin \lambda), \quad (3.69)$$

where
$$u_0 = A_1 e^{ipt} + A_2 e^{-ipt}. \quad (3.70)$$

The coefficients A_1 and A_2 are generally complex numbers whose values are determined by the initial conditions.

The easiest way of interpreting the motion of the bob is to assume momentarily that the rotation of the earth is zero. A combination of Eqs. (3.64) and (3.70) yields

$$u_0 = x_0 + iy_0 = C_1 \cos pt + iC_2 \sin pt, \quad (3.71)$$

where $C_1 = A_1 + A_2$ and $C_2 = A_1 - A_2$. Equation (3.71) indicates that both components, x_0 and y_0, perform harmonic motion with the same frequency p but different amplitudes, C_1 and C_2. It follows that the path described by the bob in the xy plane is an ellipse with the center at the origin, which is typical of the low-amplitude oscillation of a spherical pendulum. The effect of the term $\exp(-i\Omega t \sin \lambda)$ is to rotate the complex vector u_0 relative to the xy plane through an angle $-\Omega t \sin \lambda$, which increases linearly with the time t.

The motion can be summarized as follows. The complex vector u_0 performs an elliptic path around the origin with the period of motion $2\pi/p$. The effect of the earth's rotation on the pendulum is to rotate the elliptical path about the local vertical with an angular velocity $-\Omega \sin \lambda$. The rotation is clockwise in the Northern Hemisphere and counterclockwise in the Southern Hemisphere. At the North Pole, $\lambda = \pi/2$, the rotation rate of the ellipse relative to the earth's surface is $-\Omega$. On the other hand, an inertial observer will see the ellipse as stationary with the earth rotating beneath the pendulum. At the equator, $\lambda = 0$, the Coriolis effect vanishes, so that the ellipse no longer rotates relative to the earth's surface. Of course, when $C_1 = 0$ or $C_2 = 0$, the ellipse reduces to a line and the spherical pendulum becomes a simple pendulum. For the purpose of demonstrating the earth's rotation a simple pendulum is more desirable than a spherical one.

PROBLEMS

3.1 A triad x_1, x_2, x_3 is rotated by an angle ϕ about axis x_3, an angle θ about axis x_1, and an angle ψ about axis x_2. A vector **r** has constant components relative to the triad. Calculate the change $\Delta \mathbf{r}$ of the vector **r** resulting from these rotations. Express the result in terms of inertial components as well as components along the triad x_1, x_2, x_3.

3.2 For the system shown in Fig. 3.8 calculate the absolute velocity and acceleration of a point A at the tip of the bar when the bar makes an angle θ with respect to the horizontal.

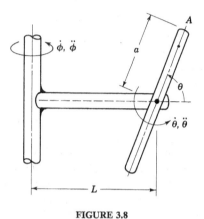

FIGURE 3.8

3.3 A particle travels along a turbine blade with velocity $\dot{s}$ and acceleration $\ddot{s}$ relative to the blade, as shown in Fig. 3.9. The radius of curvature of the blade is ρ. Calculate the absolute velocity and acceleration of the particle at the moment it leaves the blade when the wheel is in the position shown.

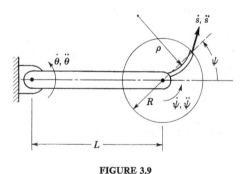

FIGURE 3.9

3.4 A particle is projected with velocity v on a smooth horizontal plane at a latitude λ in the Northern Hemisphere. If the initial direction of the particle makes an angle θ with respect to the meridian, calculate the path of the ball.

Motion Relative to Rotating Reference Frames 121

3.5 A monorail car of mass m travels east with uniform velocity v at a latitude λ in the Northern Hemisphere. Calculate the forces on the track if the track makes an angle α with the horizontal.

3.6 How should the motion of a Foucault's pendulum be initiated in order for the bob to trace a straight line?

SUGGESTED REFERENCES

1. Goldstein, H.: *Classical Mechanics*, Addison-Wesley Publishing Company, Inc., Reading, Mass., 1950.
2. Synge, J. L., and B. A. Griffith: *Principles of Mechanics*, 3d ed., McGraw-Hill Book Company, New York, 1959.

chapter four

Rigid Body Dynamics

A deformable body of finite dimensions may be regarded as being composed of an infinite number of particles, so that, in effect, it possesses an infinite number of degrees of freedom. On the other hand, by definition, a rigid body undergoes no deformations and can be shown to possess only six degrees of freedom. This results from the fact that the distance between any pair of particles in a rigid body is assumed to be constant. The concept of a rigid body is more of a mathematical idealization than a physical reality. Nevertheless, in the cases where the deformations of the body are insignificant compared to the motion of the body as a whole, the concept may be quite valid. Moreover, it is a very useful concept, as it allows us to describe the motion of a body by only six coordinates, which are generally taken as three translations and three rotations.

Many of the expressions and principles derived in connection with systems of particles are equally valid for rigid bodies. In particular, the relation between force and linear momentum, Eq. (1.59), and those between torque and angular momentum, Eqs. (1.65) and (1.66), remain unchanged for rigid bodies. The same applies to the principles of conservation of linear and angular momentum and of energy. Of course, for rigid bodies, the angular momentum and kinetic energy assume special forms, namely, in terms of moments and products of inertia. Similarly, Hamilton's principle and Lagrange's equations for rigid bodies retain the general forms derived in Chap. 2, as rigid bodies are merely special types of systems of particles. When the motion is described in terms of only six generalized coordinates, the constraint equations, resulting from the fact that the distance between any two particles in the rigid body is constant, are accounted for automatically. In this connection we may point out that when one of the generalized coordinates represents a rotation, the corresponding generalized force is actually a torque and not a force. Again the concept of center of mass plays a major role. Indeed, under certain circumstances, it enables us to separate the translational and rotational motion, so that the problem can be reduced to the investigation of the rotational motion of a body about a point.

This chapter is devoted almost entirely to the rotational motion of a rigid body about a point. First the general expressions derived for systems of particles are brought into a form suitable for the treatment of rigid bodies. The motion of a moment-free rigid body is discussed by placing the emphasis on the geometrical aspects of the solution. The phenomena of gyroscopic motion are introduced by using the symmetric top as an example of a gyroscope. Finally, the problem of a rigid body undergoing both translational and rotational motions is discussed.

4.1 KINEMATICS OF A RIGID BODY

The most general motion of a rigid body is described by six coordinates, so that a rigid body moving freely in space possesses six degrees of freedom. We may conceive of a rigid body as consisting of an infinite number of mass particles, the distance between any two particles being constant by the definition of a rigid body. It is not difficult to show that a rigid body does indeed possess only six degrees of freedom. For this purpose, let us consider one of the particles in a rigid body moving unconstrained in space, so that there are three degrees of freedom associated with that particle. A second particle within the rigid body adds only two more degrees of freedom to the count, because we must deduct from the three degrees of freedom associated with the motion of a free particle one degree of freedom due to the kinematical constraint expressing the fact that the distance between the two particles considered is constant. A third particle adds only one more degree of freedom because the distance to the first two must remain constant. No other particle can introduce an additional degree of freedom because the distance to the first three particles is constant. Thus the motion of three particles within a rigid body determines the motion completely, with the obvious conclusion that a rigid body has six degrees of freedom.

It is often convenient to choose as the six coordinates describing the motion of a rigid body three translations of a certain point within the body and three rotations about that point. To this end, a system of axes, called *body axes*, is embedded in the body, and the motion is described in terms of the translation of the origin O of the body axes as well as the rotation of these axes with respect to an inertial space. But (by design and not by coincidence) this is just how the motion was described in Chap. 3, so that the derivations there can be applied equally well here. In fact, the description of the motion of a rigid body is simpler because there is no motion of the mass points with respect to the body axes. Hence, from Eqs. (3.41), (3.43), (3.45), and (3.46) we obtain the absolute velocity of a point in a rigid body

$$\mathbf{v} = \mathbf{v}_0 + \boldsymbol{\omega} \times \mathbf{r} \tag{4.1}$$

and the absolute acceleration

$$\mathbf{a} = \mathbf{a}_0 + \dot{\boldsymbol{\omega}} \times \mathbf{r} + \boldsymbol{\omega} \times (\boldsymbol{\omega} \times \mathbf{r}) \tag{4.2}$$

after setting $\dot{\mathbf{r}}' = \ddot{\mathbf{r}}' = 0$ and letting $\mathbf{r} = \mathbf{r}'$ be the radius vector of the point in question measured with respect to the body axes.

The case in which one of the points of the body is fixed in an inertial space is of considerable interest in rigid body dynamics. The orientation of the body can be defined in terms of an orthogonal transformation given by the 3×3 matrix $[l(t)]$ relating the body axes to the inertial system at any time t. If initially the body axes were coincident with the inertial system, then $[l(0)] = [1]$, where the latter is the unit matrix.

Next we shall show that the general displacement of a rigid body with one point fixed is a rotation about some axis through the fixed point. This implies that in the inertial space there is an axis in the rigid body left unaffected by the rotation, which is equivalent to saying that the components of a vector coinciding with the axis of rotation remain the same as before the rotation. Using Eq. (3.4), this can be expressed mathematically in the matrix form

$$\{\xi\} = [l]\{x\} = \{x\}. \tag{4.3}$$

Now let us consider the general *eigenvalue problem*

$$[l]\{x\} = \lambda\{x\}, \tag{4.4}$$

of which Eq. (4.3) is a special case. The homogeneous equation (4.4) has nontrivial solutions for only certain values of λ. These constants, called *eigenvalues*, are denoted by λ_r, and the corresponding vectors, known as *eigenvectors*, are denoted by $\{x^{(r)}\}$ ($r = 1, 2, 3$) (see Ref. 2, sec. 4-3). Thus the real orthogonal matrix $[l]$, representing the rotation of a rigid body, must have an eigenvalue equal to $+1$ if the statement concerning the general displacement of a rigid body is to be true.

In general, Eq. (4.4) has three distinct eigenvalues λ_r, which can be arranged in a diagonal matrix $[\lambda]$, and three associated eigenvectors $\{x^{(r)}\}$, forming the square matrix $[x]$, so that Eq. (4.4) can be rewritten

$$[l][x] = [x][\lambda]. \tag{4.5}$$

Premultiplying Eq. (4.5) through by $[x]^{-1}$, we obtain

$$[x]^{-1}[l][x] = [\lambda], \tag{4.6}$$

so that the solution of the eigenvalue problem reduces to finding a matrix $[x]$ which transforms $[l]$ into a diagonal matrix. A transformation of the type (4.6) is known as a *similarity transformation*. Certain properties of similarity transformations will be introduced as needed.

Because $[l]$ is not symmetric, although it is real, some of the eigenvalues may be complex, from which it follows that the associated eigenvectors must also be complex. Complex eigenvectors have no meaning as far as our physical problem is concerned, but this point turns out to be irrelevant. What is relevant is that we must insist that the orthogonal transformation in question does not affect the magnitude of the vector $\{x\}$, so that denoting by $\{\xi^*\}$ and $\{x^*\}$ the complex conjugates of $\{\xi\}$ and $\{x\}$, we must have

$$\{\xi^*\}^T\{\xi\} = \{x^*\}^T\{x\}, \tag{4.7}$$

and if $\{x\}$ is an eigenvector, we must also have

$$\{x^*\}^T\{x\} = \lambda^*\lambda\{x^*\}^T\{x\}, \tag{4.8}$$

where λ^* is the complex conjugate of the eigenvalue λ. From Eq. (4.8) it follows that

$$\lambda^*\lambda = 1, \tag{4.9}$$

or the magnitude of the eigenvalues is equal to unity.

The matrices $[l]$ and $[\lambda]$ of Eq. (4.6) are said to be similar, which implies that their eigenvalues are equal and so are the values of the corresponding characteristic determinants. This allows us to write

$$|l_{ij} - \lambda\delta_{ij}| = |(\lambda_{ij} - \lambda)\delta_{ij}| = 0, \tag{4.10}$$

where δ_{ij} is the Kronecker delta and the notation $\lambda_i = \lambda_{ii}$ ($i = 1, 2, 3$) has been used. Expanding the two determinants, we conclude that

$$|l| = \lambda_1\lambda_2\lambda_3, \tag{4.11}$$

where, because all the elements l_{ij} are real, the determinant $|l|$ is real. It follows that at least one of the eigenvalues must be real and the other two are either real or complex conjugates.

In Sec. 3.1 we showed that $|l| = \pm 1$, so that the product of the three eigenvalues must equal ± 1. But the value $|l| = -1$ corresponds to an improper rotation, transforming a right-handed system into a left-handed one, and must be ignored. Thus *the real eigenvalue must be equal to* +1. It follows that the orthonormal transformation matrix [*l*], describing the rotation of a rigid body relative to an inertial space, has an eigenvalue $\lambda = 1$, so that Eq. (4.3) holds true, with the physical implication that there exists a vector which is left unaffected by the rotation. This proves the statement that *the general motion of a rigid body with one point fixed is a rotation about an axis through the point,* where the axis is called the axis of rotation. This statement is known as *Euler's theorem on the motion of rigid bodies.*

The direction of the axis of rotation can be obtained by solving the special eigenvalue problem, Eq. (4.3). The angle of rotation can also be found by means of a similarity transformation on [*l*], chosen so that one of the axes, say the x_1 axis, will coincide with the axis of rotation. But the rotation about x_1 can be written in the matrix form

$$\begin{bmatrix} 1 & 0 & 0 \\ 0 & \cos \Phi & \sin \Phi \\ 0 & -\sin \Phi & \cos \Phi \end{bmatrix}, \qquad (4.12)$$

and because a similarity transformation leaves the trace of a matrix (the sum of the matrix elements on the principal diagonal) unchanged, the trace of (4.12) must be equal to the trace of [*l*] before and after the transformation. Hence

$$1 + 2\cos \Phi = \sum_{i=1}^{3} l_{ii}, \qquad (4.13)$$

from which we obtain the value of the single rotation in terms of the three rotations defining the matrix [*l*].

The most general displacement of a rigid body consists of a translation of the body and then a rotation of the body about a line. It is possible to choose the axis of rotation parallel to the line of translation so that the motion can be envisioned as that of a screw. This statement is referred to as *Chasles' theorem.* The statement can be explained by recognizing first that a rotation about any axis is equivalent to a rotation through an equal angle about a parallel axis and a translation in a direction perpendicular to the axis.

4.2 THE LINEAR AND ANGULAR MOMENTUM OF A RIGID BODY

Let us consider a rigid body of total mass m and let x, y, and z be a set of body axes with the origin O, as shown in Fig. 4.1. The body can be assumed to be composed of a large number of particles of mass m_i, where the distance between any two particles is constant. It may also be represented as a solid

Rigid Body Dynamics

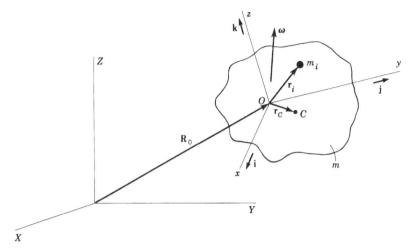

FIGURE 4.1

continuum at every point of which there is defined a differential element of mass, which can be interpreted as being the counterpart of the mass particle. The various definitions retain the same basic structure as in Sec. 1.7, and they can be expressed either in terms of an integral over the body or in terms of a summation over the system of particles. In the process, by using Eq. (4.1) for the velocity of a point in the rigid body, we take into account automatically the fact that a rigid body has only six degrees of freedom.

In view of the above discussion, the radius vector from the origin O to the mass center C is defined by

$$\mathbf{r}_C = \lim_{n \to \infty} \frac{1}{m} \sum_{i=1}^{n} m_i \mathbf{r}_i = \frac{1}{m} \int_{\text{body}} \mathbf{r} \, dm. \tag{4.14}$$

Of course, in the event that the origin coincides with the center of mass, we have $\int \mathbf{r} \, dm = \mathbf{0}$. The linear momentum of the rigid body has the expression

$$\mathbf{p} = \lim_{n \to \infty} \sum_{i=1}^{n} m_i \mathbf{v}_i = \lim_{n \to \infty} \sum_{i=1}^{n} m_i (\mathbf{v}_0 + \boldsymbol{\omega} \times \mathbf{r}_i)$$
$$= \mathbf{v}_0 \int dm + \boldsymbol{\omega} \times \int \mathbf{r} \, dm = m(\mathbf{v}_0 + \boldsymbol{\omega} \times \mathbf{r}_C), \tag{4.15}$$

where it is understood that the integration is carried out over the entire body. But the quantity in the parentheses is recognized as the velocity of the center of mass C, so that Eq. (4.15) can be written in the form

$$\mathbf{p} = m\mathbf{v}_C, \tag{4.16}$$

where $\mathbf{v}_C$ is the velocity of the center of mass. Thus the linear momentum is equal to the product of the total mass and the velocity of the center of mass. If the origin O is chosen to coincide with the center of mass C, Eq. (4.16) retains the same form but the velocity $\mathbf{v}_C$ does not involve the angular velocity $\boldsymbol{\omega}$ because $\mathbf{r}_C$ is zero.

The angular momentum of a rigid body about the origin O is defined by

$$\mathbf{L}_0 = \lim_{n \to \infty} \sum_{i=1}^{n} \mathbf{r}_i \times m_i \mathbf{v}_i = \int \mathbf{r} \times (\mathbf{v}_0 + \boldsymbol{\omega} \times \mathbf{r}) \, dm$$

$$= -\mathbf{v}_0 \times \int \mathbf{r} \, dm + \int \mathbf{r} \times (\boldsymbol{\omega} \times \mathbf{r}) \, dm. \tag{4.17}$$

But if the point O is fixed, $\mathbf{v}_0 = \mathbf{0}$, and if O coincides with the center of mass, it follows that $\int \mathbf{r} \, dm = 0$. Thus the angular momentum of a rigid body about a point O which is either *fixed* or *coincides with the center of mass* is

$$\mathbf{L}_0 = \int \mathbf{r} \times (\boldsymbol{\omega} \times \mathbf{r}) \, dm. \tag{4.18}$$

From vector algebra, however, we have the relation

$$\mathbf{r} \times (\boldsymbol{\omega} \times \mathbf{r}) = \boldsymbol{\omega}(\mathbf{r} \cdot \mathbf{r}) - \mathbf{r}(\boldsymbol{\omega} \cdot \mathbf{r}),$$

so that Eq. (4.18) reduces to

$$\mathbf{L}_0 = \int (\boldsymbol{\omega}(\mathbf{r} \cdot \mathbf{r}) - \mathbf{r}(\boldsymbol{\omega} \cdot \mathbf{r})) \, dm, \tag{4.19}$$

leading to the three components

$$\begin{aligned} L_x &= I_{xx}\omega_x - I_{xy}\omega_y - I_{xz}\omega_z, \\ L_y &= -I_{yx}\omega_x + I_{yy}\omega_y - I_{yz}\omega_z, \\ L_z &= -I_{zx}\omega_x - I_{zy}\omega_y + I_{zz}\omega_z, \end{aligned} \tag{4.20}$$

in which the three quantities

$$I_{xx} = \int (r^2 - x^2) \, dm, \quad I_{yy} = \int (r^2 - y^2) \, dm, \quad I_{zz} = \int (r^2 - z^2) \, dm \tag{4.21}$$

are called *moments of inertia* and the six quantities

$$I_{xy} = I_{yx} = \int xy \, dm, \quad I_{xz} = I_{zx} = \int xz \, dm, \quad I_{yz} = I_{zy} = \int yz \, dm \tag{4.22}$$

are called *products of inertia*. Notice that $r^2 = x^2 + y^2 + z^2$. We shall see in Sec. 4.5 that the moments and products of inertia form the components of a second-rank cartesian tensor called the *inertia tensor*. In the same section some of the properties of the inertia tensor will be examined. The inertia tensor may be conveniently displayed in the form of the symmetric matrix

$$[I] = \begin{bmatrix} I_{xx} & -I_{xy} & -I_{xz} \\ -I_{yx} & I_{yy} & -I_{yz} \\ -I_{zx} & -I_{zy} & I_{zz} \end{bmatrix}, \quad (4.23)$$

referred to as the *inertia matrix*. Introducing the column matrix $\{r_i\}$ representing the radius vector $\mathbf{r}_i$ and the diagonal matrix $[r_i^2]$ with every nonzero element equal to r_i^2, Eqs. (4.21) to (4.23) can be condensed into the compact expression

$$[I] = \lim_{n \to \infty} \sum_{i=1}^{n} ([r_i^2] - \{r_i\}(r_i)^T) m_i = \int ([r^2] - \{r\}\{r\}^T) \, dm. \quad (4.24)$$

Since, by definition, the body axes are fixed in the body, the elements of the inertia matrix $[I]$ are constant. In other cases we may choose as a reference frame a system rotating with respect to the body, in which case the elements of $[I]$ are not constant. This fact alone points to the advantage of working with a set of body axes. Sometimes, if the body exhibits axial symmetry, it may be possible to obtain constant moments of inertia even for a set of axes moving relative to the body. The moments and products of inertia are of considerable importance in the study of rigid body dynamics and will receive full attention later.

In view of Eq. (4.23), Eqs. (4.20) can be conveniently written in the compact matrix form

$$\{L\} = [I]\{\omega\}, \quad (4.25)$$

in which $\{L\}$ and $\{\omega\}$ are column matrices of the angular momentum and angular velocity components, respectively. From Eq. (4.25) we conclude that the matrix $[I]$ can be interpreted as an operator which transforms the vector $\{\omega\}$ into the vector $\{L\}$. However, unlike the operator $[l]$ of Sec. 3.1, which is nondimensional, $[I]$ has units ML^2.

The preceding results can also be presented by representing the inertia tensor in terms of a dyadic. The advantage of dyadic notation is that it enables us to operate with vectors instead of the matrix representation of vectors. Since some readers may not be as familiar with dyadic notation as with matrix notation, certain fundamental concepts are presented in Appendix A. Equation (*j*) of Appendix A indicates that the inertia tensor can be represented by the *inertia dyadic*

$$\mathbf{I} = \int ((\mathbf{r} \cdot \mathbf{r})\mathbf{1} - \mathbf{rr}) \, dm, \tag{4.26}$$

in which **1** denotes the identity dyadic. We notice that Eq. (4.26) is entirely analogous to the matrix representation of the inertia tensor, Eq. (4.24). The angular momentum vector **L** (the subscript 0 has been omitted) can be written in dyadic notation by regarding the inertia dyadic **I** as a vector operator and considering the scalar product of **I** and the angular velocity vector $\boldsymbol{\omega}$ with the result

$$\mathbf{L} = \mathbf{I} \cdot \boldsymbol{\omega} = \int ((\mathbf{r} \cdot \mathbf{r})\mathbf{1} \cdot \boldsymbol{\omega} - \mathbf{rr} \cdot \boldsymbol{\omega}) \, dm$$

$$= \int ((\mathbf{r} \cdot \mathbf{r})\boldsymbol{\omega} - \mathbf{r}(\mathbf{r} \cdot \boldsymbol{\omega})) \, dm, \tag{4.27}$$

which is identical in every respect to Eq. (4.19).

4.3 TRANSLATION THEOREM FOR THE ANGULAR MOMENTUM

We shall be interested in calculating the angular momentum about any moving point A, not necessarily within the body. In particular, we wish to express the results in terms of the translation of the center of mass and the rotation about the center of mass.

Let us refer to Fig. 4.2 and write the angular momentum about point A

$$\mathbf{L}_A = \int \mathbf{r}_A \times \mathbf{v} \, dm = \int \mathbf{r}_A \times (\mathbf{v}_C + \boldsymbol{\omega} \times \mathbf{r}) \, dm$$

$$= \int (\mathbf{r}_{AC} + \mathbf{r}) \times (\mathbf{v}_C + \boldsymbol{\omega} \times \mathbf{r}) \, dm$$

$$= \mathbf{r}_{AC} \times m\mathbf{v}_C + \int \mathbf{r} \times (\boldsymbol{\omega} \times \mathbf{r}) \, dm, \tag{4.28}$$

where we took into account the fact that the position of dm depends only on **r** and that $\int \mathbf{r} \, dm = \mathbf{0}$. But

$$m\mathbf{v}_C = \mathbf{p} \tag{4.29}$$

is the linear momentum of the body, and

$$\int \mathbf{r} \times (\boldsymbol{\omega} \times \mathbf{r}) \, dm = \mathbf{L}_C \tag{4.30}$$

is the angular momentum of the body about the center of mass. Introducing

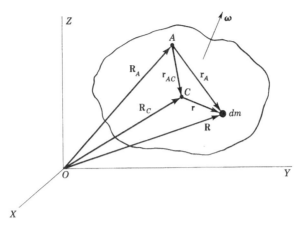

FIGURE 4.2

Eqs. (4.29) and (4.30) into Eq. (4.28), we obtain

$$\mathbf{L}_A = \mathbf{L}_C + \mathbf{r}_{AC} \times \mathbf{p}, \tag{4.31}$$

which can be regarded as a *translation theorem for the angular momentum*. In words, the angular momentum of a rigid body about any point A is equal to the sum of the angular momentum about the center of mass and the moment of the linear momentum about A, where the linear momentum vector is passing through C.

Example 4.1

As a simple illustration of the translation theorem for angular momentum, let us calculate the angular momentum of the wheel depicted in Fig. 4.3. The massless shaft of length L is rotating about the vertical axis with angular velocity Ω. It always remains in the horizontal plane normal to the wheel, where the latter has the principal moments of inertia $I_x, I_y = I_z, I_z$.

The velocity of the center C of the wheel is simply

$$\mathbf{v} = L\Omega \mathbf{j}, \tag{a}$$

so that the linear momentum vector has the value

$$\mathbf{p} = M\mathbf{v} = ML\Omega \mathbf{j}, \tag{b}$$

where M is the mass of the wheel. Assuming that the wheel is rolling without slipping, its angular velocity component about axis x is

$$\omega_x = -\frac{L}{R}\Omega \mathbf{i}, \tag{c}$$

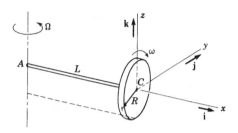

FIGURE 4.3

whereas the angular velocity components about axes y and z are 0 and Ω, respectively. Hence the angular momentum about point C is

$$\mathbf{L}_C = -I_x \frac{L}{R} \Omega \mathbf{i} + I_z \Omega \mathbf{k}. \tag{d}$$

But $\mathbf{r}_{AC} = L\mathbf{i}$ so that, using Eq. (4.31), we obtain the angular momentum about A in the form

$$\mathbf{L}_A = \mathbf{L}_C + \mathbf{r}_{AC} \times \mathbf{p} = -I_x \frac{L}{R} \Omega \mathbf{i} + (I_z + ML^2) \Omega \mathbf{k}, \tag{e}$$

where $I_z + ML^2$ is recognized as the moment of inertia of the wheel about the vertical through A.

4.4 THE KINETIC ENERGY OF A RIGID BODY

Let us refer again to Fig. 4.1 and calculate the kinetic energy of a rigid body by means of a set of body axes x, y, z with the origin at O. Using the symbols defined in Sec. 4.2, we obtain

$$T = \tfrac{1}{2} \int (\mathbf{v}_0 + \boldsymbol{\omega} \times \mathbf{r}) \cdot (\mathbf{v}_0 + \boldsymbol{\omega} \times \mathbf{r}) \, dm$$

$$= \tfrac{1}{2} m \mathbf{v}_0 \cdot \mathbf{v}_0 + \mathbf{v}_0 \cdot \boldsymbol{\omega} \times \int \mathbf{r} \, dm + \tfrac{1}{2} \int (\boldsymbol{\omega} \times \mathbf{r}) \cdot (\boldsymbol{\omega} \times \mathbf{r}) \, dm. \tag{4.32}$$

There are two cases of particular interest, namely, the case in which O is fixed relative to an inertial space and the case in which the origin O coincides with the center of mass C. When *the origin O is fixed*, $\mathbf{v}_0 = \mathbf{0}$ and Eq. (4.32) reduces to

$$T = \tfrac{1}{2} \int (\boldsymbol{\omega} \times \mathbf{r}) \cdot (\boldsymbol{\omega} \times \mathbf{r}) \, dm, \tag{4.33}$$

so that the kinetic energy can be regarded as due entirely to the rotational motion of the body about the fixed point. This is an extremely important case and will be studied in detail later. When the *origin O is chosen to coincide with the center of mass C*, we have $\int \mathbf{r}\, dm = \mathbf{0}$ and Eq. (4.32) becomes

$$T = \tfrac{1}{2} m \mathbf{v}_C \cdot \mathbf{v}_C + \tfrac{1}{2} \int (\boldsymbol{\omega} \times \mathbf{r}) \cdot (\boldsymbol{\omega} \times \mathbf{r})\, dm, \qquad (4.34)$$

where the first term in the right side of (4.34) can be recognized as the kinetic energy of translation and the second as the kinetic energy of rotation. Once again the advantage of expressing the motion in terms of a set of body axes with the origin at the center of mass is clearly demonstrated.

The translational kinetic energy can be written in the simple matrix form

$$T_{\text{tr}} = \tfrac{1}{2} m \{v_C\}^T \{v_C\}, \qquad (4.35)$$

where $\{v_C\}$ is the column matrix of the velocity components of the center of mass. The rotational kinetic energy can also be written in a relatively simple matrix form. To show this, we recall the expression of the cross product in matrix form, Eq. (3.29), and write

$$(\boldsymbol{\omega} \times \mathbf{r}) \cdot (\boldsymbol{\omega} \times \mathbf{r}) = \{r\}^T [\omega]^T [\omega] \{r\}$$

$$= \begin{Bmatrix} x \\ y \\ z \end{Bmatrix}^T \begin{bmatrix} 0 & \omega_z & -\omega_y \\ -\omega_z & 0 & \omega_x \\ \omega_y & -\omega_x & 0 \end{bmatrix} \begin{bmatrix} 0 & -\omega_z & \omega_y \\ \omega_z & 0 & -\omega_x \\ -\omega_y & \omega_x & 0 \end{bmatrix} \begin{Bmatrix} x \\ y \\ z \end{Bmatrix}$$

$$= (r^2 - x^2)\omega_x^2 + (r^2 - y^2)\omega_y^2 + (r^2 - z^2)\omega_z^2$$
$$- 2xy\omega_x\omega_y - 2xz\omega_x\omega_z - 2yz\omega_y\omega_z. \qquad (4.36)$$

Introducing Eq. (4.36) into the integral in Eq. (4.34), performing the integration, and recalling definitions (4.21) and (4.22), we obtain

$$T_{\text{rot}} = \tfrac{1}{2}(I_{xx}\omega_x^2 + I_{yy}\omega_y^2 + I_{zz}\omega_z^2 - 2I_{xy}\omega_x\omega_y - 2I_{xz}\omega_x\omega_z - 2I_{yz}\omega_y\omega_z)$$
$$= \tfrac{1}{2}\{\omega\}^T [I] \{\omega\}, \qquad (4.37)$$

where $\{\omega\}$ is the column matrix of the body angular velocity components and $[I]$ is the symmetric inertia matrix.

Equations (4.25) and (4.37) can be interpreted to imply that the kinetic energy of rotation is equal to one-half the scalar product of the angular velocity vector $\boldsymbol{\omega}$ and angular momentum vector $\mathbf{L}$. Recalling Eq. (4.27), we can write the kinetic energy of rotation in terms of the inertia dyadic $\mathbf{I}$ as

$$T_{\text{rot}} = \tfrac{1}{2}\boldsymbol{\omega} \cdot \mathbf{L} = \tfrac{1}{2}\boldsymbol{\omega} \cdot \mathbf{I} \cdot \boldsymbol{\omega}. \qquad (4.38)$$

It should be remarked that what we have done here is merely to represent Eq. (4.37), obtained by means of matrix operations, in terms of dyadic notation.

4.5 PRINCIPAL AXES

Equation (4.37) expresses the rotational kinetic energy of a rigid body in terms of the body angular velocity components and moments and products of inertia. By virtue of referring the motion to a set of body axes, the moments and products of inertia do not depend on time, so that Eq. (4.37) has the relatively simple form of a homogeneous quadratic function with constant coefficients. The direction of the set of axes used, although fixed with respect to the body, is otherwise arbitrary. It turns out that Eq. (4.37) can be further simplified by choosing a particular set of body axes, called *principal axes*. When this set of axes is used, the inertia matrix reduces to a diagonal form and the cross-product terms in the velocities disappear from the kinetic energy expression.

Before discussing the method of obtaining the principal axes, it will prove beneficial to examine the relation between the moments and products of inertia expressed in two different cartesian systems of coordinates. Hence, let us consider Fig. 4.4 and write the relation between the angular velocity components along the x_i and x_i' systems. Using the coordinate transformation described in Sec. 3.1, we can express this relation in the index notation

$$\omega_i' = \sum_{j=1}^{3} l_{ij}\omega_j, \quad i = 1, 2, 3. \tag{4.39}$$

In a three-dimensional space, a quantity with three components transforming according to Eq. (4.39) is a *tensor of first rank*. Hence a vector is a tensor of the first rank. Equation (4.39) can be written in the matrix form

$$\{\omega'\} = [l]\{\omega\}, \tag{4.40}$$

where $[l]$ is the matrix of the direction cosines between the axes x_i and x_i'. It represents an orthonormal transformation, $[l]^T = [l]^{-1}$. The kinetic energy can be expressed in either set of coordinates

$$T = \tfrac{1}{2}\{\omega\}^T[I]\{\omega\} = \tfrac{1}{2}\{\omega'\}^T[I']\{\omega'\}. \tag{4.41}$$

Introducing Eq. (4.40) into (4.41), we conclude that

$$[I'] = [l][I][l]^T, \tag{4.42}$$

Rigid Body Dynamics

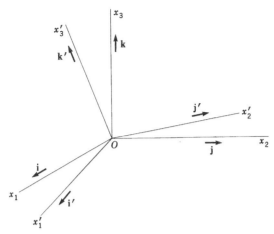

FIGURE 4.4

which represents a similarity transformation, due to the nature of $[l]$. Equation (4.42) can be expanded by means of the index notation

$$I'_{ij} = \sum_{r=1}^{3} \sum_{s=1}^{3} l_{ir} l_{js} I_{rs}, \qquad i,j = 1, 2, 3. \tag{4.43}$$

A quantity in a three-dimensional space defined by nine components transforming according to Eq. (4.43) is known as a *cartesian tensor of second rank*. Although we represented the second-rank *inertia tensors* I'_{ij} and I_{rs} by square matrices, it should be noted that a tensor has physical meaning and is restricted to similarity transformations of the special type (4.43), whereas a matrix is more of a mathematical concept not restricted to this type of transformation.

By definition, the principal axes are a set of axes for which the products of inertia are zero. Thus, to obtain the direction of the principal axes, we must seek a transformation matrix $[l]$ which, under a similarity transformation of the type (4.42), reduces $[I]$ to a diagonal form

$$[I'] = [l][I][l]^T. \tag{4.44}$$

This is equivalent to solving the eigenvalue problem

$$[I][u] = [u][\lambda] \tag{4.45}$$

associated with the symmetric inertia matrix $[I]$. This can be done by means of the method of the matrix diagonalization by successive rotations or by any other method for the solution of the eigenvalue problem, such as the characteristic determinant method (see Ref. 2, pp. 81–112). Since $[I]$ is a real symmetric matrix, all the eigenvalues λ_i ($i = 1, 2, 3$) are real. We shall not go into the details involved in obtaining the solution of the eigenvalue problem but concentrate on some of the geometrical aspects instead. To this end, let us write Eq. (4.45) in the form

$$[I]\{u\} = \lambda\{u\} \tag{4.46}$$

and premultiply through by $\{u\}^T$ to obtain

$$\{u\}^T[I]\{u\} = \lambda\{u\}^T\{u\}. \tag{4.47}$$

The vector $\{u\}$ can be normalized by writing

$$\{u\}^T\{u\} = \mathbf{u} \cdot \mathbf{u} = \sum_{i=1}^{3} u_i^2 = 1, \tag{4.48}$$

which is equivalent to saying that the length of the vector $\{u\}$ is unity. This can always be done without loss of generality because Eq. (4.46) is homogeneous. It follows that Eq. (4.47) can be written

$$\frac{1}{\lambda}(I_{11}u_1^2 + I_{22}u_2^2 + I_{33}u_3^2 - 2I_{12}u_1u_2 - 2I_{13}u_1u_3 - 2I_{23}u_2u_3) = 1, \tag{4.49}$$

which is the equation of a quadric surface with the center at the origin of the cartesian system u_i. Specifically, Eq. (4.49) represents an ellipsoid, known as the *inertia ellipsoid*.

It is not difficult to show that the vector of the direction cosines of the normal **n** to the ellipsoid at any point (u_1, u_2, u_3), if written in matrix form, is proportional to the vector $[I]\{u\}$ (see Prob. 4.5). Hence, the eigenvalue problem (4.46) can be interpreted as the problem of finding the directions for which, at a certain point, the radius vector **u** and the normal **n** to the ellipsoid are parallel to one another. These directions are the principal axes of the ellipsoid and correspond to the body axes for which the inertia matrix becomes diagonal. The equation of the ellipsoid in the canonical form is

$$\frac{1}{\lambda}(I_1 u_1'^2 + I_2 u_2'^2 + I_3 u_3'^2) = 1, \tag{4.50}$$

in which I_1, I_2, and I_3 are the *principal moments of inertia* of the body.

When two of the principal moments of inertia are equal, the quadric surface becomes an ellipsoid of revolution, and when all three moments of inertia are equal, it becomes a sphere. This is not to be interpreted, however, as an implication that the body is a body of revolution or a sphere, respectively. For example, a homogeneous cube has all three moments of inertia about a rectangular set of axes, with the origin at its center, equal to one another.

The same problem can be formulated in dyadic notation. The eigenvalue problem, Eq. (4.46), becomes

$$\mathbf{I} \cdot \mathbf{u} = \lambda \mathbf{u}, \quad (4.51)$$

where $\mathbf{I}$ is the dyadic representation of the symmetric inertia tensor. Assuming that the vector $\mathbf{u}$ is normalized according to Eq. (4.48), a scalar multiplication of both sides of Eq. (4.51) by $\mathbf{u}$ leads to

$$\frac{1}{\lambda} \mathbf{u} \cdot \mathbf{I} \cdot \mathbf{u} = 1, \quad (4.52)$$

which is the same inertia ellipsoid as the one given by Eq. (4.49). The object is to reduce Eq. (4.52) to its canonical form, which is the same as finding a set of axes, namely, the principal axes, for which the inertia tensor is diagonal and the corresponding dyadic has the particular form

$$\mathbf{I}' = I_1 \mathbf{i}'\mathbf{i}' + I_2 \mathbf{j}'\mathbf{j}' + I_3 \mathbf{k}'\mathbf{k}'. \quad (4.53)$$

When the principal axes have been found, the kinetic energy of rotation can be reduced to the simple expression

$$T_{\text{rot}} = \tfrac{1}{2}\{\omega\}^T [I']\{\omega\} = \tfrac{1}{2}\boldsymbol{\omega} \cdot \mathbf{I} \cdot \boldsymbol{\omega} = \tfrac{1}{2}(I_1 \omega_1^2 + I_2 \omega_2^2 + I_3 \omega_3^2), \quad (4.54)$$

which is a homogeneous quadratic form free of cross products. Note that for simplicity the primes on the angular velocity components have been omitted.

4.6 THE EQUATIONS OF MOTION FOR A RIGID BODY

In Sec. 1.7 we derived the equations of motion for a system of particles. These equations are perfectly valid for a rigid body, however, because a rigid body can be regarded as a system of particles. Of course, in the case of a rigid body the distance between any pair of particles is constant, and certain mass distribution characteristics are expressed by means of the position of the center of mass as well as the moments and products of inertia with respect to a set of body axes.

The force equation follows directly from Eq. (1.59)

$$\mathbf{F} = m\ddot{\mathbf{R}}_C = \frac{d}{dt}(m\mathbf{v}_C) = \dot{\mathbf{p}}, \quad (4.55)$$

where $\mathbf{p} = m\mathbf{v}_C$ = linear momentum of body,
m = total mass,
$\mathbf{v}_C$ = velocity of center of mass of body.

Equation (4.55) implies that the motion of the center of mass is the same as the motion of a particle of mass equal to the mass of the rigid body while acted upon by the same forces as the rigid body.

Similarly, the moment equation with respect to the center of mass C, Eq. (1.66), retains the form

$$\mathbf{N}_C = \dot{\mathbf{L}}_C, \tag{4.56}$$

and, in the case of a rigid body, the angular momentum $\mathbf{L}_C$ is given by Eq. (4.18). Equation (4.56), for the rotational motion about the center of mass, can be treated independently of the translational motion of the center of mass.

Equation (4.56) can be written in detail by expressing $\mathbf{L}_C$ in terms of body axes components as

$$\mathbf{L}_C = L_x \mathbf{i} + L_y \mathbf{j} + L_z \mathbf{k}, \tag{4.57}$$

in which L_x, L_y, and L_z are given by Eqs. (4.20). Because $\mathbf{L}_C$ is expressed in terms of components along a set of body axes which are rotating with angular velocity $\boldsymbol{\omega}$ relative to an inertial space, Eq. (4.56) yields

$$\mathbf{N}_C = \dot{\mathbf{L}}'_C + \boldsymbol{\omega} \times \mathbf{L}_C, \tag{4.58}$$

where, consistent with the notation of Sec. 3.3, primes denote the rate of change of a vector relative to the rotating system of axes and $\boldsymbol{\omega} \times$ denotes the rate of change resulting from the rotation of the reference frame.

Equation (4.58) can be conveniently written in matrix form by recalling Eq. (4.25) and the matrix expression corresponding to the vector $\boldsymbol{\omega} \times$ [see Eqs. (3.29) and (3.31)]. Hence

$$\{N_C\} = \{\dot{L}'_C\} + [\omega]\{L_C\} = [I]\{\dot{\omega}'\} + [\omega][I]\{\omega\}, \tag{4.59}$$

in which $[\omega]$ is the skew-symmetric matrix representing the matrix counterpart of the vector product. The matrix is obtained from Eq. (3.30) by substituting the angular velocity components ω_x, ω_y, and ω_z for ω_1, ω_2, and ω_3, respectively. The moments and products of inertia entering into the inertia matrix $[I]$ are understood to be with respect to a set of body axes with the origin at the mass center C.

If the motion takes place about a fixed point O, the torque equation for a rigid body retains the form (1.65)

$$\mathbf{N}_0 = \dot{\mathbf{L}}_0, \tag{4.60}$$

which can be expanded in terms of components along a set of body axes with the origin at O. The equations thus obtained, when written in matrix form, will have the same structure as Eq. (4.59), but, in general, the moments and products of inertia will be different.

4.7 EULER'S EQUATIONS OF MOTION

In Sec. 4.6 we stated that the rotational motion about the center of mass can be treated independently of the translational motion of the center of mass. Here the discussion will concentrate on the rotational motion about a point.

Rigid Body Dynamics

With the tacit understanding that both the torque and angular momentum are with respect to either the center of mass or a fixed point, we can drop the corresponding subscript and write the rotational equations of motion in the form

$$\{N\} = \{\dot{L}'\} + [\omega]\{L\}. \tag{4.61}$$

Although the directions of the body axes are arbitrary, the equations of motion can be considerably simplified by selecting them to coincide with the principal axes. In so doing, the products of inertia vanish, and the inertia matrix $[I]$ becomes diagonal, so that the angular momentum column matrix reduces to

$$\{L\} = [I]\{\omega\}. \tag{4.62}$$

Introducing Eq. (4.62) into (4.61), we obtain

$$\{N\} = [I]\{\dot{\omega}'\} + [\omega][I]\{\omega\}, \tag{4.63}$$

which are known as *Euler's equations of motion*. Denoting the principal axes by 1, 2, and 3 and the associated principal moments of inertia by A, B, and C, Euler's equations become

$$\begin{aligned} N_1 &= A\dot{\omega}_1 + (C - B)\omega_2\omega_3, \\ N_2 &= B\dot{\omega}_2 + (A - C)\omega_1\omega_3, \\ N_3 &= C\dot{\omega}_3 + (B - A)\omega_1\omega_2. \end{aligned} \tag{4.64}$$

It should be pointed out that ω_1, ω_2, and ω_3 are not angular velocity components in the sense that they represent time rates of change of certain angles but nonintegrable combinations of time derivatives of the angular displacements. They are sometimes referred to as the time derivatives of *quasicoordinates*.

The main reason for referring the motion to body axes is that the moments and products of inertia with respect to such axes are constant. The situation may arise, however, in which it is more advantageous to refer the motion to a coordinate system rotating with an angular velocity different from the body angular velocity, provided that the inertia coefficients with respect to the system selected are constant. This may sound like a paradox, but we must remember that in the case of a body possessing axial symmetry, for which the inertia ellipsoid is an ellipsoid of revolution, the moments of inertia corresponding to two of the principal axes are equal. These two axes constitute two orthogonal diametrical lines in the circular cross section resulting from the intersection of the inertia ellipsoid with the plane containing the two principal axes in question. Moreover, any other two diametrical lines in the same plane displaced by a certain angle with respect to the first are also principal axes, and the corresponding moments of inertia are equal to one another. In fact, the value of the moment of inertia about any diametrical line is the same. In considering such a case, we must distinguish between the angular velocity ω of the reference frame and the angular velocity Ω of the body. Whereas Eq. (4.61) retains the same form, the angular momentum matrix becomes

$$\{L\} = [I]\{\Omega\}, \qquad (4.65)$$

where two of the three diagonal elements of $[I]$ are equal. Introduction of Eq. (4.65) into (4.61) yields

$$\{N\} = [I]\{\dot{\Omega}'\} + [\omega][I]\{\Omega\}. \qquad (4.66)$$

Equation (4.66) is very useful in the study of gyroscopic motion. We shall refer to Eq. (4.66) as the *modified Euler equations*.

The motion of a system possessing inertial symmetry is discussed in Sec. 4.9, where we have an opportunity of applying the modified Euler equations.

4.8 EULER'S ANGLES

The angular velocity components ω_1, ω_2, and ω_3 about the body axes cannot be integrated to obtain angular displacements about these axes and therefore are unsatisfactory as a means of describing the orientation of a rigid body in space. Nor do the direction cosines l_{ij} meet this need, because they are not independent, which rules them out from qualifying as a set of generalized coordinates. To describe the orientation of a rigid body in space we need three independent coordinates. Such a set of coordinates is not necessarily unique, and an example of a suitable set of angular displacements was introduced in Sec. 3.2. A set which enjoys wide acceptance consists of *Euler's angles*. These are three successive angular displacements which can adequately carry out the transformation from one cartesian system of axes to another, although the rotations are not about three orthogonal axes. Moreover, the three components of the body angular velocity can be expressed in terms of Euler's angles and their time derivatives.

There is no general agreement on the notation of Euler's angles, but one of the most common, namely, ϕ, θ, and ψ, in sequence, is shown in Fig. 4.5. To demonstrate how Euler's angles provide a description of the body orientation, we shall follow a body-axes triad, initially coincident with a set of cartesian inertial axes X, Y, Z, as it assumes different orientations in space. A rotation through an angle ϕ about axis Z brings the triad into coincidence with axes ξ', η', ζ'. A further rotation θ about axis ξ' puts the body into orientation ξ, η, ζ, where axis ξ, sometimes referred to as the *nodal axis*, remains in the horizontal plane at all times. Finally, a rotation ψ about ζ makes the triad coincide with the body axes x, y, z.

Rigid Body Dynamics

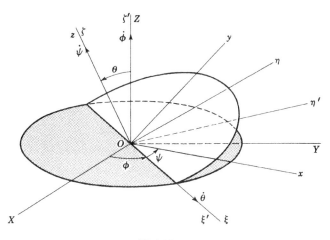

FIGURE 4.5

The coordinate transformations can be arranged in matrix form, as in Sec. 3.2. We denote the column matrix corresponding to a system of coordinates by using the first of the three coordinates of the system, so that the transformation from the X, Y, Z to the ξ', η', ζ' system is written

$$\{\xi'\} = [R_3(\phi)]\{X\}, \tag{4.67}$$

where the matrix for the rotation ϕ is

$$[R_3(\phi)] = \begin{bmatrix} \cos\phi & \sin\phi & 0 \\ -\sin\phi & \cos\phi & 0 \\ 0 & 0 & 1 \end{bmatrix}. \tag{4.68}$$

The transformation from system ξ', η', ζ' to system ξ, η, ζ has the form

$$\{\xi\} = [R_1(\theta)]\{\xi'\}, \tag{4.69}$$

where
$$[R_1(\theta)] = \begin{bmatrix} 1 & 0 & 0 \\ 0 & \cos\theta & \sin\theta \\ 0 & -\sin\theta & \cos\theta \end{bmatrix}, \tag{4.70}$$

and, finally, the transformation from ξ, η, ζ to x, y, z is

$$\{x\} = [R_3(\psi)]\{\xi\}, \tag{4.71}$$

in which
$$[R_3(\psi)] = \begin{bmatrix} \cos\psi & \sin\psi & 0 \\ -\sin\psi & \cos\psi & 0 \\ 0 & 0 & 1 \end{bmatrix}. \tag{4.72}$$

Combining Eqs. (4.67) to (4.72), we can write the transformation from the inertial axes X, Y, Z to the body axes x, y, z as

$$\{x\} = [R_3(\psi)][R_1(\theta)][R_3(\phi)]\{X\} = [R]\{X\}, \quad (4.73)$$

and we are reminded that, in general, the rotation matrices are not commutative. The matrix $[R]$, representing the triple product of the rotation matrices, is

$$[R] = \begin{bmatrix} c\phi\, c\psi - s\phi\, c\theta\, s\psi & s\phi\, c\psi + c\phi\, c\theta\, s\psi & s\theta\, s\psi \\ -c\phi\, s\psi - s\phi\, c\theta\, c\psi & -s\phi\, s\psi + c\phi\, c\theta\, c\psi & s\theta\, c\psi \\ s\phi\, s\theta & -c\phi\, s\theta & c\theta \end{bmatrix}, \quad (4.74)$$

where $s\phi = \sin\phi$, $c\phi = \cos\phi$, etc. It should be noted that $[R]$ represents an orthonormal transformation between two cartesian systems, and in view of this fact the inverse transformation giving the inertial axes in terms of the body axes is simply

$$\{X\} = [R]^T\{x\}. \quad (4.75)$$

Next we wish to establish the expressions of the angular velocity components for various systems of coordinates in terms of the angular velocities $\dot\phi$, $\dot\theta$, and $\dot\psi$, which are known as *precession*, *nutation*, and *spin*, respectively. In so doing, we shall distinguish between the angular velocity $\boldsymbol{\omega}$ of a coordinate system and the angular velocity $\boldsymbol{\Omega}$ of the body in terms of components along the same system. The discussion will concentrate on the node-axis system ξ, η, ζ and the body system x, y, z. It is easy to see from Fig. 4.5 that the *angular velocity components of system* ξ, η, ζ are

$$\omega_\xi = \dot\theta, \qquad \omega_\eta = \dot\phi \sin\theta, \qquad \omega_\zeta = \dot\phi \cos\theta. \quad (4.76)$$

On the other hand, noticing that the body moves relative to system ξ, η, ζ with an angular velocity $\dot\psi$ about ζ, we see that the *angular velocity components of the body in terms of components along system* ξ, η, ζ are

$$\Omega_\xi = \dot\theta, \qquad \Omega_\eta = \dot\phi \sin\theta, \qquad \Omega_\zeta = \dot\phi \cos\theta + \dot\psi. \quad (4.77)$$

Of course, there is no such distinction to be made for the body axes x, y, z, as both the body and the system of axes have the same velocities

$$\begin{aligned} \Omega_x &= \omega_x = \dot\phi \sin\theta \sin\psi + \dot\theta \cos\psi, \\ \Omega_y &= \omega_y = \dot\phi \sin\theta \cos\psi - \dot\theta \sin\psi, \\ \Omega_z &= \omega_z = \dot\phi \cos\theta + \dot\psi, \end{aligned} \quad (4.78)$$

which can be arranged in the matrix form

$$\begin{Bmatrix} \Omega_x \\ \Omega_y \\ \Omega_z \end{Bmatrix} = \begin{Bmatrix} \omega_x \\ \omega_y \\ \omega_z \end{Bmatrix} = \begin{bmatrix} \sin\theta \sin\psi & \cos\psi & 0 \\ \sin\theta \cos\psi & -\sin\psi & 0 \\ \cos\theta & 0 & 1 \end{bmatrix} \begin{Bmatrix} \dot\phi \\ \dot\theta \\ \dot\psi \end{Bmatrix}. \quad (4.79)$$

Rigid Body Dynamics

Notice that the transformation matrix is not orthogonal. This is to be expected because the velocities $\dot\phi$, $\dot\theta$, and $\dot\psi$ are about axes which do not form a rectangular system of coordinates.

At the point $\theta = 0$ Euler's angles can be regarded as singular, as there is no way of distinguishing between the angular velocities $\dot\phi$ and $\dot\psi$ at that point. When the interest lies in examining the motion in the neighborhood of this configuration, it may be advisable to describe it by means of a different set of angles. From Sec. 3.2 we recall that such a different system indeed exists. If the triad x, y, z is obtained by means of three successive rotations θ_1, θ_2, and θ_3 about axes X, η', and ζ, respectively, then Eq. (4.73) must be replaced by

$$\{x\} = [R_3(\theta_3)][R_2(\theta_2)][R_1(\theta_1)]\{X\} = [R]\{X\}, \tag{4.80}$$

where $[R]$ is the matrix of the direction cosines given by Eq. (3.22). The body angular velocity components can be shown to satisfy the matrix relation

$$\begin{Bmatrix} \Omega_x \\ \Omega_y \\ \Omega_z \end{Bmatrix} = \begin{bmatrix} \cos\theta_2\cos\theta_3 & \sin\theta_3 & 0 \\ -\cos\theta_2\sin\theta_3 & \cos\theta_3 & 0 \\ \sin\theta_2 & 0 & 1 \end{bmatrix} \begin{Bmatrix} \dot\theta_1 \\ \dot\theta_2 \\ \dot\theta_3 \end{Bmatrix}. \tag{4.81}$$

We shall have the opportunity to work with both sets of angles.

4.9 MOMENT-FREE INERTIALLY SYMMETRIC BODY

Let us consider a body with two of its principal moments of inertia equal, $B = A$. Although the inertia ellipsoid is an ellipsoid of revolution, with axis 3 as the symmetry axis, the body does not necessarily have to be a body of revolution. A rectangular parallelepiped may possess such inertial symmetry. Euler's equations for this special case become

$$\begin{aligned} A\dot\omega_1 + (C - A)\omega_2\omega_3 &= 0, \\ A\dot\omega_2 + (A - C)\omega_1\omega_3 &= 0, \\ C\dot\omega_3 &= 0. \end{aligned} \tag{4.82}$$

The third of Eqs. (4.82) yields immediately the integral

$$\omega_3 = \text{const.} \tag{4.83}$$

Next let

$$\frac{C - A}{A}\omega_3 = \Omega, \tag{4.84}$$

so that the first two of Eqs. (4.82) become

$$\begin{aligned} \dot\omega_1 + \Omega\omega_2 &= 0, \\ \dot\omega_2 - \Omega\omega_1 &= 0. \end{aligned} \tag{4.85}$$

Multiply the first of Eqs. (4.85) by ω_1 and the second by ω_2, add the results, and integrate with respect to time to obtain

$$\omega_1{}^2 + \omega_2{}^2 = \omega_{12}{}^2 = \text{const}, \tag{4.86}$$

where ω_{12} is the projection of the vector $\boldsymbol{\omega}$ in the 1,2 plane. It follows from Eqs. (4.83) and (4.86) that the magnitude of the body angular velocity is constant

$$|\boldsymbol{\omega}| = \text{const.} \tag{4.87}$$

For a moment-free body, $\mathbf{N} = \mathbf{0}$, the angular momentum is conserved

$$\mathbf{L} = \text{const}, \tag{4.88}$$

which implies that both the magnitude and the direction in space of $\mathbf{L}$ are constant. But the projection of $\mathbf{L}$ in the 1,2 plane is

$$L_{12} = (L_1{}^2 + L_2{}^2)^{\frac{1}{2}} = A(\omega_1{}^2 + \omega_2{}^2)^{\frac{1}{2}} = A\omega_{12}, \tag{4.89}$$

which indicates that the projections of $\mathbf{L}$ and $\boldsymbol{\omega}$ in the 1,2 plane are along the same line. Hence axes 3, $\mathbf{L}$, and $\boldsymbol{\omega}$ are in one plane.

The angles α and β between axis 3 and vectors $\mathbf{L}$ and $\boldsymbol{\omega}$, respectively, are constant, and their values are related by

$$\tan \alpha = \frac{A}{C} \tan \beta. \tag{4.90}$$

The two possible cases, $A < C$ and $A > C$, are shown in Fig. 4.6a and b, respectively. The direction of the vector $\mathbf{L}$ is fixed in space but the direction of $\boldsymbol{\omega}$ is not. Because the angle between $\mathbf{L}$ and $\boldsymbol{\omega}$ is fixed, vector $\boldsymbol{\omega}$ must rotate around $\mathbf{L}$ and in so doing describe a cone fixed in space, for which reason it is called the *space cone*. The entire 3, $\mathbf{L}$, $\boldsymbol{\omega}$ plane rotates about $\mathbf{L}$, and because the components ω_1 and ω_2 change with time, the vector $\boldsymbol{\omega}$ rotates also about axis 3 describing another cone; this one, however, is fixed in the body and called the *body cone*. This implies that the 3, $\mathbf{L}$, $\boldsymbol{\omega}$ plane rotates also with respect to the body. The body and space cones have one generatrix in common, which coincides with the vector $\boldsymbol{\omega}$, so that the motion can be visualized as the rolling of the body cone on the space cone. When $A < C$, which corresponds to a flat body we have $\alpha < \beta$ and the space cone lies inside the body cone (Fig. 4.6a). When $A > C$, corresponding to a slender body, the angles are such that $\alpha > \beta$ and the body cone lies outside the space cone (Fig. 4.6b).

To find the angular velocity of ω_{12} with respect to the body, we shall represent it as a complex vector

$$\omega_{12} = \omega_1 + i\omega_2. \tag{4.91}$$

Multiplying the second of Eqs. (4.85) by i and adding the result to the first, we obtain

$$\dot{\omega}_{12} - i\Omega\omega_{12} = 0, \tag{4.92}$$

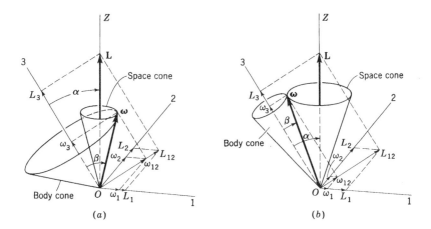

FIGURE 4.6

which has the solution

$$\omega_{12}(t) = \omega_{12}(0)e^{i\Omega t}, \tag{4.93}$$

where $\omega_{12}(0)$ is the initial value of ω_{12}. Hence, the complex vector ω_{12} rotates in the 1,2 plane with angular velocity Ω with respect to the body.

The motion is expressed in terms of body axes rather than inertial axes, so that we cannot draw conclusions concerning the absolute motion of the body in space. To complete the picture we must describe the rotational motion with respect to an inertial space, which is conveniently done by means of Euler's angles. Because the ellipsoid of inertia is symmetric with respect to the principal axis 3, we can simplify the analysis by using the modified Euler equations, Eqs. (4.66). This is frequently the case when the rotating body possesses axial symmetry. The direction of **L** is fixed in space, so that it will prove advantageous to choose the inertial axis Z to coincide with it. Considering Eqs. (4.76) and (4.77), the modified Euler equations, when written in full, assume the form

$$N_\xi = \frac{d}{dt}(A\dot{\theta}) + C(\dot{\psi} + \dot{\phi}\cos\theta)\dot{\phi}\sin\theta - A\dot{\phi}^2\sin\theta\cos\theta,$$

$$N_\eta = \frac{d}{dt}(A\dot{\phi}\sin\theta) + A\dot{\theta}\dot{\phi}\cos\theta - C(\dot{\psi} + \dot{\phi}\cos\theta)\dot{\theta}, \tag{4.94}$$

$$N_\zeta = \frac{d}{dt}[C(\dot{\psi} + \dot{\phi}\cos\theta)].$$

It may be interesting to observe at this point that the moment equations of motion can also be obtained in the form of Lagrange's equations for the generalized coordinates ϕ, θ, and ψ. Evidently the two sets of equations, Eqs. (4.94) and Lagrange's equations, must be related. Indeed it is easy to show that they are related by $N_\phi = N_\eta \sin\theta + N_\zeta \cos\theta$, $N_\theta = N_\xi$, and $N_\psi = N_\zeta$. It should be noted that the torque N_ϕ is about the inertial axis Z, so that, unlike Eqs. (4.94), Lagrange's equations are not about an orthogonal set of axes except for $\theta = \pi/2$.

To proceed with the analysis, it is advisable to take into consideration some of the results obtained by using the equations in terms of body axes. In particular, we wish to consider the fact that the angle between axis 3 (or ζ) and vector **L** (or axis Z) is constant, which leads us to the conclusion that $\theta = $ const. Of course, for a torque-free body, $N_\xi = N_\eta = N_\zeta = 0$, and the last of Eqs. (4.94) leads to the same conclusion as Eq. (4.83). On the other hand, the first of Eqs. (4.94), with $N_\xi = 0$, yields

$$\dot\phi = \frac{C}{(A-C)\cos\theta}\dot\psi, \qquad (4.95)$$

so that the precession and spin are proportional to one another. In the case in which $A < C$ (Fig. 4.6a) the sense of $\dot\phi$ is opposite to that of $\dot\psi$, for which reason this case is known as *retrograde precession*. When $A > C$ (Fig. 4.6b), the sense of $\dot\phi$ is the same as that of $\dot\psi$ and this case is called *direct precession*. From the second of Eqs. (4.94), with $N_\eta = 0$ and $\theta = $ const, we conclude that $\dot\phi = $ const, so that both the precession and spin are uniform. It is not difficult to show that the kinetic energy is also constant.

To a certain degree of approximation, the earth can be regarded as an example of a symmetric torque-free rigid body. Assuming that the earth is an oblate spheroid with the polar axis coinciding with the symmetry axis, axis 3, an observer stationed on the earth may expect to be able to detect the precession of the rotation axis $\boldsymbol{\omega}$ about axis 3. This is an extreme example of the case of Fig. 4.6a, however, since the angle β between these two axes is very small indeed. The moments of inertia of the earth are approximately related by $(C - A)/A = 0.0033$, and because the angular velocity ω_3 is for all practical purposes equal to the magnitude of $\boldsymbol{\omega}$, which is one rotation per day, it follows that

$$\Omega = \frac{C-A}{A}\omega_3 \approx 0.0033 \text{ rotation per day,}$$

or the vector $\boldsymbol{\omega}$ completes one rotation in approximately 300 days tracing a circle about the North Pole. A phenomenon resembling this, known as the *variation in latitude*, has actually been observed. The radius of the circle corresponding to the intersection of the body cone with the earth is about 13 ft, and the period is approximately 433 days rather than 300 days. The discrepancy is attributed to the fact that the earth is not perfectly rigid. As

Rigid Body Dynamics

a point of interest, it may be mentioned that the wobble is large enough that it must be considered in tracking problems involving deep space missions.

The question may be raised whether the earth is indeed a torque-free body. As it turns out, the moon and the sun do exert small torques upon the earth, causing precession and nutation of the earth's polar axis. This problem is discussed in Sec. 11.8.

4.10 GENERAL CASE OF A MOMENT-FREE BODY

The general case of a torque-free body with unequal principal moments of inertia rotating about its center of mass is considerably more complicated than the case in which two of its principal moments of inertia are equal. Because the body is free of external torques, we immediately obtain two integrals of the motion, namely, the kinetic energy T and the angular momentum **L**. To show this, we write Euler's equation for a torque-free body in terms of the body principal axes components

$$[I]\{\dot{\omega}\} + [\omega][I]\{\omega\} = \{0\}. \tag{4.96}$$

First premultiply Eq. (4.96) through by $\{\omega\}^T$ and obtain

$$\{\omega\}^T[I]\{\dot{\omega}\} + \{\omega\}^T[\omega][I]\{\omega\} = 0. \tag{4.97}$$

It is not difficult to show that the second matrix product on the left side of Eq. (4.97) is zero, whereas an integration of the first term yields

$$\{\omega\}^T[I]\{\omega\} = 2T = \text{const.} \tag{4.98}$$

Next premultiply Eq. (4.96) by $\{\omega\}^T[I]$, so that

$$\{\omega\}^T[I^2]\{\dot{\omega}\} + \{\omega\}^T[I][\omega][I]\{\omega\} = 0. \tag{4.99}$$

The second matrix product on the left side of Eq. (4.99) is zero. This can easily be seen by recognizing that the matrix product yields a scalar. The transpose of the matrix product would have to yield the same scalar, but in this case the sign changes because $[\omega]$ is skew-symmetric. It follows that the scalar must be zero. Integrating the first term on the left side of Eq. (4.99), we obtain

$$\{\omega\}^T[I^2]\{\omega\} = \{L\}^T\{L\} = L^2 = \text{const.} \tag{4.100}$$

Because the torque is zero, $\mathbf{N} = \mathbf{0}$, it follows that **L** is constant, with the implication that the direction of **L** is constant. The constancy of the magnitude of **L** is also implied, thus confirming Eq. (4.100).

The problem can be studied qualitatively by means of a very elegant geometric description due to Poinsot. To this end, let us substitute $\rho_i = \lambda^{-\frac{1}{2}} u_i$

in Eq. (4.49) and write the equation of the inertia ellipsoid in the matrix form

$$\{\rho\}^T[I]\{\rho\} = 1. \tag{4.101}$$

It was pointed out in Sec. 4.5, however, that the normal $\{n\}$ to the inertia ellipsoid is proportional to the vector $[I]\{\rho\}$, so that, denoting the proportionality factor by α, we have

$$\alpha\{n\} = [I]\{\rho\}. \tag{4.102}$$

Letting $\{\rho\} = (2T)^{-\frac{1}{2}}\{\omega\}$ in Eq. (4.102), we obtain

$$(2T)^{\frac{1}{2}}\alpha\{n\} = [I]\{\omega\} = \{L\}, \tag{4.103}$$

which indicates that the normal to the inertia ellipsoid at a point with coordinates ρ_i ($i = 1, 2, 3$) is parallel to the vector **L**. But the direction of **L** is fixed in space, so that the inertia ellipsoid must always be tangent to a plane whose normal is **L**. The distance from the center of the inertia ellipsoid to that plane is given by

$$\frac{1}{L}\{\rho\}^T\{L\} = \frac{1}{L(2T)^{\frac{1}{2}}}\{\omega\}^T[I]\{\omega\} = \frac{(2T)^{\frac{1}{2}}}{L} = \text{const}, \tag{4.104}$$

so that the plane has a fixed position in space, for which reason it is called the *invariable plane*. But the vectors $\{\rho\}$ and $\{\omega\}$ lie along the same line, which means that the point of tangency between the inertia ellipsoid and the invariable plane is on the instantaneous axis of rotation, hence at rest. It follows that the motion can be visualized geometrically as the rolling without slipping of the inertia ellipsoid on the invariable plane, while the distance between the center of the ellipsoid and the plane remains constant (Fig. 4.7).

FIGURE 4.7

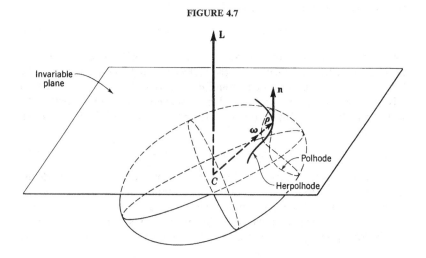

Rigid Body Dynamics

As the inertia ellipsoid rolls on the invariable plane, the locus of the points of tangency forms a curve on the ellipsoid, called the *polhode*[1] by Poinsot, and another one on the tangent plane called the *herpolhode*. Whereas the polhode is a closed curve, the herpolhode is generally not closed.

The motion of the rigid body can be regarded as consisting of three parts: a rotation of the body about the instantaneous axis ω, a precession of the instantaneous axis about **L**, and a nutation of the instantaneous axis relative to the vector **L**. The latter implies that the angle between ω and **L** is no longer constant, as it was in the case of the symmetric inertia ellipsoid.

Equation (4.96) can actually be integrated analytically and the solution expressed in terms of elliptic integrals. (The interested reader can find such a solution in Ref. 3, pp. 377–380.)

4.11 MOTION OF A SYMMETRIC TOP

A symmetric rigid body terminating in a sharp point on the symmetry axis, where the point is called the *apex* or *vertex*, is generally referred to as a top. The top is assumed to be spinning with its apex touching a plane sufficiently rough to prevent slipping, so that, in effect, the motion takes place with respect to a fixed point O coinciding with the point of contact. The problem reduces to that of determining the motion of a rigid body of revolution subjected to a uniform gravitational field when a point on the symmetry axis, generally different from the center of mass, is fixed in an inertial space. The mathematical model to a certain degree represents an entire range of physical systems, including a child's top, a variety of navigational instruments, the spinning earth, etc. These are all examples of *gyroscopic systems*, so called because they all exhibit the peculiar behavior characteristic of the spinning gyroscope. We shall choose the symmetric top to develop the mathematical theory explaining the phenomenon of gyroscopic motion.

The motion of the top can be adequately described by Euler's angles ϕ, θ, and ψ, as shown in Fig. 4.8. The inertial space is defined by the system X, Y, Z with the origin O at the apex of the top. Because the gravitational field is uniform, the center of gravity CG, at which point the resultant of the gravitational forces is acting, coincides with the center of mass at a distance l from point O along the symmetry axis z. The gravity force Mg is parallel to the inertial axis Z but in opposite direction so as to form a positive torque about the node axis ξ.

It turns out that two of the three coordinates describing the motion, namely, ϕ and ψ, are cyclic, and the motion can be described only in terms of θ. To show this, the Lagrangian approach proves more advantageous than

---
[1] Some of the curves can be seen in Ref. 4, p. 263.

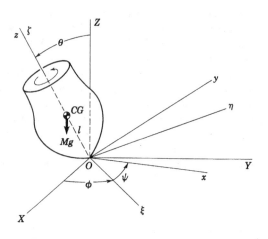

FIGURE 4.8

the use of Euler's moment equations. Because the body is symmetric, it is simpler to express the kinetic energy in terms of the body angular velocity components about the ξ, η, and ζ axes, where ζ coincides with the symmetry axis. Denoting the moment of inertia about the symmetry axis by C and that about any transverse axis through O by A, the rotational kinetic energy takes the form

$$T = \tfrac{1}{2}A(\Omega_\xi^2 + \Omega_\eta^2) + \tfrac{1}{2}C\Omega_\zeta^2, \tag{4.105}$$

where the body angular velocity components Ω_ξ, Ω_η, and Ω_ζ are given by Eqs. (4.77). It follows that

$$T = \tfrac{1}{2}A(\dot\theta^2 + \dot\phi^2 \sin^2\theta) + \tfrac{1}{2}C(\dot\phi \cos\theta + \dot\psi)^2. \tag{4.106}$$

Choosing as reference the XY plane, the potential energy is simply

$$V = Mgl \cos\theta, \tag{4.107}$$

so that the Lagrangian has the expression

$$L = T - V = \tfrac{1}{2}A(\dot\theta^2 + \dot\phi^2 \sin^2\theta) + \tfrac{1}{2}C(\dot\phi \cos\theta + \dot\psi)^2 - Mgl \cos\theta. \tag{4.108}$$

It is not difficult to see that the coordinates ϕ and ψ are cyclic, as they are absent from the Lagrangian, which provides us with two first integrals of the motion expressing the conservation of the conjugate momenta. These momenta correspond to the axes about which the rotations ϕ and ψ take place, namely, axes Z and z. The physical reason for the conservation of these momenta is that there are no torques about the associated axes. The only torque acting on the system is about the node axis ξ, and this is the axis about which the rotation θ takes place. Furthermore, because this is a natural conservative system (see Sec. 2.10), there exists a third integral of the motion consisting of the total energy E.

The momentum integrals are

$$p_\phi = \frac{\partial L}{\partial \dot\phi} = A\dot\phi \sin^2\theta + C(\dot\phi \cos\theta + \dot\psi)\cos\theta = \beta_\phi \quad (4.109)$$

and

$$p_\psi = \frac{\partial L}{\partial \dot\psi} = C(\dot\phi \cos\theta + \dot\psi) = C\omega_z = \beta_\psi, \quad (4.110)$$

where β_ϕ and β_ψ are constants depending on the initial conditions. Equation (4.110) implies that the angular velocity about the symmetry axis is constant. Using Eq. (2.134), in conjunction with Eqs. (4.108) to (4.110), the Routhian of the system becomes

$$R = L - \beta_\phi \dot\phi - \beta_\psi \dot\psi = \tfrac{1}{2} A \dot\theta^2 - \frac{(\beta_\phi - \beta_\psi \cos\theta)^2}{2A \sin^2\theta} - \frac{\beta_\psi^2}{2C} - Mgl \cos\theta, \quad (4.111)$$

which depends on the angle θ alone. The only equation of motion describing the system can be obtained from

$$\frac{d}{dt}\left(\frac{\partial R}{\partial \dot\theta}\right) - \frac{\partial R}{\partial \theta} = 0, \quad (4.112)$$

so that the system can be interpreted as a single-degree-of-freedom system with the kinetic energy $\tfrac{1}{2} A \dot\theta^2$ and the equivalent potential energy

$$\frac{(\beta_\phi - \beta_\psi \cos\theta)^2}{2A \sin^2\theta} + \frac{\beta_\psi^2}{2C} + Mgl \cos\theta.$$

At this point we shall not pursue this approach further but turn our attention to the energy integral instead. From Eqs. (4.106) and (4.107), as well as Eqs. (4.109) and (4.110), we obtain

$$E = T + V = \tfrac{1}{2} A \dot\theta^2 + \frac{(\beta_\phi - \beta_\psi \cos\theta)^2}{2A \sin^2\theta} + \frac{\beta_\psi^2}{2C} + Mgl \cos\theta, \quad (4.113)$$

which depends entirely on θ and $\dot\theta$. It is customary to introduce the constants

$$\frac{2}{A}\left(E - \frac{\beta_\psi^2}{2C}\right) = \alpha, \quad \frac{2Mgl}{A} = \beta, \quad \frac{\beta_\phi}{A} = a, \quad \frac{\beta_\psi}{A} = b, \quad (4.114)$$

and making the substitution $u = \cos\theta$, Eq. (4.113) can be reduced to

$$\dot u^2 = (\alpha - \beta u)(1 - u^2) - (a - bu)^2, \quad (4.115)$$

leading to the solution

$$t = \int_{u(0)}^{u(t)} \frac{du}{[(\alpha - \beta u)(1 - u^2) - (a - bu)^2]^{\tfrac{1}{2}}}, \quad (4.116)$$

which represents an elliptic integral. It turns out that a discussion of this solution is neither very illuminating nor necessary. Indeed, much more insight into the behavior of the system is to be gained from an examination of the cubic polynomial in u, Eq. (4.115), and in particular its roots.

Let us denote the cubic polynomial by

$$f(u) = \dot{u}^2 = (\alpha - \beta u)(1 - u^2) - (a - bu)^2 \qquad (4.117)$$

and bear in mind that, although for the sake of studying the polynomial u can assume any value from $-\infty$ to $+\infty$, only the values $-1 \leq u \leq 1$ are acceptable from a mathematical standpoint. The roots of $f(u)$ correspond to points of zero nutation, $\dot{\theta} = 0$. A brief examination of $f(u)$ yields sufficient information to plot it as a function of u. First we note that for large values of u the polynomial behaves like βu^3 and as $u \to \pm \infty$, $f(u) \to \pm \infty$ also. Moreover, for $u = \pm 1$ the first term in the cubic reduces to zero, and it follows that $f(\pm 1) \leq 0$, because the second term is never positive. A typical curve is shown in Fig. 4.9. The range of u is to be further restricted because imaginary values of u are physically unacceptable, leading us to the conclusion that only the range of u corresponding to roots between which $f(u) \geq 0$ must be considered. The physical interpretation of this is that the symmetry axis of the top must nutate in the annular region bounded by the circles $\theta_1 = \cos^{-1} u_1$ and $\theta_2 = \cos^{-1} u_2$ except when the roots u_1 and u_2 coalesce and the nutation reduces to zero.

The motion can be visualized by means of the spherical coordinates ϕ and θ of the symmetry axis. As these coordinates vary, the axis can be imagined to trace a curve on a unit sphere with the origin at O. The type of curve traced by the symmetry axis on the sphere depends not only on the nutation $\dot{\theta}$ but also on the precession $\dot{\phi}$. From Eqs. (4.109) and (4.110) and the last two of Eqs. (4.114) we obtain the precession

$$\dot{\phi} = \frac{a - bu}{1 - u^2}, \qquad (4.118)$$

FIGURE 4.9

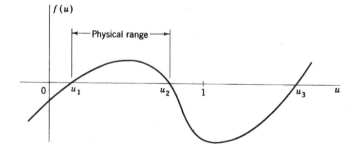

Rigid Body Dynamics

and because the denominator is in general positive, the sense of the precession depends on the sign of $a - bu$. There are three distinct cases of particular interest. They differ in the nature of the precession at the boundaries of the annular region, which, in turn, determines the type of curve traced by the symmetry axis:

1. The direction of the precession is the same at the two bounding circles, a case obtained when $a > bu_2$. This case is shown in Fig. 4.10a.

2. The precession is zero at the upper bounding circle, which is the case obtained when $a = bu_2$. The trace is in the form of cusps, as shown in Fig. 4.10b, for which reason the motion of the top is referred to as *cuspidal motion*. This case occurs when the symmetry axis of the spinning top is released from rest. Immediately after being released at $\theta = \theta_2$ the top begins to fall. As the kinetic energy increases at the expense of the potential energy, the top begins to acquire precessional motion, reaching a maximum at $\theta = \theta_1$, at which point the top begins to rise again to the initial height and the process continues to repeat itself. The case of the inverted cusps can be shown to be impossible (see Prob. 4.14).

3. The direction of the precession is different at the two bounding circles. This case, shown in Fig. 4.10c, occurs when $bu_2 > a > bu_1$. This means that at $\theta = \theta_2$ the precession is negative, whereas at $\theta = \theta_1$ it is positive. At a certain intermediate value, $\theta_2 < \theta < \theta_1$, the precession is zero, which leads us to the conclusion that the trace consists of loops. Because the average precession is not zero, the top revolves nonuniformly around the vertical axis Z.

The type of motion performed by the top depends on the initial conditions. To develop more insight into the nature of the motion, let us consider the case in which the top is imparted initially a relatively large spin, with the

FIGURE 4.10

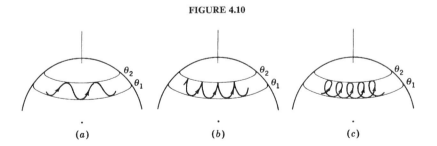

(a) (b) (c)

nutation and precession zero. From Eqs. (4.109), (4.110), and (4.114) we have

$$a = bu_0, \tag{4.119}$$

where u_0 corresponds to the initial angle θ_0. Because $\dot{\theta}_0 = 0$, it follows from Eq. (4.115) that we must also have

$$\alpha = \beta u_0, \tag{4.120}$$

so that u_0 is the root of Eq. (4.115). This enables us to write Eq. (4.115) in the form

$$f(u) = (u_0 - u)[\beta(1 - u^2) - b^2(u_0 - u)], \tag{4.121}$$

and it is obvious that the other two roots of the cubic are obtained from the quadratic equation

$$u^2 - \frac{b^2}{\beta} u + \frac{b^2 u_0}{\beta} - 1 = 0. \tag{4.122}$$

Assuming that $b^2 \gg \beta$, the roots can be shown to have the values

$$u_1 = \frac{b^2}{\beta} - u_0 + \frac{\beta}{b^2}(1 - u_0^2), \qquad u_2 = u_0 - \frac{\beta}{b^2}(1 - u_0^2). \tag{4.123}$$

It is evident that $u_1 > 1$, so that this root presents no interest. On the other hand, from the second of Eqs. (4.123), we have $u_2 < u_0$, which implies that the top must fall before it can rise again. It is also obvious that as the initial spin is increased, the annular region narrows, and because u_2 does not deviate very much from the initial value u_0, we conclude from Eq. (4.115) that $\dot{u}$ cannot be very large. To show this, let us substitute $u = u_0 - \epsilon$ in Eq. (4.115), where ϵ is small, and obtain

$$\dot{\epsilon}^2 = \epsilon[(1 - u_0^2)\beta - b^2 \epsilon]. \tag{4.124}$$

Differentiating Eq. (4.124) with respect to t, we obtain a second-order differential equation which, when we recall the initial conditions $\epsilon(0) = \dot{\epsilon}(0) = 0$, can be shown to have the solution

$$\epsilon = \frac{\beta(1 - u_0^2)}{2b^2}(1 - \cos bt), \tag{4.125}$$

whereas $\dot{u}$ is simply equal to $-\dot{\epsilon}$. Because the angle θ does not change appreciably from its initial value θ_0, and recalling that $\dot{u} = -\dot{\theta} \sin \theta = -\dot{\epsilon}$, we obtain

$$\dot{\theta} \approx \frac{\beta \sin \theta_0}{2b} \sin bt, \tag{4.126}$$

implying that the nutation is harmonic. With an increase in the initial spin the amplitude of the nutation decreases and the frequency b increases.

The precessional velocity must also be relatively small for large spin. Indeed, letting $u = u_0 - \epsilon$ in Eq. (4.118), we obtain for zero initial precession

$$\dot\phi \approx \frac{b\epsilon}{1 - u_0^2} = \frac{\beta}{2b}(1 - \cos bt), \qquad (4.127)$$

where we substituted the value of ϵ from Eq. (4.125). Thus the precession possesses, in addition to a constant term, a harmonic component. As in the case of the nutation, the amplitude of the precession decreases and the frequency of oscillation increases with an increase in the initial spin.

From the nature of Eqs. (4.126) and (4.127), we conclude that for the fast top the harmonic components of the nutation and precession are out of phase by an angle $\pi/2$. Moreover, while the average nutation is zero, for $\omega_z > 0$ the average precession has the value

$$\dot\phi_{\mathrm{av}} = \frac{\beta}{2b} = \frac{Mgl}{C\omega_z} > 0, \qquad (4.128)$$

which indicates that on balance the symmetry axis keeps revolving around the vertical axis.

The amplitudes of the nutation and precession become increasingly small as the initial spin of the top is increased; because at the same time the frequency of oscillation of the nutation and of the harmonic component of the precession becomes very large, the nutation and the harmonic component of the precession are hard to detect, with the result that the top *appears to precess uniformly.* Moreover, due to friction at the pivot O, there is a tendency for the oscillatory motion to damp out.

In the special case in which the two roots u_1 and u_0 coalesce, the angle θ assumes a constant value, say θ_s. This case, known as *steady precession*, corresponds to the conditions

$$f(u) = 0, \qquad \frac{df(u)}{du} = 0, \qquad (4.129)$$

as can be easily concluded from Fig. 4.11. Equations (4.129) can be used to

FIGURE 4.11

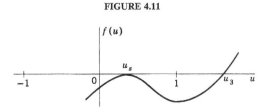

derive an explicit condition for the steady precession. It turns out, however, that this condition is more conveniently derived by means of the modified Euler equations (4.66). For the case $\ddot{\theta} = 0$ the angular velocity components of the system of axes ξ, η, ζ are

$$\omega_\xi = 0, \qquad \omega_\eta = \dot{\phi} \sin \theta, \qquad \omega_\zeta = \dot{\phi} \cos \theta, \qquad (4.130)$$

whereas the angular velocity components of the top about these axes are

$$\Omega_\xi = 0, \qquad \Omega_\eta = \dot{\phi} \sin \theta, \qquad \Omega_\zeta = \dot{\phi} \cos \theta + \dot{\psi}. \qquad (4.131)$$

The ξ component of Eqs. (4.66) yields

$$N_\xi = A\dot{\Omega}_\xi + \omega_\eta C \Omega_\zeta - \omega_\zeta A \Omega_\eta$$
$$= C\omega_z \dot{\phi} \sin \theta - A\dot{\phi}^2 \sin \theta \cos \theta = Mgl \sin \theta, \qquad (4.132)$$

leading to the quadratic equation in $\dot{\phi}$

$$\dot{\phi}^2 - \frac{C\omega_z}{A \cos \theta} \dot{\phi} + \frac{Mgl}{A \cos \theta} = 0, \qquad (4.133)$$

which has the solutions

$$\left.\begin{array}{c}\dot{\phi}_1 \\ \dot{\phi}_2\end{array}\right\} = \frac{C\omega_z}{2A \cos \theta} \pm \frac{C\omega_z}{2A \cos \theta} \left(1 - \frac{4MglA \cos \theta}{C^2 \omega_z^2}\right)^{\frac{1}{2}}, \qquad (4.134)$$

where the roots $\dot{\phi}_1$ and $\dot{\phi}_2$ are called the *fast* and *slow precession*, respectively. Of course, the solutions must be real, which is possible only if the discriminant is positive, so that the *condition for steady precession* is

$$\omega_z^2 > \frac{4MglA \cos \theta}{C^2}. \qquad (4.135)$$

For relatively large values of spin the roots can be approximated by

$$\dot{\phi}_1 = \frac{C\omega_z}{A \cos \theta}, \qquad \dot{\phi}_2 = \frac{Mgl}{C\omega_z} = \frac{\beta}{2b}, \qquad (4.136)$$

and we notice that under these circumstances the slow precession is equal to the average precession for the fast top, Eq. (4.128), which is to be expected. Due to the high kinetic energy requirement, the fast precession is generally unattainable.

There is one more special case deserving attention, namely, that in which $u_s = 1$, which corresponds to the top spinning in the upright position. This case is possible only if ω_z satisfies the condition (4.135) with $\cos \theta = 1$. As long as $\omega_z > \sqrt{4MglA/C^2}$ the top appears motionless, with the symmetry axis coinciding with the vertical, for which reason it is called the *sleeping top*. Due to friction, however, the angular velocity ω_z will fall below the critical value, and the top will begin to wobble increasingly. The same stability

Rigid Body Dynamics 157

condition will be derived in Sec. 6.9 by means of an entirely different approach, namely, the Liapunov direct method.

A physical system akin to the spinning top is the spin-stabilized artillery projectile, discussed later in the text (see Sec. 6.10).

4.12 THE LAGRANGIAN EQUATIONS FOR QUASI-COORDINATES

In Sec. 2.8 we derived Lagrange's equations of motion for the n generalized coordinates q_k ($k = 1, 2, \ldots, n$). These coordinates can be regarded as *true coordinates* in the sense that if the velocities $\dot{q}_k$ are known functions of time, an integration with respect to time yields the corresponding coordinates q_k. Lagrange's equations of motion can be written in the matrix form

$$\frac{d}{dt}\left\{\frac{\partial T}{\partial \dot{q}}\right\} - \left\{\frac{\partial T}{\partial q}\right\} = \{Q\}. \tag{4.137}$$

It may prove of interest to produce a set of differential equations of motion which is not restricted to true coordinates but uses as variables n independent linear combinations ω_s ($s = 1, 2, \ldots, n$) of the velocities $\dot{q}_k$, where, in contrast, the variables ω_s cannot be integrated to obtain true coordinates. These variables can be written in the form

$$\omega_s = \alpha_{1s}\dot{q}_1 + \alpha_{2s}\dot{q}_2 + \cdots + \alpha_{ns}\dot{q}_n$$
$$= \sum_{r=1}^{n} \alpha_{rs}\dot{q}_r, \quad s = 1, 2, \ldots, n, \tag{4.138}$$

where the coefficients α_{rs} are known functions of the generalized coordinates q_k. Let us consider a set of differentials $d\theta_s$ given by the relations

$$d\theta_s = \sum_{r=1}^{n} \alpha_{rs}\, dq_r, \quad s = 1, 2, \ldots, n, \tag{4.139}$$

where the coefficients α_{rs} are the same as in Eqs. (4.138). Unless $\partial \alpha_{js}/\partial q_k = \partial \alpha_{ks}/\partial q_j$, Eqs. (4.139) cannot be integrated to obtain the variable θ_s. We shall be interested in the very case in which Eqs. (4.139) are not integrable, so that the quantities $d\theta_s$ cannot be regarded as the differentials of the variables θ_s. The quantities $d\theta_s$ are referred to as *differentials of quasi-coordinates* although they are not differentials and the variables θ_s are undefined. Perhaps to establish the relevance of the present discussion, we may recall the statement made in Sec. 4.7 that the angular velocity components ω_i can be regarded as time derivatives of quasi-coordinates.

Equations (4.138) can be written in the matrix form

$$\{\omega\} = [\alpha]^T\{\dot{q}\}, \tag{4.140}$$

and, assuming that the matrix $[\alpha]$ is not singular, we can solve for the derivatives $\dot{q}_k$ by writing

$$\{\dot{q}\} = [\beta]\{\omega\}, \qquad [\beta][\alpha]^T = [1]. \tag{4.141}$$

This enables us to express the kinetic energy as a function of the coordinates q_k and the variables ω_k. We denote this form of the kinetic energy by $\overline{T}(\mathbf{q},\boldsymbol{\omega})$, as opposed to the original form $T(\mathbf{q},\dot{\mathbf{q}})$.

Next we proceed to replace the generalized velocities $\dot{q}_k$ by the variables ω_k in Lagrange's equations. To this end, we use Eqs. (4.138) and write the conjugate momenta

$$\frac{\partial T}{\partial \dot{q}_k} = \sum_{i=1}^{n} \frac{\partial \overline{T}}{\partial \omega_i} \frac{\partial \omega_i}{\partial \dot{q}_k} = \sum_{i=1}^{n} \alpha_{ki} \frac{\partial \overline{T}}{\partial \omega_i}, \qquad k = 1, 2, \ldots, n, \tag{4.142}$$

which can clearly be written as the column matrix

$$\left\{\frac{\partial T}{\partial \dot{q}}\right\} = [\alpha]\left\{\frac{\partial \overline{T}}{\partial \omega}\right\}, \tag{4.143}$$

from which it follows that

$$\frac{d}{dt}\left\{\frac{\partial T}{\partial \dot{q}}\right\} = [\alpha]\frac{d}{dt}\left\{\frac{\partial \overline{T}}{\partial \omega}\right\} + [\dot{\alpha}]\left\{\frac{\partial \overline{T}}{\partial \omega}\right\}. \tag{4.144}$$

Because the coefficients α_{ij} depend on coordinates alone, any element $\dot{\alpha}_{ij}$ of the square matrix $[\dot{\alpha}]$ has the form

$$\dot{\alpha}_{ij} = \sum_{r=1}^{n} \frac{\partial \alpha_{ij}}{\partial q_r} \dot{q}_r = \{\dot{q}\}^T \left\{\frac{\partial \alpha_{ij}}{\partial q}\right\} = \{\omega\}^T [\beta]^T \left\{\frac{\partial \alpha_{ij}}{\partial q}\right\}, \tag{4.145}$$

which is a scalar. It must be stressed here that the above triple matrix product does not involve summation over the indices of α_{ij}. With this idea clearly fixed, we can write the square matrix corresponding to Eq. (4.145)

$$[\dot{\alpha}] = \left[\{\omega\}^T [\beta]^T \left\{\frac{\partial \alpha}{\partial q}\right\}\right]. \tag{4.146}$$

Following the same pattern, we write

$$\frac{\partial T}{\partial q_k} = \frac{\partial \overline{T}}{\partial q_k} + \sum_{i=1}^{n} \frac{\partial \overline{T}}{\partial \omega_i} \frac{\partial \omega_i}{\partial q_k} = \frac{\partial \overline{T}}{\partial q_k} + \sum_{i=1}^{n} \frac{\partial \overline{T}}{\partial \omega_i} \sum_{j=1}^{n} \frac{\partial \alpha_{ji}}{\partial q_k} \dot{q}_j$$

$$= \frac{\partial \overline{T}}{\partial q_k} + \{\dot{q}\}^T \left[\frac{\partial \alpha}{\partial q_k}\right]\left\{\frac{\partial \overline{T}}{\partial \omega}\right\}, \tag{4.147}$$

where this time summations over both indices of α_{ij} are involved. However,

Rigid Body Dynamics

there is no summation over the index in ∂q_k. This enables us to write the column matrix

$$\left\{\frac{\partial T}{\partial q}\right\} = \left\{\frac{\partial \bar{T}}{\partial q}\right\} + \left\{\{\dot{q}\}^T \left[\frac{\partial \alpha}{\partial q}\right]\left\{\frac{\partial \bar{T}}{\partial \omega}\right\}\right\}, \tag{4.148}$$

which can be shown to be equivalent to

$$\left\{\frac{\partial T}{\partial q}\right\} = \left\{\frac{\partial \bar{T}}{\partial q}\right\} + \left[\{\omega\}^T[\beta]^T\left[\frac{\partial \alpha}{\partial q}\right]\right]\left\{\frac{\partial \bar{T}}{\partial \omega}\right\}, \tag{4.149}$$

where the triple matrix products $\{\omega\}^T[\beta]^T[\partial \alpha/\partial q]$ produce n row matrices, one for each index of ∂q. These n row matrices are arranged in the square matrix premultiplying $\{\partial \bar{T}/\partial \omega\}$. Introduction of Eqs. (4.144), (4.146), and (4.149) into Eq. (4.137) yields

$$[\alpha]\frac{d}{dt}\left\{\frac{\partial \bar{T}}{\partial \omega}\right\} + \left[\{\omega\}^T[\beta]^T\left\{\frac{\partial \alpha}{\partial q}\right\}\right]\left\{\frac{\partial \bar{T}}{\partial \omega}\right\} - \left\{\frac{\partial \bar{T}}{\partial q}\right\} - \left[\{\omega\}^T[\beta]^T\left[\frac{\partial \alpha}{\partial q}\right]\right]\left\{\frac{\partial \bar{T}}{\partial \omega}\right\} = \{Q\}. \tag{4.150}$$

Next premultiply through Eq. (4.150) by $[\beta]^T$, recall the second of Eqs. (4.141), and obtain

$$\frac{d}{dt}\left\{\frac{\partial \bar{T}}{\partial \omega}\right\} + [\beta]^T[\gamma]\left\{\frac{\partial \bar{T}}{\partial \omega}\right\} - [\beta]^T\left\{\frac{\partial \bar{T}}{\partial q}\right\} = \{N\}, \tag{4.151}$$

which are referred to as the *Lagrange equations for quasi-coordinates*, where the notation

$$[\gamma] - \left[\{\omega\}^T[\beta]^T\left\{\frac{\partial \alpha}{\partial q}\right\}\right] - \left[\{\omega\}^T[\beta]^T\left[\frac{\partial \alpha}{\partial q}\right]\right] \tag{4.152}$$

and $\quad\quad\quad \{N\} = [\beta]^T\{Q\} \tag{4.153}$

has been introduced. Again we must keep in mind Eqs. (4.145) and (4.147) for the interpretation of the matrix notation in Eq. (4.152).

As an application of this formulation, the motion of a rigid body about a point comes immediately to mind. Indeed, in this case the variables ω_k are readily identified as the body angular velocity components. The kinetic energy has the form

$$\bar{T} = \tfrac{1}{2}\{\omega\}^T[I]\{\omega\}, \tag{4.154}$$

which does not contain the coordinates explicitly, so that

$$\left\{\frac{\partial \bar{T}}{\partial q}\right\} = \{0\}. \tag{4.155}$$

Moreover, a series of algebraic manipulations leads to the simple result

$$[\beta]^T[\gamma] = [\omega], \tag{4.156}$$

where $[\omega]$ is the skew-symmetric matrix of the angular velocity components, Eq. (3.30). Introducing Eqs. (4.155) and (4.156) into (4.151), we obtain the compact form

$$\frac{d}{dt}\left\{\frac{\partial \bar{T}}{\partial \omega}\right\} + [\omega]\left\{\frac{\partial \bar{T}}{\partial \omega}\right\} = \{N\}, \tag{4.157}$$

where $\{N\}$ represents a column matrix of torque components. The advantage of Eqs. (4.157) over the commonly used Lagrange equations, Eqs. (4.137), is that Eqs. (4.157) are in terms of components about orthogonal axes whereas the Lagrange equations, when written in terms of Euler's angles ϕ, θ, and ψ, are not. The reader is urged to compare Eq. (4.157) with Eq. (4.63) and draw conclusions.

4.13 THE EQUATIONS OF MOTION REFERRED TO AN ARBITRARY SYSTEM OF AXES

In certain cases it may be desirable to refer the motion of a rigid body to a system of axes x, y, z which is neither an inertial space nor a set of body axes. Moreover, the origin A of the x, y, z system may not coincide with the center of mass C, and indeed it can move with respect to the body (see Fig. 4.12). We shall denote by **ω** the angular velocity vector of the system x, y, z and by **Ω** the angular velocity of the body. The velocity of the point A is denoted by **u** and the velocity of the center of mass of C by **v**. All these quantities will be referred to the system x, y, z.

FIGURE 4.12

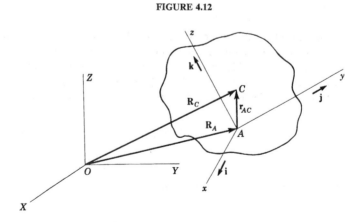

Rigid Body Dynamics

In discussing systems of particles, we derived in Sec. 1.7 the force and torque equations of motion. These equations are equally valid for rigid bodies, and indeed the force equation (1.59) can be written

$$\mathbf{F} = M\ddot{\mathbf{R}}_C = \dot{\mathbf{p}}, \tag{4.158}$$

where $\mathbf{p}$ is the linear momentum vector of the rigid body. Similarly, the torque equation about point A, Eq. (1.64), can be shown to be equivalent to

$$\mathbf{N}_A = \dot{\mathbf{L}}_A + \mathbf{u} \times \mathbf{p}, \tag{4.159}$$

where $\mathbf{L}_A$ is the angular momentum of the body about point A.

The linear and angular momenta can be written in terms of components along the moving axes x, y, z in the form

$$\mathbf{p} = M\mathbf{v} = M(v_x\mathbf{i} + v_y\mathbf{j} + v_z\mathbf{k}) = \frac{\partial T}{\partial v_x}\mathbf{i} + \frac{\partial T}{\partial v_y}\mathbf{j} + \frac{\partial T}{\partial v_z}\mathbf{k} \tag{4.160}$$

and

$$\mathbf{L}_A = L_{Ax}\mathbf{i} + L_{Ay}\mathbf{j} + L_{Az}\mathbf{k} = \frac{\partial T}{\partial \Omega_x}\mathbf{i} + \frac{\partial T}{\partial \Omega_y}\mathbf{j} + \frac{\partial T}{\partial \Omega_z}\mathbf{k}. \tag{4.161}$$

The matrix counterparts of Eqs. (4.160) and (4.161) are, respectively,

$$\{p\} = \left\{\frac{\partial T}{\partial v}\right\} \tag{4.162}$$

and

$$\{L_A\} = \left\{\frac{\partial T}{\partial \Omega}\right\}. \tag{4.163}$$

Moreover, the velocity of the origin A can be written

$$\mathbf{u} = \dot{\mathbf{R}}_A = u_x\mathbf{i} + u_y\mathbf{j} + u_z\mathbf{k}, \tag{4.164}$$

and, by analogy with the skew-symmetric matrix $[\omega]$ corresponding to the angular velocity vector $\boldsymbol{\omega}$, we can introduce the skew-symmetric matrix

$$[u] = \begin{bmatrix} 0 & -u_z & u_y \\ u_z & 0 & -u_x \\ -u_y & u_x & 0 \end{bmatrix}. \tag{4.165}$$

Recalling from Secs. 3.2 and 3.3 the matrix form of the time derivative of a vector expressed in terms of components along a moving system, Eqs. (4.158) and (4.159) assume the matrix form

$$\{F\} = \frac{d}{dt}\left\{\frac{\partial T}{\partial v}\right\} + [\omega]\left\{\frac{\partial T}{\partial v}\right\} \tag{4.166}$$

and

$$\{N_A\} = \frac{d}{dt}\left\{\frac{\partial T}{\partial \Omega}\right\} + [\omega]\left\{\frac{\partial T}{\partial \Omega}\right\} + [u]\left\{\frac{\partial T}{\partial v}\right\}, \tag{4.167}$$

where the matrix $[\omega]$ is given by Eq. (3.30). Equations (4.166) and (4.167) form a set of six coupled differential equations, which can be regarded as Lagrangian equations of motion in terms of quasi-coordinates and which are particularly suitable for the treatment of nonholonomic systems. In the special case in which *point A coincides with the center of mass C* the last matrix product in Eq. (4.167) vanishes, and the equation becomes independent of Eq. (4.166) unless the forces and moments are coupled. On the other hand, Eq. (4.166) still depends on Eq. (4.167).

In Sec. 4.14 the preceding formulation is applied to derive the equations of motion of a nonholonomic system.

4.14 THE ROLLING OF A COIN

The formulation of Sec. 4.13 is particularly suitable for the treatment of problems involving nonholonomic systems. A classical example of a nonholonomic system is the coin rolling on a rough horizontal plane, as shown in Fig. 4.13.

Let v_i denote the velocity components of the center of the coin, ω_i the angular velocity components of the system 1, 2, 3, and Ω_i the angular velocity components of a set of body axes instantaneously coinciding with the system 1, 2, 3. All these components are along axes 1, 2, 3, and the moments of inertia of the coin about these axes are A, A, and C, respectively. The coin is regarded as a thin disk. With this notation, the kinetic energy assumes the form

$$T = \tfrac{1}{2}M(v_1^2 + v_2^2 + v_3^2) + \tfrac{1}{2}A(\Omega_1^2 + \Omega_2^2) + \tfrac{1}{2}C\Omega_3^2. \quad (4.168)$$

FIGURE 4.13

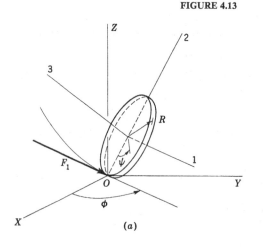
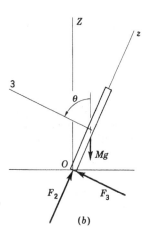

Rigid Body Dynamics 163

Introducing Eq. (4.168) into Eq. (4.166), we obtain the force equations

$$\begin{aligned} M\dot{v}_1 + M(\omega_2 v_3 - \omega_3 v_2) &= F_1, \\ M\dot{v}_2 + M(\omega_3 v_1 - \omega_1 v_3) &= F_2 - Mg\sin\theta, \\ M\dot{v}_3 + M(\omega_1 v_2 - \omega_2 v_1) &= F_3 - Mg\cos\theta, \end{aligned} \quad (4.169)$$

where F_1, F_2, and F_3 are the reactive force components along the system 1, 2, 3. Because the origin of the system 1, 2, 3 coincides with the mass center of the disk, Eq. (4.167) yields the torque equations

$$\begin{aligned} A\dot{\Omega}_1 + C\omega_2\Omega_3 - A\omega_3\Omega_2 &= -F_3 R, \\ A\dot{\Omega}_2 + A\omega_3\Omega_1 - C\omega_1\Omega_3 &= 0, \\ C\dot{\Omega}_3 + A\omega_1\Omega_2 - A\omega_2\Omega_1 &= F_1 R. \end{aligned} \quad (4.170)$$

The velocity components of the mass center and the body angular velocity components are related by

$$v_1 = -R\Omega_3, \quad v_2 = 0, \quad v_3 = R\Omega_1, \quad (4.171)$$

so that, combining Eqs. (4.169) through (4.171), we obtain

$$\begin{aligned} (A + R^2 M)\dot{\Omega}_1 + (C + R^2 M)\omega_2\Omega_3 - A\omega_3\Omega_2 &= -MgR\cos\theta, \\ A\dot{\Omega}_2 + A\omega_3\Omega_1 - C\omega_1\Omega_3 &= 0, \\ (C + R^2 M)\dot{\Omega}_3 + A\omega_1\Omega_2 - (A + R^2 M)\omega_2\Omega_1 &= 0, \end{aligned} \quad (4.172)$$

which can be recognized as the torque equations about the point of contact O.

But the system 1, 2, 3 angular velocity components ω_i can be written in terms of Euler's angles as

$$\omega_1 = \dot{\theta}, \quad \omega_2 = \dot{\phi}\sin\theta, \quad \omega_3 = \dot{\phi}\cos\theta, \quad (4.173)$$

and, in a similar fashion, the body angular velocity components Ω_i have the form

$$\Omega_1 = \dot{\theta}, \quad \Omega_2 = \dot{\phi}\sin\theta, \quad \Omega_3 = \dot{\phi}\cos\theta + \dot{\psi}. \quad (4.174)$$

Introduction of Eqs. (4.173) and (4.174) into Eqs. (4.172) yields the equations of motion

$$\begin{aligned} (A + R^2 M)\ddot{\theta} + (C + R^2 M)\dot{\phi}\sin\theta(\dot{\phi}\cos\theta + \dot{\psi}) - A\dot{\phi}^2\sin\theta\cos\theta \\ = -MgR\cos\theta, \\ A\ddot{\phi}\sin\theta + 2A\dot{\phi}\dot{\theta}\cos\theta - C\dot{\theta}(\dot{\phi}\cos\theta + \dot{\psi}) = 0, \\ (C + R^2 M)(\ddot{\phi}\cos\theta - \dot{\phi}\dot{\theta}\sin\theta + \ddot{\psi}) - MR^2\dot{\theta}\dot{\phi}\sin\theta = 0. \end{aligned} \quad (4.175)$$

Equations (4.175) are highly nonlinear, and a closed-form solution is possible only for some limiting cases, e.g., the one discussed in Example 4.2.

Example 4.2

As an application of Eqs. (4.175), let us consider the limiting case in which the coin is rolling in the vertical plane at the constant angular velocity Ω when disturbed slightly. The question posed is what magnitude of Ω ensures that the coin will not fall?

We notice that for $\theta = \pi/2$, $\phi = 0$, and $\dot\psi = \Omega_3 = $ const Eqs. (4.175) are identically satisfied. Let us assume that as a result of some disturbance the coordinate θ becomes $\theta = \pi/2 + \alpha$, where α is small, whereas ϕ is no longer zero, although it remains a small quantity. Similarly, $\dot\theta = \dot\alpha$ and $\dot\phi$ are small. In view of these assumptions, we can neglect small quantities to the second power in the third of Eqs. (4.175), integrate that equation, and obtain

$$\dot\phi \cos\theta + \dot\psi = \Omega_3 = \Omega = \text{const.} \tag{a}$$

Moreover, the first two of Eqs. (4.175) become

$$(A + R^2M)\ddot\alpha + (C + R^2M)\Omega\dot\phi - MgR\alpha = 0,$$
$$A\ddot\phi - C\Omega\dot\alpha = 0. \tag{b}$$

The second of Eqs. (b) can be integrated immediately, with the result

$$\dot\phi = \frac{C\Omega}{A}\alpha. \tag{c}$$

Introducing Eq. (c) into the first of Eqs. (b), we arrive at

$$(A + R^2M)\ddot\alpha + \left[(C + R^2M)\frac{C\Omega^2}{A} - MgR\right]\alpha = 0. \tag{d}$$

Equation (d) has an oscillatory nondivergent solution only if the coefficient of α is positive. This yields the condition

$$\Omega^2 > \frac{AMgR}{C(C + R^2M)}, \tag{e}$$

which the angular velocity Ω must satisfy for the rolling motion to be stable. More rigorous treatment of stability problems will be presented in Chaps. 5 to 7.

PROBLEMS

4.1 Calculate the angular momentum about point O of the uniform thin bar of mass m per unit length, as shown in Fig. 4.14. Disregard the mass of the horizontal bar.

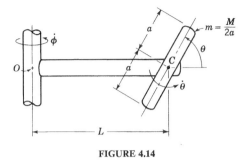

FIGURE 4.14

4.2 Calculate the kinetic energy of the bar of Prob. 4.1.

4.3 A homogeneous right circular cone of mass m rolls on a rough horizontal plane with its apex fixed at point O (see Fig. 4.15). If the cone revolves about the vertical through O with angular velocity ω, calculate the moment of momentum about O and the kinetic energy of the cone.

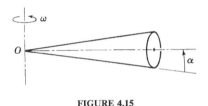

FIGURE 4.15

4.4 A homogeneous circular cylinder of radius r and mass m rolls inside a rough circular track of radius R, as shown in Fig. 4.16. If the cylinder is

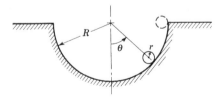

FIGURE 4.16

released from rest at point $\theta = \pi/2$, use the principle of conservation of energy to determine the angular velocity $\dot{\theta}$ of the cylinder when in position θ.

4.5 Prove that the vector of the direction cosines of the normal **n** to the inertia ellipsoid at any point (u_1, u_2, u_3) on the ellipsoid is proportional to the vector $[I]\{u\}$, where $[I]$ is the symmetric matrix of the moments of inertia.

4.6 Consider Fig. 4.17 and obtain the principal moments of inertia of the system consisting of a thin disk of mass M and a rod of distributed mass m when (a) the origin is at A and (b) the origin is at the center of mass C. The normal to the disk makes an angle θ with respect to the rod.

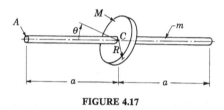

FIGURE 4.17

4.7 Derive the moment equations of the rod of Fig. 4.14 with respect to the center of mass C and with respect to point O. What torque components must be applied to produce the motion?

4.8 Two masses are symmetrically attached to a shaft, as shown in Fig. 4.18. Calculate the bearing reactions if the shaft rotates with the constant angular velocity ω.

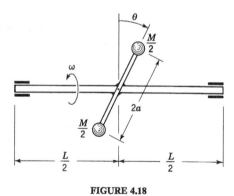

FIGURE 4.18

4.9 A disk of moment of inertia I about the symmetry axis is spinning at the constant angular velocity Ω, as shown in Fig. 4.19. If the entire system is

Rigid Body Dynamics

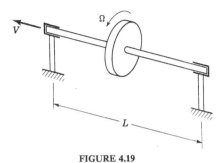

FIGURE 4.19

mounted on a vehicle which travels with velocity v in the direction of the shaft and the vehicle makes a turn of radius R, what are the bearing reactions?

4.10 A thin disk is thrown spinning into the air. If the disk has a slight wobble, derive an expression for the ratio between the wobble angular velocity and the spin angular velocity. Are the wobble and spin angular velocities in the same or opposite sense?

4.11 Derive the Lagrange equation of motion for the cylinder of Prob. 4.4. (See the introductory remarks at the beginning of this chapter.) Assume small angles θ and derive the frequency of oscillation about the position $\theta = 0$.

4.12 Two equal uniform links hinged at point O lie at rest on a smooth horizontal table when struck by an impulsive force, as shown in Fig. 4.20. Derive the Lagrange equations of motion (see the introductory remarks at the beginning of this chapter) and determine the subsequent motion of the system. What should be the value of the distance D if the link not struck is to remain at rest?

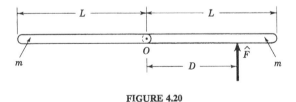

FIGURE 4.20

4.13 Formulate the problem of a moment-free body of Sec. 4.10 in dyadic notation.

4.14 Consider the spinning symmetric top and show that motion in which the symmetry axis describes inverted cusps is not possible.

4.15 Prove that for the sleeping top the cubic polynomial in u, Eq. (4.117), becomes
$$f(u) = (1-u)^2 \left[\frac{2Mgl}{A}(1+u) - \frac{C^2}{A^2}\omega_z^2 \right].$$
Whereas $u = 1$ is a double root, the third root is greater or less than unity, depending on the system parameters. The two cases are shown in solid lines in Fig. 4.21a and b. If the top is disturbed slightly, we no longer have a sleeping top and the corresponding curves are as shown in dashed lines. Discuss the motion associated with each case and, in particular, the stability of motion. How does your conclusion agree with the conclusion reached in Sec. 4.11?

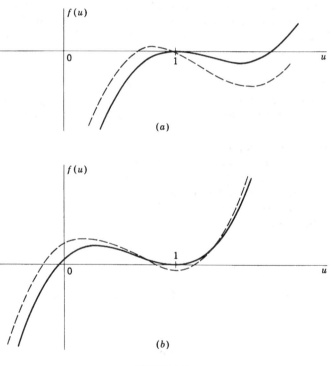

FIGURE 4.21

4.16 Prove Eq. (4.156).

Rigid Body Dynamics

4.17 Consider Eqs. (4.175) for the general motion of a coin and derive a stability condition for the case in which the coin is rotating about a diametrical axis in the upright position.

4.18 Consider Eqs. (4.175) for the general motion of a coin and discuss the case in which the coin is rotating with axis 3 nearly vertical.

SUGGESTED REFERENCES

1. Goldstein, H.: *Classical Mechanics*, Addison-Wesley Publishing Company, Inc., Reading, Mass., 1950.
2. Meirovitch, L.: *Analytical Methods in Vibrations*, The Macmillan Company, New York, 1967.
3. Synge, J. L., and B. A. Griffith: *Principles of Mechanics*, 3d ed., McGraw-Hill Book Company, New York, 1959.
4. Webster, A. G.: *The Dynamics of Particles*, Dover Publications, Inc., New York, 1959.
5. Whittaker, E. T.: *Analytical Dynamics*, Cambridge University Press, London, 1937.

chapter five

Behavior of Dynamical Systems. Geometric Theory

The first four chapters have been devoted almost exclusively to fundamental principles and techniques leading to convenient formulations of problems associated with dynamical systems. The formulations, for the most part, consist of sets of simultaneous nonlinear differential equations. To gain insight into the behavior of a physical system, however, we must integrate these equations and examine their solution. This turns out to be a formidable task when the equations are nonlinear, which is almost the rule with dynamical systems. In Chap. 2 we showed that under certain circumstances the system possesses first integrals of the motion. Such integrals, if they exist, may go a long way toward revealing some of the motion properties of a system. In general, however, the question of extracting this information from the differential equations of the system is still with us. In this chapter and in the following four we discuss methods suitable for investigating the behavior of dynamical systems.

A large class of problems in dynamics deals with the small motions of elastic systems generally described by sets of simultaneous linear equations with constant coefficients. This class of problems led to the development of the vibration theory of linear systems, which is a study in itself. In this text, however, we shall concentrate primarily on the dynamical behavior of rigid bodies and, in particular, rotational motion. Such motion is generally governed by sets of simultaneous nonlinear differential equations, which often require entirely different methods of attack. The theory for the treatment of nonlinear systems is not nearly as fully developed as that for linear systems. In fact, a common procedure in the study of nonlinear systems is to examine the motion in the neighborhood of known motions, which amounts to representing the nonlinear system by its linear approximation. Of course, this is not always advisable, as we shall see.

The geometric theory of sets of first-order differential equations plays an important role in the study of nonlinear systems. We may recall that Hamilton's canonical equations are of this type. Of particular interest is the representation of the solution of a system of $2n$ first-order equations in a $2n$-dimensional phase space or a $(2n + 1)$-dimensional motion space. Quite often, in the case of nonlinear systems of equations, no explicit solutions in the form of time-dependent functions can be expected and we must be content with a statement concerning the stability characteristics of the system in the neighborhood of known motions. Although the discussion of systems of first-order differential equations is motivated by our interest in the Hamilton canonical equations, the theory presented in this text is not restricted to Hamiltonian systems except when a specific statement is made to that effect.

In this chapter we first introduce the concepts associated with the geometric theory of dynamical systems defined by sets of $2n$ simultaneous first-order differential equations and then give precise mathematical definitions for various types of stability. Subsequently, the discussion is confined to second-order autonomous systems and the representation of solutions for such systems in the phase plane. The subject of system singularities receives considerable attention, and ways of identifying them are presented. Finally interest is focused on the question of periodic solutions, and an exposition of the important Poincaré-Bendixson theory for the existence of cyclic trajectories in the phase plane is given.

5.1 FUNDAMENTAL CONCEPTS

The differential equations describing the motion of an n-degree-of-freedom holonomic dynamical system can be written in the form of Lagrange's equations

$$\frac{d}{dt}\left(\frac{\partial L}{\partial \dot{q}_k}\right) - \frac{\partial L}{\partial q_k} = 0, \qquad k = 1, 2, \ldots, n, \qquad (5.1)$$

where L is the Lagrangian and q_k ($k = 1, 2, \ldots, n$) are the generalized coordinates. Equations (5.1) constitute a set of simultaneous second-order equations which are generally nonlinear.

We indicated in Sec. 2.1 that it is possible to associate with the generalized coordinates q_k ($k = 1, 2, \ldots, n$), representing the solution of the dynamical problem, an n-dimensional space.[1] This space will be referred to as the *Lagrangian configuration space*. The coordinates q_k can be regarded as the rectangular coordinates of a *representative point P* in a Euclidean n-space, denoted by E^n, so that the motion of the system can be visualized geometrically in terms of the motion of point P along a given curve in that space. At first we may be tempted to conclude that a geometrical representation in terms of rectangular coordinates cannot be satisfactory in view of the fact that some of the generalized coordinates may be curvilinear, which implies a distortion of the actual space. Indeed *metrical properties* such as angles, distances, and areas are changed by this mapping. On the other hand, *topological properties* do not change, because a point, the neighborhood of a point, and continuous and differentiable curves remain so in a Euclidean space. It turns out that the topological properties, rather than the metrical ones, are essential to our geometrical representation of the dynamical problems.[2] The use of such an abstract n-space enables us to extend the concepts of the geometry of motion of a single particle in a physical space to the motion of a complicated dynamical system in an n-space.

In Sec. 2.13 we showed that the n second-order Lagrangian differential equations can be replaced by $2n$ first-order Hamiltonian equations

$$\dot{q}_k = \frac{\partial H}{\partial p_k}, \qquad \dot{p}_k = -\frac{\partial H}{\partial q_k}, \qquad k = 1, 2, \ldots, n. \tag{5.2}$$

The advantage of Hamilton's equations is that time derivatives appear only on the left side of Eqs. (5.2), because the Hamiltonian H contains only the variables q_k and p_k and not their time derivatives. This may help, at times, to uncover first integrals of motion. Furthermore, the initial conditions can be regarded as being of one kind only, as they involve the initial values of q_k and p_k but not the initial values of their time derivatives. The advantage of these equations in representing the solution geometrically will soon become apparent.

Letting

$$q_k = x_k, \qquad p_k = x_{n+k},$$
$$\frac{\partial H}{\partial p_k} = X_k, \qquad -\frac{\partial H}{\partial q_k} = X_{n+k}, \qquad k = 1, 2, \ldots, n, \tag{5.3}$$

Eqs. (5.2) can be written

$$\dot{x}_i = X_i(x_1, x_2, \ldots, x_{2n}, t), \qquad i = 1, 2, \ldots, 2n. \tag{5.4}$$

---

[1] For simplicity in future discussions we shall omit wherever appropriate, the word "dimensional" and refer to an n-dimensional space or an n-dimensional vector as simply an n-space or an n-vector, respectively.

[2] The reader unfamiliar with this terminology is advised to refer to Appendix B, where many of these concepts are discussed.

Regarding the quantities x_i and X_i as the components of $2n$-dimensional vectors, or simply $2n$-vectors, **x** and **X**, and representing these vectors by column matrices $\{x\}$ and $\{X\}$, respectively, Eqs. (5.4) can be written in the matrix form

$$\{\dot{x}\} = \{X\}. \tag{5.5}$$

From Eqs. (5.4) or Eq. (5.5) we conclude that in the Hamiltonian mechanics the motion of the dynamical system can be represented geometrically by the motion of a representative point P in a $2n$-space defined by the n Lagrangian coordinates q_k and the n conjugate momenta p_k. The $2n$-space defined by the variables q_k and p_k, referred to by Gibbs as the *phase space*, can be taken, for convenience, as a Euclidean space E^{2n}. There is no need to place undue stress upon the physical difference between the variables q_k and p_k, as they can all be regarded as having equal status (see Sec. 9.3).

Both in the Lagrangian configuration space and the Hamiltonian phase space the representative point P traces a path corresponding to a particular solution of the dynamical problem. The real advantage of the $2n$-dimensional representation becomes evident when we consider not only one path but the *totality of paths*, known as the *phase portrait*, representing all possible solutions. In the Lagrangian configuration space the paths may intersect one another. For a given point in the configuration space the motion may start in every direction with arbitrary initial velocity, so that the picture of the totality of paths appears one of confusion. On the other hand, for any given point in the phase space Hamilton's equations determine uniquely the velocity of that point, thus determining uniquely the motion of the representative point. Hence, given the initial conditions $x_i(t_0) = \alpha_i$ ($i = 1, 2, \ldots, 2n$), we can write the unique solution

$$x_i(t) = \psi_i(\alpha_1, \alpha_2, \ldots, \alpha_{2n}, t), \quad i = 1, 2, \ldots, 2n. \tag{5.6}$$

By contrast with the solution in the configuration space, the solution in the phase space has an orderly appearance. The conditions that the functions $X_i(x_1, x_2, \ldots, x_{2n}, t)$ ($i = 1, 2, \ldots, 2n$) must satisfy to ensure uniqueness will be examined shortly.

When the time t is introduced as an additional coordinate, it is possible to define a $(2n + 1)$-space $(\mathbf{x}, t)$, referred to as the *motion space*, which can be regarded as the product of a $2n$-vector space and the time line. The totality of solutions can be visualized geometrically as an infinite set of curves in the $(2n + 1)$-space. Because Hamilton's equations determine the tangent at any point of the motion space uniquely, these curves never intersect. Equations (5.4) can be regarded as describing the motion of a fluid in a $2n$-space so that the fluid velocity at any point $(\mathbf{x}, t)$ is given by the vector **X**. Each streamline represents the motion of the dynamical system for a given set of initial conditions. This fluid flow analogy corresponds to the field description of the fluid motion in which the velocity vector is determined at any point **x** and time t. The *integral curves* x_i, as given by Eqs. (5.6), in the $(2n + 1)$-space $(\mathbf{x},t)$ are called *characteristics*.

To ensure the uniqueness of the solution (5.6), certain restrictions must be placed on the functions X_i. For the purpose of discussing these restrictions, let us consider the scalar function $f = f(x_1, x_2, \ldots, x_{2n}, t)$ defined at each point of a domain D of the $(2n + 1)$-space $(\mathbf{x},t)$. The function f is said to satisfy a *Lipschitz condition* with respect to **x** in D if there exists a positive constant k such that the inequality

$$|f(x_1', x_2', \ldots, x_{2n}', t) - f(x_1'', x_2'', \ldots, x_{2n}'', t)| \leq k \sum_{i=1}^{2n} |x_i' - x_i''| \quad (5.7)$$

is satisfied for every $(\mathbf{x}',t)$ and $(\mathbf{x}'',t)$ in D. The constant k for which the above inequality is true is called a *Lipschitz constant*, and the function f is said to be a *Lipschitz function* in the domain D. It can be shown that $f(\mathbf{x},t)$ is a Lipschitz function if the domain D is convex and the partial derivatives $\partial f/\partial x_i$ exist and are bounded in D (see Ref. 3, p. 26). Hence, the function f must be of class C_1 in D (see Sec. 1.2 for definition) if it is to be Lipschitzian.

A vector function $\mathbf{f}(\mathbf{x},t)$ satisfies Lipschitz conditions if every component $f_i(\mathbf{x},t)$ satisfies a Lipschitz condition, where the Lipschitz constant may be different for each component.

Next let us consider $2n$ functions $X_i(x_1, x_2, \ldots, x_{2n}, t)$, where the variables $x_1, x_2, \ldots, x_{2n}, t$ have the initial values $\alpha_1, \alpha_2, \ldots, \alpha_{2n}, t_0$. According to the *Cauchy-Lipschitz theorem* (see Ref. 4, p. 31), the set of first-order differential equations (5.4) has a unique solution, Eqs. (5.6), if the functions X_i satisfy Lipschitz conditions in D. This theorem asserts not only that a solution exists but that the solution is uniquely determined by the initial conditions. In our discussions we shall be concerned only with Lipschitzian differential equations.

When the vector **X** does not depend explicitly on t, the system is said to be *autonomous*; otherwise it is *nonautonomous*. In the autonomous case the fluid flow is *steady*. The paths resulting from the projection of the characteristic curves onto the phase space are called *trajectories*. They represent the paths traced by the representative points without regard to time. In this case Eqs. (5.6) can be regarded as providing a parametric representation of the trajectories with the time t playing the role of parameter. Assuming that the trajectories are initiated at time $t = t_0$, the trajectories corresponding to $t \geq t_0$ are called *positive half-trajectories*, whereas those corresponding to $t \leq t_0$ are known as *negative half-trajectories*.

A point for which $\{X\}^T\{X\} > 0$ is said to be an *ordinary* or *regular point*. On the other hand, a point for which $\{X\} = \{0\}$ is called a *singular point* or an *equilibrium point*. A trajectory may approach a singular point only for

$t \to \pm \infty$. Another interpretation of an equilibrium point is a point at which the solution is constant, $\varphi_i(\boldsymbol{\alpha},t) = \alpha_i$ ($i = 1, 2, \ldots, 2n$). If $\boldsymbol{\alpha}$ is the only constant solution in some neighborhood of the point $\boldsymbol{\alpha}$, then the constant solution is referred to as an *isolated point*. In this text we shall be concerned only with isolated singular points.

The *Euclidean length*, or *Euclidean norm*, of the vector $\mathbf{x}$ is defined by $\|\mathbf{x}\| = (\{x\}^T\{x\})^{\frac{1}{2}} = (\sum_{i=1}^{2n} x_i^2)^{\frac{1}{2}}$. A sphere of radius r and with the center at the origin of the Euclidean space E^{2n} can be written in terms of the Euclidean length in the form $\|\mathbf{x}\| = r$, whereas the open domain enclosed by that sphere is defined by $\|\mathbf{x}\| < r$. If the vector $\mathbf{x}(t)$ represents an integral curve of system (5.5), then the integral curve is said to be *bounded* if $\|\mathbf{x}(t)\| \leq r$ for some r. Hence an integral curve $\mathbf{x}(t)$ is bounded if it remains at all times inside the spherical region of radius r of the $2n$-dimensional phase space or inside the cylindrical region of radius r of the $(2n + 1)$-dimensional motion space, where the cylinder axis coincides with the t axis.

Next let us consider a particular solution $x_i = \varphi_i(t)$ ($i = 1, 2, \ldots, 2n$) of Eqs. (5.4) and examine the motion in the neighborhood of that solution. We shall refer to the particular solution $\varphi_i(t)$ as the *unperturbed solution* and to the motion $x_i(t)$ in the neighborhood of $\varphi_i(t)$ as the *perturbed motion*. Denoting the *perturbations* by $y_i(t)$, we have

$$y_i(t) = x_i(t) - \varphi_i(t), \qquad i = 1, 2, \ldots, 2n, \tag{5.8}$$

so that, introducing Eqs. (5.8) into (5.4), we have

$$\dot{y}_i(t) = Y_i(y_1, y_2, \ldots, y_{2n}, t), \qquad i = 1, 2, \ldots, 2n, \tag{5.9}$$

where

$Y_i(y_1, y_2, \ldots, y_{2n}, t)$
$= X_i(y_1 + \varphi_1, y_2 + \varphi_2, \ldots, y_{2n} + \varphi_{2n}, t) - X_i(\varphi_1, \varphi_2, \ldots, \varphi_{2n}, t).$ (5.10)

Equations (5.9) are referred to as the *differential equations of the perturbed motion*, and they are generally more complicated than the original equations (5.4). For example, a situation may arise in which Eqs. (5.4) are autonomous and Eqs. (5.9) are nonautonomous. Equations (5.9) can be written in the vector form

$$\dot{\mathbf{y}}(t) = \mathbf{Y}(\mathbf{y},t), \tag{5.11}$$

and we notice that Eq. (5.11) has the *trivial solution* $\mathbf{y} = \mathbf{0}$ as an equilibrium point because $\mathbf{Y}(\mathbf{0},t) = \mathbf{0}$.

In view of the above discussion, we lose no generality by concerning ourselves with the dynamical system

$$\dot{\mathbf{x}} = \mathbf{X}(\mathbf{x},t), \tag{5.12}$$

and, assuming that the origin is a singular point, we have $\mathbf{X}(\mathbf{0},t) = \mathbf{0}$. The vector $\mathbf{x} = \mathbf{0}$ is called the *null solution* of the system (5.12). The functions $X_i(x_1, x_2, \ldots, x_{2n}, t)$ are assumed to satisfy the conditions of the Cauchy-Lipschitz existence theorem in a connected open set D of the $2n$-vector space $\mathbf{x}$ for all values of $t \geq t_0$, where D contains the origin.

One of the basic questions in the study of dynamics is the stability of an equilibrium point. Specifically, the question is whether a slight perturbation of a dynamical system from an equilibrium state will produce a motion confined to the neighborhood of the equilibrium point or a motion tending to leave that neighborhood. For simplicity, let us consider the case in which the equilibrium point is at the origin and denote the integral curve at a given time $t_0 > 0$ by $\mathbf{x}(t_0) = \mathbf{x}_0$. Then, assuming that the origin is an isolated singularity, introduce the following definitions due to Liapunov:

1. The null solution is *stable in the sense of Liapunov* if for any arbitrary positive ϵ and time t_0 there exists a $\delta = \delta(\epsilon, t_0) > 0$ such that if the inequality

$$\|\mathbf{x}_0\| < \delta \tag{5.13}$$

is satisfied, then the inequality

$$\|\mathbf{x}(t)\| < \epsilon, \quad t_0 \leq t < \infty, \tag{5.14}$$

is implied. If δ is independent of t_0, the stability is said to be *uniform*.

2. The null solution is *asymptotically stable* if it is Liapunov stable and in addition

$$\lim_{t \to \infty} \|\mathbf{x}(t)\| = 0. \tag{5.15}$$

If Eq. (5.15) holds, then a uniformly stable solution is said to be *uniformly asymptotically stable*.

The part of the $\mathbf{x}$-space characterized by the fact that every motion originated in it is asymptotically stable to $\mathbf{x} = \mathbf{0}$ is called the *domain of attraction* of the point $\mathbf{x} = \mathbf{0}$. If the domain of attraction of the point $\mathbf{x} = \mathbf{0}$ includes the entire $\mathbf{x}$-space, the null solution is said to be *globally stable*.

3. The null solution is said to be *unstable* if for any arbitrarily small δ and any time t_0 such that

$$\|\mathbf{x}_0\| < \delta, \tag{5.16}$$

we have at some other finite time t_1 the situation

$$\|\mathbf{x}(t_1)\| = \epsilon, \quad t_1 > t_0. \tag{5.17}$$

This is equivalent to the statement that a motion initiated inside an open sphere of radius δ and with the center at the origin reaches the boundary of the sphere of radius ϵ in finite time.

When the functions X_i do not depend on time, i.e., when the dynamical system is autonomous, Eq. (5.12) reduces to

$$\dot{x} = X(x). \tag{5.18}$$

In such cases the time t_0 may be shifted to zero for convenience, from which we conclude that for autonomous systems stability is always uniform.

The preceding definitions of stability are unduly restrictive, and they preclude cases which should be considered stable, such as periodic phenomena. To examine such motions, a new concept, known as *orbital stability* or *stability in the sense of Poincaré*, must be introduced. For simplicity, let us restrict ourselves to an autonomous system and consider a solution of Eq. (5.18) in the form of the closed trajectory C in the phase space. Using the notation of set theory (see Appendix B), we denote a point on C by x' and an arbitrary point near C by x and write the distance from x to the closed trajectory C in the form

$$d(x,C) = \inf \{d(x,x'): x' \in C\}. \tag{5.19}$$

Expression (5.19) defines the distance from a point x to a set of points belonging to the trajectory C as the infimum of all distances from x to any point x' on C. With this in mind, we can introduce the following definitions:

4. The closed trajectory C is said to be *orbitally stable* if, given any $\epsilon > 0$, there exists a $\delta > 0$ such that every solution $x(t)$ passing through the point $x = x_0$ of the phase space at time $t = t_0$, where x_0 is such that

$$d(x_0,C) < \delta, \tag{5.20}$$

has the property

$$d(x,C) < \epsilon \tag{5.21}$$

for all $t \geq t_0$.

5. The same trajectory C is said to be *asymptotically orbitally stable* if it is orbitally stable and there is an $\epsilon_0 > 0$ such that for every solution $x(t)$ which passes at $t = t_0$ through a point x_0 of the phase space the inequality

$$d(x_0,C) < \epsilon_0 \tag{5.22}$$

implies that

$$\lim_{t \to \infty} d(x(t),C) = 0. \tag{5.23}$$

Physically, a closed trajectory C is orbitally stable if any trajectory Γ passing sufficiently close to C at a certain time remains close to C for ever after. The closed trajectory is asymptotically orbitally stable if the trajectory Γ tends to C as $t \to \infty$.

On the other hand, another type of stability definition, requiring only the *boundedness* of the solutions for any time $t > t_0$, appears to be too general for practical purposes. Such stability is called *stability in the sense of Lagrange* by some and *in the sense of Laplace* by others.

On occasion it may be possible to approximate the behavior of a nonlinear system in the neighborhood of an equilibrium point by the behavior of the associated linearized system. In this case we speak of *infinitesimal stability* rather than stability.

An alternate formulation of the problems of dynamics, also in terms of $2n$ first-order differential equations of the type (5.4), is obtained by using as an auxiliary set of n coordinates the generalized velocities, $x_{n+k} = \dot{q}_k$ ($k = 1, 2, \ldots, n$), as discussed in Sec. 2.13. This results in a phase space defined by q_k and $\dot{q}_k$, and the discussion of such a phase space parallels that of the phase space defined by q_k and p_k.

A relatively complete geometrical theory of differential equations exists only for second-order autonomous systems of the type (5.18). This theory was developed largely by Poincaré. In the rest of this chapter we confine ourselves to second-order autonomous systems, emphasizing the geometrical aspects of the solution. The dynamical behavior of higher-order systems, autonomous as well as nonautonomous, will be discussed in subsequent chapters.

5.2 MOTION OF SINGLE-DEGREE-OF-FREEDOM AUTONOMOUS SYSTEMS ABOUT EQUILIBRIUM POINTS

Geometric analysis of integral curves provides a useful tool in the study of phenomena associated with autonomous dynamical systems. In particular, it provides considerable insight into the nature of the motion, and such an analysis should prove of great value when the interest lies in the qualitative aspects of the motion rather than in solutions in the form of explicit functions of time. The usefulness of the geometric investigation of integral curves is limited to autonomous systems of low order. Although the concepts can be extended to systems of higher order, many of the concepts and results presented in the balance of this chapter reflect geometrical and topological properties peculiar to the Euclidean two-space E^2 not shared by higher-order spaces.

We shall be concerned with the motion of a single-degree-of-freedom autonomous system defined by the two first-order differential equations

$$\dot{x}_1 = X_1(x_1,x_2), \qquad \dot{x}_2 = X_2(x_1,x_2), \tag{5.24}$$

where X_1 and X_2 are real functions in a certain domain D of the phase plane defined by the coordinates x_1 and x_2 and satisfy Lipschitz conditions on both

x_1 and x_2 (see the definition in Sec. 5.1) in some neighborhood of every point of D.

The differential equation of the trajectories in the phase plane can be obtained immediately by eliminating the explicit time dependence from Eqs. (5.24) and writing

$$\frac{dx_2}{dx_1} = \frac{X_2(x_1,x_2)}{X_1(x_1,x_2)}, \qquad X_1(x_1,x_2) \neq 0. \tag{5.25}$$

Of course, this elimination is possible because the system is autonomous. Equations (5.25) give the tangent to the trajectory curves at any point in the phase plane without reference to time, with the exception of points at which expression (5.25) is indeterminate. Such points are referred to as *critical points*, and they coincide with the singular points of the system. The converse is not necessarily true, however. The difference between critical and singular points can best be demonstrated by considering the case in which X_1 and X_2 contain a common factor which vanishes at some particular point. Such a point is a singular point of system (5.24) but is not a critical point because the common factor cancels and the tangent at that point is not indeterminate. Whereas a singular point represents a dynamical property of the system, namely, an equilibrium point, a critical point represents a geometrical property. Because we are interested primarily in the dynamical properties of the system (and, moreover, later in the section attribute to the term critical a different meaning), we do not insist on that distinction and refer to all such points as singular. By contrast with Eqs. (5.25), the original equations (5.24) contain additional information because they provide the direction in which a representative point moves in the motion space and, obviously, also in the phase plane.

As a result of the Cauchy-Lipschitz theorem, we conclude that *through any regular point of the phase plane there passes at most one integral curve* corresponding to Eqs. (5.24). It follows that two integral curves have no point in common. If Eqs. (5.24) are regarded as representing a flow in the phase plane, then a singular point (α_1,α_2) must be interpreted as a stationary point of the flow, $x_1 = \alpha_1$, $x_2 = \alpha_2$. This indicates that the integral curve passing through a singular point consists only of the point itself. A trajectory passing through an ordinary point cannot approach a singular point in finite time. In fact, there are cases in which the trajectories never approach a singular point. When we recall that half of the variables defining a phase space denote coordinates and the other half reflect velocities, the name equilibrium point for a singular point appears fully justified.

It will prove of interest to examine the motion of the system in the neighborhood of an equilibrium point. The values of the variables at an equilibrium point, $x_1 = \alpha_1$ and $x_2 = \alpha_2$, are obtained by setting $X_1(\alpha_1,\alpha_2) = X_2(\alpha_1,\alpha_2) = 0$. Without any loss of generality, the origin can be made to coincide with an equilibrium point by means of a coordinate transformation representing simple translation. We shall be interested only in isolated singular points, so that the Jacobian $\partial(X_1,X_2)/\partial(x_1,x_2)$ must be different from zero at that point. Moreover, assuming that X_1 and X_2 are real analytic functions in the neighborhood of the origin, it follows that they can be expanded in power series about the origin, so that Eqs. (5.24) can be written in the form

$$\dot{x}_1 = a_{11}x_1 + a_{12}x_2 + \epsilon_1(x_1,x_2), \qquad \dot{x}_2 = a_{21}x_1 + a_{22}x_2 + \epsilon_2(x_1,x_2), \quad (5.26)$$

where $a_{ij} = \partial X_i(0,0)/\partial x_j$ are constant coefficients and the functions ϵ_1 and ϵ_2 have, in the neighborhood of the origin, power series expansions containing terms of at least second order in x_1 and x_2. This implies that $\lim_{r \to 0} \epsilon_i/r = 0$ ($i = 1, 2$), where $r = \sqrt{x_1^2 + x_2^2}$. Equations (5.26) can be written in the matrix form

$$\{\dot{x}\} = [a]\{x\} + \{\epsilon\}, \qquad (5.27)$$

and to ensure that the origin is an isolated singularity we must have $|a| \neq 0$, or the determinant of the coefficients must be different from zero. This implies that the matrix $[a]$ is nonsingular.

Ignoring the higher-order terms in Eq. (5.27), we obtain

$$\{\dot{x}\} = [a]\{x\}, \qquad (5.28)$$

where the equations represented by (5.28) are known as the *first-approximation* or *linearized equations*, in contrast with the *complete equations* (5.26). It is reasonable to expect that, under certain circumstances, the linearized equations can provide some guidance concerning the quality of motion of the complete system in the neighborhood of the equilibrium point. This is not always the case, however, as pointed out at the end of this section.

Next we show that the nature of the motion in the neighborhood of the equilibrium point depends on the eigenvalues of the matrix $[a]$. These eigenvalues coincide with the roots λ_1 and λ_2 of the characteristic equation, where the characteristic equation can be written in the determinant form

$$|a_{ij} - \lambda \delta_{ij}| = 0. \qquad (5.29)$$

This can be demonstrated to be the case by noticing that the solution of Eq. (5.28) has the form $\{x\} = e^{(t-t_0)[a]}\{x_0\}$, where the column matrix $\{x_0\}$ has constant elements and is equal to the initial conditions at $t = t_0$. (In the sequel we shall assume, without loss of generality, that $t_0 = 0$.) Moreover, the matrix $[a]$ is similar to a matrix whose main diagonal elements consist of the eigenvalues λ_1 and λ_2.

To justify the preceding statement, we express Eq. (5.28) in a different set of coordinates such that the square matrix of coefficients corresponding to

Behavior of Dynamical Systems. Geometric Theory

[a] assumes a simpler form, preferably diagonal. To this end, we introduce the linear transformation

$$\{x\} = [b]\{u\}, \tag{5.30}$$

where [b] is a nonsingular matrix of constant coefficients. A transformation of this type does not alter the topological features of system (5.28). Indeed it will be noticed that the origin remains the same, straight lines map into straight lines, and parallel lines map into parallel lines, with the spacing between the parallel lines remaining proportional. Equation (5.30) represents a special case of an *affine transformation*.

Substituting Eq. (5.30) into (5.28) and premultiplying the result by $[b]^{-1}$, we obtain

$$\{\dot{u}\} = [c]\{u\}, \tag{5.31}$$

where
$$[c] = [b]^{-1}[a][b]. \tag{5.32}$$

Because Eq. (5.32) represents a similarity transformation, matrices [a] and [c] have the same eigenvalues (see Sec. 6.2). Moreover, the transformation matrix [b] can be chosen so that the matrix [c] assumes a simple form, known as the *Jordan canonical form*. First we shall review the possible Jordan forms corresponding to a second-order system[1] and then discuss the implication of these forms with respect to the nature of motion in the neighborhood of the equilibrium point.

There are essentially three cases.

1. When the characteristic roots λ_1 and λ_2 are real and distinct, the Jordan form is a diagonal one

$$[c] = \begin{bmatrix} \lambda_1 & 0 \\ 0 & \lambda_2 \end{bmatrix}. \tag{5.33}$$

The type of motion obtained depends on the signs of λ_1 and λ_2.

2. When the eigenvalues are real and equal, two Jordan forms are possible, namely,

$$[c] = \begin{bmatrix} \lambda_1 & 0 \\ 0 & \lambda_1 \end{bmatrix} \tag{5.34}$$

and
$$[c] = \begin{bmatrix} \lambda_1 & 1 \\ 0 & \lambda_1 \end{bmatrix}, \tag{5.35}$$

and the motion depends on the particular Jordan form obtained.

---

[1] A more extensive coverage of the various types of Jordan forms, applicable to higher-order systems, is given in Sec. 6.2.

3. If the roots are complex, they must be complex conjugates, $\lambda_2 = \lambda_1^*$, so that

$$[c] = \begin{bmatrix} \lambda_1 & 0 \\ 0 & \lambda_1^* \end{bmatrix}. \tag{5.36}$$

In this case the motion is controlled by the real part of the roots.

Next we turn our attention to the nature of motion associated with each case, for which purpose we introduce the various types of Jordan forms, Eqs. (5.33) to (5.36), in Eq. (5.31) and examine the corresponding solutions.

Case 1. Introducing Eq. (5.33) into (5.31), we obtain

$$\dot{u}_1 = \lambda_1 u_1, \qquad \dot{u}_2 = \lambda_2 u_2, \tag{5.37}$$

which have the solutions

$$u_1 = u_{10} e^{\lambda_1 t}, \qquad u_2 = u_{20} e^{\lambda_2 t}. \tag{5.38}$$

There are two possibilities. If both roots are of the same sign, we obtain a *node*. Figure 5.1a shows the phase portrait for $\lambda_2 < \lambda_1 < 0$. Because all half-trajectories approach the origin O, the node is said to be *stable*. The tangent to the trajectories has the equation

$$\frac{du_2}{du_1} = \frac{\dot{u}_2}{\dot{u}_1} = \frac{u_{20} \lambda_2}{u_{10} \lambda_1} e^{(\lambda_2 - \lambda_1)t}, \tag{5.39}$$

and we notice that all trajectories tend to have a horizontal slope as $t \to \infty$, provided $u_{10} \neq 0$. When $u_{10} = 0$, the half-trajectories coincide with axis u_2. If $\lambda_2 > \lambda_1 > 0$, the arrowheads change directions and the node is said to be *unstable*. In this case we can regard the trajectories as approaching the origin as $t \to -\infty$.

When the roots are real but of different sign, say $\lambda_2 < 0 < \lambda_1$, the singularity is a saddle point and the trajectories are as shown in Fig. 5.1b. Because the trajectories do not tend to the origin except for the case corresponding to $u_{10} = 0$ or $u_{20} = 0$, the saddle point is an *unstable* singularity.

Case 2. A substitution of the form (5.34) into Eq. (5.31) yields

$$\dot{u}_1 = \lambda_1 u_1, \qquad \dot{u}_2 = \lambda_1 u_2, \tag{5.40}$$

leading to

$$u_1 = u_{10} e^{\lambda_1 t}, \qquad u_2 = u_{20} e^{\lambda_1 t}, \tag{5.41}$$

so that the trajectories are the straight lines

$$u_2 = \frac{u_{20}}{u_{10}} u_1 \tag{5.42}$$

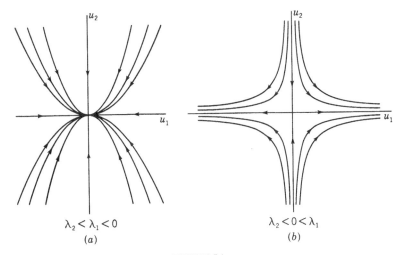

FIGURE 5.1

and they all pass through the origin. Figure 5.2a shows the case $\lambda_1 < 0$ for which a *stable node* occurs.

In the event the second Jordan form, Eq. (5.35), is obtained, Eq. (5.31) leads to

$$\dot{u}_1 = \lambda_1 u_1 + u_2, \qquad \dot{u}_2 = \lambda_1 u_2, \tag{5.43}$$

which have the solutions

$$u_1 = (u_{10} + u_{20} t)e^{\lambda_1 t}, \qquad u_2 = u_{20} e^{\lambda_1 t}, \tag{5.44}$$

and if $\lambda_1 < 0$, the singularity is a stable node of the form shown in Fig. 5.2b. A node of this type is said to be *degenerate*. The slope of the trajectories has the expression

$$\frac{du_2}{du_1} = \frac{\lambda_1 u_{20}}{\lambda_1 u_{10} + (\lambda_1 t + 1)u_{20}}, \tag{5.45}$$

and it is not difficult to see that any half-trajectory initiated in the upper half-plane, $u_{20} > 0$, will approach the origin from the right with zero slope as $t \to \infty$.

Case 3. When the eigenvalues are complex conjugates, $\lambda_1 = \alpha + i\beta$ and $\lambda_2 = \lambda_1^* = \alpha - i\beta$, Eqs. (5.31) and (5.36) yield

$$\dot{u}_1 = (\alpha + i\beta)u_1, \qquad \dot{u}_2 = (\alpha - i\beta)u_2. \tag{5.46}$$

Introducing the notation

$$v_1 = (1 + i)u_1 + (1 - i)u_2, \qquad v_2 = (1 - i)u_1 + (1 + i)u_2, \tag{5.47}$$

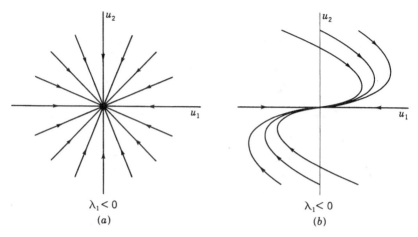

FIGURE 5.2

Eqs. (5.46) can be shown to lead to

$$\dot{v}_1 = \alpha v_1 - \beta v_2, \qquad \dot{v}_2 = \beta v_1 + \alpha v_2. \tag{5.48}$$

Next define a complex vector by

$$v = v_1 + iv_2, \tag{5.49}$$

so that, multiplying the second of Eqs. (5.48) by i and adding the result to the first, we obtain

$$\dot{v} = (\alpha + i\beta)v, \tag{5.50}$$

which has the solution

$$v = (v_0 e^{\alpha t})e^{i\beta t}. \tag{5.51}$$

Equation (5.51) represents a logarithmic spiral, and the singular point is known as a *spiral point* or *focus*. When $\alpha < 0$, the focus is *stable*, and when $\alpha > 0$, it is *unstable*. Hence, the real part of the eigenvalues controls the system stability. On the other hand, the value of β merely gives the direction of the trajectories in the phase plane. A typical trajectory is shown in Fig. 5.3a for $\alpha < 0$ and $\beta > 0$.

When $\alpha = 0$, the magnitude of the radius vector becomes constant and the trajectories reduce to circles with the center at the origin (see Fig. 5.3b). In this case the representative point performs periodic motion about the singular point, which is called a *center* or *vortex point*. The center must be regarded as a limiting case between stable and unstable spirals.

It should be mentioned at this point that although affine transformations of the type (5.30) or (5.47) do not alter the quality of the singularity, they may

Behavior of Dynamical Systems. Geometric Theory

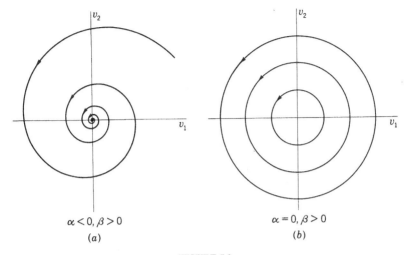

FIGURE 5.3

distort the phase portrait. As an example, the circles of Fig. 5.3*b* become ellipses when expressed in a different coordinate system.

It will prove of interest to relate the coefficients a_{ij} of Eq. (5.28) with the various types of integral curves obtained. To this end, we expand the characteristic equation (5.29) and obtain

$$|a_{ij} - \lambda \delta_{ij}| = \lambda^2 - \lambda(a_{11} + a_{22}) + a_{11}a_{22} - a_{12}a_{21} = 0. \quad (5.52)$$

Next introduce the parameters

$$\begin{aligned} a_{11} + a_{22} &= \text{tr } [a] = p, \\ a_{11}a_{22} - a_{12}a_{21} &= |a| = q, \end{aligned} \quad (5.53)$$

where p is recognized as the trace and q as the determinant of the matrix $[a]$. This enables us to write the characteristic equation in the form

$$\lambda^2 - p\lambda + q = 0, \quad (5.54)$$

from which we obtain the eigenvalues

$$\left.\begin{aligned}\lambda_1 \\ \lambda_2\end{aligned}\right\} = \frac{p}{2} \pm \sqrt{\left(\frac{p}{2}\right)^2 - q}. \quad (5.55)$$

We wish to distinguish the following cases:

1. Case $p^2 > 4q$ points to real distinct roots. If q is positive, both roots are of the same sign, which implies a node. The node is stable (*SN*) when p is negative and unstable (*UN*) when p is positive. If q is

negative, the roots are of different sign and the singularity is a saddle point (*SP*) regardless of the sign of p. The corresponding domains are shown in the parameter plane of Fig. 5.4.

2. Case $p^2 = 4q$ implies that both roots are equal. This singularity must be regarded as a borderline node and is possible only if a_{12} and a_{21} are of opposite signs. It represents the parabola separating the nodes from the foci in Fig. 5.4.

3. Case $p^2 < 4q$ indicates complex roots provided $q > 0$. The singularity is a focus. It is a stable focus (*SF*) when p is negative and an unstable focus (*UF*) when p is positive. When $p = 0$, we obtain a center (*C*), which is the border line between the stable and unstable foci. This corresponds to the q axis in Fig. 5.4.

From Fig. 5.4 we notice that the centers lie on the boundary separating two domains. This underscores the fact that the center represents a mathematical concept more than a physical reality. Nevertheless such mathematical models are commonly assumed, the most frequent example being the simple oscillator. In practice, however, it is more likely that the system possesses a small amount of damping, resulting in a weak focus rather than a center. The domain in the right half-plane bounded by the parabola $p^2 = 4q$ in Fig. 5.4 is characterized by oscillatory motion, stable or unstable. On the other hand, in the entire left half-plane the motion is unstable. The domains of oscillatory motion and the domain of instability of the saddle-point type are separated by domains of nodal points representing aperiodic motion.

FIGURE 5.4

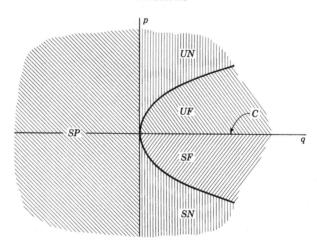

The above results, pertinent to the linear equations (5.28), can be summarized as follows:

1. If λ_1 and λ_2 are real and of the same sign, the singularity is a nodal point. The equilibrium point is asymptotically stable if λ_1 and λ_2 are negative and unstable if they are positive.
2. If λ_1 and λ_2 are real and of opposite signs, the singularity is a saddle point, hence unstable.
3. If λ_1 and λ_2 are purely imaginary, the equilibrium point is a center which is stable.
4. If λ_1 and λ_2 are complex conjugates, the singular point is a focus. The equilibrium is asymptotically stable if the real part of λ_1 and λ_2 is negative and unstable if the real part is positive.

The case in which λ_1 and λ_2 are real and equal leads to nodes, but this case is not very common.

In conclusion, the necessary and sufficient condition for the equilibrium of the linear system to be stable is that both eigenvalues λ_1 and λ_2 be real and negative or that they be complex conjugates with either negative or zero real parts.

The natural question to pose is to what extent the linear approximation provides an accurate classification of the singular points associated with a general nonlinear system. We shall not go into the details of the proof but advance the following theorem: *If the origin is an isolated singularity and the nonlinear terms* ϵ_1 *and* ϵ_2 *are such that* $\lim_{r \to 0} \epsilon_i/r = 0$ *(i = 1, 2), where* $r = \sqrt{x_1^2 + x_2^2}$, *the character of the singularity of the linear first-approximation equations is the same as for the complete equations except when the singularity of the linear equations is a center.* In this case the system is said to possess *critical behavior*, and the singularity of the complete equations may be either a center or a focal point.

Hence when the eigenvalues of the matrix $[a]$ are complex conjugates with zero real part, the linear approximation is inconclusive in determining the nature of the singularity. The theorem has intuitive appeal in view of the statement concerning centers made in the discussion of Fig. 5.4. (For a proof of the theorem see Ref. 7, pp. 174–177.) We shall return to this subject in Sec. 6.5, when higher-order systems are discussed.

Example 5.1

Consider a simple pendulum in a resisting medium where the resisting force is proportional to the velocity. It is not difficult to show that the differential equation of motion is

$$mL^2\ddot{\theta} + cL^2\dot{\theta} + mgL \sin \theta = 0, \quad (a)$$

where m = mass of bob,
L = length of massless rod connecting bob,
c = damping coefficient.

Introducing the notation $c/m = 2\zeta\omega$ and $g/L = \omega^2$, Eq. (a) reduces to

$$\ddot{\theta} + 2\zeta\omega\dot{\theta} + \omega^2 \sin \theta = 0. \quad (b)$$

Letting $\theta = x_1$, $\dot{\theta} = x_2$, Eq. (b) can be replaced by two first-order differential equations

$$\begin{aligned}\dot{x}_1 &= x_2, \\ \dot{x}_2 &= -\omega^2 \sin x_1 - 2\zeta\omega x_2.\end{aligned} \quad (c)$$

There are two equilibrium positions to consider: $x_1 = x_2 = 0$ and $x_1 = \pi$, $x_2 = 0$.

In the first case the equations of the linear approximation assume the matrix form

$$\begin{Bmatrix}\dot{x}_1 \\ \dot{x}_2\end{Bmatrix} = \begin{bmatrix} 0 & 1 \\ -\omega^2 & -2\zeta\omega \end{bmatrix}\begin{Bmatrix}x_1 \\ x_2\end{Bmatrix}, \quad (d)$$

and the corresponding characteristic equation becomes

$$\begin{vmatrix} -\lambda & 1 \\ -\omega^2 & -2\zeta\omega - \lambda \end{vmatrix} = \lambda^2 + 2\zeta\omega\lambda + \omega^2 = 0. \quad (e)$$

The two eigenvalues are

$$\lambda_1 = -\zeta\omega + (\zeta^2 - 1)^{\frac{1}{2}}\omega, \quad \lambda_2 = -\zeta\omega - (\zeta^2 - 1)^{\frac{1}{2}}\omega. \quad (f)$$

First we shall assume that $\zeta > 1$, in which case the roots λ_1 and λ_2 are real. Because they are both negative, the singular point $x_1 = x_2 = 0$ is a stable node and the motion is aperiodic. To obtain the equation of the trajectories, we use Eqs. (5.37) and write

$$\frac{du_2}{du_1} = \frac{\lambda_2 u_2}{\lambda_1 u_1}, \quad (g)$$

which has the solution

$$\frac{u_1}{u_{10}} = \left(\frac{u_2}{u_{20}}\right)^{[-\zeta + (\zeta^2 - 1)^{\frac{1}{2}}]^2}, \quad \zeta > 1, \quad (h)$$

where u_{10} and u_{20} are the coordinates of the point at which the trajectory was initiated. If $\zeta < 1$, Eqs. (5.31) and (5.36) lead to

$$\dot{u}_1 = [-\zeta\omega + i(1 - \zeta^2)^{\frac{1}{2}}\omega]u_1, \quad \dot{u}_2 = [-\zeta\omega - i(1 - \zeta^2)^{\frac{1}{2}}\omega]u_2, \quad (i)$$

and, making the substitution $u_1 = re^{i\theta}$, $u_2 = re^{-i\theta}$, either of Eqs. (i) yields

$$\dot{r} = -\zeta\omega r, \qquad \dot{\theta} = (1 - \zeta^2)^{\frac{1}{2}}\omega. \qquad (j)$$

Assuming that $\theta(0) = 0$, Eqs. (j) lead to the trajectory equation

$$r = r_0 e^{-\zeta\theta/(1-\zeta^2)^{\frac{1}{2}}}, \qquad (k)$$

which represents a logarithmic spiral. For positive ζ the singular point is a stable focus, and the motion is a damped oscillation. Of course, for $\zeta = 0$ the spiral reduces to the circle $r = r_0$, and the singular point becomes a center. If the trajectory is expressed in terms of the coordinates x_1 and x_2 rather than u_1 and u_2, it assumes the form of the ellipse

$$x_1^2 + \frac{x_2^2}{\omega^2} = x_{10}^2 + \frac{x_{20}^2}{\omega^2}. \qquad (l)$$

Hence for any value $\zeta \geq 0$ the singular point is stable.

The pendulum exhibits an entirely different behavior in the neighborhood of the second equilibrium point, $x_1 = \pi$, $x_2 = 0$. The corresponding first-order differential equations have the matrix form

$$\begin{Bmatrix} \dot{x}_1 \\ \dot{x}_2 \end{Bmatrix} = \begin{bmatrix} 0 & 1 \\ \omega^2 & -2\zeta\omega \end{bmatrix} \begin{Bmatrix} x_1 \\ x_2 \end{Bmatrix}, \qquad (m)$$

and the eigenvalues for this case are

$$\lambda_1 = -\zeta\omega + (1 + \zeta^2)^{\frac{1}{2}}\omega, \qquad \lambda_2 = -\zeta\omega - (1 + \zeta^2)^{\frac{1}{2}}\omega. \qquad (n)$$

From Eqs. (n) we conclude that, independently of the value of ζ, the eigenvalues are real and of opposite sign. It follows that the singular point is a saddle point, hence unstable. The trajectory equation can be shown to be

$$\frac{u_1}{u_{10}}\left(\frac{u_2}{u_{20}}\right)^{[\zeta-(1+\zeta^2)^{\frac{1}{2}}]^2} = 1, \qquad (o)$$

representing hyperbolas.

5.3 CONSERVATIVE SYSTEMS. MOTION IN THE LARGE

The discussion of single-degree-of-freedom autonomous systems in Sec. 5.2 concentrated on the motion in the neighborhood of singular points. It will prove of interest to investigate now the behavior of the integral curves in the entire phase plane. We shall confine ourselves to the simple case of an autonomous conservative system described by

$$\ddot{x} = f(x), \qquad (5.56)$$

where $f(x)$ is the force per unit mass, which is in general a nonlinear analytic function. If we recall that $\ddot{x} = \dot{x}\, d\dot{x}/dx$, Eq. (5.56) becomes

$$\dot{x}\, d\dot{x} = f(x)\, dx, \qquad (5.57)$$

which yields the energy integral

$$\tfrac{1}{2}\dot{x}^2 + V(x) = E = \text{const}, \qquad (5.58)$$

where $V(x) = \int_x^0 f(x)\, dx =$ potential energy per unit mass,

$\tfrac{1}{2}\dot{x}^2 =$ kinetic energy per unit mass,

$E =$ total energy per unit mass.

Notice that Eq. (5.58) formed the basis of the energy diagrams of Sec. 1.6. We can complement the analysis of that section by means of phase portraits.

Let us introduce the notation $x = x_1$, $\dot{x} = x_2$, so that Eq. (5.58) becomes

$$\tfrac{1}{2}x_2^2 + V(x_1) = E, \qquad (5.59)$$

which represents a family of integral curves in the phase plane, where the parameter of the family is E. These integral curves are symmetric with respect to the x_1 axis. We can envision an axis E normal to the phase plane and interpret the integral curves geometrically as *level curves* resulting from the intersection of a surface $E = E(x_1, x_2)$, as given by Eq. (5.59), and the planes $E =$ const. This interpretation of the integral curves as level curves helps us rule out singularities such as nodal points and focal points. Because Eq. (5.59) implies that the value of the integral is the same for any point on a given level curve, and, moreover, since all integral curves corresponding to nodal and focal points tend to these singular points asymptotically as $t \to \infty$ (or $t \to -\infty$), we must conclude that $E(x_1, x_2)$ has the same value at every regular point surrounding the singular point. This, however, contradicts the definition of level curves, so that the only possible elementary singular points are the center and the saddle point, which precludes asymptotic stability.

Before examining the *motion in the large*, i.e., the motion at points other than the immediate neighborhood of the singular points, let us investigate the effect of nonlinear terms on the trajectories in the vicinity of singularities. Consider a singular point with coordinates $x_1 = \alpha_1$, $x_2 = \alpha_2 = 0$, introduce the transformation $x_1 = \alpha_1 + y_1$, $x_2 = y_2$, and expand the function $f(x_1)$ in the Taylor's series

$$f(x_1) = \sum_{n=1}^{\infty} \frac{a_n}{n!} y_1^n, \qquad a_n = \left.\frac{\partial^n f}{\partial x_1^n}\right|_{x_1 = \alpha_1}, \qquad (5.60)$$

from which it follows that

$$V(x_1) = V(\alpha_1) - \sum_{n=1}^{\infty} \frac{a_n}{(n+1)!} y_1^{n+1}. \qquad (5.61)$$

Behavior of Dynamical Systems. Geometric Theory

But $V(\alpha_1) = E(\alpha_1,0)$ so that Eq. (5.59) can be rewritten as

$$\tfrac{1}{2} y_2^2 - \sum_{n=1}^{\infty} \frac{a_n}{(n+1)!} y_1^{n+1} = E(x_1,x_2) - E(\alpha_1,0) = \Delta E, \quad (5.62)$$

where ΔE represents the energy difference between a given level curve and the equilibrium point.

If only the linear term is retained in expansion (5.60), and if $V(\alpha_1)$ represents the minimum value of $V(x_1)$, then $a_1 < 0$, in which case Eq. (5.62) yields ellipses as integral curves and the equilibrium point $x_1 = \alpha_1$, $x_2 = 0$ is a center. If, on the other hand, $V(x_1)$ has a maximum at the point $x_1 = \alpha_1$, then $a_1 > 0$, from which it follows that the level curves are hyperbolas and the singular point is a saddle point. The asymptotes of the hyperbolas are the straight lines $y_2 = \pm \sqrt{a_1} y_1$.

In the event that $a_n = 0$ $(n = 1, 2, \ldots, k-1)$ and $a_k < 0$, where k is odd, the function $V(x_1)$ has a minimum at the singular point and it makes a contact of kth order with the horizontal line $E = V(\alpha_1)$. The level curves have the equation

$$\tfrac{1}{2} y_2^2 + \frac{|a_k|}{(k+1)!} y_1^{k+1} = \Delta E, \quad (5.63)$$

which represents closed curves but not ellipses. The corresponding motion is periodic but not harmonic, and the singular point is obviously a center (see Fig. 5.5). Moreover, in contrast with linear systems, for which the period is

FIGURE 5.5

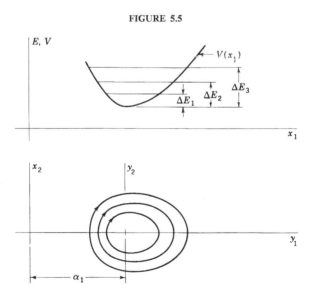

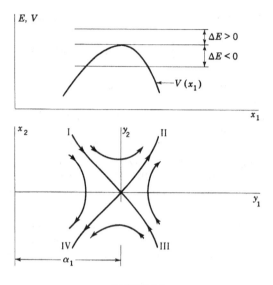

FIGURE 5.6

constant, the period of nonlinear systems is not constant but depends on the value of ΔE. If $a_k > 0$, where k is odd, however, $V(x_1)$ has a maximum at $x_1 = \alpha_1$ and the equation of the trajectories assumes the form

$$\tfrac{1}{2} y_2^2 - \frac{a_k}{(k+1)!} y_1^{k+1} = \Delta E, \tag{5.64}$$

which yields a family of level curves similar in shape to branches of hyperbolas but having curvilinear asymptotes, as shown in Fig. 5.6. The singular point is a saddle point, which is unstable. For $\Delta E = 0$ we obtain the curvilinear asymptotes I, II, III, and IV. The level curves between I and II, as well as between III and IV, correspond to $\Delta E > 0$, whereas the level curves between I and IV and between II and III correspond to $\Delta E < 0$. The meaning of these domains will become evident shortly, when motion in the large is discussed.

When $V(x_1)$ has a stationary value at $x_1 = \alpha_1$ which is neither a maximum nor a minimum, the singular point can be regarded as the coalescence of a center and a saddle point. This corresponds to the case in which the potential energy has an inflection point with zero slope at $x_1 = \alpha_1$. This case is obtained when the first coefficient different from zero in expansion (5.60) is a_k, where k is even. There are two trajectories tending to the singular point, one as $t \to \infty$ and the other as $t \to -\infty$, as shown in Fig. 5.7. The singular point is said to be *nonelementary* and must be regarded as unstable.

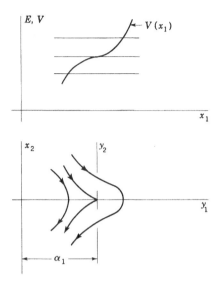

FIGURE 5.7

This simple example, Eq. (5.56), illustrates a theorem due to Lagrange which can be stated as follows: *An isolated equilibrium position corresponding to a minimum value of the potential energy is stable.* This theorem is also associated with the name of Dirichlet. The converse theorem, attributed to Liapunov, reads: *If the potential energy corresponding to an equilibrium point is not a minimum, the equilibrium is unstable.* We shall return to these theorems in Sec. 6.9 when multi-degree-of-freedom canonical systems are discussed.

There is no difficulty in obtaining the complete phase portrait of the conservative system (5.56) for a given analytic function $V(x_1)$. Using Eq. (5.59), we write simply

$$x_2 = \pm \sqrt{2[E - V(x_1)]}, \tag{5.65}$$

which enables us to plot level curves corresponding to a series of values for E. In this manner the entire phase plane is divided into domains possessing different motion characteristics. The trajectories bounding these domains are called *separatrices*. As an illustration, we consider the ball on a frictionless track similar to that in Sec. 1.6. In this case the potential energy is

$$V(x_1) = mgh, \tag{5.66}$$

where h is the height above a reference line, as shown in Fig. 5.8a. Figure

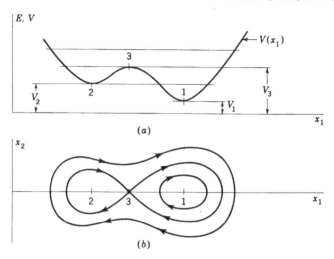

FIGURE 5.8

5.8b shows the phase portrait in the form of level curves. It will be noticed that no motion is possible for $E < V_1$. At $E = V_1$ there is a center, and as the energy E is increased, periodic motion takes place about center 1. For $V_2 < E < V_3$ there can be periodic motion about either center 1 or center 2. The energy level $E = V_3$ corresponds to a separatrix associated with the saddle point 3. The motion in the neighborhood of point 3 is unstable. The motion in the large, however, is periodic with the trajectories surrounding both centers and the saddle point. It is a characteristic of conservative systems that the closed paths enclose an odd number of singularities, with the number of centers exceeding the number of saddle points by 1 (see Prob. 5.8).

It may prove of interest to discuss the motion of conservative systems in terms of the Jordan canonical form. This implies that in the neighborhood of a singularity it is possible to approximate the restoring force by the expression $f_1 \approx a_1 y_1$. Using the developments of Sec. 5.2, we conclude that there are two cases possible: (1) for $a_1 > 0$ the Jordan form is according to Eq. (5.33), in which $\lambda_1 = \sqrt{a_1}$ and $\lambda_2 = -\sqrt{a_1}$, and (2) for $a_1 < 0$ the Jordan form is of the type (5.36), in which $\lambda_1 = i\sqrt{|a_1|}$ and $\lambda_1^* = -i\sqrt{|a_1|}$. In terms of the characteristic equation (5.54), we have $p = 0$ and $q = -a_1$, so that the roots lie on the q axis of Fig. 5.4. Of course, for $a_1 > 0$ the singular points are saddle points, and for $a_1 < 0$ they are centers. In both cases the domains reduce to straight lines, underscoring the degree of idealization of conservative systems.

5.4 THE INDEX OF POINCARÉ

The index of Poincaré provides a characteristic of a singular point which is easy to calculate and proves useful in establishing the existence, or rather lack of existence, of *closed trajectories*. Because the term *closed* has a special topological connotation (see Appendix B), we shall use this term primarily to denote topological closure and refer to closed trajectories as *cyclic trajectories*. Such trajectories, if they exist, are an indication of periodic motion.

Let us consider again the second-order autonomous system

$$\dot{x}_1 = X_1(x_1, x_2), \qquad \dot{x}_2 = X_2(x_1, x_2), \qquad (5.67)$$

where X_1 and X_2 are real continuous functions of class C_1 in a certain domain D of the Euclidean plane defined by the coordinates x_1 and x_2, and denote by $\mathbf{X}(P)$ a vector function whose initial point is $P(x_1, x_2)$ and whose components are X_1 and X_2. The totality of such vectors constitutes a vector field defined by Eqs. (5.67). The vector $\mathbf{X}(P)$ is tangent to the trajectories at any point P and it vanishes if and only if P is a singular point.

Next we consider a *Jordan curve*, namely, a simple closed curve satisfying the conditions of the *Jordan curve theorem* (see Appendix B). Letting the point P be on a Jordan curve J (see Fig. 5.9), we can examine the behavior of the vector $\mathbf{X}(P)$ as the point P traverses the curve J in the counterclockwise sense. If the point P traverses J once, the vector $\mathbf{X}(P)$ undergoes an angle change $2\pi j$ with respect to a fixed direction in the $x_1 x_2$ plane. The integer j, which may be positive, negative, or zero, is called the *index of the curve J for the vector field* $\mathbf{X}(P)$, and is often referred to simply as the *index of Poincaré*.

It is not difficult to see that the *index j does not change if the curve J is altered without crossing singular points*. Since the index takes only discrete values, if the value of the index is to vary continuously with a continuous deformation of J, it must not vary at all. This leads to the theorem: *The index of a curve surrounding no singularities is zero*, because we can deform J continuously until it becomes a point.

FIGURE 5.9

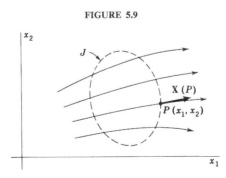

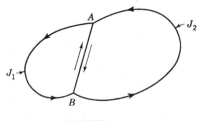

FIGURE 5.10

If J_1 and J_2 represent two Jordan curves with a common arc AB, as shown in Fig. 5.10, and J_3 is the Jordan curve obtained by deleting the arc AB, then

$$j_3 = j_1 + j_2. \tag{5.68}$$

This enables us to state the theorem: *The index of a Jordan curve surrounding a finite number r of isolated singularities S_i is*

$$j = \sum_{i=1}^{r} j_i. \tag{5.69}$$

This theorem becomes evident if we consider Fig. 5.11. The proof is similar to that for the residue theorem in the theory of complex variables.

Next we state the following theorem: *If J is a Jordan curve with a continuously turning tangent vector v in the $x_1 x_2$ plane, and if v does not vanish at any point on J, then its index is* 1. (For a proof of this theorem see Ref. 2, p. 399.)

The preceding theorem leads to the following consequence: *The index of a cyclic trajectory is* 1. Moreover, we have two corollaries: (1) *a cyclic trajectory encloses at least one singular point*, and (2) *if J is a cyclic trajectory and the singular points of the system are isolated, then J encloses a finite number of singularities with the sum of the indices equal to* 1.

FIGURE 5.11

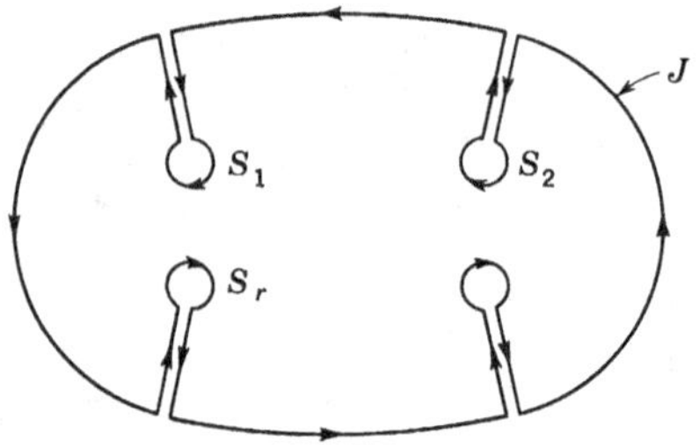

The mathematical definition of the index is

$$j = \frac{1}{2\pi}\oint_J d\left(\tan^{-1}\frac{X_2}{X_1}\right) = \frac{1}{2\pi}\oint_J \frac{X_1\,dX_2 - X_2\,dX_1}{X_1^2 + X_2^2}. \tag{5.70}$$

Equation (5.70) can be used to evaluate the index of curves surrounding singular points of the elementary type. We shall calculate such indices for the linear system (5.28), in which case Eq. (5.70) becomes

$$j = \frac{1}{2\pi}\oint_J \frac{(a_{11}x_1 + a_{12}x_2)d(a_{21}x_1 + a_{22}x_2) - (a_{21}x_1 + a_{22}x_2)d(a_{11}x_1 + a_{12}x_2)}{(a_{11}x_1 + a_{12}x_2)^2 + (a_{21}x_1 + a_{22}x_2)^2}. \tag{5.71}$$

Choosing the Jordan curve as the ellipse

$$(a_{11}x_1 + a_{12}x_2)^2 + (a_{21}x_1 + a_{22}x_2)^2 = 1, \tag{5.72}$$

it follows that

$$j = \frac{q}{2\pi}\oint_J (x_1\,dx_2 - x_2\,dx_1) = \frac{q}{\pi}A_e, \tag{5.73}$$

where q is the determinant of the coefficients, namely, the second of Eqs. (5.53), and A_e is the area of the ellipse. The transformation

$$\xi_1 = a_{11}x_1 + a_{12}x_2, \quad \xi_2 = a_{21}x_1 + a_{22}x_2 \tag{5.74}$$

maps the ellipse (5.72) into the circle $\xi_1^2 + \xi_2^2 = 1$. But the area of the circle is related to the area of the ellipse by the Jacobian of the transformation

$$A_c = \left|J\!\left(\frac{\xi_1, \xi_2}{x_1, x_2}\right)\right| A_e = |q|A_e, \tag{5.75}$$

where the absolute value of the determinant must be taken, to account for the fact that areas are, by definition, positive. Moreover, the area of the circle is π, so that, introducing Eq. (5.75) into (5.73), we obtain

$$j = \frac{q}{|q|}. \tag{5.76}$$

From Fig. 5.4 we conclude that *the index of a Jordan curve surrounding a singularity is equal to* 1 *if the singularity is a nodal point, a focal point, or a center and is equal to* -1 *if the singularity is a saddle point.*

If the system is described by the nonlinear equations (5.27), the question can be asked whether the above results, obtained for the linear approximation (5.28), are valid for the complete nonlinear system. The answer is again provided by the theorem of Sec. 5.2, which states the conditions under which the linear approximation can be used to reach conclusions concerning the characteristics of singular points associated with the complete nonlinear system.

5.5 LIMIT CYCLES OF POINCARÉ

From the results of Sec. 5.2 we could reach the conclusion that cyclic trajectories in the phase plane are obtained only when the singular point is a center. Moreover, from Sec. 5.3 we could receive the impression that this phenomenon occurs only in conservative systems, with the cyclic trajectories reflecting periodic motion. Yet conservative systems are more of a mathematical concept than a common physical occurrence, and, as pointed out in Sec. 5.2, centers must be regarded as borderline cases separating weakly stable focal points from weakly unstable ones (see Fig. 5.4). Hence we are tempted to conclude that cyclic trajectories are not physically possible. This turns out not to be the case, however, as cyclic trajectories reflecting periodic motion are indeed a physical possibility, and, furthermore, they are a phenomenon associated with nonlinear nonconservative systems rather than conservative ones.

Let us consider the set of differential equations

$$\dot{x}_1 = X_1(x_1, x_2), \qquad \dot{x}_2 = X_2(x_1, x_2), \qquad (5.77)$$

where X_1 and X_2 are real continuous functions defined over the domain D of the real Euclidean plane E^2 defined by the coordinates x_1 and x_2. Under certain circumstances system (5.77) admits periodic solutions which can be represented in the phase plane as cyclic trajectories, known as *limit cycles of Poincaré* or simply *limit cycles*.

The limit cycles, which are also called *cyclic characteristics*, are isolated closed trajectories representing stationary motion, in contrast with singular points representing equilibrium states. If such a periodic solution exists, then every trajectory Γ in the neighborhood of the limit cycle C tends to C asymptotically as $t \to \infty$, or as $t \to -\infty$, so that the limit cycle is said to be *stable*, or *unstable*, accordingly. To illustrate the concept, consider the system

$$\dot{x}_1 = x_2 + x_1(1 - x_1^2 - x_2^2), \qquad \dot{x}_2 = -x_1 + x_2(1 - x_1^2 - x_2^2), \qquad (5.78)$$

introduce the polar coordinates $x_1 = r \cos \theta$, $x_2 = r \sin \theta$, and transform Eqs. (5.78) into

$$\dot{r} = r(1 - r^2), \qquad \dot{\theta} = -1, \qquad (5.79)$$

which have the solution

$$r = (1 + Ae^{-2t})^{-\frac{1}{2}}, \qquad \theta = -t, \qquad (5.80)$$

where A is a constant of integration depending on the initial condition. For simplicity, θ and t are measured from zero. Equations (5.80) represent a family of curves tending to the circle $r = 1$ as $t \to \infty$. When $A > 0$, the curves are spirals approaching the circle $r = 1$ from inside as $t \to \infty$ and the

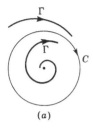 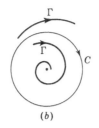 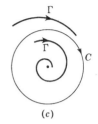

(a) (b) (c)

FIGURE 5.12

origin as $t \to -\infty$. On the other hand, when $A < 0$, the curves approach the circle $r = 1$ from outside as $t \to \infty$ (see Fig. 5.12a). Hence, in this case, the origin is an unstable focal point, and the limit cycle C is *stable*. Figure 5.12b and c shows other possibilities. The first illustrates an *unstable* limit cycle with a stable focal point, and the second shows a *semistable* limit cycle enclosing a stable focal point. A semistable limit cycle is unstable for all practical purposes.

The remarkable feature of the systems exhibiting limit cycles is that the amplitude of the stationary motion does not depend on the initial conditions but on the system parameters, in contrast with the conservative systems, for which the amplitude of the closed trajectories surrounding centers does depend on initial conditions. As an illustration we notice that the system (5.78) achieves ultimately stationary motion on the limit cycle $r = 1$ regardless of the initial condition. Stable limit cycles generally enclose a singular point representing a position of unstable equilibrium, so that oscillations build up and sustain themselves in the absence of external forces. For this reason such oscillations are called *self-excited* or *self-sustained oscillations*. This type of behavior is encountered in the case of an oscillator possessing nonlinear amplitude-dependent damping forces, which tend to increase the amplitude of oscillations for small motions and decrease the amplitude when the motions become large. A state of stationary motion is achieved in which the system gains energy during part of the cycle and dissipates energy during the remaining part, so that at the end of each cycle the net energy exchange is zero.

Next we turn our attention to ways of establishing the existence of limit cycles. We shall discuss the classical Poincaré-Bendixson theorem, the exposition of which follows closely that given by Coddington and Levinson (Ref. 2, secs. 16.1–16.3). For the topological concepts used see Appendix B.

If a solution $\varphi_1(t) = x_1(t)$, $\varphi_2(t) = x_2(t)$ of Eqs. (5.77) exists for all $t_0 \leq t < \infty$, or $-\infty < t \leq t_0$, or $-\infty < t < \infty$, then the solution is said to represent a positive half-trajectory C^+, or a negative half-trajectory C^-, or a full trajectory C, respectively. Hence, C^+ is the set of all points $P(t)$

of the domain D with coordinates $\varphi_1(t)$ and $\varphi_2(t)$, where $t_0 \leq t < \infty$. A point Q in the Euclidean plane E^2 is said to be a *limit point* of C^+ if there exists a sequence of real numbers t_n ($n = 1, 2, \ldots$) such that $t_n \to \infty$ and $P(t_n) \to Q$ as $n \to \infty$. The set of limit points of C^+ in E^2 is referred to as a *limiting set* for the half-trajectory in question and denoted by $L(C^+)$. A similar definition exists for $L(C^-)$. The full trajectory C can be regarded as the sum of C^+ and C^-. Denoting the limiting sets $L(C^+)$ and $L(C^-)$ by $L^+(C)$ and $L^-(C)$, respectively, the limiting set of the full trajectory can be written as the union of the two sets, namely, $L(C) = L^+(C) \cup L^-(C)$. For example, the origin in Figs. 5.1a, 5.2a and b, and 5.3a is the limiting set $L^+(C)$ of all positive half-trajectories C^+. The origin in Fig. 5.1b is the limiting set $L^+(C)$ of two positive half-trajectories C^+ and the limiting set $L^-(C)$ of two negative half-trajectories C^-, whereas all other trajectories have empty limiting sets. Each closed trajectory in Fig. 5.3b is the limiting set for the trajectory in question.

One of the few general theorems asserting the existence of periodic solutions of Eqs. (5.77) is the Poincaré-Bendixson theorem. It follows from the investigation of limiting sets corresponding to a half-trajectory of system (5.77), which trajectory remains inside a compact, namely, a bounded and closed subset R of D. In the sequel we shall be concerned primarily with just such a positive half-trajectory and denote it simply by C unless special emphasis is necessary.

As background to the Poincaré-Bendixson theorem, several lemmas are needed. Two of these lemmas can be stated as follows:

1. *The limiting set $L(C)$ of C is nonempty, closed, and connected.*

2. *If $L(C)$ contains a regular point Q, then the trajectory Γ through Q is a full trajectory and Γ is a subset of $L(C)$, $\Gamma \subseteq L(C)$, which is the same as saying that Γ lies entirely in $L(C)$.*

(Proofs of these lemmas can be found in Ref. 2, pp. 390–391.)

Several other lemmas are peculiar to the plane and are based on the Jordan curve theorem (see Appendix B). Before proceeding with the discussion, it will prove convenient to introduce a new concept referred to as a *transversal* by some and as a *segment without contact* by others. Let $\mathbf{X}$ be the vector whose components in D are the real continuous functions $X_1(x_1,x_2)$ and $X_2(x_1,x_2)$. Then a transversal with respect to $\mathbf{X}$ is defined as a finite closed segment l of a straight line in the plane E^2 such that every point of l is a regular point of $\mathbf{X}$ and at every point of l the direction of l is different from that of $\mathbf{X}$. Hence a transversal passes through no singularity of $\mathbf{X}$ and is tangent to no trajectory of system (5.77). One of the important properties of a transversal is that every trajectory which meets a transversal must cross it and all such trajectories cross the transversal in the same direction. This property is a result of the continuity of the vector field **X** in the neighborhood of the intersection points. The concept of transversal enables us to state the following lemma:

3. *If a finite closed arc γ of a trajectory C crosses a transversal l, it does so in a finite number of points, whose order on γ is the same as the order on l. If C is periodic, then it intersects l in only one point.*

To prove this lemma, we consider an arc γ intersecting the transversal l at two successive points $P_1 = P(t_1)$ and $P_2 = P(t_2)$, where $t_2 > t_1$ (see Fig. 5.13a and b). Let us denote by $\gamma_0 = \widehat{P_1 P_2}$ the open arc from P_1 to P_2 on γ and by $l_0 = \overline{P_2 P_1}$ the closed segment from P_2 to P_1 on l, such that $J = \gamma_0 + l_0$ represents a Jordan curve defining the boundary of the region R. The two complementary arcs γ_1 and γ_2 of γ, the first corresponding to $t < t_1$ and the second to $t > t_2$, are such that one is inside and one is outside R. Moreover, the complementary segments l_1 and l_2 of l are also one inside and one outside R, respectively. Figure 5.13a and b illustrates the two possible cases. Considering the case of Fig. 5.13a, we conclude that γ_2 must remain inside R for all $t > t_2$ because by uniqueness it cannot cross γ_0 and by continuity it cannot cross l_0 in the wrong direction. Hence, the next intersection P_3 must be different from P_1 and P_2 and lie inside R, that is, after P_2. A similar argument leads us to the conclusion that the order of the intersections is preserved also in the case of Fig. 5.13b. It also follows that C is periodic if and only if $P_1 = P_2$.

Next let us consider a point P_1 which is a common point of both C and $L(C)$. If P_1 is a singular point, then $C = L(C) = P_1$ and C is a degenerate cyclic trajectory. If P_1 is a regular point, it can be made an interior point of a transversal l. It can be shown that the successive intersections of C with l

FIGURE 5.13

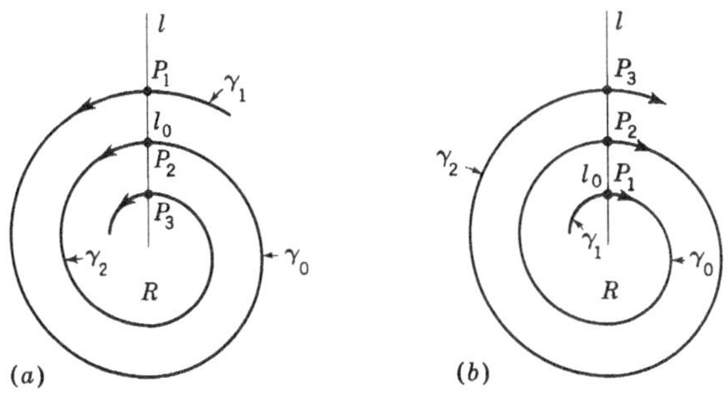

form a monotone sequence of points P_n ($n = 2, 3$ . . .) which tends away from P_1 if $P_2 \neq P_1$. Hence P_1 cannot be a limit point of C, so that P_1 is not in $L(C)$. This contradiction leads to the following lemmas:

4. *If C and $L(C)$ have a regular point P in common, then $C = L(C)$ and C is periodic.*

5. *A transversal l cannot meet $L(C)$ in more than one point.*

Finally we state one more lemma:

6. *If $L(C)$ contains a cyclic trajectory Γ, then $L(C) = \Gamma$.*

The proof of this lemma can be effected by showing that the assumption $L(C) - \Gamma \neq 0$ leads to a contradiction of Lemma 5 (see Ref. 2, p. 394).

We are now in a position to consider the *Poincaré-Bendixson theorem*, which can be stated as follows:

Theorem 5.5.1 *If C is a positive half-trajectory contained in a closed bounded region $R \subset D$, and if $L(C)$ consists only of regular points, then either C [$= L(C)$] is a cyclic trajectory or $L(C)$ is a cyclic trajectory.*

The implication of the second statement of the theorem is that C approaches a cyclic trajectory, namely, $L(C)$, asymptotically either from the inside or from the outside. Following is a proof of the theorem. If C is a cyclic trajectory, then clearly $C = L(C)$. If C is not a cyclic trajectory, because $L(C)$ is not empty and consists of regular points only, by virtue of Lemma 2 there exists a full trajectory $\Gamma \subset L(C) \subset R$. But Γ is contained in R, which implies that the positive half-trajectory Γ^+ has a limit point P_0 and $P_0 \in L(C)$. Let l be a transversal through P_0. Clearly both P_0 and Γ^+ are contained in $L(C)$, so that, by Lemma 5, l can meet $L(C)$ only at P_0. Since P_0 is a limit point of Γ^+, l must meet Γ^+ in some point, which must by necessity be P_0. It follows that Γ^+ and $L(\Gamma^+)$ have the point P_0 in common. By Lemma 4, Γ^+ is periodic, which implies, by Lemma 6, that $\Gamma = L(C)$.

A corollary of the Poincaré-Bendixson theorem can be stated as follows: *If C is a positive half-trajectory contained in a compact set $R \subset D$ in which $\mathbf{X}$ has no singularities, then R contains a cyclic trajectory.*

It remains to characterize the behavior of limiting sets in the presence of singularities. To this end we state the following theorem.

Theorem 5.5.2 *Let the positive half-trajectory C of the system (5.77) be contained in a closed set $R \subset D$, and let the domain D possess only a finite number of singular points. Then the limiting set $L(C)$ of C is either:*

1. A single singular point which C approaches as $t \to \infty$.
2. A cyclic trajectory.
3. Or a finite number of singular points with connecting trajectories (non-periodic) each of which tends to one of the singular points as $t \to \pm\infty$.

If $L(C)$ consists only of singular points, then $L(C)$ is just a single point since $L(C)$ is connected and there is only a finite number of singular points. The half-trajectory C must approach this point as $t \to \infty$. If $L(C)$ has regular points, then it is the union of singular points and a set of limit trajectories $\Gamma \subset L(C)$. But by the argument used in proving the Poincaré-Bendixson theorem, Γ can have a regular limit only if it is periodic. Hence, Γ is either periodic, in which case $L(C) = \Gamma$ by Lemma 6, or Γ has no regular limit points, in which case all trajectories in $L(C)$ are not periodic and have only singular points as limit points. Thus if Γ is such a trajectory, and hence by Lemma 2 is a full trajectory, then both $L^+(\Gamma)$ and $L^-(\Gamma)$ are single singular points, which may be distinct or coincident.

The Poincaré-Bendixson theorem indicates that if a bounded region R contains a half-trajectory but no singular points, then it must also contain a cyclic trajectory. If there is a bounded region whose boundaries consist of two concentric Jordan curves, and if the vector field $\mathbf{X}$ is directed inward at every point of the boundaries (or directed outward at every point of the boundaries), then the annular region enclosed by the two Jordan curves contains at least one cyclic trajectory (see Fig. 5.14).

On the other hand, if a Jordan curve encloses exactly one unstable (stable) focal point or nodal point, and if the vector field $\mathbf{X}$ is directed inward (outward) at every point of the Jordan curve, then there must be a cyclic trajectory interior to the Jordan curve enclosing the singularity (see Fig. 5.15). If there is more than one trajectory, the trajectories must be concentric. If there is just one trajectory, then it is asymptotically orbitally stable (unstable).

If there are two cyclic trajectories C_1 and C_2 one of which, say C_1, is interior to the other, and if in the annular region R between C_1 and C_2 there are no singular points or other cyclic trajectories, then C_1 and C_2 are said to

FIGURE 5.14

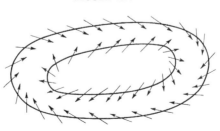

FIGURE 5.15

be *adjacent*. Every trajectory inside R has one of the cyclic trajectories for its positive limit cycle and the other for its negative limit cycle. Hence, two adjacent cyclic trajectories cannot be both positively stable on the sides facing each other, and the interior trajectories represent motion from one limit cycle to another as $t \to \infty$ (see Fig. 5.16).

The main drawback of the Poincaré-Bendixson theorem is the difficulty of determining the region R for which the conditions of the theorem are applicable. One approach is to choose the two Jordan curves of Fig. 5.14 in the form of two concentric circles, check the direction of the vector field **X** at every point of the boundaries, and investigate whether there are any singular points inside the annular region R. We also recall that the index of Poincaré can be used to check whether a cyclic trajectory is possible or not. But the index establishes only a necessary and not a sufficient condition for the existence of a cyclic trajectory, so that the index can prove conclusively only that a system does not admit a periodic solution.

The proof of the Poincaré-Bendixson theorem is based on Lemma 4, which, in turn, is based on the Jordan curve theorem. As the Jordan curve theorem applies only to the plane, we conclude that the Poincaré-Bendixson theory is also so restricted. No such complete geometric theory for spaces of higher dimension exists.

FIGURE 5.16

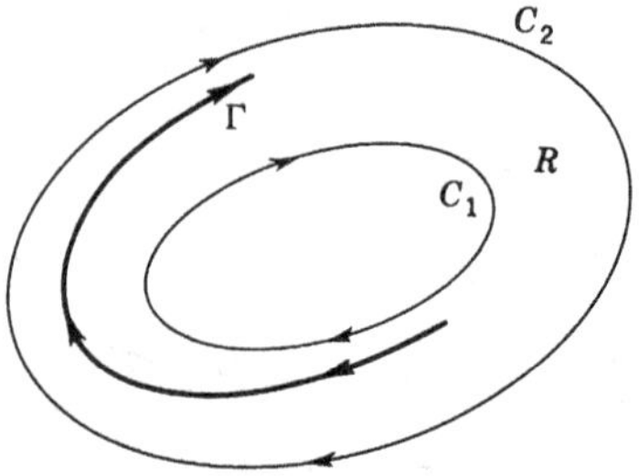

A criterion due to Bendixson, based on the properties of the divergence of the vector field **X**, gives a sufficient condition for the *absence* of a cyclic trajectory, for which reason it is referred to as the *Bendixson negative criterion*. Introducing the notation div $\mathbf{X} = \partial X_1/\partial x_1 + \partial X_2/\partial x_2$, the criterion can be stated as follows:

Bendixon's criterion *If in a region R of the phase plane*, div **X** *has a constant sign (zero excluded), then system* (5.77) *admits no solution in the form of a cyclic trajectory.*

Let us assume that there is a cyclic trajectory C for system (5.77) in the region $R \subset D$ and denote the period of motion by T. If R is taken as the simple Jordan region whose boundary is C, then, by Green's theorem, we have

$$\iint_R \text{div } \mathbf{X}\, dx_1\, dx_2 = \oint_C (X_1\, dx_2 - X_2\, dx_1) = \int_0^T (\dot{x}_1 \dot{x}_2 - \dot{x}_2 \dot{x}_1)\, dt = 0. \tag{5.81}$$

It follows from Eq. (5.81) that div **X** cannot have a constant sign in R. Since this contradicts the hypothesis, we must conclude that C cannot represent a periodic solution, which is the essence of Bendixson's criterion. On the other hand, if div **X** does change sign in R, we still cannot conclude that a cyclic trajectory does indeed exist for system (5.77).

The existence of cyclic trajectories can also be checked geometrically by means of a graphical construction due to Liénard. (For a description of this method see Ref. 5, chap. 4.)

The limit cycle is a phenomenon typical of nonlinear nonconservative systems and entirely eludes an analysis confined to the linear approximation. A classical example of a physical system exhibiting a cyclic trajectory is the van der Pol's oscillator described by the differential equation

$$\ddot{x} + \mu(x^2 - 1)\dot{x} + x = 0, \qquad \mu > 0. \tag{5.82}$$

The term $\mu(x^2 - 1)$ can be regarded as a nonlinear damping coefficient. The term is negative for $|x| < 1$ and positive for $|x| > 1$, so that a limit cycle is anticipated and indeed obtained.

Equations (5.82) can be replaced by the two first-order differential equations

$$\dot{x}_1 = x_2, \qquad \dot{x}_2 = -x_1 + \mu(1 - x_1^2)x_2, \tag{5.83}$$

where $x_1 = x$. It is obvious that the origin is a singular point. The differential equation of the trajectories is

$$\frac{dx_2}{dx_1} = \mu(1 - x_1^2) - \frac{x_1}{x_2}, \tag{5.84}$$

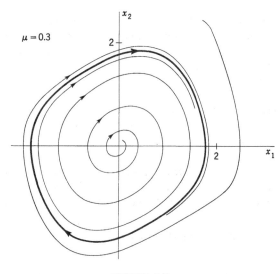

FIGURE 5.17

which can be integrated graphically by the isoclines method. A cyclic trajectory enclosing the origin is obtained. The shape of the cyclic trajectory depends on the parameter μ. For small values of μ the limit cycle resembles a circle, and the corresponding motion is nearly harmonic. For $0 < \mu < 2$ the origin is an unstable focus, and for $\mu > 2$ it is an unstable node. Figure 5.17 shows the phase portrait for $\mu = 0.3$. It is easy to see from Eqs. (5.83) that the linear approximation would entirely ignore the existence of a limit cycle and judge the system to be unstable.

The case in which $\mu < 0$ can be easily treated by substituting $t = -\tau$ in Eqs. (5.82) and regarding τ as the independent variable. Since the equation in terms of τ with $\mu < 0$ is equivalent to the equation in terms of t with $\mu > 0$, the preceding discussion is entirely applicable. Of course, when $\mu < 0$, the limit cycle is unstable because it is approached as $\tau \to \infty$, which is the same as $t \to -\infty$.

PROBLEMS

5.1 Consider the system of Example 2.2, let the equilibrium angle be $\theta = \theta_0$, and derive the set of differential equations about the equilibrium position in the form (5.27). Solve the eigenvalue problem associated with the linearized system and determine the nature of the singularity.

5.2 Consider the system of Prob. 2.2 and derive the equation of motion of the system. Determine the equilibrium positions of the system, derive the differential equation of motion about each of these positions, and check the stability of each equilibrium point.

5.3 Plot the phase portraits for the system of Prob. 5.2 corresponding to the two cases $\omega = 2\sqrt{g/R}$ and $\omega = 0$. Let $-\pi < \theta < \pi$.

5.4 Plot the phase portrait for the spherical pendulum of Example 2.9 for the case $\beta_\phi = 2mL^2$. Let $-\pi < \theta < \pi$.

5.5 Plot the phase portrait for a mass-spring system. The spring is nonlinear with the restoring force per unit mass given by

$$f(x) = \omega^2(x - \epsilon x^3), \qquad \epsilon > 0.$$

Choose your own values of ω and ϵ. Note that such a spring is referred to as a "soft" spring.

5.6 With reference to Prob. 5.5, if the restoring force has the form

$$f(x) = \omega^2(x + \epsilon x^3), \qquad \epsilon > 0,$$

the spring is said to be "hard." Derive an expression for the period of oscillation for such a case and draw conclusions concerning its dependence on the system energy.

5.7 Obtain the index of a curve enclosing the origin for the vector field $\mathbf{X}(P)$ whose components are given by Eq. (d) of Example 5.1. Calculate the index by using the definition (5.70).

5.8 Use the theorems of Sec. 5.4 and show that for a conservative system a closed trajectory encloses an odd number of singularities, with the number of centers exceeding the number of saddle points by 1.

5.9 Use the method of isoclines to plot the phase portrait for the system given by Eqs. (5.78) and thus establish the existence of a limit cycle.

5.10 Consider the system (5.78), choose an annular region R defined by two appropriate concentric circles C_1 and C_2 with the center at the origin, and establish the existence of a limit cycle by checking the direction of the vector field $\mathbf{X}$ at the boundaries of R.

SUGGESTED REFERENCES

1. Cesari, L.: *Asymptotic Behavior and Stability Problems in Ordinary Differential Equations*, Springer-Verlag OHG, Berlin, 1963.
2. Coddington, E. A., and N. Levinson: *Theory of Ordinary Differential Equations*, McGraw-Hill Book Company, New York, 1955.
3. Hurewicz, W.: *Lectures on Ordinary Differential Equations*, John Wiley & Sons, Inc., New York, 1958.
4. Lefschetz, S.: *Differential Equations: Geometric Theory*, 2d ed., John Wiley & Sons, Inc., New York, 1963.
5. Minorsky, N.: *Nonlinear Oscillations*, D. Van Nostrand Company, Inc., Princeton, N.J., 1962.
6. Nemytskii, V. V., and V. V. Stepanov: *Qualitative Theory of Differential Equations*, Princeton University Press, Princeton, N.J., 1960.
7. Struble, R. A.: *Nonlinear Differential Equations*, McGraw-Hill Book Company, New York, 1962.

chapter six

Stability of Multi-Degree-of-Freedom Autonomous Systems

The motion of a large number of dynamical systems is described by sets of differential equations in which the time appears not in an explicit form but implicitly through the variables $x_i(t)$. Such systems are said to be *autonomous*.

Whereas in the case of linear differential equations with constant coefficients it is possible to obtain closed-form solutions, at present there are no general methods for obtaining such solutions for nonlinear systems. When no general solutions are possible, sufficient information concerning the behavior of the system may be obtained by examining the motion in the neighborhood of equilibrium points. This amounts to investigating the behavior of a linear set of equations, known as the *variational equations*, corresponding to the differential equations of the perturbed motion. If the original equations are autonomous and the perturbed motion takes place in the neighborhood of a constant solution of the original equations, then the variational system is autonomous and its solution presents no particular difficulty. The Liapunov direct method, on the other hand, provides an entirely different approach from that based on the variational equations. The Liapunov direct method is very powerful and has two important features: (1) the method is not restricted to linear systems or a small neighborhood of the equilibrium, and (2) it does not require the solution of the differential equations for an investigation of the system's stability. Its main drawback, however, is that it necessitates constructing a testing function, known as a *Liapunov function*, which may not always be possible. In fact the Liapunov direct method should be regarded as more of a philosophy of approach than a method since at present there are no general procedures for constructing Liapunov functions except for linear autonomous systems.

In this chapter the general theory of linear systems is first introduced. Discussion of the linear theory is then concentrated on autonomous systems for which explicit solutions are possible. The linear theory can be used to study the infinitesimal stability of systems in the neighborhood of singular points. To this end, the variational equations corresponding to the perturbed motion are derived. When the original system of equations is canonical, the variational equations are shown to lead to interesting results. An exposition of the salient features of the Liapunov direct method is given, together with the more important theorems concerning the stability and instability of autonomous systems and a geometrical interpretation of the theorems. Due to our special interest in canonical systems, the application of the Liapunov direct method to the stability analysis of such systems is discussed in detail. Finally the procedure for constructing the Liapunov functions for linear autonomous systems of equations is presented.

6.1 GENERAL LINEAR SYSTEMS

Let us consider a general dynamical system, not necessarily Hamiltonian. Under certain circumstances, the equations describing the motion of the dynamical system can be written in the form of the *homogeneous linear system of equations*

$$\dot{x}_i = \sum_{j=1}^{m} A_{ij}(t) x_j, \qquad i = 1, 2, \ldots, m. \tag{6.1}$$

Half the x_i variables in Eqs. (6.1) represent generalized coordinates, and the remaining half represent generalized momenta (or velocities). Hence m is an even number, $m = 2n$, where n is the number of degrees of freedom of the system. It should be mentioned, however, that the present discussion of linear systems is general in scope and in no way restricted to canonical systems or systems of even order. Canonical systems are discussed later.

The coefficients $A_{ij}(t)$ $(i, j = 1, 2, \ldots, m)$ are assumed to be continuous functions of t over an open interval $t_1 < t < t_2$. The interval $t_1 < t < t_2$ will often be referred to as simply the interval I. Given a set of initial conditions $x_i(t_0) = \alpha_i$ $(i = 1, 2, \ldots, m)$, we conclude from Sec. 5.1 that Eqs. (6.1) have a unique solution

$$x_i(t) = \varphi_i(\alpha_1, \alpha_2, \ldots, \alpha_m, t), \qquad i = 1, 2, \ldots, m, \tag{6.2}$$

defined over the interval I. In the sequel we shall discuss the form of this solution.

The solution of Eqs. (6.1) is discussed most conveniently by expressing the equations in the compact matrix notation

$$\{\dot{x}\} = [A(t)]\{x\}, \qquad (6.3)$$

where $\{x\}$ is an $m \times 1$ column matrix, which is the equivalent of a vector **x** (these two notations will be used interchangeably) with components x_i in the Euclidean m-space E^m, and $[A(t)]$ is an $m \times m$ matrix continuous in time but generally nonsymmetrical. System (6.3) is completely determined by the matrix $[A(t)]$.

Next we shall examine certain properties of the solution of Eq. (6.3). Writing this solution in the matrix form $\{\varphi(t)\}$, we observe that the null solution $\{\varphi(t)\} \equiv \{0\}$ is always a solution. It is often referred to as the *trivial solution*. If $\{\varphi(t_0)\} = \{0\}$ is a solution for some number t_0 in the interval I ($t_0 \in I$), then, by uniqueness, $\{\varphi(t)\} \equiv \{0\}$ is a solution in that interval.

Since system (6.3) is homogeneous, if $\{\varphi(t)\}$ is a solution, then $c\{\varphi(t)\}$ is also a solution, where c is any arbitrary constant. Moreover, the system is linear, so that if $\{\varphi^{(1)}\}$ and $\{\varphi^{(2)}\}$ are solutions, then $c_1\{\varphi^{(1)}\} + c_2\{\varphi^{(2)}\}$ is also a solution, from which we conclude that the solutions of (6.3) form a *linear manifold*; i.e., any linear combination of solutions is itself a solution.

A set of k m-vectors $\{u^{(1)}\}, \{u^{(2)}\}, \ldots, \{u^{(k)}\}$ is said to be *linearly independent* over the interval I if the statement

$$\sum_{i=1}^{k} c_i \{u^{(i)}(t)\} = \{0\}, \qquad t \in I, \qquad (6.4)$$

where c_i are constants, implies $c_i = 0$ ($i = 1, 2, \ldots, k$). If the constants c_i are not all zero, the set of vectors is said to be *linearly dependent*.

With this definition in mind, we can state the following basic theorem on linear systems.

Theorem *The set of all solutions of Eq. (6.3) in the interval I form an m-vector space.*

To prove this theorem we must show that it is possible to produce a set of m independent solutions $\{\varphi^{(i)}\}$ ($i = 1, 2, \ldots, m$) of Eq. (6.3) such that every other solution of (6.3) is a linear combination of the vectors $\{\varphi^{(i)}\}$. For this purpose we consider a set of m linearly independent vectors in the Euclidean m-space **x**. As such a set we can take the *unit vectors* $\{e_i\}$ ($i = 1, 2, \ldots, m$), where a typical vector $\{e_i\}$ is defined as having all its components zero except for the ith component, which is equal to 1. Note that the vectors $\{e_i\}$ are simply the corresponding columns of the identity matrix [1] of order m. If these unit vectors are regarded as representing initial conditions, $\{\varphi^{(i)}(t_0)\} = \{e_i\}$, then, by the existence theorem, it follows that system (6.3) admits m solutions $\{\varphi^{(i)}(t)\}$. Moreover, it is not difficult to show that these solutions are independent. If the set of vectors $\{\varphi^{(i)}(t)\}$ is to be dependent, there must be m constants c_i, not all zero, such that the relation

$$\sum_{i=1}^{m} c_i \{\varphi^{(i)}(t)\} = \{0\}, \qquad t \in I, \tag{6.5}$$

is satisfied. But for $t_0 \in I$, Eq. (6.5) becomes

$$\sum_{i=1}^{m} c_i \{\varphi^{(i)}(t_0)\} = \sum_{i=1}^{m} c_i \{e_i\} \neq \{0\}, \tag{6.6}$$

so that the solutions $\{\varphi^{(i)}\}$ ($i = 1, 2, \ldots, m$) must be linearly independent.

If $\{\varphi(t)\}$ is any solution of system (6.3) in the interval I, then its initial value $\{\varphi(t_0)\}$ can be written in the form

$$\{\varphi(t_0)\} = \sum_{i=1}^{m} c_i \{e_i\}, \qquad t_0 \in I, \tag{6.7}$$

where the constants c_i are unique, because the vectors $\{e_i\}$ span an m-vector space. It follows, by uniqueness, that the solution $\{\varphi(t)\}$ has the expression

$$\{\varphi(t)\} = \sum_{i=1}^{m} c_i \{\varphi^{(i)}(t)\}, \qquad t \in I. \tag{6.8}$$

Hence, every solution $\{\varphi(t)\}$ can be written as a unique linear combination of the vectors $\{\varphi^{(i)}\}$, which is another way of stating the basic theorem on linear systems.

A set of m linearly independent solutions $\{\varphi^{(i)}\}$ ($i = 1, 2, \ldots, m$) of Eq. (6.3) is known as a *fundamental set* or *linear basis*. An $m \times m$ matrix $[\varphi(t)]$ with its columns consisting of a set of linearly independent solutions of (6.3) is called a *fundamental matrix* for the homogeneous system (6.3). The fundamental matrix satisfies the matrix equation

$$[\dot{\varphi}(t)] = [A(t)][\varphi(t)], \qquad t \in I. \tag{6.9}$$

Next let us imagine m solutions $\{x^{(i)}\}$ of Eq. (6.3). If these solutions are taken as the columns of an $m \times m$ matrix $[x(t)]$, the corresponding m differential equations can be written in the compact form

$$[\dot{x}(t)] = [A(t)][x(t)], \quad t \in I, \tag{6.10}$$

which is sometimes referred to as the *matrix differential equation associated with* (6.3) on *I*. Clearly [φ] is a solution of Eq. (6.10), so that the fundamental matrix provides complete knowledge of the set of solutions of Eq. (6.3).

By the theorem on linear systems, however, any solution of Eq. (6.3) can be expressed as a linear combination of the fundamental set of solutions of (6.3). This implies that any solution of (6.3) can be written in the form

$$\{x(t)\} = [\varphi(t)]\{c\}, \tag{6.11}$$

where $\{c\}$ is a column matrix of unique constants c_i ($i = 1, 2, \ldots, m$), not all zero. Denoting the solution at a particular time t_0 by $\{x_0\}$ Eq. (6.11) leads to

$$\{x(t_0)\} = \{x_0\} = [\varphi(t_0)]\{c\}, \tag{6.12}$$

which can be regarded as a set of m simultaneous equations in the unknowns c_i. Equation (6.12) has a unique solution if and only if det $[\varphi(t_0)] \neq 0$. Since t_0 is arbitrary, it follows that a *necessary and sufficient condition for the solutions* $\{\varphi^{(i)}(t)\}$ ($i = 1, 2, \ldots, m$) *to form a fundamental set is that the determinant of the associated matrix* $[\varphi(t)]$ *be different from zero for all* $t \in I$. This is equivalent to the requirement that the columns $\{\varphi^{(i)}(t)\}$ of $[\varphi(t)]$ be linearly independent for all times in *I*.

If [φ] is a fundamental matrix of (6.3) and [c] is an $m \times m$ nonsingular constant matrix, then [φ][c] is also a fundamental matrix of (6.3). This can be easily seen by postmultiplying Eq. (6.9) by [c] and obtaining

$$\frac{d}{dt}([\varphi][c]) = [A]([\varphi][c]). \tag{6.13}$$

Clearly det ([φ][c]) $\neq 0$ because det [φ] $\neq 0$ and [c] is nonsingular. If the matrix solution of Eq. (6.10) is such that it satisfies the initial condition

$$[\varphi(t_0)] = [1], \tag{6.14}$$

namely, that at t_0 the solution is equal to the identity matrix, the solution [$\varphi(t)$] is said to be a *principal matrix* for the system (6.3).

From the formula for the derivative of a determinant it is possible to extract a very basic result. To show this, we recall that Eq. (6.9) represents a set of m^2 simultaneous scalar differential equations of the type

$$\dot{\varphi}_{ij}(t) = \sum_{k=1}^{m} A_{ik}(t)\varphi_{kj}(t), \quad i, j = 1, 2, \ldots, m, \tag{6.15}$$

and consider

$$\frac{d}{dt} \det [\varphi(t)] = \begin{vmatrix} \dot{\varphi}_{11} & \dot{\varphi}_{12} & \cdots & \dot{\varphi}_{1m} \\ \varphi_{21} & \varphi_{22} & \cdots & \varphi_{2m} \\ \cdots & \cdots & \cdots & \cdots \\ \varphi_{m1} & \varphi_{m2} & \cdots & \varphi_{mm} \end{vmatrix} + \begin{vmatrix} \varphi_{11} & \varphi_{12} & \cdots & \varphi_{1m} \\ \dot{\varphi}_{21} & \dot{\varphi}_{22} & \cdots & \dot{\varphi}_{2m} \\ \cdots & \cdots & \cdots & \cdots \\ \varphi_{m1} & \varphi_{m2} & \cdots & \varphi_{mm} \end{vmatrix} + \cdots$$

$$+ \begin{vmatrix} \varphi_{11} & \varphi_{12} & \cdots & \varphi_{1m} \\ \varphi_{21} & \varphi_{22} & \cdots & \varphi_{2m} \\ \cdots & \cdots & \cdots & \cdots \\ \dot{\varphi}_{m1} & \dot{\varphi}_{m2} & \cdots & \dot{\varphi}_{mm} \end{vmatrix}. \tag{6.16}$$

Using Eqs. (6.15) and operating on the rows of the first determinant on the right side of (6.16), we can show that this first determinant yields

$$\begin{vmatrix} \sum_k A_{1k}\varphi_{k1} & \sum_k A_{1k}\varphi_{k2} & \cdots & \sum_k A_{1k}\varphi_{km} \\ \varphi_{21} & \varphi_{22} & \cdots & \varphi_{2m} \\ \cdots & \cdots & \cdots & \cdots \\ \varphi_{m1} & \varphi_{m2} & \cdots & \varphi_{mm} \end{vmatrix} = \begin{vmatrix} A_{11}\varphi_{11} & A_{11}\varphi_{12} & \cdots & A_{11}\varphi_{1m} \\ \varphi_{21} & \varphi_{22} & \cdots & \varphi_{2m} \\ \cdots & \cdots & \cdots & \cdots \\ \varphi_{m1} & \varphi_{m2} & \cdots & \varphi_{mm} \end{vmatrix}$$

$$= A_{11} \det [\varphi(t)]. \tag{6.17}$$

Similar expressions can be derived for the remaining $m - 1$ determinants in (6.16), which leads us to the result

$$\frac{d}{dt} \det [\varphi(t)] = \left(\sum_{i=1}^{m} A_{ii} \right) \det [\varphi(t)] = \text{tr} [A(t)] \det [\varphi(t)], \tag{6.18}$$

where tr $[A]$ denotes the trace of the matrix $[A]$. Equation (6.18) is a scalar differential equation, which has the solution

$$\det [\varphi(t)] = \det [\varphi(t_0)] \exp \left(\int_{t_0}^{t} \text{tr} [A(s)] \, ds \right), \tag{6.19}$$

where s is a dummy variable of integration. Relation (6.19), sometimes referred to as the *Jacobi-Liouville formula*, implies that in order to obtain a fundamental matrix it is necessary to specify the initial fundamental matrix $[\varphi(t_0)]$ so that its determinant is different from zero, because only then is the fundamental matrix $[\varphi(t)]$ nonsingular. In view of Eq. (6.14), the principal matrix is always nonsingular, since det $[\varphi(t_0)] = 1$.

The fundamental matrix $[\varphi(t)]$ possesses an inverse $[\varphi(t)]^{-1}$, by definition. The product of the two matrices is equal to the identity matrix, from which it follows that

$$[0] = \frac{d}{dt} ([\varphi(t)][\varphi(t)]^{-1}) = \frac{d}{dt} [\varphi(t)][\varphi(t)]^{-1} + [\varphi(t)] \frac{d}{dt} [\varphi(t)]^{-1}. \tag{6.20}$$

Stability of Multi-Degree-of-Freedom Autonomous Systems

In view of Eq. (6.9), we conclude that $[\varphi(t)]^{-1}$ satisfies the matrix equation

$$\frac{d}{dt}[\varphi(t)]^{-1} = -[\varphi(t)]^{-1}[A(t)]. \tag{6.21}$$

If $\{y\}$ is a column matrix corresponding to the transpose of a row matrix in $[\varphi(t)]^{-1}$, then Eq. (6.21) leads to

$$\{\dot{y}\}^T = -\{y\}^T[A(t)], \tag{6.22}$$

which is referred to as the *adjoint equation* of (6.3). If the matrix $[A(t)]$ is skew-symmetric, Eq. (6.22) is equivalent to Eq. (6.3) and the system is said to be *self-adjoint*.

Under certain circumstances the solution of Eq. (6.3) can be obtained in explicit form. To verify this statement, let us introduce a matrix $[B(t)]$ defined by the integral

$$[B(t)] = \int_{t_0}^{t} [A(s)]\, ds. \tag{6.23}$$

We shall show that when the matrices $[A(t)]$ and $[B(t)]$ commute, it is possible to obtain an explicit form for the fundamental matrix. A proof of this statement can be effected by the method of successive approximations.

Let us consider the case in which $[A(t)]$ and $[B(t)]$ commute, in which case we can use Eq. (6.23) and write

$$\frac{d}{dt}[B]^k = k[B]^{k-1}\frac{d}{dt}[B] = k[A][B]^{k-1}, \qquad k = 1, 2, \ldots . \tag{6.24}$$

Next assume that system (6.3) is subject to the initial condition $\{x(t_0)\} = \{x_0\}$ and let this initial value be a first approximation for the solution $\{x(t)\}$. Then a second approximation, $\{x_1\}$, satisfies the equation

$$\{\dot{x}_1\} = [A]\{x_0\}, \qquad \{x_1(t_0)\} = \{x_0\}, \tag{6.25}$$

which has the solution

$$\{x_1(t)\} = \{x_0\} + \int_{t_0}^{t} [A]\{x_0\}\, ds = ([1] + [B])\{x_0\}. \tag{6.26}$$

The subsequent approximations satisfy

$$\{\dot{x}_k\} = [A]\{x_{k-1}\}, \qquad \{x_k(t_0)\} = \{x_0\}, \qquad k = 2, 3, \ldots \tag{6.27}$$

so that, using Eqs. (6.24) and (6.26), it follows by induction that

$$\{x_k(t)\} = \left([1] + [B] + \frac{1}{2!}[B]^2 + \cdots + \frac{1}{k!}[B]^k\right)\{x_0\}, \qquad k = 2, 3, \ldots . \tag{6.28}$$

According to the Picard-Lindelöf theorem,[1] the successive approximations $\{x_k(t)\}$ exist and converge uniformly in a sufficiently small neighborhood of t_0 to the solution $\{x(t)\}$ of Eq. (6.3) as $k \to \infty$. Hence, the solution of Eq. (6.3), subject to the initial condition $\{x_0\}$, can be written in the form

$$\{x(t)\} = e^{[B]}\{x_0\} = \exp\left(\int_{t_0}^{t} [A(s)]\,ds\right)\{x_0\}, \qquad (6.29)$$

where $e^{[B]}$ represents the *exponential* of the matrix $[B]$, namely, the matrix series

$$e^{[B]} = [1] + [B] + \frac{1}{2!}[B]^2 + \frac{1}{3!}[B]^3 + \cdots. \qquad (6.30)$$

The series converges uniformly over any closed interval in which the elements of the matrix $[A(t)]$ are continuous.

It is easy to see that the matrices $[A]$ and $[B]$ commute if $[A]$ is constant or if $[A(t)]$ is diagonal. Hence, in either of these two cases $e^{[B]}$ represents a fundamental matrix.

Next we wish to consider the *nonhomogeneous* problem associated with system (6.3). If the system is acted upon by a set of time-dependent forces which can be written in the form of the vector $\{f(t)\}$, Eq. (6.3) must be replaced by

$$\{\dot{x}(t)\} = [A(t)]\{x(t)\} + \{f(t)\}. \qquad (6.31)$$

Premultiplying Eq. (6.31) by $[\varphi(t)]^{-1}$, postmultiplying Eq. (6.21) by $\{x(t)\}$, and adding the results, we arrive at

$$\frac{d}{dt}([\varphi(t)]^{-1}\{x(t)\}) = [\varphi(t)]^{-1}\{f(t)\}, \qquad (6.32)$$

which can be integrated to obtain

$$[\varphi(t)]^{-1}\{x(t)\} = [\varphi(t_0)]^{-1}\{x(t_0)\} + \int_{t_0}^{t} [\varphi(s)]^{-1}\{f(s)\}\,ds. \qquad (6.33)$$

If $[\varphi(t)]$ is the principal matrix, then $[\varphi(t_0)] = [1]$, so that when Eq. (6.33) is premultiplied by $[\varphi(t)]$, it becomes

$$\{x(t)\} = [\varphi(t)]\{x(t_0)\} + [\varphi(t)]\int_{t_0}^{t} [\varphi(s)]^{-1}\{f(s)\}\,ds. \qquad (6.34)$$

Finally, noticing that $[\varphi(t)]\{x(t_0)\}$ is the solution $\{\varphi(t)\}$ of the homogeneous equation (6.3) but subject to the same initial conditions as the vector $\{x(t)\}$

---

[1] Actually the Picard-Lindelöf theorem is concerned with the general nonlinear system $\{\dot{x}\} = \{X\}$, and it requires that the column vector $\{X\}$ be Lipschitzian. This certainly is the case with system (6.3). For a discussion of the theorem and the method of successive approximations, see Ref. 5, sec. 1.3.

of Eq. (6.31), that is, $\{\varphi(t_0)\} = \{x(t_0)\}$, $t_0 \in I$, we see that Eq. (6.34) reduces to

$$\{x(t)\} = \{\varphi(t)\} + [\varphi(t)] \int_{t_0}^{t} [\varphi(s)]^{-1}\{f(s)\}\, ds, \qquad t \in I. \tag{6.35}$$

Hence, the solution of the linear system reduces to the determination of a fundamental matrix. When the elements of the matrix $[A]$ are constant or periodic, additional information concerning the nature of the solution can be obtained. The case of a linear system with constant coefficients is discussed in Sec. 6.2 and that of a system with periodic coefficients in Sec. 7.1.

6.2 LINEAR AUTONOMOUS SYSTEMS

A case of particular interest is that in which the dynamical system is autonomous. In this case the matrix $[A]$ in Eq. (6.3) is constant. Hence, we shall be concerned with the linear homogeneous system

$$\{\dot{x}\} = [A]\{x\}, \tag{6.36}$$

in which $\{x\}$ is an $m \times 1$ column matrix and $[A]$ is an $m \times m$ generally nonsymmetrical matrix whose elements are real constants. For canonical systems the order m is an even number, $m = 2n$, where n is the number of degrees of freedom of the system. The solutions of Eq. (6.36) are defined for all values of t, and these solutions possess derivatives of all orders.

In contrast with the nonautonomous system, it is easily verified that $\{\dot{x}\}$ is also a solution of Eq. (6.36). Indeed for constant $[A]$ we can write

$$\frac{d}{dt}\{\dot{x}\} = \frac{d}{dt}([A]\{x\}) = [A]\{\dot{x}\}. \tag{6.37}$$

Similarly, derivatives of any order or linear combinations of derivatives are also solutions.

In Sec. 6.1 we showed that when $[A]$ is constant $e^{[B]}$ is a fundamental matrix. In view of Eq. (6.23), we conclude that

$$[\varphi(t)] = e^{[B]} = e^{(t-t_0)[A]}. \tag{6.38}$$

It follows that the solution of Eq. (6.36), valid for any time t, can be written in the form

$$\{x(t)\} = e^{(t-t_0)[A]}\{x_0\}, \tag{6.39}$$

where $\{x_0\}$ is a column matrix representing the initial conditions at $t = t_0$. Hence, the behavior of the dynamical system is completely determined by the behavior of the matrix $e^{(t-t_0)[A]}$, which, in turn, is governed by the eigenvalues of the matrix $[A]$. In the sequel we examine how the matrix $[A]$ controls the behavior of the system (6.36), and for this purpose we shall review some preliminary definitions and notations from matrix algebra.

The *characteristic polynomial* associated with the $m \times m$ matrix $[A]$ is defined by the *characteristic determinant* $\det([A] - \lambda[1]) = |[A] - \lambda[1]|$. The roots of the characteristic polynomial, denoted by λ_j ($j = 1, 2, \ldots, m$), are called the *eigenvalues* of the matrix $[A]$. To a certain eigenvalue λ_j belongs the eigenvector $\{u^{(j)}\}$ satisfying the equation

$$[A]\{u^{(j)}\} = \lambda_j \{u^{(j)}\}. \tag{6.40}$$

By expanding the characteristic determinant, we obtain the characteristic polynomial

$$\det([A] - \lambda[1]) = a_0 \lambda^m + a_1 \lambda^{m-1} + \cdots + a_m = \prod_{j=1}^{m} (\lambda_j - \lambda), \tag{6.41}$$

where, in our case, m is an even integer and a_m is the determinant of the matrix $[A]$. In the event that a_m is zero, one of the characteristic roots is zero, from which it follows that for the matrix $[A]$ to be nonsingular all the eigenvalues must be different from zero.

Next let us consider the case in which there is a repeated root, say λ_i, so that Eq. (6.41) assumes the form

$$\det([A] - \lambda[1]) = (\lambda_i - \lambda)^{\mu_i} P_i(\lambda), \tag{6.42}$$

where $(\lambda_i - \lambda)^{\mu_i}$ is called an *elementary divisor* of the characteristic polynomial and $P_i(\lambda)$ is a polynomial in λ such that $P_i(\lambda_i) \neq 0$. The integer μ_i is known as the *multiplicity* of λ_i, and the number of linearly independent eigenvectors associated with λ_i is called the *nullity* of λ_i and denoted by ν_i. In general $\nu_i \leq \mu_i$. If λ_i is an eigenvalue of the matrix $[A]$ and $\nu_i = \mu_i$, then $(\lambda_i - \lambda)^{\mu_i}$ is said to be a *simple elementary divisor*.

Two $m \times m$ matrices $[A]$ and $[C]$ are said to be similar if there exists a nonsingular matrix $[P]$ such that

$$[C] = [P]^{-1}[A][P]. \tag{6.43}$$

The characteristic polynomial associated with $[C]$ is

$$\begin{aligned}\det([C] - \lambda[1]) &= \det([P]^{-1}([A] - \lambda[1])[P]) \\ &= \det[P]^{-1} \det([A] - \lambda[1]) \det[P] \\ &= \det([A] - \lambda[1]),\end{aligned} \tag{6.44}$$

from which we conclude that *two similar matrices have the same characteristic polynomial*. It follows that *the coefficients a_k ($k = 1, 2, \ldots, m$) in Eq. (6.41) are invariant under similarity transformations*.

Stability of Multi-Degree-of-Freedom Autonomous Systems

We shall be interested in a similarity transformation such that the matrix $[C]$ assumes as simple a form as possible. In this connection we state the following theorem.

Theorem *Every $m \times m$ matrix $[A]$ is similar to a matrix of the form*

$$[J] = \begin{bmatrix} [J_0] & [0] & \cdots & [0] \\ [0] & [J_1] & \cdots & [0] \\ \cdots & \cdots & \cdots & \cdots \\ [0] & [0] & \cdots & [J_s] \end{bmatrix}, \quad (6.45)$$

where $[J_0]$ is a $q \times q$ diagonal matrix with elements $\lambda_1, \lambda_2, \ldots, \lambda_q$ and

$$[J_i] = \begin{bmatrix} \lambda_{q+i} & 1 & 0 & \cdots & 0 & 0 \\ 0 & \lambda_{q+i} & 1 & \cdots & 0 & 0 \\ \cdots & \cdots & \cdots & \cdots & \cdots & \cdots \\ 0 & 0 & 0 & \cdots & \lambda_{q+i} & 1 \\ 0 & 0 & 0 & \cdots & 0 & \lambda_{q+i} \end{bmatrix}, \quad i = 1, 2, \ldots, s. \quad (6.46)$$

The quantities λ_j ($j = 1, 2, \ldots, q + s$), which are the eigenvalues of the matrix $[A]$, are not necessarily distinct.

The matrix $[J_i]$ has r_i rows and columns so that

$$q + \sum_{i=1}^{s} r_i = m. \quad (6.47)$$

The *block diagonal* matrix $[J]$, Eq. (6.45), is said to be a *Jordan canonical form*. The Jordan form was briefly discussed in Sec. 5.2. The matrices $[J_i]$, called *companion matrices* associated with the eigenvalues λ_{q+i} ($i = 1, 2, \ldots, s$), are diagonal only when the elementary divisors are all simple. Hence, *if the elementary divisors of the characteristic polynomial associated with the matrix $[A]$ are all simple, the corresponding Jordan canonical form is diagonal.* It should be pointed out that the Jordan form associated with a given matrix $[A]$ is unique except for a permutation of rows and columns.

Of the invariant coefficients of the characteristic polynomial, the coefficients corresponding to the trace and determinant of the matrix $[A]$ are of particular significance. These coefficients have the values

$$-a_1 = \operatorname{tr}[A] = \sum_i \lambda_i, \quad a_m = \det[A] = \prod_i \lambda_i, \quad (6.48)$$

where the sum and product are taken over all roots and each multiple root is counted as many times as its multiplicity.

To investigate the behavior of the fundamental matrix (6.38), we introduce the linear transformation

$$\{x\} = [P]\{u\}, \tag{6.49}$$

where $[P]$ is an $m \times m$ nonsingular matrix of constants. Introducing Eq. (6.49) into (6.36) and premultiplying the result by $[P]^{-1}$, we obtain

$$\{\dot{u}\} = [P]^{-1}[A][P]\{u\}. \tag{6.50}$$

We shall assume that the elements of $[P]$ are so chosen that the triple matrix product $[P]^{-1}[A][P]$ is the Jordan canonical form associated with $[A]$, so that Eq. (6.50) can be written

$$\{\dot{u}\} = [J]\{u\}, \tag{6.51}$$

where the matrix $[J]$ is of the block diagonal form (6.45). By analogy with Eqs. (6.36) and (6.38), a fundamental matrix associated with system (6.51) can be written in the form

$$[\psi(t)] = e^{(t-t_0)[J]}. \tag{6.52}$$

Since $e^{t_0[J]}$ is a constant matrix, it follows that the investigation of the solution of Eq. (6.36) reduces to the investigation of the matrix

$$e^{t[J]} = \begin{bmatrix} e^{t[J_0]} & [0] & \cdots & [0] \\ [0] & e^{t[J_1]} & \cdots & [0] \\ \cdots & \cdots & \cdots & \cdots \\ [0] & [0] & \cdots & e^{t[J_s]} \end{bmatrix}. \tag{6.53}$$

But $[J_0]$ is a diagonal matrix with its elements consisting of the first q eigenvalues of matrix $[A]$, so that

$$e^{t[J_0]} = \begin{bmatrix} e^{t\lambda_1} & 0 & \cdots & 0 \\ 0 & e^{t\lambda_2} & \cdots & 0 \\ \cdots & \cdots & \cdots & \cdots \\ 0 & 0 & \cdots & e^{t\lambda_q} \end{bmatrix}. \tag{6.54}$$

Moreover, from Eqs. (6.46), we can write the $r_i \times r_i$ matrix $[J_i]$ in the form

$$[J_i] = \lambda_{q+i}[1] + [Z_i], \tag{6.55}$$

where
$$[Z_i] = \begin{bmatrix} 0 & 1 & 0 & \cdots & 0 & 0 \\ 0 & 0 & 1 & \cdots & 0 & 0 \\ 0 & 0 & 0 & \cdots & 0 & 0 \\ \cdots & \cdots & \cdots & \cdots & \cdots & \cdots \\ 0 & 0 & 0 & \cdots & 0 & 1 \\ 0 & 0 & 0 & \cdots & 0 & 0 \end{bmatrix}. \tag{6.56}$$

This is a square matrix with all elements zero except those immediately above the main diagonal, which are equal to 1. It follows that

$$e^{t[J_i]} = e^{t\lambda_q + i e^{t[Z_i]}} = e^{t\lambda_{q+i}} \begin{bmatrix} 1 & t & t^2/2! & \cdots & t^{r_i-1}/(r_i-1)! \\ 0 & 1 & t & \cdots & t^{r_i-2}/(r_i-2)! \\ 0 & 0 & 1 & \cdots & t^{r_i-3}/(r_i-3)! \\ \cdots & \cdots & \cdots & \cdots & \cdots \\ 0 & 0 & 0 & \cdots & 1 \end{bmatrix}. \tag{6.57}$$

It should be pointed out that for a solution of Eq. (6.51), although it is desirable that the matrix $[J]$ be diagonal, it is not necessary. Indeed, for an explicit solution of Eq. (6.51) it is sufficient for the fundamental matrix (6.52) to be triangular, and this is certainly the case, as can be concluded from Eqs. (6.53) and (6.57).

From Eqs. (6.50) and (6.51) we obtain a relation between $[A]$ and $[J]$ which enables us to write

$$e^{(t-t_0)[A]} = e^{(t-t_0)[P][J][P]^{-1}} = [P]e^{(t-t_0)[J]}[P]^{-1}, \tag{6.58}$$

so that, introducing Eq. (6.58) into (6.38) and recalling Eq. (6.52), we obtain

$$[\varphi] = e^{(t-t_0)[A]} = [P]e^{(t-t_0)[J]}[P]^{-1} = [P][\psi][P]^{-1}. \tag{6.59}$$

Thus the behavior of the fundamental matrix $[\varphi]$, corresponding to the system (6.36), is known if the behavior of the fundamental matrix $[\psi]$ of the system (6.51) is known.

Next let us consider the nonhomogeneous system (6.31), in which the matrix $[A]$ is constant. In this case, formula (6.35) can be written in an explicit form. To show this, we recall that the postmultiplication of a fundamental matrix by a constant matrix yields another fundamental matrix, as shown in Eq. (6.13). Regarding the matrix $[\varphi(s)]^{-1}$ in Eq. (6.35) as constant, we can write a fundamental matrix associated with $[A]$ in the form $[\Phi(t)]_1 = [\varphi(t)][\varphi(s)]^{-1}$ and note that $[\Phi(t)]_1$ is the principal matrix corresponding to the initial time s, since $[\Phi(s)]_1 = [1]$. On the other hand, $[\Phi(t)]_2 = [\varphi(t-s)]e^{t_0[A]}$ is also a principal matrix for $[A]$ corresponding to the initial time s. But, by uniqueness, the matrix $[A]$ admits only one principal matrix, from which it follows that $[\varphi(t)][\varphi(s)]^{-1} = [\varphi(t-s)]e^{t_0[A]}$. This enables us to write Eq. (6.35) as

$$\{x(t)\} = \{\varphi(t)\} + e^{t_0[A]} \int_{t_0}^{t} [\varphi(t-s)]\{f(s)\}\,ds, \qquad t \in I, \qquad (6.60)$$

where t_0 is in the interval I. In view of Eq. (6.38) and recalling that $\{\varphi(t)\} = [\varphi(t)]\{x(t_0)\}$, we see that Eq. (6.60) assumes the explicit form

$$\{x(t)\} = e^{(t-t_0)[A]}\{x(t_0)\} + \int_{t_0}^{t} e^{(t-s)[A]}\{f(s)\}\,ds, \qquad t \in I. \qquad (6.61)$$

6.3 STABILITY OF LINEAR AUTONOMOUS SYSTEMS. ROUTH-HURWITZ CRITERION

Quite frequently there is no particular interest in producing the fundamental matrix $[\varphi(t)]$ of the system (6.36), and a statement concerning the stability of the system suffices. In Sec. 6.2 we saw that the behavior of the matrix $[\varphi(t)]$ is known if the fundamental matrix $[\psi(t)]$ of the system (6.51) is known. From Eqs. (6.52) to (6.54) and (6.57), however, we conclude that the behavior of $[\psi(t)]$ depends on the eigenvalues λ_j of the matrix $[A]$ defining system (6.36). This enables us to make the following statements concerning the stability of system (6.36):

1. When all the eigenvalues of $[A]$ have negative real parts, the system is *asymptotically stable*.
2. When all the eigenvalues of $[A]$ have nonpositive real parts but some of the eigenvalues have vanishing real parts, the system is said to possess *critical behavior*. In this case, if the elementary divisors corresponding to the eigenvalues with zero real parts are all simple, then the system is *stable*. This is true because if all the elementary divisors are simple, the matrix $[J_i]$ reduces to a diagonal form.
3. If at least one of the eigenvalues of $[A]$ has a positive real part, the system is *unstable*.

When the system exhibits a behavior in accordance with either statement 1 or 3, the system is said to possess *significant behavior*.

The eigenvalues λ_j can be produced by solving the *characteristic equation* obtained by setting the characteristic determinant, Eq. (6.41), equal to zero

$$a_0\lambda^m + a_1\lambda^{m-1} + a_2\lambda^{m-2} + \cdots + a_m = 0. \qquad (6.62)$$

At times the solution of Eq. (6.62) may prove to be a formidable task, particularly for higher-order systems. In these instances it is desirable to be able to make a statement about the system stability without actually solving the characteristic equation. Because the imaginary parts of the eigenvalues have no effect upon the stability of the system, the information concerning the real parts of the eigenvalues, in particular the sign of the real parts, suffices. The *Routh-Hurwitz criterion* serves precisely this purpose. To illustrate the criterion, let us form the array

$$\begin{matrix} a_1 & a_0 & 0 & 0 & \cdots & 0 \\ a_3 & a_2 & a_1 & a_0 & \cdots & 0 \\ a_5 & a_4 & a_3 & a_2 & \cdots & 0 \\ a_7 & a_6 & a_5 & a_4 & \cdots & 0 \\ \cdots & \cdots & \cdots & \cdots & \cdots & \cdots \\ a_{2m-1} & a_{2m-2} & a_{2m-3} & \cdots & \cdots & a_m \end{matrix}$$

where a_i ($i = 0, 1, 2, \ldots, m$) are the coefficients in Eq. (6.62), and construct the determinants

$$\Delta_1 = a_1,$$

$$\Delta_2 = \begin{vmatrix} a_1 & a_0 \\ a_3 & a_2 \end{vmatrix},$$

$$\Delta_3 = \begin{vmatrix} a_1 & a_0 & 0 \\ a_3 & a_2 & a_1 \\ a_5 & a_4 & a_3 \end{vmatrix},$$

$$\cdots \cdots \cdots \cdots \cdots \quad (6.63)$$

$$\Delta_m = \begin{vmatrix} a_1 & a_0 & 0 & 0 & \cdots & 0 \\ a_3 & a_2 & a_1 & a_0 & \cdots & 0 \\ a_5 & a_4 & a_3 & a_2 & \cdots & 0 \\ \cdots & \cdots & \cdots & \cdots & \cdots & \cdots \\ a_{2m-1} & a_{2m-2} & \cdots & \cdots & \cdots & a_m \end{vmatrix}.$$

All the elements corresponding to subscripts r such that $r > m$ or $r < 0$ in the above are to be replaced by zero.

The Routh-Hurwitz criterion states that *the necessary and sufficient conditions for all the roots λ_j of the characteristic polynomial to have negative real parts is that all the determinants $\Delta_1, \Delta_2, \ldots, \Delta_m$ be positive*, provided that Eq. (6.62) is put in such a form that $a_0 > 0$. (The proof of the criterion can be found in Ref. 4, pp. 75–76.)

We notice, however, that the last two determinants are related by

$$\Delta_m = a_m \Delta_{m-1}, \tag{6.64}$$

so that, in effect, we need only check the sign of the first $m - 1$ determinants and of the coefficient a_m.

Example 6.1

As an application of the Routh-Hurwitz criterion, let us check the stability of the system shown in Fig. 6.1, assuming that the angle θ is restricted to small values.

The equations of motion for small θ can be shown to have the form

$$\begin{aligned}(M + m)\ddot{x} + c\dot{x} + kx + mL\ddot{\theta} &= 0, \\ mL\ddot{x} + mL^2\ddot{\theta} + mgL\theta &= 0,\end{aligned} \tag{a}$$

leading to the characteristic equation

$$\begin{aligned}a_0 \lambda^4 &+ a_1 \lambda^3 + a_2 \lambda^2 + a_3 \lambda + a_4 \\ &= ML\lambda^4 + cL\lambda^3 + [(M + m)g + kL]\lambda^2 + cg\lambda + kg = 0.\end{aligned} \tag{b}$$

Since $a_0 = ML > 0$, we can proceed to check the sign of the determinants Δ_i ($i = 1, 2, 3, 4$). The determinants have the values

$$\begin{aligned}\Delta_1 &= a_1 = cL > 0, \\ \Delta_2 &= a_1 a_2 - a_0 a_3 = cL(mg + kL) > 0, \\ \Delta_3 &= a_3 \Delta_2 - a_1^2 a_4 = c^2 mLg^2 > 0, \\ \Delta_4 &= a_4 \Delta_3 = c^2 mLkg^3 > 0,\end{aligned} \tag{c}$$

which are all positive. Hence, according to the Routh-Hurwitz criterion, all the roots of the polynomial (b) have negative real parts, with the implication that the motion is asymptotically stable.

FIGURE 6.1

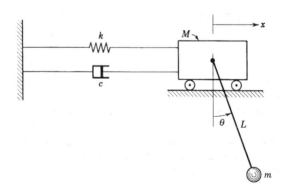

6.4 THE VARIATIONAL EQUATIONS

The motion of an n-degree-of-freedom dynamical system can be described by a set of $2n$ first-order differential equations of the Hamiltonian type, such as Eqs. (5.4). For an autonomous system Eqs. (5.4) reduce to the form

$$\dot{x}_i = X_i(x_1, x_2, \ldots, x_m), \qquad i = 1, 2, \ldots, m, \qquad (6.65)$$

where $m = 2n$. The functions X_i, which for an autonomous system do not depend explicitly on time, are assumed to satisfy Lipschitz conditions in a domain D, ensuring that Eqs. (6.65) have unique solutions

$$x_i(t) = \varphi_i(\alpha_1, \alpha_2, \ldots, \alpha_m, t), \qquad i = 1, 2, \ldots, m, \qquad (6.66)$$

where the α_i's represent the initial conditions.

The above statements cannot be construed, however, to mean that a solution of Eqs. (6.65) can always be found. Indeed, general solutions of the nonlinear system (6.65) are difficult, if not impossible, to obtain. Quite frequently, meaningful information about the system (6.65) can be obtained by studying the motion characteristics in the neighborhood of known motions. The cases in which the motion takes place in the vicinity of either singular points or closed trajectories are of particular importance. In Chap. 5 we discussed the behavior of system (6.65), and in particular its motion stability in the neighborhood of equilibrium points or stationary motions, by using a geometric approach. In this chapter we shall use analytical methods to investigate the stability of the system.

Recalling some of the definitions of Sec. 5.1, we shall refer to a known solution of Eqs. (6.65) as the *unperturbed motion* and define the solution of these equations in the neighborhood of the unperturbed motion as the *perturbed motion*. The latter motion can be expressed in the form

$$x_i(t) = \varphi_i(t) + y_i(t), \qquad i = 1, 2, \ldots, m, \qquad (6.67)$$

where the functions $\varphi_i(t)$ constitute a known solution of Eqs. (6.65) and the functions $y_i(t)$ are referred to as *perturbations*. We shall be interested in the cases in which the functions $\varphi_i(t)$ are either constant or periodic. When the functions $\varphi_i(t)$ are constant, the perturbed motion takes place in the neighborhood of equilibrium points, and when they are periodic, the perturbed motion is confined to the neighborhood of cyclic trajectories representing stationary motions. Introducing Eqs. (6.67) into (6.65) and recalling that the functions $\varphi_i(t)$ satisfy Eqs. (6.65), we obtain

$$\dot{y}_i(t) = X_i(\varphi_1 + y_1, \varphi_2 + y_2, \ldots, \varphi_m + y_m) - X_i(\varphi_1, \varphi_2, \ldots, \varphi_m),$$
$$i = 1, 2, \ldots, m, \qquad (6.68)$$

which are known as the *differential equations of the perturbed motion*. The origin $y_i = 0$ is a trivial solution of these equations. Expanding Eqs. (6.68) about the origin, we obtain

$$\dot{y}_i = \sum_{j=1}^{m} a_{ij}(t) y_j + \epsilon_i(y_1, y_2, \ldots, y_m, t), \quad i = 1, 2, \ldots, m, \quad (6.69)$$

where
$$a_{ij}(t) = \frac{\partial X_i(x_1, x_2, \ldots, x_m)}{\partial x_j}\bigg|_{x_i = \varphi_i} \quad (6.70)$$

are either constant or periodic, as the case may be, and the functions ϵ_i are power series in $y_1, y_2, \ldots, y_m$ containing terms of second and higher powers in those variables.

The perturbations y_i generally result from small disturbances of unknown source. Assuming that the perturbations are sufficiently small to permit second-order terms in y_j to be ignored, Eqs. (6.69) can be approximated by

$$\dot{y}_i = \sum_{j=1}^{m} a_{ij}(t) y_j, \quad i = 1, 2, \ldots, m. \quad (6.71)$$

Equations (6.71), representing the first-approximation equations, are called the *variational equations* of Poincaré. Because Eqs. (6.71) are confined to the neighborhood of states of equilibrium or stationary motion, such that the equations represent a linear approximation, the stability must be regarded as an *infinitesimal stability*. Whereas the stability definitions remain the same as in Sec. 5.1, the question whether the linearized system exhibits significant or critical behavior (see Sec. 6.3) has important implications.

It should be emphasized at this point that although the complete system of equations (6.65) is autonomous, the variational equations may be either autonomous or nonautonomous, depending on whether the coefficients a_{ij} in Eqs. (6.71) are constant or time-dependent. Note that these coefficients are denoted by lowercase letters in Eqs. (6.71), as opposed to the uppercase letters used in Eqs. (6.1), to indicate that Eqs. (6.71) represent only a linear approximation corresponding to an otherwise nonlinear system.

By contrast with the variational equations, which enable us to examine the behavior of the system in the neighborhood of known motions, later in this chapter we shall discuss the *direct, or second, method of Liapunov*, which is suitable for testing system *stability in the large*.

6.5 THEOREM ON THE FIRST-APPROXIMATION STABILITY

We first turn our attention to the case in which the coefficients a_{ij} in the variational equations (6.71) are constant, so that the set of equations representing the linear approximation is autonomous. This also implies that the

Stability of Multi-Degree-of-Freedom Autonomous Systems

motion takes place in the neighborhood of an equilibrium point at the origin. Because in this case the known solutions of Eqs. (6.65) are constant, $x_i(t) = \varphi_i(t) = \text{const}$ ($i = 1, 2, \ldots, m$), Eqs. (6.67) represent a simple translation of the origin making it coincide with a singular point. The constants φ_i are obtained by replacing the variables $x_1, x_2, \ldots, x_m$ by $\varphi_1, \varphi_2, \ldots, \varphi_m$, respectively, in the functions X_i of Eqs. (6.65) and solving the algebraic equations

$$X_i(\varphi_1, \varphi_2, \ldots, \varphi_m) = 0, \qquad i = 1, 2, \ldots, m. \tag{6.72}$$

In problems involving rotational motion Eqs. (6.72) are generally transcendental.

The variational equations

$$\dot{y}_i = \sum_{j=1}^{m} a_{ij} y_j, \qquad i = 1, 2, \ldots, m, \tag{6.73}$$

can be regarded as representing a linear autonomous system similar in form to system (6.36). It follows that the behavior of system (6.73) in the neighborhood of the origin can be investigated by the methods of Sec. 6.3. In contrast with Sec. 6.3, however, if the eigenvalues associated with the matrix [a] of the coefficients a_{ij} possess nonpositive real parts, then the system is said to be *infinitesimally stable* rather than merely stable. Furthermore, the stability conclusions reached by using the variational equations (6.73), instead of the complete equations (6.65), must be qualified according to the following:

Theorem *If the variational system of equations possesses significant behavior, then the stability characteristics of the linear approximation are the same as for the complete nonlinear equations.*

This theorem is often referred to as *Liapunov's theorem on the stability in the first approximation*.

If, on the other hand, the variational system possesses critical behavior, the stability analysis based on the linear approximation must be regarded as inconclusive, and the higher-order terms contained in the nonlinear functions ϵ_i of Eqs. (6.69) should be considered. This statement can easily be visualized in terms of a single-degree-of-freedom system for which the case of critical behavior was shown in Sec. 5.2 to correspond to the line separating the regions of asymptotic stability and asymptotic instability in the parameter plane (Fig. 5.4).

The case in which the coefficients a_{ij} are periodic functions of time will be discussed in Chap. 7.

Example 6.2

Let us consider the stability of rotation about the center of mass of a torque-free body with moments of inertia A, B, and C about the principal axes 1, 2, and 3, respectively.

The rotational motion is governed by Euler's moment equations

$$\begin{aligned} A\dot{\omega}_1 + (C - B)\omega_2\omega_3 &= 0, \\ B\dot{\omega}_2 + (A - C)\omega_1\omega_3 &= 0, \\ C\dot{\omega}_3 + (B - A)\omega_1\omega_2 &= 0. \end{aligned} \quad (a)$$

As unperturbed motion, we consider the steady rotation about axis 3 defined by

$$\omega_1 = \omega_2 = 0, \qquad \omega_3 = \omega_0 = \text{const}, \quad (b)$$

so that the perturbed motion can be written

$$\omega_1 = \epsilon_1, \qquad \omega_2 = \epsilon_2, \qquad \omega_3 = \omega_0 + \epsilon_3, \quad (c)$$

where the ϵ_i ($i = 1, 2, 3$) are small quantities. Introducing Eqs. (c) into (a), we obtain the equations of the perturbed motion

$$\begin{aligned} A\dot{\epsilon}_1 + (C - B)(\omega_0 + \epsilon_3)\epsilon_2 &= 0, \\ B\dot{\epsilon}_2 + (A - C)(\omega_0 + \epsilon_3)\epsilon_1 &= 0, \\ C\dot{\epsilon}_3 + (B - A)\epsilon_1\epsilon_2 &= 0. \end{aligned} \quad (d)$$

Ignoring the cross products in the perturbations ϵ_1, ϵ_2, and ϵ_3, we obtain the variational equations

$$\begin{aligned} \dot{\epsilon}_1 + \frac{(C - B)\omega_0}{A}\epsilon_2 &= 0, \\ \dot{\epsilon}_2 + \frac{(A - C)\omega_0}{B}\epsilon_1 &= 0, \\ \dot{\epsilon}_3 &= 0, \end{aligned} \quad (e)$$

leading to the characteristic equation

$$\lambda\left[\lambda^2 - \frac{(C - B)(A - C)\omega_0^2}{AB}\right] = 0. \quad (f)$$

One root of the equation is zero, whereas the remaining two roots are

$$\left.\begin{aligned}\lambda_1 \\ \lambda_2\end{aligned}\right\} = \pm i\omega_0\left[\frac{(C - B)(C - A)}{AB}\right]^{\frac{1}{2}}. \quad (g)$$

From Eq. (g) we conclude that there are two cases in which the body exhibits significant behavior, namely, when $C > B$ and $C < A$ or when $C < B$ and $C > A$. Both cases represent rotation about the axis of intermediate moment of inertia, and this rotation is *unstable*, as the roots λ_1 and λ_2 are real and of opposite sign. On the other hand, when $C > B$ and $C > A$, or when $C < B$ and $C < A$, the body exhibits critical behavior because in both these cases the roots are purely imaginary. The cases represent rotations about the axes of maximum and minimum moments of inertia, respectively. In both cases the analysis based on the linear approximation is inconclusive. We shall return to this example in Sec. 6.7.

6.6 VARIATION FROM CANONICAL SYSTEMS. CONSTANT COEFFICIENTS

The preceding discussion of the theory of linear equations with constant coefficients was of a perfectly general nature and applicable to systems from a large variety of fields. Yet our interest lies primarily not in a general system but in a particular one, namely, in a system of variational equations corresponding to a set of Hamilton's canonical equations. It turns out that in this particular case additional information concerning the system behavior can be gained without actually solving the characteristic equation. In this section we discuss the autonomous case, and in Sec. 7.4 we consider a variational system with periodic coefficients.

Let us consider the Hamilton canonical equations corresponding to an n-degree-of-freedom holonomic conservative system

$$\dot{q}_i = \frac{\partial H}{\partial p_i}, \qquad \dot{p}_i = -\frac{\partial H}{\partial q_i}, \qquad i = 1, 2, \ldots, n, \tag{6.74}$$

where $H = H(q_1, q_2, \ldots, q_n, p_1, p_2, \ldots, p_n)$ is the Hamiltonian function, which does not depend explicitly on time. Let us denote the n generalized coordinates q_i and n generalized momenta p_i as

$$q_i = x_i, \qquad p_i = x_{n+i}, \qquad i = 1, 2, \ldots, n, \tag{6.75}$$

so that Eqs. (6.74) reduce to

$$\dot{x}_i = \frac{\partial H}{\partial x_{n+i}}, \qquad \dot{x}_{n+i} = -\frac{\partial H}{\partial x_i}, \qquad i = 1, 2, \ldots, n. \tag{6.76}$$

Introducing the $2n \times 2n$ skew-symmetric matrix

$$[Z] = \begin{bmatrix} [0] & [1_n] \\ -[1_n] & [0] \end{bmatrix}, \tag{6.77}$$

in which $[1_n]$ is the $n \times n$ unit matrix, Eqs. (6.76) can be written in the form

$$\{\dot{x}\} = [Z]\left\{\frac{\partial H}{\partial x}\right\}, \tag{6.78}$$

where the column matrices $\{x\}$ and $\{\partial H/\partial x\}$ represent $2n$-vectors.

Next we consider the constant solution $x_i = \varphi_i$ ($i = 1, 2, \ldots, 2n$) and denote the perturbed motion by

$$x_i = \varphi_i + y_i, \qquad i = 1, 2, \ldots, 2n, \tag{6.79}$$

where the y_i's represent small perturbations. Introducing Eqs. (6.79) into (6.78) and retaining only the first-order terms in y_i ($i = 1, 2, \ldots, 2n$), we obtain the variational equations with constant coefficients

$$\{\dot{y}\} = [Z][\mathscr{H}]\{y\}, \tag{6.80}$$

in which $[\mathscr{H}]$ is the Hessian matrix associated with the Hamiltonian, namely, a $2n \times 2n$ symmetric matrix whose elements are

$$\mathscr{H}_{ij} = \left.\frac{\partial^2 H}{\partial x_i\, \partial x_j}\right|_{x_i = \varphi_i}, \qquad i,j = 1, 2, \ldots, 2n. \tag{6.81}$$

From Sec. 6.2 we conclude that the behavior of system (6.80) depends on the roots of the characteristic equation

$$\det\left([Z][\mathscr{H}] - \lambda[1_{2n}]\right) = 0, \tag{6.82}$$

where $[1_{2n}]$ is the $2n \times 2n$ unit matrix. The matrix $[Z]$ possesses certain interesting properties. For example, on the one hand we note that $[Z][Z] = -[1_{2n}]$ and $[Z]^T[Z] = [1_{2n}]$, and on the other hand that $\det[Z] = \det([Z][Z]) = \det([Z]^T[Z]) = 1$. But the characteristic determinant corresponding to the transpose of $[Z][\mathscr{H}]$ must also be zero

$$\det\left([\mathscr{H}][Z]^T - \lambda[1_{2n}]\right) = 0. \tag{6.83}$$

Recalling the rule on the determinant of a product of matrices, we conclude from Eq. (6.82) and the properties of $[Z]$ that

$$\det[Z]\det\left([Z][\mathscr{H}] - \lambda[1_{2n}]\right)\det[Z]^T$$
$$= \det\left([Z]([Z][\mathscr{H}] - \lambda[1_{2n}])[Z]^T\right)$$
$$= \det\left(-[\mathscr{H}][Z]^T - \lambda[1_{2n}]\right) = \det\left([\mathscr{H}][Z]^T + \lambda[1_{2n}]\right) = 0, \tag{6.84}$$

from which it follows that if λ is an eigenvalue of the variational system (6.80), then $-\lambda$ is also an eigenvalue. Hence, the characteristic equation (6.82) must contain only even powers of λ. If there is an eigenvalue λ with negative real part, then there must be another one with positive real part, so that the system is unstable. Stability is possible only when all the roots are purely imaginary, in which case they occur in pairs of complex conjugates. This is consistent with the fact that the system is conservative, for which no asymptotic stability is possible. However, since the case of purely imaginary roots implies that the variational system possesses critical behavior, it follows that system stability is determined by the higher-order terms in the differential equations of the perturbed motion. Of course, if the system possessing purely imaginary roots is linear, so that there are no higher-order terms, then it must be regarded as stable.

6.7 THE LIAPUNOV DIRECT METHOD

The *Liapunov direct method*, also known as *Liapunov's second method*, provides an approach to the question of the stability of dynamical systems entirely different from the analysis based on the variational equations. The idea behind this approach is to answer the stability question by utilizing the differential equations governing system behavior *without actually solving these equations*. The method consists of devising for the dynamical system a suitable scalar function defined in the phase space or motion space and using this function in conjunction with the differential equations in an attempt to test system stability. Such testing functions, if they can be found, are referred to as *Liapunov functions*. As noted in the introductory remarks to this chapter, the Liapunov direct method must be regarded at present as more of a philosophy of approach than a method, since there is no systematic way of producing a Liapunov function for a general dynamical system. The approach can truly be called a method only in the case of linear autonomous systems, for which a Liapunov function can be produced by solving a set of simultaneous algebraic equations of order $n(2n + 1)$, where n is the number of degrees of freedom of the system. When a Liapunov function can be found for the system, the stability criteria are derived on the basis of the sign-definiteness of the function and its total time derivative, where the latter is evaluated along a trajectory of the system of differential equations. In this manner, the problem of integrating the differential equations of the system is circumvented. By contrast with the method based on the variational equations, the Liapunov direct method is by no means restricted to linear systems, and, provided that a Liapunov function is found, the method is suitable for testing the stability of dynamical systems in the large. The method was inspired by Dirichlet's proof of Lagrange's theorem on the stability of equilibrium of a system (see Sec. 5.3), so that, in effect, the Liapunov method can be regarded as an extension of the energy concept, the Liapunov function being the counterpart of the energy function. The Liapunov direct method is on a higher level of abstraction, however, as the Liapunov function represents a mathematical concept not necessarily possessing physical meaning, and, as such, it can be applied to stability problems from a large variety of fields. There is no unique Liapunov function for a given system, and indeed there is a large degree of flexibility in the selection of a Liapunov function. The method represents a basic tool in the treatment of stability problems.

Let us confine ourselves to an n-degree-of-freedom autonomous system governed by a set of m first-order differential equations, $m = 2n$, which can be written in the vector form

$$\dot{\mathbf{x}} = \mathbf{X}(\mathbf{x}), \tag{6.85}$$

where $\mathbf{x}$ and $\mathbf{X}$ are real m-vectors. The vector $\mathbf{X}$ is assumed to be continuous and satisfy Lipschitz conditions in a spherical domain D_h: $\|\mathbf{x}\| \le h$ with the center at the origin of the Euclidean m-space E^m, where h is a positive constant. In addition $\mathbf{X}$ satisfies the relation $\mathbf{X}(\mathbf{0}) = \mathbf{0}$, so that the origin is a singular point of the system. For any initial vector $\mathbf{x}(t_0) = \mathbf{x}_0$ in D_h, Eq. (6.85) has a unique solution $\mathbf{x}(\mathbf{x}_0, t)$, $t \ge t_0$. We shall be concerned with the stability of that solution in the neighborhood of the origin.

Next let us consider a real continuous function $V(\mathbf{x})$ possessing continuous first partial derivatives with respect to the variables x_i ($i = 1, 2, \ldots, m$) in D_h and vanishing at the origin, $V(\mathbf{0}) = 0$. For such a function, we introduce the following definitions:

1. The function $V(\mathbf{x})$ is called *positive (negative) definite* in a certain domain D_h if $V(\mathbf{x}) > 0$ (< 0) for all $\mathbf{x} \ne \mathbf{0}$ and $V(\mathbf{0}) = 0$.
2. The function $V(\mathbf{x})$ is said to be *positive (negative) semidefinite* in a certain domain D_h if $V(\mathbf{x}) \ge 0$ (≤ 0), and it can vanish also for some $\mathbf{x} \ne \mathbf{0}$ in D_h.
3. The function $V(\mathbf{x})$ is called *indefinite* if it can assume both positive and negative values in the domain D_h no matter how small the value of h.

The positive (negative) definite and semidefinite functions are sometimes referred to as *constant with respect to sign*, or *sign-constant*, whereas indefinite functions are called *variable with respect to sign*, or *sign-variable*.

If $V(\mathbf{x}) = V(x_1, x_2, \ldots, x_m)$ is a homogeneous function of order p in the variables x_i, then, for any arbitrary λ, we have

$$V(\lambda x_1, \lambda x_2, \ldots, \lambda x_m) = \lambda^p V(x_1, x_2, \ldots, x_m), \tag{6.86}$$

so that if p is an odd integer, the function V is indefinite. When $p = 2$, the function V has the quadratic form

$$V = \sum_{i=1}^{m} \sum_{j=1}^{m} \alpha_{ij} x_i x_j = \{x\}^T [\alpha]\{x\}, \tag{6.87}$$

where $[\alpha]$ is the symmetric matrix of the coefficients α_{ij}. It may be possible to check the sign of V by using a linear transformation reducing $[\alpha]$ to a diagonal form. The positive definiteness of the quadratic form V can more readily be checked by means of *Sylvester's theorem*, which can be stated as follows: *The necessary and sufficient conditions for the quadratic form* (6.87) *to be positive definite are that all the principal minor determinants corresponding to the symmetric matrix of the coefficients* $[\alpha]$ *be positive.* Mathematically, the conditions can be written

$$|\alpha_{qr}| > 0, \quad \begin{array}{l} q, r = 1, 2, \ldots, k, \\ k = 1, 2, \ldots, m. \end{array} \quad (6.88)$$

The principal minor determinants $|\alpha_{qr}|$ are called *discriminants* of the quadratic form. (A proof of the theorem is given in Ref. 4, Sec. 20.) The theorem is often referred to as *Sylvester's criterion*. If the quadratic form V is positive definite, then the matrix of the coefficients $[\alpha]$ is said to be a *positive definite matrix*.

It will prove of interest to examine the geometrical nature of positive definite functions. To this end, we consider a positive definite function $V(\mathbf{x})$ in a given domain D_h of the Euclidean m-space. For any positive constant c, the equation

$$V(\mathbf{x}) = V(x_1, x_2, \ldots, x_m) = c \quad (6.89)$$

represents an m-dimensional surface in the Euclidean space. By the definition of a positive definite function, when $c = 0$, the surface reduces to a point at the origin of the space, $V(\mathbf{0}) = 0$. For small values of c, say $c = c_1$, $V(\mathbf{x}) = c_1$ represents a closed surface enclosing the origin (see Fig. 6.2). For any other value $c_2 > c_1$ the surface $V(\mathbf{x}) = c_2$ also represents a closed surface enclosing the surface $V(\mathbf{x}) = c_1$ without intersecting it. Hence, $V(\mathbf{x}) = c$ represents a family of closed nonintersecting surfaces in the neighborhood of the origin. These surfaces increase in size with c and shrink to a point at the origin for $c = 0$. Next let us consider the surface $V = k$ and denote by $\|\mathbf{x}\| = \epsilon$ the smallest sphere enclosing V and by $\|\mathbf{x}\| = \delta$ the

FIGURE 6.2

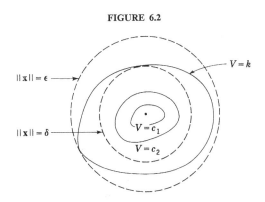

largest sphere enclosed by V, where ϵ and δ are positive numbers such that $\delta < \epsilon$. For a given ϵ, δ depends on ϵ, $\delta = \delta(\epsilon)$. The infimum of V on the sphere $\|\mathbf{x}\| = \epsilon$ will be designated by $V_\epsilon = \inf V(\mathbf{x})$, on $\|\mathbf{x}\| = \epsilon$, and the supremum of V on the sphere $\|\mathbf{x}\| = \delta$ by $V_\delta = \sup V(\mathbf{x})$, on $\|\mathbf{x}\| = \delta$. We note that for $\|\mathbf{x}\| > \epsilon$ we have $V(\mathbf{x}) > V_\epsilon$, whereas for $\|\mathbf{x}\| < \delta$ we have $V(\mathbf{x}) < V_\delta$. Hence for $V = k$ to lie entirely in D_h we must have $\epsilon \leq h$.

The total derivative of the function $V(\mathbf{x})$ with respect to time, evaluated along a trajectory Γ of system (6.85), is defined as

$$\dot{V} = \frac{dV}{dt} = \sum_{i=1}^{m} \frac{\partial V}{\partial x_i} \frac{dx_i}{dt} = \sum_{i=1}^{m} \frac{\partial V}{\partial x_i} X_i = \nabla V \cdot \mathbf{X}, \tag{6.90}$$

where ∇V is the gradient of the scalar function V.

Now we are in the position to consider the *Liapunov stability theorems*. The first theorem can be stated as follows:

Theorem 6.7.1 *If there exists for the system* (6.85) *a positive* (*negative*) *definite function* $V(\mathbf{x})$ *whose total time derivative* $\dot{V}(\mathbf{x})$ *is negative* (*positive*) *semidefinite along every trajectory of* (6.85), *then the trivial solution* $\mathbf{x} = \mathbf{0}$ *is stable.*

To prove the theorem, we assume that the function V is positive definite in the domain D_h and, moreover, that the inequality

$$\dot{V} = \nabla V \cdot \mathbf{X} \leq 0 \tag{6.91}$$

is satisfied in the same domain. Let ϵ be an arbitrarily small positive number such that $\epsilon < h$ and denote by $\|\mathbf{x}\| = \epsilon$ the sphere centered at the origin and of radius ϵ. If $V_\epsilon = \inf V(\mathbf{x})$ on the sphere $\|\mathbf{x}\| = \epsilon$, it follows that

$$V(\mathbf{x}) \geq V_\epsilon \quad \text{on} \quad \|\mathbf{x}\| = \epsilon. \tag{6.92}$$

We note that $V_\epsilon > 0$ because V is positive definite.

If the motion is initiated at a point $\mathbf{x}(t_0) = \mathbf{x}_0$ inside a spherical domain

$$\|\mathbf{x}\| \leq \delta, \tag{6.93}$$

where $\delta = \delta(\epsilon)$ is a positive number, we must have

$$V(\mathbf{x}_0) < V_\epsilon. \tag{6.94}$$

Such a number δ exists by virtue of the fact that $V(\mathbf{x})$ is a continuous function and $V(\mathbf{0}) = 0$.

Substituting the solution $\mathbf{x}(t)$, corresponding to the initial condition $\mathbf{x}_0$, in the function V, we obtain a function of time which, in view of (6.91), cannot increase in the domain D_h. It follows that

$$V(\mathbf{x}) \le V(\mathbf{x}_0) < V_\epsilon, \qquad t > t_0, \tag{6.95}$$

from which we conclude that

$$\|\mathbf{x}\| \le \epsilon, \qquad t > t_0, \tag{6.96}$$

at least for values of t close to t_0. Assuming for the moment that at a certain time $t = T$ the solution of Eq. (6.85) is such that

$$\|\mathbf{x}(T)\| = \epsilon, \tag{6.97}$$

then, because of (6.92), we must have

$$V(\mathbf{x}(T)) \ge V_\epsilon. \tag{6.98}$$

This is impossible, however, because $\epsilon < h$ and the set $\|\mathbf{x}\| = \epsilon$ lies inside the domain D_h in which the inequality (6.95) must be satisfied. Hence, if the solution $\mathbf{x}$ of Eq. (6.85) is such that the inequality (6.95) is satisfied, the motion will remain in the domain $\|\mathbf{x}\| < \epsilon$ for all $t > t_0$, so that the equilibrium is stable.

The second stability theorem of Liapunov is concerned with the asymptotic stability of a system in the neighborhood of the origin.

Theorem 6.7.2 *If there exists for the system* (6.85) *a positive* (*negative*) *definite function* $V(\mathbf{x})$ *whose total time derivative* $\dot{V}(\mathbf{x})$ *is negative* (*positive*) *definite along every trajectory of* (6.85), *then the trivial solution is asymptotically stable.*

The proof of Theorem 6.7.1 can be used to prove that any motion initiated in the domain D_δ, $\|\mathbf{x}\| \le \delta$ is at least stable. But in the present case $\dot{V}$ is negative definite in D_h rather than merely negative semidefinite, so that V must be a monotonically decreasing function in time for all $\mathbf{x} \ne \mathbf{0}$. We shall show that V is never stalled above a certain value α, where $\alpha > 0$, but tends to zero as $t \to \infty$.

Since $\dot{V}$ is a negative definite function in D_h, we assume that there is a positive number b such that

$$\dot{V}(\mathbf{x}) \le -b. \tag{6.99}$$

It follows that for all $t > t_0$ we must have

$$V(\mathbf{x}) = V(\mathbf{x}_0) + \int_{t_0}^{t} \dot{V}(\mathbf{x}) \, dt \le V(\mathbf{x}_0) - b(t - t_0). \tag{6.100}$$

But (6.100) implies that for a value of t sufficiently large $V(\mathbf{x})$ becomes negative in D_h, which contradicts the condition of the theorem that $V(\mathbf{x})$ be positive definite in D_h. Hence, we are led to the conclusion that

$$\lim_{t \to \infty} V(\mathbf{x}) = 0, \qquad (6.101)$$

from which it follows that

$$\lim_{t \to \infty} \mathbf{x}(t) = \mathbf{0}, \qquad (6.102)$$

or the motion initiated inside D_δ is asymptotically stable. We conclude from the above that the *domain of attraction* of the point $\mathbf{x} = \mathbf{0}$ is no smaller than D_δ.

Similarly, there are two *Liapunov instability theorems*.

Theorem 6.7.3 *If there exists for the system* (6.85) *a function* $V(\mathbf{x})$ *whose total time derivative* $\dot{V}(\mathbf{x})$ *is positive (negative) definite along every trajectory of* (6.85) *and the function itself can assume positive (negative) values for arbitrarily small values of* $\mathbf{x}$, *then the trivial solution is unstable.*

Theorem 6.7.4 *If there exists for the system* (6.85) *a function* $V(\mathbf{x})$ *such that*

$$\dot{V}(\mathbf{x}) = \lambda V(\mathbf{x}) + W(\mathbf{x}) \qquad (6.103)$$

along every trajectory of (6.85), *where* λ *is a positive constant and* $W(\mathbf{x})$ *is either identically zero or is a positive (negative) function, and if in the latter case the function* $V(\mathbf{x})$ *is not negative (positive), then the trivial solution is unstable.*

Liapunov's instability theorem, Theorem 6.7.3, requires that $\dot{V}$ be sign-definite in an entire spherical domain D_h: $\|\mathbf{x}\| \leq h$ of the origin and that V itself be not sign-definite of opposite sign in the same domain. It turns out that these conditions are unduly restrictive. Indeed, as demonstrated by Chetayev, conditions on $\dot{V}$ and V need be satisfied only in an arbitrarily small subdomain D_1 of D_h. *Chetayev's instability theorem* can be stated as follows:

Theorem 6.7.5 *If there exists for the system* (6.85) *a function* $V(\mathbf{x})$ *such that* (1) *in any arbitrarily small neighborhood of the origin there is a region* D_1 *in which* $V > 0$ *and on whose boundaries* $V = 0$, (2) *at all points of the region in which* $V > 0$ *the total time derivative* $\dot{V}$ *assumes positive values along every trajectory of* (6.85), *and* (3) *the origin is a boundary point of* D_1, *then the trivial solution is unstable.*

Chetayev's instability theorem represents an important generalization of Liapunov's instability theorem, Theorem 6.7.3.

We shall not attempt to prove the preceding instability theorems, but in Sec. 6.8 we present a geometric interpretation of them instead. The interested reader can find proofs of the theorems in Ref. 12, pp. 37–43 and 195–196.

Perhaps of no less importance are two other generalizations of the Liapunov theorems which are designed to enhance the usefulness of the Liapunov direct method by removing certain restrictions on the Liapunov functions.

The first generalization, proved by Barbasin and Krasovskii, concerns the asymptotic stability of a system and can be stated as follows:

Theorem 6.7.6 *If there exists for the system (6.85) a positive (negative) definite function $V(\mathbf{x})$ whose total time derivative $\dot{V}(\mathbf{x})$ is negative (positive) semidefinite along every trajectory of (6.85), and if the set of points S at which $\dot{V}(\mathbf{x})$ is zero contains no nontrivial positive half-trajectory $\mathbf{x}(t)$, $t \geq t_0$, then the trivial solution is asymptotically stable.*

The proof of the theorem is based on the fact that the positive half-trajectory must contain points for which $\dot{V}(\mathbf{x}) < 0$, so that there must exist a time $t_1 > t_0$ for which $V(\mathbf{x}(t_1)) < V(\mathbf{x}(t_0))$. This, in turn, is an indication that there cannot be a trajectory along which $\|\mathbf{x}\| > \eta > 0$ for all $t \geq t_0$.

The second generalization, due to Krasovskii, is concerned with the instability of a system and reads as follows:

Theorem 6.7.7 *Suppose that there exists for the system (6.85) a function $V(\mathbf{x})$ which is continuous at the origin and vanishing there, $V(\mathbf{0}) = 0$, and whose total time derivative $\dot{V}(\mathbf{x})$ is positive (negative) semidefinite along every trajectory of (6.85). Moreover, the function $V(\mathbf{x})$ can assume positive (negative) values, and the set of points S at which $\dot{V}(\mathbf{x})$ is zero contains no nontrivial positive half-trajectory $\mathbf{x}(t)$, $t \geq t_0$. Suppose further that in every neighborhood of the origin there is a point $\mathbf{x}(t_0) = \mathbf{x}_0$ such that for arbitrary $t_0 \geq 0$ we have $V(\mathbf{x}_0) > 0$ (< 0). Then the trivial solution is unstable, and the trajectories $\mathbf{x}(\mathbf{x}_0, t_0, t)$ for which $V(\mathbf{x}_0) > 0$ (< 0) must leave the open domain $\|\mathbf{x}\| < \epsilon$ as the time t increases.*

Theorems 6.7.6 and 6.7.7 are special cases of more general theorems, applicable to both autonomous systems and systems for which the vector $\mathbf{X}(\mathbf{x},t)$ is periodic, the proofs of which can be found in Ref. 9, secs. 14 and 15. The theorems are designed to replace Theorems 6.7.2 and 6.7.3 of Liapunov. There appears no reason, however, why Theorem 6.7.7 cannot be generalized to become the counterpart of Chetayev's instability theorem, Theorem 6.7.5. A geometric interpretation of Theorems 6.7.6 and 6.7.7 is provided in Sec. 6.8. These theorems have significant implications, and their practical value is amply demonstrated in problems involving the stability analysis of damped mechanical systems, as we shall see in Sec. 6.9.

The main drawback of the Liapunov direct method lies in the fact that there are no established criteria for the selection of a Liapunov function except for the linear autonomous case. A certain Liapunov function may provide a strong sufficient condition for stability which may not be a necessary condition. This implies that an inappropriate choice of the Liapunov function may lead to unduly restrictive stability conditions. The fact that a Liapunov function cannot be found gives no indication of system stability or its lack. Although the construction of Liapunov functions leaves the impression of being more an art than a scientific endeavor, the situation is not altogether bad. This is particularly true in the case of mechanical systems for which clues are available for the construction of Liapunov functions in the form of motion integrals, such as the Jacobi integral and momentum integrals or a combination thereof. Moreover, for linear autonomous systems the construction of a Liapunov function reduces to the solution of a set of algebraic equations. This case is discussed in Sec. 6.11. If precise rules for the construction of Liapunov functions existed for any system, the Liapunov second method could achieve the status of a unifying principle for the general treatment of stability problems. A great deal of effort is constantly being directed toward developing such rules for broad classes of systems.

Example 6.3

Now let us return to Example 6.2, from which we obtain the equations for the perturbed motion

$$A\dot{\epsilon}_1 + (C - B)(\omega_0 + \epsilon_3)\epsilon_2 = 0,$$
$$B\dot{\epsilon}_2 + (A - C)(\omega_0 + \epsilon_3)\epsilon_1 = 0, \qquad (a)$$
$$C\dot{\epsilon}_3 + (B - A)\epsilon_1\epsilon_2 = 0.$$

Multiplying the first of Eqs. (a) by $(A - C)\epsilon_1$ and the second one by $(C - B)\epsilon_2$, and subtracting the one from the other, we obtain

$$\frac{d}{dt}[\tfrac{1}{2}A(A - C)\epsilon_1^2 - \tfrac{1}{2}B(C - B)\epsilon_2^2] = 0, \qquad (b)$$

so that the function

$$V_1(\epsilon_1,\epsilon_2) = A(A - C)\epsilon_1^2 - B(C - B)\epsilon_2^2 \qquad (c)$$

appears as a suitable prospect for consideration as a Liapunov function since its time derivative is zero. If V_1 is positive definite, by virtue of Liapunov's Theorem 6.7.1, the rotational motion is stable. Hence, for stability we must have

$$A > C \quad \text{and} \quad C < B. \qquad (d)$$

This implies that *the rotational motion about the axis of minimum moment of inertia is stable.*

Choosing the Liapunov function in the form

$$V_2(\epsilon_1,\epsilon_2) = B(C - B)\epsilon_2{}^2 - A(A - C)\epsilon_1{}^2 \qquad (e)$$

and noting that $\dot{V}_2 = 0$, we conclude that the motion is stable when

$$C > B \quad \text{and} \quad A < C, \qquad (f)$$

so that *the rotational motion about the axis of maximum moment of inertia is also stable.*

To investigate the stability of motion about the axis of intermediate moment of inertia, we select the Liapunov function

$$V_3 = \epsilon_1 \epsilon_2, \qquad (g)$$

and using Eqs. (*a*), we can write

$$\dot{V}_3 = (\omega_0 + \epsilon_3)\left(\frac{C - A}{B}\epsilon_1{}^2 + \frac{B - C}{A}\epsilon_2{}^2\right). \qquad (h)$$

When $B > C > A$, $\dot{V}_3$ is positive definite, and because V_3 is positive in the region $\epsilon_1, \epsilon_2 > 0$, by Chetayev's instability theorem, we conclude that *the rotation about the axis of intermediate moment of inertia is unstable.*

6.8 GEOMETRIC INTERPRETATION OF THE LIAPUNOV DIRECT METHOD

The theorems of the preceding section lend themselves to a geometric interpretation which deserves special consideration. We shall find it convenient to introduce the concepts by means of a second-order system, $m = 2$, for which the phase space reduces to the phase plane $x_1 x_2$. The function $V(x_1,x_2)$ can be plotted in three dimensions by introducing a third coordinate z, normal to both x_1 and x_2, and letting $z = V(x_1,x_2)$. For the case in which V is positive definite the surface $z = V(x_1,x_2)$ resembles a cup tangent to the phase plane at the origin, as shown in Fig. 6.3a. The intersections of the surface with the planes $z = $ const consist of level curves which appear in the phase plane as closed curves surrounding the origin. Similarly, any path on the surface $z = V(x_1,x_2)$ has a projection on the phase plane. The trajectories I, II, and III in the phase plane represent solutions of Eqs. (6.85) illustrating Theorems 6.7.1, 6.7.2, and 6.7.3, respectively. The curves have corresponding projections on the surface $z = V(x_1,x_2)$. Curve I corresponds to a negative semidefinite $\dot{V}$, and although its trend is downward, the possibility exists that the motion may become stalled on a level curve at some distance away from the origin, which implies mere stability. Curve II, on the other hand, corresponds to a negative definite $\dot{V}$. The curve approaches

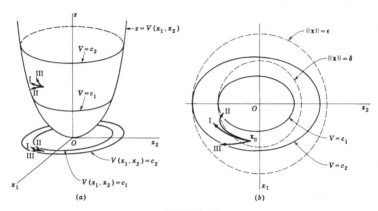

FIGURE 6.3

the origin, indicating that the system is asymptotically stable. By contrast, curve III, for which $\dot{V}$ is positive definite, represents instability. Figure 6.3b shows the associated phase plane. A trajectory from any arbitrary point on the circle $\|\mathbf{x}\| = \epsilon$ to the origin will intersect a closed curve $V = c$ provided that the circle encloses that curve. When the system is asymptotically stable, a trajectory originating at a point $\mathbf{x} = \mathbf{x}_0$, inside the circle $\|\mathbf{x}\| = \delta$ but outside the curve $V = c_1$ enclosed by that circle, will cross the curve $V = c_1$ from the exterior toward the interior.

Although the above representation corresponds to a second-order system, the geometric interpretation remains essentially the same for higher-order systems. For systems of higher order the functions $V = c$ can be regarded as representing closed nonintersecting surfaces lying one inside the other in the phase space and shrinking to the origin for $V = 0$, as indicated in Sec. 6.7. For mere stability a trajectory initiated at a point in the space enclosed by two properly chosen surfaces $V = c_1$ and $V = c_2$, $c_2 > c_1$, will remain in the enclosed space, hence will remain inside the domain $\|\mathbf{x}\| < \epsilon$. In the case of asymptotic stability a solution curve will cross the surface $V = c_1$ on its way to the origin, whereas a trajectory corresponding to an unstable system will cross the surface $V = c_2$ from the inside with the representative point increasing its distance from the origin. This again is easier to visualize by considering the phase plane (Fig. 6.3b). Hence, the Liapunov function can be regarded as a measure of the distance of the representative point from the origin, with $\dot{V}$ being a measure of the rate at which the representative point approaches the origin.

It is to be noted, however, that Theorem 6.7.3 does not require that V be sign-definite. To investigate the case in which V is not constant with respect to sign, we consider again a second-order system. The curve $V = 0$ may

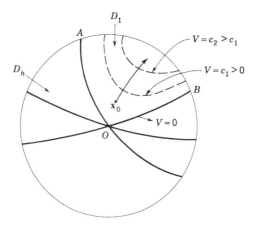

FIGURE 6.4

have one or more branches passing through the origin, as shown in Fig. 6.4. We shall assume that $\dot{V}$ is positive definite and that there is at least one region, say the region D_1 enclosed by the curves OA and OB, in which $V > 0$. Any trajectory initiated at a point $\mathbf{x}_0$ inside D_1, reasonably close to the origin, will intersect the curves $V = c > 0$ in the direction of increasing c, thus increasing the distance from the origin. On the boundaries OA and OB the function is zero, $V = 0$. A trajectory will never cross the boundaries OA or OB because this would imply that $\dot{V}$ is negative at the crossing points. Hence, a motion initiated in D_1 inside D_h will finally leave D_h.

The above geometric interpretation of Liapunov's Theorem 6.7.3 does not actually require that $\dot{V}$ be positive definite everywhere in D_h. Indeed the conclusions would be the same if $\dot{V}$ assumed positive values only at every point in D_1, instead of the entire D_h. This observation led Chetayev to the formulation of his instability theorem, Theorem 6.7.5. Hence, the preceding geometric interpretation is equally valid for Chetayev's theorem with the exception that, in this case, $\dot{V}$ need not be positive everywhere in D_h but only in the entire region in which $V > 0$, namely, in D_1. In view of this, Liapunov's Theorem 6.7.3 can be regarded as a special case of the more general Chetayev's instability theorem.

Figure 6.3 can also be used to interpret Theorems 6.7.6 and 6.7.7 geometrically. The implication of these theorems is that asymptotic stability and instability, respectively, are possible provided no positive half-trajectory coincides with a level curve or a given point other than the origin, for all times $t \geq t_0$.

Example 6.4

Let us consider the stability of the system of differential equations

$$\dot{x}_1 = -(x_1 - \beta^2 x_2)(1 - a^2 x_1^2 - b^2 x_2^2),$$
$$\dot{x}_2 = -(x_2 + \alpha^2 x_1)(1 - a^2 x_1^2 - b^2 x_2^2), \quad (a)$$

where α, β, a, and b are positive constants such that $\alpha < \beta$ and $a < b$. Clearly the origin is a trivial solution of the system of equations.

As a Liapunov function we consider the positive definite function

$$V = \alpha^2 x_1^2 + \beta^2 x_2^2. \quad (b)$$

The total time derivative of V is

$$\dot{V} = 2\alpha^2 x_1 \dot{x}_1 + 2\beta^2 x_2 \dot{x}_2, \quad (c)$$

and upon introducing Eqs. (a) into (c) we obtain

$$\dot{V} = -2(\alpha^2 x_1^2 + \beta^2 x_2^2)(1 - a^2 x_1^2 - b^2 x_2^2), \quad (d)$$

which is negative definite sufficiently close to the origin, so that the trivial solution $x_1 = x_2 = 0$ is asymptotically stable.

To prove this we consider Fig. 6.5. The circle

$$\|\mathbf{x}\| = \frac{1}{b} \quad (e)$$

encloses the circle

$$\|\mathbf{x}\| = A \quad (f)$$

FIGURE 6.5

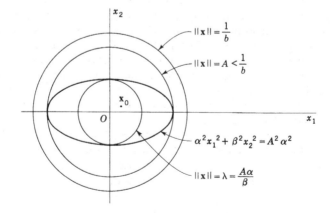

provided $A < 1/b$. Within the circle $\|\mathbf{x}\| = A$ we can inscribe the ellipse

$$\alpha^2 x_1{}^2 + \beta^2 x_2{}^2 = A^2\alpha^2, \tag{g}$$

which is tangent to the circle $\|\mathbf{x}\| = A$ at the points $x_1 = \pm A$, $x_2 = 0$. This is another way of saying that the function $V = \alpha^2 x_1{}^2 + \beta^2 x_2{}^2$ has an infimum on the circle $\|\mathbf{x}\| = A$ at either of these two points. On the other hand, V has a supremum on the circle

$$\|\mathbf{x}\| = \lambda = \frac{A\alpha}{\beta} \tag{h}$$

at either of the two points $x_1 = 0$, $x_2 = \pm A\alpha/\beta$. No motion initiated at a point $\mathbf{x}_0$ inside the circle $\|\mathbf{x}\| = \lambda$ will ever reach the ellipse $\alpha^2 x_1{}^2 + \beta^2 x_2{}^2 = A^2\alpha^2$ because $\dot{V}$ is negative definite inside that region. Hence, the representative point will cross the ellipses $V = \text{const}$, enclosed by the circle $\|\mathbf{x}\| = \lambda$, from the outside toward the inside and ultimately will reach the origin as $t \to \infty$. But for any small value $A < 1/b$ we can always find a quantity $\lambda = A\alpha/\beta$, so that the trivial solution of Eqs. (a) is asymptotically stable. It follows that the domain of attraction of the origin is no smaller than the circle $\|\mathbf{x}\| = \lambda$.

6.9 STABILITY OF CANONICAL SYSTEMS

Lagrange's theorem is a generalization of Torricelli's theorem on the stability of equilibrium of mechanical systems. We encountered Lagrange's theorem when discussing the geometric theory associated with a single-degree-of-freedom autonomous system. This theorem will be shown to be a mere example of Liapunov's Theorem 6.7.1 on the stability of trivial solutions discussed in Sec. 6.7. In doing so we shall extend the theorem to multi-degree-of-freedom systems.

Let us consider the Hamilton canonical equations associated with the n-degree-of-freedom holonomic system

$$\dot{q}_i = \frac{\partial H}{\partial p_i}, \qquad \dot{p}_i = -\frac{\partial H}{\partial q_i}, \qquad i = 1, 2, \ldots, n, \tag{6.104}$$

where q_i = generalized coordinates of system,
p_i = generalized momenta of system,
H = Hamiltonian of system.

We shall consider the case in which the origin is a trivial solution of Eqs. (6.104), so that the point $q_i = p_i = 0$ represents an equilibrium position of the system; moreover, we assume that the singularity is isolated.

For a *holonomic conservative system*, in which the Hamiltonian does not depend explicitly on time, we obtain from Eq. (2.179) $\dot{H} = 0$. This points to the Hamiltonian as a possible Liapunov function. Indeed the Hamiltonian will be a suitable Liapunov function provided it is sign-definite. This we proceed to investigate. To this end, we consider a natural system for which the kinetic energy is a homogeneous quadratic function

$$T = \frac{1}{2} \sum_{i=1}^{n} \sum_{j=1}^{n} \alpha_{ij} \dot{q}_i \dot{q}_j. \tag{6.105}$$

Recalling that the Lagrangian has the form $L = T - V$, where V is the potential energy, and using Eq. (2.165), we obtain the Hamiltonian

$$H = \sum_{i=1}^{n} \frac{\partial L}{\partial \dot{q}_i} \dot{q}_i - L = T + V = h = \text{const}, \tag{6.106}$$

so that the Hamiltonian is equal to the Jacobi integral h, which is further equal to the total energy for a natural system. But, by definition, the kinetic energy is a positive definite function of the variables p_i. If the potential energy is a positive definite function in a small neighborhood of the equilibrium, then the Hamiltonian is positive definite, hence a suitable Liapunov function. Recalling Liapunov's first stability theorem, we can state *Lagrange's theorem:*

Theorem 6.9.1 *If the potential energy has an isolated minimum in the equilibrium position, then the equilibrium is stable.*

Next we shall prove the converse of Lagrange's theorem, for which purpose we select a Liapunov function in the form

$$U = \sum_{i=1}^{n} p_i q_i. \tag{6.107}$$

Considering system (6.104), we can write

$$\dot{U} = \sum_{i=1}^{n} (p_i \dot{q}_i + \dot{p}_i q_i) = \sum_{i=1}^{n} \left(p_i \frac{\partial H}{\partial p_i} - q_i \frac{\partial H}{\partial q_i} \right), \tag{6.108}$$

where the Hamiltonian has the form (6.106).

For a natural system, the kinetic energy can be written in terms of the generalized momenta in the form

$$T = \frac{1}{2} \sum_{i=1}^{n} \sum_{j=1}^{n} \beta_{ij} p_i p_j, \tag{6.109}$$

Stability of Multi-Degree-of-Freedom Autonomous Systems

which is a positive definite function, by definition. Furthermore, we shall assume that the potential energy can be written

$$V = \sum_{r=2}^{s} V_r(q_1, q_2, \ldots, q_n) \qquad (6.110)$$

and that it is a negative definite function in the variables q_i, where V_r are homogeneous functions of degree r. The integer s represents the extent to which higher-order terms are included in V. In the neighborhood of the origin the dominant term is V_2 which obviously must be negative definite.

Introducing Eqs. (6.106), (6.109), and (6.110) into (6.108) and recalling Euler's theorem on homogeneous functions, we obtain

$$\begin{aligned} \dot{U} &= 2T - \sum_{r=2}^{s} rV_r - \sum_{k=1}^{n} \frac{\partial T}{\partial q_k} q_k \\ &= 2T - \sum_{r=2}^{s} rV_r - \sum_{k=1}^{n} q_k \sum_{i=1}^{n} \sum_{j=1}^{n} \frac{\partial \beta_{ij}}{\partial q_k} p_i p_j, \end{aligned} \qquad (6.111)$$

where a new dummy index k has been introduced to avoid confusion. But the second expression on the right side of Eq. (6.111) (minus sign included) is positive definite by virtue of the fact that the dominant term in the series is negative definite, whereas the last expression in (6.111) can be seen to be of higher order in the variables q_i and p_i. It follows that $\dot{U}$ is a positive definite function in the variables q_i and p_i. On the other hand, the function U is sign-variable, so that using Liapunov's Theorem 6.7.3, we conclude that the trivial solution is unstable. Hence, we arrive at *Liapunov's theorem:*

Theorem 6.9.2 *If for the system* (6.104) *the potential energy has a maximum in the equilibrium position, the equilibrium is unstable.*

Actually for the equilibrium to be unstable it is necessary only that the potential energy have no minimum in the equilibrium position. To show this, we consider a Liapunov function in the form

$$U = -H \sum_{i=1}^{n} p_i q_i. \qquad (6.112)$$

Because $H(q_1, q_2, \ldots, q_n, 0, 0, \ldots, 0) = V(q_1, q_2, \ldots, q_n)$, it follows that when V has no minimum in the neighborhood of the origin, by necessity there must exist in that neighborhood a region in which $H < 0$ for sufficiently small p_i ($i = 1, 2, \ldots, n$), so that in this region $V < 0$. The portion of this region in which, in addition to $H < 0$, we have $\sum_i p_i q_i > 0$ will be designated by D_1

$$D_1: H < 0 \text{ and } \sum_{i=1}^{n} p_i q_i > 0. \tag{6.113}$$

Hence $U > 0$ in D_1, and, moreover, $U = 0$ on the boundaries of D_1 because either $H = 0$ or $\sum_i p_i q_i = 0$ on these boundaries. It is also obvious that the origin is a point on the boundary.

We shall assume that the system is governed by Eqs. (6.104) and the kinetic and potential energy are given by Eqs. (6.109) and (6.110), where the latter is no longer assumed negative definite. In view of the fact that $\dot{H} = 0$ and considering Eq. (6.111), we see that the total time derivative of U becomes

$$\dot{U} = -H\left(2T - \sum_{r=2}^{s} rV_r - \sum_{k=1}^{n} q_k \sum_{i=1}^{n} \sum_{j=1}^{n} \frac{\partial \beta_{ij}}{\partial q_k} p_i p_j\right). \tag{6.114}$$

Because H is negative in D_1 and T is positive definite by definition, it follows that V must be negative. Furthermore, the last term in the parentheses is of higher order, indicating that the expression in the parentheses is positive in D_1; and because H is negative in that region, it follows that $\dot{U} > 0$ in D_1. Hence, U satisfies the conditions of Chetayev's instability theorem, Theorem 6.7.5, which leads to the *extended Liapunov theorem*.

Theorem 6.9.3 *If for the system* (6.104) *the potential energy has no minimum in the equilibrium position, the equilibrium is unstable.*

It may prove of interest to examine the motion stability of a nonnatural system, namely, a system for which the kinetic energy contains linear terms in the velocities. It will be recalled from Sec. 2.13 that when the motion is expressed in terms of rotating coordinates, the kinetic energy can be separated into terms of second, first, and zero order in the generalized velocities $\dot{q}_i$ as follows:

$$T = \frac{1}{2} \sum_{i=1}^{n} \sum_{j=1}^{n} \alpha_{ij} \dot{q}_i \dot{q}_j + \sum_{i=1}^{n} \beta_i \dot{q}_i + \gamma = T_2 + T_1 + T_0, \tag{6.115}$$

where α_{ij}, β_i, and γ are functions of the generalized coordinates q_i but not of the generalized velocities $\dot{q}_i$. If the system is governed by the differential equations (6.104), we still have a Jacobi integral in the form of the Hamiltonian, but for the nonnatural case the Hamiltonian is no longer equal to $T + V$. In fact, it was shown in Sec. 2.13 that for a nonnatural system the Hamiltonian takes the form

$$H = T_2 - T_0 + V = \text{const}, \tag{6.116}$$

which contains no linear terms in the generalized velocities. Because the Hamiltonian is constant, its total derivative is zero. Hence the Hamiltonian may again prove to be a suitable Liapunov function for testing the system stability. It turns out that T_2 is positive definite, so that we are able to formulate a *stability theorem for nonnatural systems*, which may be regarded as the counterpart of Lagrange's theorem, Theorem 6.9.1. The theorem can be stated as follows:

Theorem 6.9.4 *If for a nonnatural system the expression $V - T_0$ has an isolated minimum in the equilibrium position, then the equilibrium is stable.*

A theorem similar to Theorem 6.9.2 or Theorem 6.9.3 can be advanced with regard to the instability of nonnatural systems.

Theorems 6.9.1 and 6.9.4 give conditions under which the system is merely stable. Since system (6.104) is conservative, asymptotic stability is precluded. Although the concept of a conservative system is a valuable one, physical systems are seldom truly conservative. Hence it is natural to raise a question as to the manner in which nonconservative forces affect the system stability.

From Sec. 2.13 we see that the Hamilton canonical equations associated with an n-degree-of-freedom *holonomic nonconservative system* have the form

$$\dot{q}_i = \frac{\partial H}{\partial p_i}, \qquad \dot{p}_i = -\frac{\partial H}{\partial q_i} + Q_i, \qquad i = 1, 2, \ldots, n, \qquad (6.117)$$

where Q_i are the generalized nonconservative forces. In contrast with system (6.104), however, for system (6.117) the Hamiltonian H is not constant, so that we no longer possess a motion integral in the form of the Jacobi integral. Since the Hamiltonian does not depend explicitly on time, use of Eqs. (6.117) leads to

$$\dot{H} = \sum_{i=1}^{n} \left(\frac{\partial H}{\partial q_i} \dot{q}_i + \frac{\partial H}{\partial p_i} \dot{p}_i \right) = \sum_{i=1}^{n} Q_i \dot{q}_i, \qquad (6.118)$$

so that the sign-definiteness of $\dot{H}$ depends on the nature of the nonconservative forces Q_i.

We have considerable interest in the case in which the quantities Q_i represent damping forces. These forces are such that $\dot{H}$ is never positive; it can be zero, however. In the sequel we introduce several definitions concerning the damping of a system.

First we consider the case in which $\sum_{i=1}^{n} Q_i \dot{q}_i$, hence $\dot{H}$, is a negative definite function of the generalized velocities $\dot{q}_i$. In this case system (6.117) is said to possess *complete damping*. If H is positive definite, we consider the

Liapunov function $U = H + \beta \sum_i q_i \dot{q}_i$, where β is a sufficiently small positive constant for U to be positive definite and $\dot{U}$ to be negative definite, so that by Liapunov's second stability theorem the trivial solution is asymptotically stable. A similar proposition shows that an unstable trivial solution is not affected by complete damping. (For details concerning both these statements, see Ref. 4, sec. 39.) Hence, complete damping does not alter the nature of the equilibrium in a meaningful way: a stable system becomes asymptotically stable, whereas an unstable one remains unstable.

There are cases in which $\sum_{i=1}^{n} Q_i \dot{q}_i$, hence $\dot{H}$, is only a negative semi-definite function of the generalized velocities $\dot{q}_i$. Then, if the set of points for which $\dot{H} = 0$ contains no nontrivial positive half-trajectory of system (6.117), the system is said to possess *pervasive damping*. The implication is that for a system with pervasive damping $\dot{H}$ does not reduce to zero at a certain time and remains zero ever after at any point of the domain other than the origin. Complete damping clearly represents a special case of the more general pervasive damping. Hence, if we recall Theorems 6.7.6 and 6.7.7, we can state the following *general theorems:*

Theorem 6.9.5 *If for the system* (6.117) *the Hamiltonian is positive definite, and if the system possesses pervasive damping, then the trivial solution is asymptotically stable.*

Theorem 6.9.6 *If for the system* (6.117) *the Hamiltonian can assume negative values in the neighborhood of the origin, and if the system possesses pervasive damping, then the trivial solution is unstable.*

As an illustration of Theorems 6.9.5 and 6.9.6 we consider a system subjected to nonconservative forces of the Rayleigh type. From Sec. 2.12, we recall that these forces have the expression

$$Q_i = -\frac{\partial F}{\partial \dot{q}_i}, \qquad i = 1, 2, \ldots, n, \qquad (6.119)$$

where

$$F = \frac{1}{2} \sum_{i=1}^{n} \sum_{j=1}^{n} c_{ij} \dot{q}_i \dot{q}_j \qquad (6.120)$$

is Rayleigh's dissipation function, in which c_{ij} are damping coefficients. In view of Eqs. (6.119) and (6.120) and again using Euler's theorem on homogeneous functions, we see that Eq. (6.118) becomes

$$\dot{H} = \sum_{i=1}^{n} Q_i \dot{q}_i = -\sum_{i=1}^{n} \frac{\partial F}{\partial \dot{q}_i} \dot{q}_i = -2F. \qquad (6.121)$$

If F is positive definite in the generalized velocities $\dot{q}_i$, then $\dot{H}$ is negative definite in these velocities. The possibility exists, however, that some generalized velocities are absent from F, so that F itself may be only positive semidefinite in $\dot{q}_i$. But when the generalized coordinates q_i are constant, F reduces to zero independently of $\dot{q}_i$, so that $\dot{H}$ is negative semidefinite. In order for the system to satisfy the definition of pervasive damping, however, $\dot{H}$ must reduce to zero if and only if all the coordinates q_i vanish at the same time, as all the velocities $\dot{q}_i$ do. This turns out to be the case if system (6.117) admits no solution in which one (or more) of the generalized coordinates q_i does not reduce to zero as $t \to \infty$. If for such a system H is positive definite, the system is asymptotically stable; and if H can assume negative values in the neighborhood of the origin, the system is unstable.

Let us consider again a holonomic conservative system and recall from Sec. 2.10 that when certain coordinates do not appear in the Hamiltonian, the coordinates in question are ignorable and the corresponding momenta are conserved. Indeed, if for an n-degree-of-freedom system l coordinates q_s are ignorable, we obtain l motion integrals in the form of the conserved momenta

$$\frac{\partial L}{\partial \dot{q}_s} = \beta_s = \text{const}, \qquad s = n - l + 1, \ldots, n. \tag{6.122}$$

Moreover, in Sec. 2.11 we showed that Eqs. (6.122) can be used to eliminate the ignorable coordinates from the formulation of the problem and to obtain the system of equations of motion

$$\frac{d}{dt}\left(\frac{\partial R}{\partial \dot{q}_k}\right) - \frac{\partial R}{\partial q_k} = 0, \qquad k = 1, 2, \ldots, n - l, \tag{6.123}$$

where R is the Routhian of the system

$$R = R(q_1, q_2, \ldots, q_{n-l}, \dot{q}_1, \dot{q}_2, \ldots, \dot{q}_{n-l}, \beta_{n-l+1}, \ldots, \beta_n). \tag{6.124}$$

Under these circumstances, we obtain the Jacobi integral

$$H = \sum_{k=1}^{n-l} \frac{\partial R}{\partial \dot{q}_k} \dot{q}_k - R = h = \text{const}, \tag{6.125}$$

which appears as a good prospect for a Liapunov function. It turns out, however, that when all the nonignorable coordinates and velocities are zero, in general, H does not reduce to zero but assumes the constant value H_0. Thus, we may consider a Liapunov function in the form

$$U = H - H_0 \tag{6.126}$$

and, because $\dot{U} = 0$, regard the system as stable if U is sign-definite, provided the integrals β_s are not perturbed. Hence, *in the presence of ignorable coordinates*, Lagrange's theorem, Theorem 6.9.1, is to be replaced by what is sometimes referred to as *Routh's theorem*.

Theorem 6.9.7 *If for the system* (6.123) *the function* $U = H - H_0$ *is sign-definite, then the equilibrium is stable.*

The fact that in a certain system $U = H - H_0$ is not sign-definite cannot be construed as an indication that the system is unstable. Indeed, if ignorable coordinates are present, it may be possible to prove stability by choosing a Liapunov function as a suitable combination of the motion integrals.

In the sequel two relatively simple illustrative examples are shown. Further applications of the theory to the stability analysis of canonical systems are presented later in the book.

Example 6.5

As a simple illustration, we consider the stability of the system of Example 6.1 in the neighborhood of $\theta = 0$. It is not difficult to show that the expressions for the kinetic energy, potential energy, and Rayleigh's dissipation function are

$$T = \tfrac{1}{2}[(M + m)\dot{x}^2 + 2mL\dot{x}\dot{\theta}\cos\theta + mL^2\dot{\theta}^2], \quad (a)$$

$$V = \tfrac{1}{2}kx^2 + mgL(1 - \cos\theta), \quad (b)$$

and $\qquad F = \tfrac{1}{2}c\dot{x}^2, \quad (c)$

respectively. The system is a natural one, so that the Hamiltonian has the form

$$H = T + V. \quad (d)$$

Considering the Hamiltonian as a Liapunov function and confining the motion to the neighborhood of $\theta = 0$, we obtain

$$H = \tfrac{1}{2}[(M + m)\dot{x}^2 + 2mL\dot{x}\dot{\theta} + mL^2\dot{\theta}^2] + \tfrac{1}{2}kx^2 + \tfrac{1}{2}mgL\theta^2, \quad (e)$$

which can easily be shown to be positive definite. On the other hand, F is positive semidefinite, so that

$$\dot{H} = -2F \leq 0. \quad (f)$$

Moreover, examining the equations of motion, Eqs. (*a*) of Example 6.1, we conclude that there is no way of solving for either x or θ independently of one another. Indeed the system does not admit a constant solution other than the trivial one, so that the system possesses pervasive damping and, by Theorem 6.9.5, the equilibrium is asymptotically stable. The conclusion agrees with that reached by the Routh-Hurwitz criterion.

Example 6.6

An ingenious construction of Liapunov function is given by Chetayev (Ref. 4, sec. 10) in connection with the stability problem of a sleeping top (see Sec. 4.11). We shall follow his exposition closely.

Referring to Fig. 4.8, we denote the direction cosines between the fixed vertical direction Z and the body axes x, y, and z by l, m, and n, respectively. The angular velocity components of the top about the body axes are ω_x, ω_y, and ω_z so that the unperturbed motion of the sleeping top is defined by

$$\omega_x = \omega_y = 0, \qquad \omega_z = \omega_0 = \text{const}, \qquad l = m = 0, \qquad n = 1. \qquad (a)$$

We shall study the stability of the sleeping top by using ω_x, ω_y, ω_z, l, m, and n as the variables of the system.

We recall that for the symmetric top we have three motion integrals, the Hamiltonian, which is equal to the total energy E, the angular momentum p_Z about the vertical direction, and the angular momentum p_z about the symmetry axis. Furthermore, the sum of the direction cosines squared is equal to 1, so that all these relations can be written

$$\begin{aligned} E &= \tfrac{1}{2}A(\omega_x^2 + \omega_y^2) + \tfrac{1}{2}C\omega_z^2 + mgLn, \\ p_Z &= A(l\omega_x + m\omega_y) + Cn\omega_z, \\ p_z &= C\omega_z, \\ 1 &= l^2 + m^2 + n^2, \end{aligned} \qquad (b)$$

where L is the distance between the center of mass of the top and the pivot. No confusion will arise from the fact that m denotes both the mass of the top and one of the direction cosines.

The perturbed motion can be defined as

$$\begin{aligned} \omega_x &= \epsilon_x, & \omega_y &= \epsilon_y, & \omega_z &= \omega_0 + \epsilon_z, \\ l &= \delta_x, & m &= \delta_y, & n &= 1 + \delta_z, \end{aligned} \qquad (c)$$

where ϵ_x, ϵ_y, ϵ_z, δ_x, δ_y, and δ_z are small perturbations. It follows that the equations of the perturbed motion have the integrals

$$\begin{aligned} V_1 &= \tfrac{1}{2}A(\epsilon_x^2 + \epsilon_y^2) + \tfrac{1}{2}C\epsilon_z(\epsilon_z + 2\omega_0) + mgL\delta_z, \\ V_2 &= A(\delta_x\epsilon_x + \delta_y\epsilon_y) + C(\omega_0\delta_z + \epsilon_z + \delta_z\epsilon_z), \\ V_3 &= C\epsilon_z, \\ V_4 &= \delta_x^2 + \delta_y^2 + \delta_z^2 + 2\delta_z. \end{aligned} \qquad (d)$$

We seek a Liapunov function as a combination of the above integrals chosen to yield a quadratic form. Hence, let

$$\begin{aligned} V &= 2V_1 + 2\lambda V_2 + \mu V_3^2 - 2(\omega_0 + \lambda)V_3 - (mgL + \lambda C\omega_0)V_4 \\ &= A(\epsilon_x^2 + \epsilon_y^2) + C(1 + \mu C)\epsilon_z^2 - (mgL + \lambda C\omega_0)(\delta_x^2 + \delta_y^2 + \delta_z^2) \\ &\quad + 2\lambda A(\epsilon_x\delta_x + \epsilon_y\delta_y) + 2\lambda C\epsilon_z\delta_z, \end{aligned} \qquad (e)$$

so that the matrix of the coefficients becomes

$$[\alpha] = \begin{bmatrix} A & 0 & 0 & \lambda A & 0 & 0 \\ 0 & A & 0 & 0 & \lambda A & 0 \\ 0 & 0 & C + \mu C^2 & 0 & 0 & \lambda C \\ \lambda A & 0 & 0 & -N & 0 & 0 \\ 0 & \lambda A & 0 & 0 & -N & 0 \\ 0 & 0 & \lambda C & 0 & 0 & -N \end{bmatrix}, \quad (f)$$

where
$$N = mgL + \lambda C \omega_0. \quad (g)$$

It is not difficult to show by means of Sylvester's theorem (see Sec. 6.7) that by choosing

$$\mu = \frac{C - A}{AC} \quad (h)$$

the condition for the matrix $[\alpha]$ to be positive definite reduces to

$$A\lambda^2 + C\omega_0\lambda + mgL < 0. \quad (i)$$

The inequality can be satisfied only when the roots λ_1 and λ_2 of the associated polynomial are real and distinct. This leads to the requirement

$$\omega_0^2 > \frac{4AmgL}{C^2}, \quad (j)$$

which coincides with the stability condition obtained in Sec. 4.11.

Because the mathematical model does not include damping forces, no asymptotic stability is expected.

6.10 STABILITY IN THE PRESENCE OF GYROSCOPIC AND DISSIPATIVE FORCES

Let us consider a holonomic conservative system for which the Hamiltonian is given by

$$H = \frac{1}{2}\sum_{i=1}^{n} p_i^2 + \frac{1}{2}\sum_{i=1}^{n} \Lambda_i q_i^2 = T + V = h = \text{const.} \quad (6.127)$$

The system has an equilibrium point at the origin. Since the Hessian matrix is diagonal, it is easy to see that the system is stable if all Λ_i ($i = 1, 2, \ldots, n$) are positive. If at least one of the quantities Λ_i is negative, the potential energy has no minimum in the equilibrium position and by Liapunov's theorem, Theorem 6.9.3, the equilibrium is unstable. The question is

Stability of Multi-Degree-of-Freedom Autonomous Systems

whether there are any forces which when applied to the system can stabilize the equilibrium. In Sec. 6.9 we showed that dissipative forces do not alter the nature of the equilibrium in a meaningful way, so that such forces are not satisfactory.

It turns out that there indeed exists a class of forces which can stabilize an unstable equilibrium of system (6.127). To show this, we consider the set of forces

$$Q_i = \sum_{j=1}^{n} g_{ij}\dot{q}_j, \quad g_{ij} = -g_{ji}, \quad i = 1, 2, \ldots, n, \quad (6.128)$$

where the coefficients g_{ij} are skew-symmetric, as indicated. The forces Q_i can be identified as gyroscopic forces, and it can be shown without difficulty that they perform no work. In the presence of such forces the Hamiltonian equations of motion become

$$\dot{q}_i = \frac{\partial H}{\partial p_i} = p_i,$$
$$\dot{p}_i = -\frac{\partial H}{\partial q_i} + Q_i = -\Lambda_i q_i + \sum_{j=1}^{n} g_{ij}\dot{q}_j, \quad i = 1, 2, \ldots, n. \quad (6.129)$$

Using Eqs. (6.129), the time derivative of the Hamiltonian becomes

$$\dot{H} = \sum_{i=1}^{n} \frac{\partial H}{\partial q_i}\dot{q}_i + \sum_{i=1}^{n} \frac{\partial H}{\partial p_i}\dot{p}_i = \sum_{i=1}^{n}\sum_{j=1}^{n} g_{ij}\dot{q}_i\dot{q}_j = 0. \quad (6.130)$$

The result can be easily explained by the fact that the coefficients g_{ij} are skew-symmetric.

Equations (6.129) can be reduced to a set of n second-order differential equations of the form

$$\ddot{q}_i - \sum_{j=1}^{n} g_{ij}\dot{q}_j + \Lambda_i q_i = 0, \quad i = 1, 2, \ldots, n, \quad (6.131)$$

leading to the characteristic equation

$$\Delta(\lambda) = |(\lambda^2 + \Lambda_i)\delta_{ij} - \lambda g_{ij}| = 0. \quad (6.132)$$

Because the determinant of a matrix and that of the transposed matrix are equal, Eq. (6.132) can also be written in the form

$$\Delta(\lambda) = |(\lambda^2 + \Lambda_i)\delta_{ij} - \lambda g_{ji}|$$
$$= |(\lambda^2 + \Lambda_i)\delta_{ij} + \lambda g_{ij}| = 0. \quad (6.133)$$

Hence, it follows from Eqs. (6.132) and (6.133) that if λ_r is a characteristic root of the system (6.131), then $-\lambda_r$ is also a root. It remains to show that the forces (6.128) can under certain circumstances stabilize an otherwise unstable equilibrium.

First we consider the case in which the gyroscopic forces cannot stabilize an unstable equilibrium. This is the case in which the original system, namely, the system without the gyroscopic forces, has an odd number of negative Λ_i. To show this, we observe from Eq. (6.132) that

$$\Delta(0) = \prod_{i=1}^{n} \Lambda_i \tag{6.134}$$

and
$$\lim_{\lambda \to \infty} \Delta(\lambda) > 0. \tag{6.135}$$

If there is an odd number of negative Λ_i's, then $\Delta(0)$ will be negative and $\Delta(\infty)$ positive. Hence Eq. (6.132) must have at least one positive root for λ, which implies instability.

If, on the other hand, the number of negative Λ_i's is even, gyroscopic forces can render the unstable equilibrium stable. Letting $n = 2$, Eq. (6.132) reduces to

$$\Delta(\lambda) = \begin{vmatrix} \lambda^2 + \Lambda_1 & -\lambda g_{12} \\ \lambda g_{12} & \lambda^2 + \Lambda_2 \end{vmatrix} = \lambda^4 + (g_{12}{}^2 + \Lambda_1 + \Lambda_2)\lambda^2 + \Lambda_1\Lambda_2 = 0, \tag{6.136}$$

where Λ_1 and Λ_2 are assumed to be negative. The gyroscopic forces render the system stable if all the roots of Eq. (6.136) are purely imaginary. It follows that for stability the coefficient g_{12} must satisfy the conditions

$$g_{12}{}^2 + \Lambda_1 + \Lambda_2 > 0, \qquad (g_{12}{}^2 + \Lambda_1 + \Lambda_2)^2 - 4\Lambda_1\Lambda_2 > 0, \tag{6.137}$$

which implies that stability is possible provided g_{12} is sufficiently large. We note that whereas $\dot{H} = 0$, H is sign-variable, so that the Hamiltonian itself is not a Liapunov function.

As it turns out, gyroscopic stabilization is possible only if the original system is conservative. Indeed in the presence of dissipative forces this type of stabilization is lost. For dissipative forces of the Rayleigh type, Eqs. (6.128) assume the form

$$Q_i = -\frac{\partial F}{\partial \dot{q}_i} + \sum_{j=1}^{n} g_{ij}\dot{q}_j, \qquad g_{ij} = -g_{ji}, \tag{6.138}$$

where
$$F = \frac{1}{2} \sum_{i=1}^{n} \sum_{j=1}^{n} c_{ij}\dot{q}_i\dot{q}_j \tag{6.139}$$

is the Rayleigh dissipation function. We shall assume that the system possesses complete damping, so that F is positive definite in the generalized

Stability of Multi-Degree-of-Freedom Autonomous Systems 255

velocities. Whereas the Hamiltonian remains in the form (6.127), its time derivative along a trajectory of the system becomes $\dot{H} = -2F$, which is negative definite in the velocities $\dot{q}_i$. Considering a Liapunov function in the form

$$U = -H - \beta \sum_{i=1}^{n} \Lambda_i q_i \dot{q}_i, \qquad (6.140)$$

where β is a positive constant, we can choose β sufficiently small for $\dot{U}$ to be positive definite. On the other hand, since for a system which was originally unstable at least one of the eigenvalues Λ_i is negative, a subdomain of the phase space can be found in which U is positive, so that by Chetayev's instability theorem the system is unstable. Hence, a point of unstable equilibrium for the original system remains unstable in the presence of both gyroscopic and dissipative forces.

The above results are due to Lord Kelvin, who referred to stability due to gyroscopic forces in the absence of dissipative forces as *temporary stability* (see Ref. 4, Sec. 40) because such stability is lost as soon as the system becomes completely damped. Actually from Sec. 6.9 we conclude that damping need not be complete but only pervasive.

Example 6.7

Let us concern ourselves with the stability of the rotational motion of an artillery shell, in the form of an axially symmetric elongated body, moving in a resisting medium. The shell is assumed to be subjected to aerodynamic forces with a resultant **R** acting at the point P, known as the *center of pressure*, a distance e ahead of the center of mass C. The force **R** is tangent to the trajectory and opposed to the direction of motion. For any deviation of the shell axis from alignment with the direction of motion, the force **R** gives rise to a torque about the center of mass, causing the projectile to tumble.

Let us denote by ξ, η, ζ a set of axes with the origin at the center of mass of the shell and with axis ζ along the symmetry axis. The center of mass C moves with velocity **v** relative to an inertial space x, y, z, as shown in Fig. 6.6. A rotation θ_1 about x takes a triad from an original position parallel to system x, y, z to a position coincident with a set of axes ξ', η', ζ', and a rotation θ_2 about axis ξ' defines the system ξ, η, ζ. The shell rotates relative to system ξ, η, ζ with angular velocity $\dot{\theta}_3$ about the symmetry axis ζ.

We shall assume that the projectile, envisioned as a rigid body, travels with uniform velocity in a flat trajectory, which is equivalent to ignoring the effect of gravity and the retarding force due to air resistance on the motion of the center of mass. Under these assumptions, **R** is regarded as constant, and the problem reduces to that of stability of the spinning shell under the effect of aerodynamic torques.

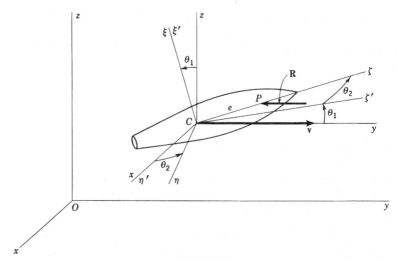

FIGURE 6.6

The body angular velocity components about axes ξ, η, ζ can be shown to be

$$\Omega_\xi = \dot\theta_2, \qquad \Omega_\eta = \dot\theta_1 \cos\theta_2, \qquad \Omega_\zeta = \dot\theta_3 - \dot\theta_1 \sin\theta_2, \tag{a}$$

where the angles θ_1, θ_2, and θ_3 play the role of Euler's angles. Denoting the moments of inertia of the shell about axes ξ, η, ζ by A, A, C, respectively, and taking advantage of symmetry, we write the rotational kinetic energy expression

$$T = \tfrac{1}{2}[A(\dot\theta_2^{\,2} + \dot\theta_1^{\,2}\cos^2\theta_2) + C(\dot\theta_3 - \dot\theta_1\sin\theta_2)^2]. \tag{b}$$

The aerodynamic torques can be conveniently derived from the potential energy

$$V = Re\cos\theta_1\cos\theta_2, \tag{c}$$

where R is the magnitude of the aerodynamic force.

First we assume that the shell has zero spin, $\dot\theta_3 = 0$, so that the Hamiltonian reduces to the Jacobi integral

$$H = T + V = \tfrac{1}{2}[A(\dot\theta_2^{\,2} + \dot\theta_1^{\,2}\cos^2\theta_2) + C\dot\theta_1^{\,2}\sin^2\theta_2] + Re\cos\theta_1\cos\theta_2$$
$$= h = \text{const}, \tag{d}$$

because H is not an explicit function of time and the system is natural. It is not difficult to see that $\theta_1 = \theta_2 = \dot\theta_1 = \dot\theta_2 = 0$ represents an equilibrium point. Since V has a maximum at this point, we conclude by Liapunov's theorem, Theorem 6.9.2, that the equilibrium is unstable.

Stability of Multi-Degree-of-Freedom Autonomous Systems

The motion can be stabilized by imparting to the projectile a certain spin $\dot{\theta}_3$ about axis ζ. Indeed let us consider the case in which $\dot{\theta}_3 \neq 0$, where it is not difficult to see that θ_3 is a cyclic coordinate, so that the angular momentum about ζ is conserved

$$L_\zeta = C(\dot{\theta}_3 - \dot{\theta}_1 \sin \theta_2) = Cn = \text{const.} \quad (e)$$

Moreover, the aerodynamic force being parallel to the y axis, the angular momentum about y is conserved

$$L_y = -A\dot{\theta}_2 \sin \theta_1 + (Cn - A\dot{\theta}_1 \sin \theta_2) \cos \theta_1 \cos \theta_2 = \text{const.} \quad (f)$$

The third integral of motion is the Hamiltonian, which coincides with the total energy of the system

$$H = T + V = \tfrac{1}{2}A(\dot{\theta}_2^2 + \dot{\theta}_1^2 \cos^2 \theta_2) + \tfrac{1}{2}Cn^2 + Re \cos \theta_1 \cos \theta_2 = \text{const.} \quad (g)$$

The system has an equilibrium position defined by $\theta_1 = \theta_2 = \dot{\theta}_1 = \dot{\theta}_2 = 0$, in which position the rotational motion of the projectile consists of pure spin about the ζ axis with the aerodynamic forces being zero. We shall check the stability of motion about this equilibrium position.

As in Example 6.6, we select a Liapunov function U as a combination of the motion integrals having the form

$$U = \tfrac{1}{2}A(\dot{\theta}_2^2 + \dot{\theta}_1^2 \cos^2 \theta_2) + Re \cos \theta_1 \cos \theta_2$$
$$+ \lambda[A\dot{\theta}_2 \sin \theta_1 - (Cn - A\dot{\theta}_1 \sin \theta_2) \cos \theta_1 \cos \theta_2] = \text{const,} \quad (h)$$

where λ is determined from the condition that U be sign-definite with respect to the variables $\theta_1, \theta_2, \dot{\theta}_1$, and $\dot{\theta}_2$. Because $\dot{U}$ is zero, we can prove at most mere stability. To test U for sign-definiteness, we use Sylvester's criterion (see Sec. 6.7) and write the Hessian matrix $[\mathcal{H}]_E$, corresponding to the equilibrium position $\theta_1 = \theta_2 = \dot{\theta}_1 = \dot{\theta}_2 = 0$,

$$[\mathcal{H}]_E = \begin{bmatrix} A & 0 & 0 & \lambda A \\ 0 & A & \lambda A & 0 \\ 0 & \lambda A & \lambda Cn - Re & 0 \\ \lambda A & 0 & 0 & \lambda Cn - Re \end{bmatrix}. \quad (i)$$

By definition, $A > 0$, so that the first two principal minor determinants are positive. The condition that the third principal minor determinant be positive can be reduced to

$$A\lambda^2 - Cn\lambda + Re < 0, \quad (j)$$

which is satisfied only when the associated quadratic equation has simple real roots. This in turn leads to the condition

$$(Cn)^2 - 4ARe > 0. \quad (k)$$

The fourth determinant leads to no new condition, so that the rotational motion of the projectile is stable if the spin satisfies the condition

$$n^2 > \frac{4ARe}{C^2}. \tag{l}$$

As a matter of interest, it should be pointed out that Eq. (l) would be equivalent to the stability condition for the sleeping top obtained in Sec. 4.11 if the aerodynamic torque Re were to be replaced by the gravitational torque mgl on the spinning top. In view of the assumptions made here, the two systems are similar.

A more realistic mathematical model can be obtained by assuming that the motions θ_i give rise to damping torques proportional to the corresponding angular velocities. To study such a model, we assume that these nonconservative torques are derivable from the Rayleigh dissipation function

$$F = \frac{1}{2} \sum_{i=1}^{3} \sum_{j=1}^{3} c_{ij} \dot{\theta}_i \dot{\theta}_j, \tag{m}$$

where the damping coefficients c_{ij} ($i, j = 1, 2, 3$) are such that F is a positive definite quadratic form in the angular velocities $\dot{\theta}_i$. In this case the Hamiltonian H is the sum of T and V, as given by Eqs. (b) and (c), but is no longer constant. In fact, $\dot{H} = -2F$, which is negative definite in the angular velocities $\dot{\theta}_i$. Choosing a Liapunov function in the form (6.140), it can be shown that the equilibrium is unstable (see Ref. 4, sec. 40).

6.11 CONSTRUCTION OF LIAPUNOV FUNCTIONS FOR LINEAR AUTONOMOUS SYSTEMS

When the system behavior is described by a set of $2n$ first-order linear differential equations with constant coefficients, there is no longer need for ingenuity in constructing a Liapunov function.

Let us consider the system of equations

$$\{\dot{x}\} = [A]\{x\}, \tag{6.141}$$

where $\{x\}$ is an m-dimensional column matrix of variables and $[A]$ is an $m \times m$ matrix of constant real coefficients, $m = 2n$. A Liapunov function will be sought as the quadratic form

$$V = \{x\}^T[B]\{x\}, \tag{6.142}$$

where we assume, without loss of generality, that $[B]$ is a real symmetric matrix with constant elements which have yet to be determined. The total derivative of V, along a trajectory of Eq. (6.141), becomes

$$\dot{V} = \{\dot{x}\}^T[B]\{x\} + \{x\}^T[B]\{\dot{x}\} = \{x\}^T([A]^T[B] + [B][A])\{x\}, \tag{6.143}$$

and we shall be interested in the case in which $\dot{V}$ is negative definite. This is ensured by assuming that we can determine a matrix $[B]$ such that the equation

$$[A]^T[B] + [B][A] = -[1] \tag{6.144}$$

is satisfied, where $[1]$ is the identity matrix of order m. Equation (6.144) represents a set of linear algebraic equations in $m(m + 1)/2$ unknowns B_{ij}. We must still answer the question whether the system of equations possesses a unique solution. To this end, we refer to a theorem given in Ref. 2, p. 231:

Theorem *The necessary and sufficient condition for the system of equations*

$$[E][G] + [G][F] = [C] \tag{6.145}$$

to have a solution for all $[C]$ is that $\lambda_i + \mu_j \neq 0$, where λ_i are the eigenvalues of the matrix $[E]$ and μ_i are the eigenvalues of $[F]$.

Since in our case $[F] = [E]^T = [A]$, it follows that Eq. (6.144) has a unique solution provided $\lambda_i + \lambda_j \neq 0$ $(i, j = 1, 2, \ldots, m)$, where λ_i are the eigenvalues of $[A]$. The question remains whether the solution $[B]$ represents a positive definite matrix or not. To answer this question, we designate a matrix $[A]$ as a *stability matrix* if all its eigenvalues λ_i have negative real parts and refer to another theorem given in Ref. 2, p. 245:

Theorem *The necessary and sufficient condition that the real matrix $[A]$ be a stability matrix is that $[B]$ be a positive definite matrix, where $[B]$ is determined by Eq. (6.144).*

Now we are in a position to state the following:

Theorem *If there exists a symmetric positive definite matrix $[B]$ which is the unique solution of Eq. (6.144), then the trivial solution of Eq. (6.141) is asymptotically stable.*

We can test the matrix $[B]$ for positive definiteness by using Sylvester's theorem (see Sec. 6.7).

If there are eigenvalues λ_i of the matrix $[A]$ with positive real parts, then the matrix $[B]$, if it exists, is indefinite, and if all the eigenvalues have positive real parts, then $[B]$ is a negative definite matrix and, by virtue of Liapunov's Theorem 6.7.3, we conclude that the system (6.141) is *unstable*.

When the matrix $[A]$ is not in a Jordan form, the solution of Eq. (6.144) is very difficult. A note by Barnett and Storey (Ref. 1) discusses some approaches to this problem.

PROBLEMS

6.1 Consider the system of Example 6.1 and use the Routh-Hurwitz criterion to determine the system stability in the neighborhood of the equilibrium point $x = 0$, $\theta = \pi$.

6.2 For the differential equation

$$\ddot{x} = -k\left(x - \frac{1}{4a^2}x^3\right),$$

representing the equation of motion of a mass-spring system, where the mass is equal to unity and the spring is "soft," (a) determine the equilibrium points, (b) obtain the variational equations associated with these points, and (c) devise a linear transformation leading to the Jordan canonical form for each set of variational equations.

6.3 Derive the differential equations of motion for the system shown in Fig. 6.7. The force in the spring k_1 has the expression

$$k_1\left(x_1 - \frac{1}{4a^2}x_1^3\right),$$

whereas spring k_2 is linear. The corresponding masses are m_1 and m_2, and the damping is viscous with damping coefficients c_1 and c_2. For the derived system of equations (a) establish the equilibrium positions, and (b) derive the variational equations and test their stability by means of the Routh-Hurwitz criterion.

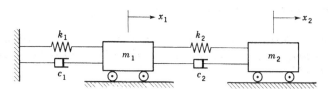

FIGURE 6.7

6.4 Derive Hamilton's canonical equations for the system of Example 6.1 by allowing large angles θ. Obtain the variational equations corresponding to the equilibrium point $x = \theta = 0$ and verify Eq. (6.78).

6.5 The system of Fig. 6.8 represents a gyropendulum with the rotor shaft locked and the outer gimbal rotating at a constant angular velocity Ω.

Derive the equation of motion for the system and test the stability of the system for all the equilibrium positions. The rotor consists of a thin disk of diameter D.

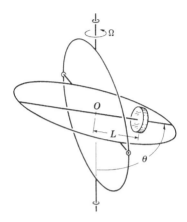

FIGURE 6.8

6.6 Test the stability of the equilibrium positions of Prob. 6.3 by means of the Liapunov direct method.

6.7 Consider the van der Pol's oscillator of Sec. 5.5, use

$$V(x_1,x_2) = \tfrac{1}{2}(x_1{}^2 + x_2{}^2)$$

as a Liapunov function, and determine the domain of attraction for the *negative* half-trajectories.

6.8 Solve Prob. 6.1 by the Liapunov direct method.

6.9 For the system of Prob. 2.2 test the system stability by the Liapunov direct method for each of the equilibrium positions.

6.10 Consider a two-body system subjected to the attractive central force of magnitude $f(r)$ and use Routh's method (Sec. 6.9) to derive the condition for the stability of the circular orbit $r = r_0$. (Note that here H_0 represents the value of the Hamiltonian corresponding to $r = r_0$.) If the force is proportional to the radial distance to the power n, use the stability condition and determine the values of n for which the orbit is stable. Note that n is an integer.

6.11 Let the system of Prob. 6.2 be viscously damped and denote the damping coefficient by c. Using the developments of Sec. 6.11, construct Liapunov functions for the sets of variational equations associated with each equilibrium position and test the stability of these positions.

SUGGESTED REFERENCES

1. Barnett, S., and C. Storey: "Solution of Lyapunov Matrix Equation," *Electron. Letters*, vol. 2, no. 12, pp. 466–467, 1966.
2. Bellman, R.: *Introduction to Matrix Analysis*, McGraw-Hill Book Company, New York, 1960.
3. Cesari, L.: *Asymptotic Behavior and Stability Problems in Ordinary Differential Equations*, Springer-Verlag OHG, Berlin, 1963.
4. Chetayev, N. G.: *The Stability of Motion*, Pergamon Press, New York, 1961.
5. Coddington, E. A., and N. Levinson: *Theory of Ordinary Differential Equations*, McGraw-Hill Book Company, New York, 1955.
6. Hahn, W.: *Theory and Application of Liapunov's Direct Method*, Prentice-Hall, Inc., Englewood Cliffs, N.J., 1963.
7. Hurewicz, W.: *Lectures on Ordinary Differential Equations*, John Wiley & Sons, Inc., New York, 1958.
8. Kalman, R. E., and J. E. Bertram: "Control System Analysis and Design via the 'Second Method' of Lyapunov," pt. I, *J. Basic Eng.*, pp. 371–393, June, 1960.
9. Krasovskii, N. N.: *Stability of Motion*, Stanford University Press, Stanford, Calif., 1963.
10. LaSalle, J., and S. Lefschetz: *Stability by Liapunov's Direct Method*, Academic Press, Inc., New York, 1961.
11. Lefschetz, S.: *Differential Equations: Geometric Theory*, 2d ed., John Wiley & Sons, Inc., New York, 1963.
12. Malkin, I. G.: *Theory of Stability of Motion*, U.S. Atomic Energy Commission Translation, AEC-tr-3352, Department of Commerce, Washington, D.C., 1958.
13. Nemytskii, V. V., and V. V. Stepanov: *Qualitative Theory of Differential Equations*, Princeton University Press, Princeton, N.J., 1960.

chapter seven

Nonautonomous Systems

When the differential equations governing the behavior of a physical system contain time-dependent coefficients, the system is said to be *nonautonomous*. Such equations arise when some of the system parameters are explicit functions of time. As a typical illustration we may mention a simple pendulum whose length changes with time. Nonautonomous systems can also arise in the form of variational equations associated with the perturbation of a known periodic motion of an autonomous system.

The mathematical treatment of nonautonomous systems of equations is considerably more complicated than the treatment of autonomous ones. In fact, much of the theory is restricted to linear systems with coefficients in the form of periodic functions of time. Furthermore, the theory is concerned primarily with the form of the solution rather than exact solutions in terms of explicit functions of time. When the variational system of equations is the result of a perturbation of a set of Hamilton's canonical equations, the situation is appreciably improved because in this case it is possible to establish in advance some of the properties of the characteristic exponents of the system. In general, however, a measure of success in solving nonautonomous problems analytically has been achieved only for low-order systems, particularly second-order systems such as governed by Hill's equation.

An interesting aspect of the stability of a second-order system is unveiled when we study the change in the stability characteristics of the motion brought about by a variation in the system parameters. We can envision a parameter plane divided into regions of stability and instability with the periodic solutions of period T and $2T$ defining the boundary curves. This approach circumvents the often impossible task of determining the characteristic exponents.

As for autonomous systems, the Liapunov direct method can be regarded as a unified approach to the problem of nonlinear nonautonomous systems. Again the main difficulty lies in constructing a Liapunov function for the system, but in the case of nonautonomous systems the prospects are far less promising than for autonomous ones.

7.1 LINEAR SYSTEMS WITH PERIODIC COEFFICIENTS. FLOQUET'S THEORY

The general theory of homogeneous linear systems of equations was introduced in Chap. 6. As can be concluded from that chapter, in the important case of linear autonomous systems, namely, systems with constant coefficients, certain interesting results can be obtained. Of the nonautonomous dynamical systems, the homogeneous linear system for which the coefficients $A_{ij}(t)$ of Eqs. (6.1) are periodic is of particular significance. The theory of such systems of equations is frequently referred to as *Floquet's theory*.

We shall be concerned with a system of m first-order differential equations, which can be written in the matrix form

$$\{\dot{x}\} = [A(t)]\{x\}, \tag{7.1}$$

where the elements $A_{ij}(t)$ $(i, j = 1, 2, \ldots, m)$ of the matrix $[A(t)]$ are continuous periodic functions of time with period T

$$[A(t + T)] = [A(t)]. \tag{7.2}$$

We shall be interested in the stability of the trivial solution of Eq. (7.1).

System (7.1) represents a special case of the general homogeneous linear systems discussed in Sec. 6.1. Many of the definitions and concepts introduced in that section are equally valid here and will not be repeated.

A system described by Eqs. (7.1) and (7.2) is said to be periodic. It turns out that a fundamental matrix for such systems can be represented as the product of a periodic matrix of period T and a solution matrix for a system with constant coefficients. This is stated by the following theorem.

Theorem 7.1.1 *If $[\varphi(t)]$ is a fundamental matrix of the system described by Eqs. (7.1) and (7.2), then $[\varphi(t + T)]$ is also a fundamental matrix. Moreover, for every such $[\varphi(t)]$ there exists a periodic nonsingular matrix $[Q(t)]$ with period T and a constant matrix $[R]$ such that*

$$[\varphi(t)] = [Q(t)]e^{t[R]}, \qquad [Q(t)] = [Q(t + T)]. \tag{7.3}$$

The first part of Theorem 7.1.1 is relatively easy to prove. To this end, we recall from Sec. 6.1 that a fundamental matrix satisfies the matrix equation

$$[\dot{\varphi}(t)] = [A(t)][\varphi(t)]. \tag{7.4}$$

In view of Eq. (7.2), we can write

$$[\dot{\varphi}(t + T)] = [A(t + T)][\varphi(t + T)] = [A(t)][\varphi(t + T)], \tag{7.5}$$

from which it follows that $[\varphi(t + T)]$ is also a fundamental matrix of system (7.1).

Because both $[\varphi(t)]$ and $[\varphi(t + T)]$ represent fundamental solutions, there exists a nonsingular constant matrix $[C]$ such that

$$[\varphi(t + T)] = [\varphi(t)][C], \tag{7.6}$$

where the matrix $[C]$ is sometimes referred to as the *monodromy matrix* of the fundamental matrix $[\varphi(t)]$. But a matrix $[R]$ can be found (see Ref. 3, sec. 5.1) which satisfies

$$[C] = e^{T[R]}, \tag{7.7}$$

so that, combining Eqs. (7.6) and (7.7), we obtain

$$[\varphi(t + T)] = [\varphi(t)]e^{T[R]}. \tag{7.8}$$

Next let us define a matrix $[Q(t)]$ by

$$[Q(t)] = [\varphi(t)]e^{-t[R]} \tag{7.9}$$

and use Eq. (7.8) to obtain

$$[Q(t + T)] = [\varphi(t + T)]e^{-(t+T)[R]} = [\varphi(t)]e^{T[R]}e^{-(t+T)[R]}$$
$$= [\varphi(t)]e^{-t[R]} = [Q(t)], \tag{7.10}$$

from which we conclude that $[Q(t)]$ is a periodic matrix with period T. Because $[\varphi(t)]$ and $\exp(-t[R])$ are nonsingular, $[Q(t)]$ is also nonsingular. As Eqs. (7.9) and (7.10) are equivalent to Eqs. (7.3), the second part of Theorem 7.1.1 has just been proved.

The implication of Theorem 7.1.1 is that the determination of $[\varphi(t)]$ over one period is equivalent to the determination of $[\varphi(t)]$ for all values of time. Let us assume that $[\varphi(t)]$ is determined over the period $0 \leq t \leq T$. Then, setting $t = 0$ in Eq. (7.6), we conclude that

$$[C] = [\varphi(0)]^{-1}[\varphi(T)], \qquad (7.11)$$

and from Eq. (7.7) we obtain

$$[R] = \frac{1}{T}\log[C]. \qquad (7.12)$$

Hence, by Eq. (7.9), $[Q(t)]$ is determined over the period $0 \leq t \leq T$. Since $[Q(t)]$ is periodic, however, it is determined for $-\infty < t < \infty$, so that, from Eqs. (7.3), it follows that $[\varphi(t)]$ also is determined for $-\infty < t < \infty$.

The matrix $[C]$ is not unique but depends on the particular fundamental matrix $[\varphi(t)]$. Although this is so, the eigenvalues associated with the monodromy matrix $[C]$ are unique, because any other fundamental matrix corresponding to $[A(t)]$ possesses a monodromy matrix which is similar to $[C]$. To show this, let us consider another fundamental matrix $[\Phi(t)]$ related to $[\varphi(t)]$ by

$$[\Phi(t)] = [\varphi(t)][B], \qquad (7.13)$$

where $[B]$ is a constant nonsingular matrix. Introducing Eq. (7.13) into (7.6) and postmultiplying the result by $[B]$, we obtain

$$[\Phi(t + T)] = [\Phi(t)][B]^{-1}[C][B]. \qquad (7.14)$$

Hence, any fundamental matrix $[\Phi(t)]$ has a monodromy matrix $[B]^{-1}[C][B]$ which is clearly similar to $[C]$. It follows that all the fundamental matrices of (7.1), and hence $[A(t)]$, determine uniquely all quantities associated with $[C]$ which are invariant under a similarity transformation, such as the eigenvalues and the elementary divisors. This implies that all the monodromy matrices can be reduced to the same Jordan canonical form $[J]$. The eigenvalues of the matrix $[C]$, denoted by $\lambda_1, \lambda_2, \ldots, \lambda_m$, are called the *characteristic multipliers* associated with the periodic matrix $[A(t)]$. None of the multipliers vanishes, for

$$\det[C] = \prod_{j=1}^{m} \lambda_j \neq 0. \qquad (7.15)$$

The eigenvalues of $[R]$, denoted by $\rho_1, \rho_2, \ldots, \rho_m$, are called the *characteristic exponents* associated with the periodic matrix $[A(t)]$ and are related to the characteristic multipliers by

$$\lambda_j = e^{T\rho_j}, \qquad j = 1, 2, \ldots, m. \tag{7.16}$$

Whereas the characteristic multipliers λ_j are uniquely defined, only the real parts of the characteristic exponents ρ_j are defined uniquely, as can be seen from

$$\rho_j = \frac{1}{T}(\log |\lambda_j| + i \arg \lambda_j), \qquad j = 1, 2, \ldots, m. \tag{7.17}$$

The imaginary parts are determined up to an integral multiple of $2\pi/T$. A characteristic multiplier is said to be *simple* if its multiplicity is 1.

To explore the form of solution of system (7.1), we express the fundamental matrix $[\Phi(t)]$ in the form

$$[\Phi(t)] = [P(t)]e^{t[J]}, \qquad [P(t)] = [P(t + T)], \tag{7.18}$$

where $[P(t)]$ is a periodic matrix and $[J]$ is the Jordan form associated with the matrix $[A(t)]$. In a manner analogous to the case of linear equations with constant coefficients (see Sec. 6.2), we can write

$$e^{t[J]} = \begin{bmatrix} e^{t[J_0]} & [0] & \cdots & [0] \\ [0] & e^{t[J_1]} & \cdots & [0] \\ \cdots & \cdots & \cdots & \cdots \\ [0] & [0] & \cdots & e^{t[J_s]} \end{bmatrix}, \tag{7.19}$$

in which

$$e^{t[J_0]} = \begin{bmatrix} e^{t\rho_1} & 0 & \cdots & 0 \\ 0 & e^{t\rho_2} & \cdots & 0 \\ \cdots & \cdots & \cdots & \cdots \\ 0 & 0 & \cdots & e^{t\rho_q} \end{bmatrix} \tag{7.20}$$

and

$$e^{t[J_i]} = e^{t\rho_{q+i}} \begin{bmatrix} 1 & t & \frac{t^2}{2!} & \cdots & \frac{t^{r_i-1}}{(r_i-1)!} \\ 0 & 1 & t & \cdots & \frac{t^{r_i-2}}{(r_i-2)!} \\ 0 & 0 & 1 & \cdots & \frac{t^{r_i-3}}{(r_i-3)!} \\ \cdots & \cdots & \cdots & \cdots & \cdots \\ 0 & 0 & 0 & \cdots & 1 \end{bmatrix}, \quad \begin{array}{l} i = 1, 2, \ldots, s; \\ q + \sum_{i=1}^{s} r_i = m. \end{array} \tag{7.21}$$

The m linearly independent columns $\{\Phi^{(j)}(t)\}$ of the fundamental matrix

[Φ(t)], Eqs. (7.18), can be obtained by means of Eqs. (7.19) to (7.21) in the form

$$\{\Phi^{(1)}(t)\} = e^{t\rho_1}\{P^{(1)}(t)\},$$

$$\{\Phi^{(2)}(t)\} = e^{t\rho_2}\{P^{(2)}(t)\},$$

$$\cdots\cdots\cdots\cdots,$$

$$\{\Phi^{(q)}(t)\} = e^{t\rho_q}\{P^{(q)}(t)\},$$

$$\{\Phi^{(q+1)}(t)\} = e^{t\rho_{q+1}}\{P^{(q+1)}(t)\},$$

$$\{\Phi^{(q+2)}(t)\} = e^{t\rho_{q+1}} \sum_{k=0}^{1} \frac{t^k}{k!}\{P^{(q+2-k)}(t)\},$$

$$\{\Phi^{(q+3)}(t)\} = e^{t\rho_{q+1}} \sum_{k=0}^{2} \frac{t^k}{k!}\{P^{(q+3-k)}(t)\},$$

$$\cdots\cdots\cdots\cdots\cdots\cdots\cdots,$$ (7.22)

$$\{\Phi^{(q+r_1)}(t)\} = e^{t\rho_{q+1}} \sum_{k=0}^{r_1-1} \frac{t^k}{k!}\{P^{(q+r_1-k)}(t)\},$$

$$\cdots\cdots\cdots\cdots\cdots\cdots\cdots,$$

$$\{\Phi^{(m-r_s+1)}(t)\} = e^{t\rho_{q+s}}\{P^{(m-r_s+1)}(t)\},$$

$$\cdots\cdots\cdots\cdots\cdots\cdots\cdots,$$

$$\{\Phi^{(m)}(t)\} = e^{t\rho_{q+s}} \sum_{k=0}^{r_s-1} \frac{t^k}{k!}\{P^{(m-k)}(t)\},$$

where $\{P^{(j)}(t)\}$ are the columns of the periodic matrix $[P(t)]$.

Hence the solution consists of polynomials in t multiplying exponential terms of the form $\exp(t\rho_j)$. The polynomials possess periodic coefficients in t of period T. When all the characteristic exponents are distinct (with the exception of the case when the imaginary parts alone differ by an integral multiple of $2\pi/T$), the Jordan form becomes purely diagonal and the terms multiplying $\exp(t\rho_j)$ become strictly periodic. This enables us to reach the following conclusions:

1. If all the characteristic exponents have negative real parts, all solutions of Eq. (7.1) are *asymptotically stable*

$$\lim_{t\to\infty} \{\Phi^{(j)}(t)\} = 0, \quad \text{Re}(\rho_j) < 0, \quad j = 1, 2, \ldots, m. \quad (7.23)$$

Note from Eqs. (7.17) that $\text{Re}(\rho_j) < 0$ implies $|\lambda_j| < 1$.

2. If at least one of the characteristic exponents has a positive real part, system (7.1) is *unstable*.

3. If some of the characteristic exponents have negative real parts and the remaining ones have zero real parts, and if the elementary divisors corresponding to the latter are simple, the solution is *stable*; if the said divisors are not simple, the solution is *unstable*.

4. *A purely periodic solution* is possible only when one of the characteristic exponents is identically zero.

The question remains of how to determine the characteristic exponents. Unfortunately there are no general methods available for the exact determination of the characteristic exponents, so that, in effect, the above theory determines only the form of the solution rather than the solution itself. It is possible, however, to obtain some information concerning the characteristic exponents. To this end, we recall the theory of Sec. 6.1, let $t_0 = 0$ in the Jacobi-Liouville formula (6.19), and obtain

$$\det [\varphi(t)] = \det [\varphi(0)] \exp \left(\int_0^t \text{tr } [A(s)] \, ds \right), \tag{7.24}$$

where s is a dummy variable of integration and $\det [\varphi(0)]$ is the determinant of the fundamental matrix evaluated at $t = 0$. Assuming that the initial conditions are such that

$$[\varphi(0)] = [1], \tag{7.25}$$

and using Eq. (7.11), the monodromy matrix reduces to

$$[C] = [\varphi(T)], \tag{7.26}$$

which is simply the principal matrix for the system (7.1) evaluated at the end of the period, $t = T$. Recalling Eq. (7.15), setting $t = T$ in Eq. (7.24), and using Eq. (7.26), we obtain

$$\det [C] = \prod_{j=1}^m \lambda_j = \exp \left(\int_0^T \text{tr } [A(s)] \, ds \right). \tag{7.27}$$

Hence, if $m - 1$ of the characteristic multipliers are known, Eq. (7.27) yields the remaining one.

We are primarily interested in the case in which the elements of the matrix $[A(t)]$ are all real. Then if λ_j is a characteristic multiplier corresponding to the system (7.1), the complex conjugate λ_j^* is also a characteristic multiplier. Furthermore, if $[\varphi(t)]$ is a real fundamental matrix of that system such that Eqs. (7.25) and (7.26) hold, then the characteristic equation $\det ([C] - \lambda[1]) = 0$ consists of a polynomial in λ with real coefficients.

If there exists a matrix $[K(t)]$ which, together with the determinant of its reciprocal, det $[K(t)]^{-1}$, is bounded and the matrix is such that the linear transformation

$$\{x\} = [K(t)]\{y\} \tag{7.28}$$

reduces system (7.1) to a system of linear equations with constant coefficients, then system (7.1) is said to be *reducible*. In this connection we can state the following theorem, due to Liapunov:

Theorem 7.1.2 *The system defined by Eqs. (7.1) and (7.2) is reducible by means of a periodic matrix.*

Furthermore, we have the following theorem about the characteristic exponents of the original system with periodic coefficients, Eqs. (7.1) and (7.2):

Theorem 7.1.3 *The eigenvalues of the transformed system with constant coefficients constitute the characteristic exponents of the original system with periodic coefficients.*

A more detailed discussion of reducible systems can be found in Ref. 10, pp. 159–164.

It should be noted at this point that the above considerations, regarding a system with periodic coefficients, are valid even when the matrix $[A(t)]$ is constant rather than periodic. Then the matrix $[A]$ can be regarded as pertaining to the reduced system with constant coefficients and, by Theorem 7.1.3, it is possible to obtain the characteristic exponents by solving the eigenvalue problem associated with $[A]$.

A semianalytical method of determining the characteristic multipliers is based on Eq. (7.26). It follows from this equation that the monodromy matrix $[C]$ is equal to the principal matrix for system (7.1) evaluated at the end of the period. Hence the matrix $[C]$ can be produced by integrating Eq. (7.1) numerically m times, in sequence, obtaining each time a solution $\{\varphi^{(j)}(t)\}$ corresponding to a column matrix of initial conditions with all its elements equal to zero except the element in the jth row, which is equal to 1, and evaluating the solutions at $t = T$. The monodromy matrix is the matrix having as its columns the solutions $\{\varphi^{(j)}(T)\}$, and the characteristic multipliers can be obtained by solving the eigenvalue problem associated with that matrix.

The application of the theory of linear differential equations with periodic coefficients has been fruitful in certain cases, particularly in the case of second-order systems governed by Hill's equation. There has also been success in treating general systems defined by a matrix $[A(t,\epsilon)]$ whose elements $A_{ij}(t,\epsilon)$ consist of constant terms and periodic terms multiplied by a small parameter ϵ.

7.2 STABILITY OF VARIATIONAL EQUATIONS WITH PERIODIC COEFFICIENTS

We shall be concerned with n-degree-of-freedom nonautonomous dynamical systems governed by the m nonlinear first-order differential equations

$$\dot{x}_i = X_i(x_1, x_2, \ldots, x_m, t), \qquad i = 1, 2, \ldots, m, \qquad (7.29)$$

where $m = 2n$. The functions X_i are real, continuous, and possess first-order derivatives with respect to the variables x_i. In particular, we wish to explore the behavior of the system in the vicinity of the real periodic solution $p_i(t)$ ($i = 1, 2, \ldots, m$) of Eqs. (7.29). For this purpose we let

$$x_i(t) = p_i(t) + y_i(t), \qquad i = 1, 2, \ldots, m, \qquad (7.30)$$

where the periodic solution $p_i(t)$ has period T, $p_i(t) = p_i(t + T)$, and $y_i(t)$ are small perturbations from the periodic solution. Introducing Eqs. (7.30) into (7.29), we obtain

$$\begin{aligned}\dot{y}_i(t) &= X_i(p_1 + y_1, p_2 + y_2, \ldots, p_m + y_m, t) - X_i(p_1, p_2, \ldots, p_m, t) \\ &= \sum_{j=1}^{m} a_{ij}(t) y_j(t) + \epsilon_i(y_1, y_2, \ldots, y_m, t),\end{aligned} \qquad (7.31)$$

where

$$a_{ij}(t) = \left.\frac{\partial X_i}{\partial x_j}\right|_{x_i = p_i}, \qquad i, j = 1, 2, \ldots, m, \qquad (7.32)$$

are periodic coefficients of period T, $a_{ij}(t) = a_{ij}(t + T)$, and ϵ_i are higher-order terms in the perturbations y_j. Assuming that the perturbations are sufficiently small for the terms ϵ_i to be neglected, we obtain the first-approximation equations, or variational equations

$$\dot{y}_i(t) = \sum_{j=1}^{m} a_{ij}(t) y_j(t), \qquad i = 1, 2, \ldots, m. \qquad (7.33)$$

We shall be interested in establishing the conditions under which the trivial solution of Eqs. (7.33) yields valid information concerning the stability of the periodic solution of Eqs. (7.29).

Equations (7.33) constitute a set of simultaneous linear equations with periodic coefficients of the type discussed in Sec. 7.1, where criteria for the stability of linear equations with periodic coefficients were derived. Under certain circumstances these criteria are applicable to the variational equations (7.33). To establish when this is so, we consider the following theorems:

Theorem 7.2.1 *If all the characteristic exponents associated with the variational equations* (7.33) *possess negative real parts, then the periodic solution $p_i(t)$ of the complete system* (7.29) *is asymptotically stable.*

Theorem 7.2.2 *If there is at least one characteristic exponent associated with Eqs. (7.33) with a positive real part, then the periodic solution $p_i(t)$ of the complete system (7.29) is unstable.*

The conditions to be satisfied by the functions ϵ_i for Theorem 7.2.1 or 7.2.2 to be applicable, as well as proofs of both these theorems, are given in Ref. 8, pp. 272–275.

When system (7.33) satisfies either Theorem 7.2.1 or 7.2.2, it is said to possess *significant behavior*. On the other hand, when some of the characteristic exponents associated with Eqs. (7.33) have negative real parts and the remaining exponents have zero real parts, the system is said to exhibit *critical behavior*. In this case system (7.31) can be either stable or unstable, depending on the higher-order terms (see Ref. 8, p. 275).

7.3 ORBITAL STABILITY

A case of considerable interest arises when the complete nonlinear dynamical system

$$\dot{x}_i = X_i(x_1, x_2, \ldots, x_m), \qquad i = 1, 2, \ldots, m, \tag{7.34}$$

has a periodic solution $p_i(t)$ ($i = 1, 2, \ldots, m$) with period T. Note that, by contrast with the system of the preceding section, the original system (7.34) is autonomous. The variational equations remain in the form

$$\dot{y}_i(t) = \sum_{j=1}^{m} a_{ij}(t) y_j(t), \qquad i = 1, 2, \ldots, m, \tag{7.35}$$

where the periodic coefficients $a_{ij}(t)$ are given by Eqs. (7.32).

The periodic solution $p_i(t)$ satisfies Eqs. (7.34), so that we can write

$$\dot{p}_i(t) = X_i(p_1, p_2, \ldots, p_m), \qquad i = 1, 2, \ldots, m. \tag{7.36}$$

Differentiating Eqs. (7.36) with respect to time and recalling Eqs. (7.32), we obtain

$$\frac{d}{dt}(\dot{p}_i) = \sum_{j=1}^{m} \frac{\partial X_i}{\partial p_j} \dot{p}_j = \sum_{j=1}^{m} a_{ij}(t) \dot{p}_j, \qquad i = 1, 2, \ldots, m, \tag{7.37}$$

which indicates that $\dot{p}_i(t)$ ($i = 1, 2, \ldots, m$) is a solution of the variational equations. This implies that the variational equations have a purely periodic solution and the characteristic exponent associated with that solution must be zero. Hence Eqs. (7.35) can have at most $m - 1$ characteristic exponents with negative real parts, which rules out asymptotic stability.

The solution $x_i(t) = p_i(t)$ $(i = 1, 2, \ldots, m)$ can be interpreted as representing a cyclic trajectory in the phase space defined by the coordinates x_i. If $m - 1$ characteristic exponents associated with the system (7.35) have negative real parts, then the cyclic trajectory is asymptotically stable in the sense that any trajectory in the neighborhood of the cyclic trajectory tends to it as $t \to \infty$. This implies that the periodic motion possesses *asymptotic orbital stability*. This can be summarized by the following theorem:

Theorem *A periodic solution of the dynamical system (7.34) is asymptotically orbitally stable if $m - 1$ characteristic exponents associated with the variational equations (7.35) possess negative real parts.*

(A proof of this theorem is given in Ref. 3, pp. 323–327.)

7.4 VARIATION FROM CANONICAL SYSTEMS. PERIODIC COEFFICIENTS

Our discussion of variational equations with periodic coefficients was perhaps a little too general to suit out purposes, since our interest lies in variational equations from Hamiltonian systems. In view of our experience in Sec. 6.6 with such systems, we have every reason to expect that in the case in which the coefficients are periodic rather than constant, it is also possible to gain additional information by taking advantage of the special form of the variational equations. This indeed turns out to be the case.

Let us again consider the Hamilton equations for an n-degree-of-freedom holonomic conservative system

$$\dot{q}_i = \frac{\partial H}{\partial p_i}, \qquad \dot{p}_i = -\frac{\partial H}{\partial q_i}, \qquad i = 1, 2, \ldots, n, \tag{7.38}$$

where $H = H(q_1, q_2, \ldots, q_n, p_1, p_2, \ldots, p_n)$ is the Hamiltonian function. We have shown in Sec. 6.6 that introducing the notation $q_i = x_i$, $p_i = x_{n+i}$ $(i = 1, 2, \ldots, n)$, Eqs. (7.38) can be written in the matrix form

$$\{\dot{x}\} = [Z]\left\{\frac{\partial H}{\partial x}\right\}, \tag{7.39}$$

where the column matrices $\{x\}$ and $\{\partial H/\partial x\}$ represent $2n$-vectors and $[Z]$ is the skew-symmetric matrix defined by Eq. (6.77).

Next we consider the solution of Eq. (7.39) in the form

$$\{x(t)\} = \{P(t)\} + \{y(t)\}, \tag{7.40}$$

in which the $2n$-vectors $\{P(t)\}$ and $\{y(t)\}$ represent a periodic solution of Eq. (7.39) and a small perturbation from that periodic solution, respectively. Introducing Eq. (7.40) into (7.39) and retaining only first-order terms in the perturbations y_i, we obtain a set of variational equations having the matrix form

$$\{\dot{y}\} = [Z][\mathcal{H}]\{y\} = [A(t)]\{y\}, \tag{7.41}$$

where $[\mathcal{H}]$ is the Hessian matrix associated with the Hamiltonian, namely, a $2n \times 2n$ symmetric periodic matrix whose elements are defined by

$$\mathcal{H}_{ij}(t) = \frac{\partial^2 H}{\partial x_i \, \partial x_j}, \qquad i, j = 1, 2, \ldots, 2n, \tag{7.42}$$

in which the periodic solution $x_i(t) = P_i(t)$ ($i = 1, 2, \ldots, 2n$) is substituted after differentiation.

Let us assume that $\{y^{(r)}\}$ and $\{y^{(s)}\}$ are two independent solutions of Eq. (7.41) and introduce a scalar K defined by

$$K = \{y^{(r)}\}^T [Z]\{y^{(s)}\}. \tag{7.43}$$

Differentiating Eq. (7.43) with respect to time, using Eq. (7.41), and recalling the properties of $[Z]$ from Sec. 6.6 as well as the fact that $[\mathcal{H}]$ is symmetric, we obtain

$$\dot{K} = \{\dot{y}^{(r)}\}^T [Z]\{y^{(s)}\} + \{y^{(r)}\}^T [Z]\{\dot{y}^{(s)}\}$$
$$= \{y^{(r)}\}^T [\mathcal{H}]^T [Z]^T [Z]\{y^{(s)}\} + \{y^{(r)}\}^T [Z][Z][\mathcal{H}]\{y^{(s)}\} = 0. \tag{7.44}$$

Hence K is constant, and it follows that the relation

$$\{y^{(r)}(t)\}^T [Z]\{y^{(s)}(t)\} = \{y^{(r)}(0)\}^T [Z]\{y^{(s)}(0)\} \tag{7.45}$$

must hold true. If $[\varphi(t)]$ is a fundamental matrix of the variational equations (7.41), then any solution $\{y(t)\}$ of these equations can be written as

$$\{y(t)\} = [\varphi(t)]\{y_0\}, \tag{7.46}$$

where $\{y_0\}$ is a suitable $2n$-vector. In particular, if

$$[\varphi(0)] = [1_{2n}], \tag{7.47}$$

where $[1_{2n}]$ is the unit matrix of order $2n$, then Eq. (7.46) becomes

$$\{y(t)\} = [\varphi(t)]\{y(0)\}, \tag{7.48}$$

which gives the solution $\{y(t)\}$ in terms of the initial conditions $\{y(0)\}$. But if (7.47) holds true, then

$$[\varphi(T)] = [C], \tag{7.49}$$

where $[C]$ is the monodromy matrix associated with the fundamental system $[\varphi(t)]$ (see Sec. 7.1). Letting $t = T$ in Eqs. (7.45) and (7.48) and taking into consideration Eq. (7.49), we obtain

$$\{y^{(r)}(0)\}^T[C]^T[Z][C]\{y^{(s)}(0)\} = \{y^{(r)}(0)\}^T[Z]\{y^{(s)}(0)\}, \qquad (7.50)$$

and, since this must be true for any arbitrary set of initial conditions, it follows that the monodromy matrix must satisfy the relation

$$[C]^T[Z][C] = [Z]. \qquad (7.51)$$

A matrix satisfying Eq. (7.51) is said to be *symplectic*.

In Sec. 7.1, however, it was indicated that the characteristic multipliers λ_i are the eigenvalues of the monodromy matrix; hence they satisfy the characteristic equation

$$\det\left([C]^T - \lambda[1_{2n}]\right) = 0, \qquad (7.52)$$

from which it follows that

$$\det\left([C]^T[Z][C] - \lambda[Z][C]\right) = 0. \qquad (7.53)$$

Using Eq. (7.51), Eq. (7.53) reduces to

$$\det\left([Z] - \lambda[Z][C]\right) = 0, \qquad (7.54)$$

leading to

$$\det\left([C] - \lambda^{-1}[1_{2n}]\right) = 0, \qquad (7.55)$$

so that λ_i^{-1} are also characteristic multipliers of the system (7.41). But the characteristic exponents ρ_i are such that $\lambda_i = \exp(T\rho_i)$, which indicates that the characteristic exponents occur in pairs, ρ_i and $-\rho_i$. Hence, since the number of characteristic exponents is even, if there is a periodic solution of the original canonical system, then two of the characteristic exponents of the corresponding variational equations are zero. Furthermore, the monodromy matrix $[C]$ is real, so that if λ_i is a characteristic multiplier, then the complex conjugate λ_i^* must also be a characteristic multiplier.

The above discussion enables us to state the following:

Theorem *If ρ_i is a characteristic exponent of a canonical system with periodic coefficients and ρ_i is neither real nor purely imaginary, then $-\rho_i$, ρ_i^*, and $-\rho_i^*$ are also characteristic exponents. If the original Hamiltonian equations have a periodic solution, then the variational equations possess two characteristic exponents equal to zero.*

The interested reader can find an extensive discussion of canonical systems with periodic coefficients in Ref. 4.

Example 7.1

Let us consider the stability of motion of a particle in a central force field for the special case in which the orbit is circular. Using as generalized coordinates q_r and q_θ the polar coordinates r and θ, respectively, we can write the kinetic energy

$$T = \tfrac{1}{2}m(\dot{r}^2 + r^2\dot\theta^2). \tag{a}$$

Denoting the potential energy by $V(r)$ and recalling that the Lagrangian of the system is $L = T - V$, we see that the generalized momenta become

$$p_r = \frac{\partial L}{\partial \dot r} = m\dot r, \quad p_\theta = \frac{\partial L}{\partial \dot\theta} = mr^2\dot\theta. \tag{b}$$

This enables us to write the Hamiltonian in the form

$$H = \frac{\partial L}{\partial \dot r}\dot r + \frac{\partial L}{\partial \dot\theta}\dot\theta - L = \frac{1}{2m}\left(p_r^2 + \frac{p_\theta^2}{r^2}\right) + V(r), \tag{c}$$

from which we derive the Hamilton canonical equations

$$\begin{aligned}
\dot r &= \frac{\partial H}{\partial p_r} = \frac{p_r}{m}, \\
\dot\theta &= \frac{\partial H}{\partial p_\theta} = \frac{p_\theta}{mr^2}, \\
\dot p_r &= -\frac{\partial H}{\partial r} = \frac{p_\theta^2}{mr^3} - V'(r), \\
\dot p_\theta &= -\frac{\partial H}{\partial \theta} = 0,
\end{aligned} \tag{d}$$

where the prime denotes differentiation with respect to r. Introducing the notation

$$x_1 = r, \quad x_2 = \theta, \quad x_3 = p_r, \quad x_4 = p_\theta, \tag{e}$$

the Hamiltonian becomes

$$H = \frac{1}{2m}\left(x_3^2 + \frac{x_4^2}{x_1^2}\right) + V(x_1). \tag{f}$$

The circular orbit is defined by

$$x_1 = a, \quad \dot x_2 = \omega, \quad x_4 = ma^2\omega, \tag{g}$$

where a is the orbit radius and ω the orbital angular velocity, so that, introducing Eq. (f) into expression (7.42), we obtain the Hessian matrix

$$[\mathscr{H}] = \begin{bmatrix} 3m\omega^2 + V''(a) & 0 & 0 & -2\omega a^{-1} \\ 0 & 0 & 0 & 0 \\ 0 & 0 & m^{-1} & 0 \\ -2\omega a^{-1} & 0 & 0 & m^{-1}a^{-2} \end{bmatrix}. \tag{h}$$

Upon use of Eq. (7.41) the matrix of the coefficients becomes

$$[A(t)] = [Z][\mathcal{H}] = \begin{bmatrix} 0 & 0 & m^{-1} & 0 \\ -2\omega a^{-1} & 0 & 0 & m^{-1}a^{-2} \\ -3m\omega^2 - V''(a) & 0 & 0 & 2\omega a^{-1} \\ 0 & 0 & 0 & 0 \end{bmatrix}, \quad (i)$$

which turns out to be constant rather than periodic.

Recalling Theorem 7.1.3, we obtain the characteristic exponents by solving the characteristic equation

$$\det([A] - \rho[1]) = \rho^2 \left[\rho^2 + 3\omega^2 + \frac{1}{m} V''(a) \right]. \quad (j)$$

As expected, two of the characteristic exponents are zero, $\rho_1 = \rho_2 = 0$. To obtain the remaining two we must specify the form of the potential energy. Letting

$$V(x_1) = -\alpha \frac{x_1^{k+1}}{k+1}, \quad (k)$$

where k is an integer and α a constant, and because for a circular orbit

$$-m a \omega^2 = F(a) = \alpha a^k, \quad (l)$$

the remaining characteristic exponents become

$$\left. \begin{array}{c} \rho_3 \\ \rho_4 \end{array} \right\} = \pm i\omega(3+k)^{\frac{1}{2}}. \quad (m)$$

For $k > -3$ the characteristic exponents are purely imaginary, so that the orbit is merely orbitally stable. This includes the orbit under the inverse square law, for which $k = -2$ and

$$\left. \begin{array}{c} \rho_3 \\ \rho_4 \end{array} \right\} = \pm i\omega. \quad (n)$$

Note that this is really a case in which the system exhibits critical behavior. For $k < -3$ the characteristic exponents are real, and since one of them is positive, the orbit is unstable.

7.5 SECOND-ORDER SYSTEMS WITH PERIODIC COEFFICIENTS

A large number of important dynamical systems lead to mathematical formulations in terms of a single second-order linear differential equation with periodic coefficients. Of these, the most commonly encountered are the

Mathieu equation and Hill's equation, where the first may be regarded as a special case of the second. Both equations can be written in the general form

$$\ddot{x} + p(t)x = 0, \tag{7.56}$$

where $p(t)$ is a periodic function of period T, $p(t) = p(t + T)$. Equation (7.56) can be reduced to two first-order equations by introducing the notation $x = x_1$, $\dot{x} = x_2$, which leads to

$$\begin{aligned}\dot{x}_1 &= x_2, \\ \dot{x}_2 &= -p(t)x_1,\end{aligned} \tag{7.57}$$

or in matrix form

$$\{\dot{x}\} = [A(t)]\{x\}, \tag{7.58}$$

where $\{x\}$ is the corresponding 2×1 column matrix of the variables and

$$[A(t)] = \begin{bmatrix} 0 & 1 \\ -p(t) & 0 \end{bmatrix} \tag{7.59}$$

is the 2×2 periodic matrix of the coefficients.

The 2×2 fundamental matrix $[\varphi(t)]$ associated with system (7.58) satisfies the equation

$$[\varphi(t + T)] = [\varphi(t)][C], \tag{7.60}$$

where the monodromy matrix $[C]$ is chosen so that

$$[\varphi(0)] = [1]. \tag{7.61}$$

From Eq. (7.59), however, we observe that tr $[A(t)] = 0$, and in view of Eq. (7.27) we conclude that

$$\det [C] = \lambda_1 \lambda_2 = e^0 = 1. \tag{7.62}$$

It follows that the characteristic equation associated with the monodromy matrix reduces to

$$\det ([C] - \lambda[1]) = \lambda^2 - 2A\lambda + 1 = 0, \tag{7.63}$$

where the notation

$$2A = C_{11} + C_{22} = \varphi_{11}(T) + \varphi_{22}(T) \tag{7.64}$$

has been introduced. Solving Eq. (7.63), we obtain the characteristic multipliers

$$\left.\begin{matrix}\lambda_1 \\ \lambda_2\end{matrix}\right\} = A \pm \sqrt{A^2 - 1}. \tag{7.65}$$

For $A^2 < 1$ the characteristic multipliers λ_1 and λ_2 are complex conjugates. Because, by Eq. (7.62), $\lambda_1\lambda_2 = 1$, their absolute values are equal to 1, $|\lambda_1| = |\lambda_2| = 1$. This implies that both characteristic exponents, ρ_1 and ρ_2, have zero real parts, so that all solutions are bounded. If $A^2 > 1$, both characteristic multipliers are real and such that $0 < \lambda_2 < 1 < \lambda_1$ when A is positive. It follows from Eqs. (7.17) that the characteristic exponent ρ_1 has a positive real part; hence the solution is unbounded.

When $A^2 = 1$, the characteristic multipliers assume the values $\lambda_1 = \lambda_2 = 1$ or $\lambda_1 = \lambda_2 = -1$, and in either of these cases Eq. (7.56) has at least one periodic solution. In the first case the period is T, and in the second the period is $2T$. These cases are very infrequent, but they are interesting from a mathematical standpoint, as they represent the dividing lines between bounded and unbounded solutions.

It may prove of interest to consider a graphic representation in the complex plane of the various possible cases of characteristic multipliers (see Fig. 7.1). The point on the unit circle intersecting the positive real axis in Fig. 7.1 represents periodic motion with period T, whereas the intersection with the negative real axis represents periodic motion with period $2T$. Considering the case in which A is positive, a slight variation from the value $A = 1$ will cause the two multipliers to split. For $0 < A < 1$ the multipliers λ_1 and λ_2 become complex conjugates, of magnitude equal to 1, moving in opposite directions on the unit circle, as shown in Fig. 7.1. For $A > 1$ the multipliers move in opposite directions along the real axis, their values being the reciprocals of each other. We shall return to this subject in Sec. 7.7.

The preceding analysis yields the form of solution of system (7.58) provided the characteristic multipliers λ_1 and λ_2 are known. This, however,

FIGURE 7.1

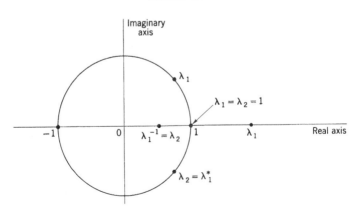

requires the elements $C_{11} = \varphi_{11}(T)$ and $C_{22} = \varphi_{22}(T)$ of the monodromy matrix, and there is no simple way of obtaining them.

Under certain circumstances it may be possible to test the boundedness of the solution without actually having the monodromy matrix. To this end, we state the following theorem:

Theorem *If $p(t)$ is a continuous periodic function of period T and the function satisfies the conditions*

1. $\int_0^T p(t)\, dt \geq 0, \qquad p(t) \neq 0,$ (7.66)

2. $\int_0^T |p(t)|\, dt \leq \dfrac{4}{T},$ (7.67)

then all solutions of system (7.58) are bounded as $t \to \pm \infty$.

(A proof of this theorem can be found in Ref. 1, pp. 124–125.)

7.6 HILL'S INFINITE DETERMINANT

In the preceding section we employed Floquet's theory to discuss the solution of a second-order system with periodic coefficients. As we pointed out, the theory provides us with the form of the solution rather than the solution itself. For example, the solution of the second-order equation

$$\ddot{x} + p(t)x = 0, \qquad (7.68)$$

in which $p(t)$ is periodic with period T, has the form

$$x(t) = e^{\rho t}\phi(t), \qquad (7.69)$$

where $\phi(t)$ is a periodic function of period T and ρ is the characteristic exponent, which depends on the system parameters. In general determination of ρ is not an easy matter.

In some cases, such as Hill's equation, the form of the solution alone enables us to construct a solution. In the case of Hill's equation the periodic coefficient $p(t)$ has the expression

$$p(t) = \theta_0 + 2\sum_{r=1}^{\infty} \theta_r \cos 2rt, \qquad (7.70)$$

which has the period $T = \pi$. The coefficients θ_r are known, and the series in (7.70) converges absolutely. In accordance with Eq. (7.69) a solution of

Nonautonomous Systems

Hill's equation is assumed in the form

$$x(t) = e^{\rho t} \sum_{r=-\infty}^{\infty} b_r e^{2rit}, \quad i = \sqrt{-1}. \tag{7.71}$$

Introducing Eqs. (7.71) into (7.68), with $p(t)$ in the form (7.70), we obtain

$$\sum_{r=-\infty}^{\infty} (\rho + 2ri)^2 b_r e^{(\rho + 2ri)t} + \left(\sum_{s=-\infty}^{\infty} \theta_s e^{2sit} \right) \left(\sum_{r=-\infty}^{\infty} b_r e^{(\rho + 2ri)t} \right) = 0, \tag{7.72}$$

where $\theta_{-s} = \theta_s$. Equating the coefficients of every integral power of e^{2it} to zero, we obtain the infinite set of algebraic equations

$$(\rho + 2ri)^2 b_r + \sum_{s=-\infty}^{\infty} \theta_s b_{r-s} = 0, \quad r = \ldots, -2, -1, 0, 1, 2, \ldots \tag{7.73}$$

The linear homogeneous system of equations (7.73) has a solution, other than the trivial one, only if the determinant of the coefficients is equal to zero. This is also true in the case in which the number of unknowns is infinite. Dividing every one of Eqs. (7.73) by $\theta_0 - 4r^2$ $(r = \ldots, -2, -1, 0, 1, 2, \ldots)$, correspondingly, to ensure convergence, we obtain Hill's determinantal equation

$$\Delta(i\rho) = \begin{vmatrix} \cdots & \cdots & \cdots & \cdots & \cdots & \cdots & \cdots \\ \cdots & \dfrac{(i\rho+4)^2 - \theta_0}{4^2 - \theta_0} & \dfrac{-\theta_1}{4^2 - \theta_0} & \dfrac{-\theta_2}{4^2 - \theta_0} & \dfrac{-\theta_3}{4^2 - \theta_0} & \dfrac{-\theta_4}{4^2 - \theta_0} & \cdots \\ \cdots & \dfrac{-\theta_1}{2^2 - \theta_0} & \dfrac{(i\rho+2)^2 - \theta_0}{2^2 - \theta_0} & \dfrac{-\theta_1}{2^2 - \theta_0} & \dfrac{-\theta_2}{2^2 - \theta_0} & \dfrac{-\theta_3}{2^2 - \theta_0} & \cdots \\ \cdots & \dfrac{-\theta_2}{-\theta_0} & \dfrac{-\theta_1}{-\theta_0} & \dfrac{(i\rho)^2 - \theta_0}{-\theta_0} & \dfrac{-\theta_1}{-\theta_0} & \dfrac{-\theta_2}{-\theta_0} & \cdots \\ \cdots & \dfrac{-\theta_3}{2^2 - \theta_0} & \dfrac{-\theta_2}{2^2 - \theta_0} & \dfrac{-\theta_1}{2^2 - \theta_0} & \dfrac{(i\rho-2)^2 - \theta_0}{2^2 - \theta_0} & \dfrac{-\theta_1}{2^2 - \theta_0} & \cdots \\ \cdots & \dfrac{-\theta_4}{4^2 - \theta_0} & \dfrac{-\theta_3}{4^2 - \theta_0} & \dfrac{-\theta_2}{4^2 - \theta_0} & \dfrac{-\theta_1}{4^2 - \theta_0} & \dfrac{(i\rho-4)^2 - \theta_0}{4^2 - \theta_0} & \cdots \\ \cdots & \cdots & \cdots & \cdots & \cdots & \cdots & \cdots \end{vmatrix} = 0. \tag{7.74}$$

It can be shown (see Ref. 11, pp. 415–416), however, that the determinant $\Delta(i\rho)$ can be written in the form

$$\Delta(i\rho) = \Delta(0) - \sin^2\left(\tfrac{1}{2}\pi i\rho\right) \csc^2\left(\tfrac{1}{2}\pi\sqrt{\theta_0}\right), \tag{7.75}$$

so that the characteristic exponents can be obtained by solving the transcendental equation

$$\sin^2\left(\tfrac{1}{2}\pi i\rho\right) = \Delta(0) \sin^2\left(\tfrac{1}{2}\pi\sqrt{\theta_0}\right). \tag{7.76}$$

The values of ρ thus obtained, when introduced in Eqs. (7.73), enable us to evaluate the coefficients b_r in terms of b_0 [because Eq. (7.68) is homogeneous, the amplitude of solution (7.71) is arbitrary]. This again necessitates the evaluation of infinite determinants, which raises the question of convergence of such determinants. Infinite determinants and tests for their convergence are discussed in Ref. 11, pp. 36–37.

In some instances a good approximation for ρ can be obtained by considering only the three central rows and columns of Hill's determinant [enclosed by the dashed rectangle in (7.74)].

Actually we shall be interested in an approach which circumvents the difficult task of evaluating characteristic exponents. From Eq. (7.74) we conclude that the characteristic exponents depend on the coefficients θ_r in Eq. (7.70), which are assumed to be known quantities. There are instances where the quantities θ_r can be regarded as parameters, in which case we could conceive of a parameter space defined by these quantities. The parameter space can be divided into regions of bounded and unbounded motions with the periodic motions providing the surfaces separating these regions. The concept is particularly useful when the system is defined by only two parameters, say θ_0 and θ_1, in which case the parameter space reduces to a parameter plane and the boundary surfaces become boundary curves. Some of these ideas will be applied in Sec. 7.7 in connection with the study of Mathieu's equation.

7.7 MATHIEU'S EQUATION

A very important and frequently encountered ordinary differential equation in mathematical physics is Mathieu's equation. A simple illustration of a dynamical system whose behavior is governed by the Mathieu equation consists of an oscillator with its natural frequency varying harmonically with time. The differential equation of the oscillator is

$$\ddot{x} + (\delta + 2\epsilon \cos 2t)x = 0. \tag{7.77}$$

The quantities δ and ϵ may be regarded as parameters reflecting the system properties. It is not difficult to see that Eq. (7.77) is a special case of Hill's equation, Eqs. (7.68) and (7.70). Mathieu's equation has been amply explored, and numerous publications discuss its solution in terms of Mathieu's functions (for a comprehensive coverage see, for example, Ref. 9). We shall not go into the details of deriving these solutions but concentrate on an interesting aspect of the problem, namely, how the nature of the solution changes as the system parameters δ and ϵ vary.

Depending on the parameters δ and ϵ, the solution of Eq. (7.77) can be bounded or unbounded. For certain combinations of the parameters at least one of the solutions becomes periodic. (Both solutions can be periodic only if the Jordan form associated with the system is diagonal.) In particular, when the characteristic multipliers are equal to 1, there is a periodic solution with period $T = \pi$, whereas when the characteristic multipliers are equal to -1, there is a periodic solution with period $2T = 2\pi$, i.e., $x(t + T) = -x(t)$. The periodic solutions appear in the parameter plane (δ, ϵ) in the form of curves separating the regions of bounded motion, for which the characteristic exponents ρ_1 and ρ_2 have zero real parts, from the regions of unbounded motion, for which the characteristic exponent ρ_1 has a positive real part. The preceding results were discussed in Sec. 7.5 by means of a graphical representation of the characteristic multipliers λ_1 and λ_2 in the complex plane. This graphical representation acquires added meaning when used in conjunction with the representation of the solution in the parameter plane.

We recall from Sec. 7.5 that the characteristic multipliers have the expressions

$$\lambda_1 = A + \sqrt{A^2 - 1}, \qquad \lambda_2 = A - \sqrt{A^2 - 1}, \tag{7.78}$$

where
$$2A = \varphi_{11}(\pi) + \varphi_{22}(\pi), \tag{7.79}$$

so that $2A$ represents the trace of the fundamental matrix $[\varphi(t)]$ evaluated at the end of one period, $T = \pi$. For the sake of this discussion we shall not be concerned with the fundamental solution. It is evident, however, that this solution, hence A, must depend on the system parameters δ and ϵ. As the parameters vary, A varies and so do the characteristic multipliers λ_1 and λ_2. Returning to the complex plane representation (Fig. 7.2), we see that for $A = 1$ we obtain a periodic solution with period $T = \pi$. Let us hold the corresponding value of ϵ fixed and vary δ in such a way that A increases. It follows from Eqs. (7.78) that the characteristic multipliers remain real, with

FIGURE 7.2

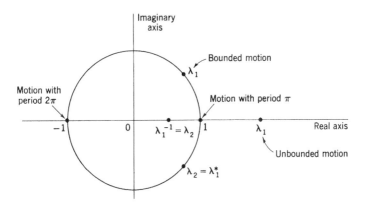

λ_1 moving in the positive direction of the real axis and λ_2 in the negative direction. Since, by Eq. (7.62), $\lambda_1\lambda_2 = 1$, it follows that λ_2 is the reciprocal of λ_1. But λ_1 cannot become infinite, nor can λ_2 become zero. Hence at a certain point the roots λ_1 and λ_2 must conceivably reverse their directions and return to the point $\lambda_1 = \lambda_2 = 1$. As $\lambda_1 > 1$, in the interval bounded by these two sets of values of δ and ϵ the motion is unstable. If, on the other hand, we hold ϵ constant and vary δ in such a way that A decreases, then λ_1 and λ_2 become complex conjugates of magnitude 1 and the roots will move along the unit circle, in opposite directions, until they coalesce at the point $\lambda_1 = \lambda_2 = -1$, at which point the motion is periodic with period $2T = 2\pi$. The above discussion enables us to make the following statements: *If the periodic motion at both ends of an interval in the parameter plane possesses the same period T or 2T, then the enclosed interval is characterized by unbounded motion. If the periodic motion at one end of an interval in the parameter plane possesses period T (2T) and at the other end possesses period 2T (T), then the motion in that interval is bounded.*

It turns out that by keeping ϵ fixed and increasing δ we obtain a sequence of discrete points at which the motion is periodic. These points occur in alternating pairs with period T and $2T$, so that the intervals enclosed are characterized by unbounded motion (corresponding to end points of equal period) and bounded motion (for end points of different period), alternately. By also allowing ϵ to vary the boundary points become boundary curves separating the regions of unbounded and bounded motion. Hence, we can replace in the above statements the words "ends" and "interval" by "boundary curves" and "region," respectively.

In the above discussion the tacit assumption was made that the characteristic multipliers behave as if they were continuous functions of the system parameters. This indeed turns out to be the case. (For a justification of the assumption see Ref. 2, p. 59.)

To establish the regions of bounded and unbounded motions in the parameter plane we need only obtain the boundary curves. Since on the boundary curves the motion is periodic, we can obtain the boundaries corresponding to the period $2T$ by representing the solution by the Fourier series

$$x(t) = \sum_{r=1}^{\infty} [a_r \sin(2r-1)t + b_r \cos(2r-1)t]. \qquad (7.80)$$

Substituting Eq. (7.80) into (7.77) and equating the coefficients of the trigonometric terms $\sin(2r-1)t$ to zero, we obtain the following homogeneous algebraic equations

$$\begin{aligned}(\delta - 1 - \epsilon)a_1 + \epsilon a_2 &= 0, \\ \epsilon(a_r + a_{r+2}) + [\delta - (2r+1)^2]a_{r+1} &= 0, \quad r = 1, 2, \ldots,\end{aligned} \qquad (7.81)$$

and, similarly, for the terms $\cos(2r - 1)t$ we obtain

$$(\delta - 1 + \epsilon)b_1 + \epsilon b_2 = 0,$$
$$\epsilon(b_r + b_{r+2}) + [\delta - (2r + 1)^2]b_{r+1} = 0, \quad r = 1, 2, \ldots. \quad (7.82)$$

We note that the coefficients of a_k are the same as the ones of b_k with the exception of the coefficients of a_1 and b_1, which differ only in the sign of ϵ. It follows that the conditions for the two sets (7.81) and (7.82) to have a nontrivial solution can be combined into one determinantal equation

$$\begin{vmatrix} \delta - 1 \pm \epsilon & \epsilon & 0 & 0 & \cdots \\ \epsilon & \delta - 3^2 & \epsilon & 0 & \cdots \\ 0 & \epsilon & \delta - 5^2 & \epsilon & \cdots \\ \cdots & \cdots & \cdots & \cdots & \cdots \end{vmatrix} = 0, \quad (7.83)$$

which, of course, involves an infinite determinant. Equation (7.83) represents curves ϵ versus δ in the parameter plane corresponding to boundary curves on which the motion is periodic with period $2T$.

Similarly, we can derive an expression for the boundary curves on which the motion is periodic with period T. To this end, we represent the solution of Eq. (7.77) by the Fourier series

$$x(t) = \sum_{r=0}^{\infty} (a_r \sin 2rt + b_r \cos 2rt). \quad (7.84)$$

Introducing Eq. (7.84) into (7.77) and equating the coefficients of $\sin 2rt$ and $\cos 2rt$ to zero, we obtain

$$(\delta - 4)a_1 + \epsilon a_2 = 0,$$
$$[\delta - 4(r + 1)^2]a_{r+1} + \epsilon(a_r + a_{r+2}) = 0, \quad r = 1, 2, \ldots, \quad (7.85)$$

and

$$\delta b_0 + \epsilon b_1 = 0,$$
$$2\epsilon b_0 + (\delta - 4)b_1 + \epsilon b_2 = 0, \quad (7.86)$$
$$[\delta - 4(r + 1)^2]b_{r+1} + \epsilon(b_r + b_{r+2}) = 0, \quad r = 1, 2, \ldots,$$

respectively, which lead to the determinantal equations

$$\begin{vmatrix} \delta - 2^2 & \epsilon & 0 & 0 & \cdots \\ \epsilon & \delta - 4^2 & \epsilon & 0 & \cdots \\ 0 & \epsilon & \delta - 6^2 & \epsilon & \cdots \\ \cdots & \cdots & \cdots & \cdots & \cdots \end{vmatrix} = 0 \quad (7.87)$$

and

$$\begin{vmatrix} \delta & \epsilon & 0 & 0 & 0 & \cdots \\ 2\epsilon & \delta - 2^2 & \epsilon & 0 & 0 & \cdots \\ 0 & \epsilon & \delta - 4^2 & \epsilon & 0 & \cdots \\ 0 & 0 & \epsilon & \delta - 6^2 & \epsilon & \cdots \\ \cdots & \cdots & \cdots & \cdots & \cdots & \cdots \end{vmatrix} = 0, \quad (7.88)$$

where again the determinants are infinite.

The preceding analysis enables us to divide the parameter plane ϵ versus δ into regions of bounded and unbounded motions. As it turns out, the regions are symmetric with respect to the axis δ, as can be seen in Fig. 7.3. The parameter plane of Fig. 7.3 is often referred to as a *Strutt diagram*.

FIGURE 7.3

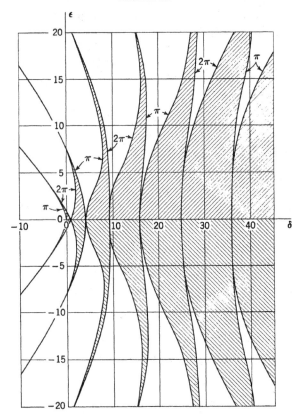

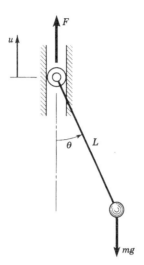

FIGURE 7.4

Quite frequently, if only an approximate solution is desired, the problem of evaluating infinite determinants can be avoided by considering in their place the associated principal minor determinants of finite order. We can begin with the principal minor determinant of second order and continuously increase the order of the determinant considered until sufficient accuracy in the determination of the first few boundary curves is achieved. This procedure is particularly suitable when ϵ is a small parameter relative to δ. This subject will be discussed in Chap. 8.

Example 7.2

The pendulum whose support is undergoing oscillatory motion provides an example of a system that can be described by Mathieu's equation. An interesting aspect of this problem is the manner in which the system parameters affect its stability characteristics.

Let us consider the pendulum illustrated in Fig. 7.4 and denote by u the vertical motion of the support when acted upon by the force F. We describe the motion of the system by the coordinates θ and u, although we shall assume later than u has a known harmonic expression. The kinetic energy, potential energy, and virtual work of the system are, respectively,

$$T = \tfrac{1}{2}m[(L\dot\theta \cos\theta)^2 + (\dot u + L\dot\theta \sin\theta)^2] = \tfrac{1}{2}m(L^2\dot\theta^2 + 2L\dot\theta\dot u \sin\theta + \dot u^2), \quad (a)$$
$$V = mg[L(1 - \cos\theta) + u], \quad (b)$$
and
$$\overline{\delta W} = F\,\delta u. \quad (c)$$

It is not difficult to show that the Lagrange equations of motion for the coordinates θ and u are

$$mL^2\ddot{\theta} + mL\ddot{u}\sin\theta + mgL\sin\theta = 0,$$
$$mL\ddot{\theta}\sin\theta + mL\dot{\theta}^2\cos\theta + m\ddot{u} + mg = F. \quad (d)$$

In the neighborhood of $\theta = 0$ Eqs. (d) reduce to

$$\ddot{\theta} + \left(\frac{g}{L} + \frac{\ddot{u}}{L}\right)\theta = 0,$$
$$\ddot{u} + g = \frac{F}{m}. \quad (e)$$

If the motion of the support is assumed to be harmonic and of the form $u = A\cos\omega t$, then the first of Eqs. (e) becomes

$$\ddot{\theta} + \left(\frac{g}{L} - \frac{A\omega^2}{L}\cos\omega t\right)\theta = 0, \quad (f)$$

which is recognized as Mathieu's equation, whereas the second of Eqs. (e) gives simply the force necessary to produce the harmonic motion of the support.

In view of Fig. 7.3 we conclude that the stability characteristics of a pendulum with the oscillating support are substantially different from those of a simple pendulum. Whereas a simple pendulum is stable in the neighborhood of $\theta = 0$, the pendulum with the oscillating support can be rendered unstable for certain combinations of the parameters g/L and $A\omega^2/L$.

In the neighborhood of $\theta = \pi$, the first of Eqs. (e) becomes

$$\ddot{\theta} - \left(\frac{g}{L} - \frac{A\omega^2}{L}\cos\omega t\right)\theta = 0. \quad (g)$$

We note that the same equation can be obtained by assuming that the gravity force is acting upward (the sign of the harmonic term is immaterial, as can be observed from Fig. 7.3). A simple pendulum is known to be unstable in the upright position. On the other hand, from Eq. (g) and Fig. 7.3 we conclude that for the right choice of the system parameters the equilibrium of the pendulum can be rendered stable by imparting an oscillatory motion to its support.

7.8 THE LIAPUNOV DIRECT METHOD

In Sec. 6.7 we discussed the Liapunov direct method in connection with the stability analysis of autonomous systems. The method is by no means restricted to autonomous systems, however, although its application to nonautonomous systems remains scarce. Except for slight modifications, the basic ideas are the same as for autonomous systems. The application of the Liapunov direct method to nonautonomous systems will necessitate the introduction of certain new definitions.

We shall be concerned with an n-degree-of-freedom nonautonomous system described by a set of m first-order differential equations, $m = 2n$, having the vector form

$$\dot{\mathbf{x}} = \mathbf{X}(\mathbf{x},t), \tag{7.89}$$

where $\mathbf{x}$ and $\mathbf{X}$ are real m-vectors. The vector $\mathbf{X}$ is Lipschitzian in the half-cylindrical neighborhood D_{H,t_0}: $t \geq t_0 > 0$, $\|\mathbf{x}\| \leq H$, of the t axis of the motion space, where t_0 and H are positive constants. It will be assumed that $\mathbf{X}(\mathbf{0},t) = \mathbf{0}$, so that Eq. (7.89) admits the trivial solution $\mathbf{x} = \mathbf{0}$.

Next consider the scalar function $V(\mathbf{x},t)$ defined in the neighborhood D_{h,t_0}, $0 < h \leq H$, and assumed to have the following properties: (1) V possesses in domain D_{h,t_0} continuous partial derivatives $\partial V/\partial t$, $\partial V/\partial x_i$ ($i = 1, 2, \ldots, m$), and (2) $V(\mathbf{0},t) = 0$ for all $t \geq t_0$. With this in mind, we introduce the following definitions:

1. A function $V(\mathbf{x},t)$, defined in the domain D_{h,t_0} of the motion space, is said to be *positive* (*negative*) *semidefinite* if $V(\mathbf{x},t) \geq 0 \, (\leq 0)$ in D_{h_1,t_0}, for suitable $h_1 \leq h$.

2. A function $V(\mathbf{x},t)$ is said to be *positive* (*negative*) *definite* if there exists in a domain D_{h_1}, $h_1 \leq h$, of the phase space a positive definite function $W(\mathbf{x})$ such that $V(\mathbf{x},t) \geq W(\mathbf{x}) \, [\leq -W(\mathbf{x})]$.

As an example, the function $V = e^{-t} \sum_{i=1}^{m} x_i^2$ is positive semidefinite, as it becomes zero for $t \to \infty$ regardless of the values x_i. On the other hand, $V = (2 + \sin t) \sum_{i=1}^{m} x_i^2$ is positive definite because $W = \sum_{i=1}^{m} x_i^2$ is positive definite.

Furthermore, we introduce additional definitions concerning the boundedness of a function:

3. A scalar function $V(\mathbf{x},t)$, as considered above, is said to be *bounded* if there is a positive constant M such that $|V(\mathbf{x},t)| < M$ in a given domain.

4. A bounded function $V(\mathbf{x},t)$ is said to have an *infinitesimal upper bound* if for any positive number ϵ there exists another positive number h such that $|V(\mathbf{x},t)| \leq \epsilon$ in the domain D_{h,t_0}. This implies that as $\|\mathbf{x}\|$ approaches zero V approaches zero uniformly in t.

As an illustration, the function $V = \sin(t \sum_{i=1}^{m} x_i)$ is merely bounded, whereas the function $V = (\sum_{i=1}^{m} x_i) \sin t$ admits an infinitesimal upper bound. In technical literature, the terminology "V is decrescent" is sometimes used instead of "V admits an infinitesimal upper bound."

It may prove of interest to explore the geometric interpretation of a positive definite function $V(\mathbf{x},t)$. For this purpose we consider the surface $V(\mathbf{x},t) = c$ and regard t as a parameter. For a certain value of t the equation $V = c_1$ represents in the phase space a closed surface surrounding the origin, where c_1 represents an arbitrarily small value of c. If t is allowed to vary, then the surface $V = c_1$ becomes a surface in motion. Next we consider the stationary surface $W(\mathbf{x}) = c_1$ and assume that $V(\mathbf{x},t) \geq W(\mathbf{x})$, from which we conclude that the surface $V = c_1$ will remain within the surface $W = c_1$ at all times. This can be best visualized by considering Fig. 6.3a and b.

If a positive definite function $V(\mathbf{x},t)$ admits an infinitesimal upper bound, then the moving surface $V(\mathbf{x},t) = c_1$ will remain at all times outside a certain sufficiently small neighborhood of the origin. This is so because it is possible to find a spherical region $\|\mathbf{x}\| < \mu$, where μ is sufficiently small for the surface $V = c_1$ to lie outside the spherical region.

The total derivative of the function $V(\mathbf{x},t)$ in the domain D_{h,t_0}, where the vector $\mathbf{x}$ satisfies Eq. (7.89), is given by

$$\dot{V}(\mathbf{x},t) = \frac{dV(\mathbf{x},t)}{dt} = \sum_{i=1}^{m} \frac{\partial V}{\partial x_i} \dot{x}_i + \frac{\partial V}{\partial t} = \nabla V \cdot \mathbf{X} + \frac{\partial V}{\partial t}, \quad (7.90)$$

where ∇V is the gradient of the scalar function V.

The definitions above enable us to state the *Liapunov stability theorems*:

Theorem 7.8.1 *If there exists a positive (negative) definite function $V(\mathbf{x},t)$ whose total derivative $\dot{V}(\mathbf{x},t)$ for the system (7.89) is negative (positive) semi-definite, then the trivial solution $\mathbf{x} = \mathbf{0}$ is stable.*

Theorem 7.8.2 *If there exists a positive (negative) definite function $V(\mathbf{x},t)$ which admits an infinitesimal upper bound and whose total derivative $\dot{V}(\mathbf{x},t)$ for the system (7.89) is negative (positive) definite, then the trivial solution $\mathbf{x} = \mathbf{0}$ is asymptotically stable.*

In a similar way, we can state the *Liapunov instability theorem*:

Theorem 7.8.3 *If there exists a function $V(\mathbf{x},t)$ which admits an infinitesimal upper bound, if its total derivative $\dot{V}(\mathbf{x},t)$ for the system (7.89) is positive (negative) definite, and if $V(\mathbf{x},t)$ itself can assume positive (negative) values for arbitrarily small values of $\mathbf{x}$ and for arbitrarily large values of t, then the trivial solution $\mathbf{x} = \mathbf{0}$ is unstable.*

A less restrictive *instability theorem due to Chetayev* makes requirements on the function $V(\mathbf{x},t)$ and its total derivative $\dot{V}(\mathbf{x},t)$ only in certain parts of the neighborhood of the origin rather than the entire neighborhood. The theorem has essentially the same form as the theorem for autonomous systems presented in Sec. 6.7.

Proofs of the above theorems can be found in Ref. 8, pp. 186–196. We shall present here, however, a geometric interpretation of Theorem 7.8.1. We consider the spherical region $\|\mathbf{x}\| \leq \epsilon$ and select a number c sufficiently small for the closed surface $W(\mathbf{x}) = c$ to lie entirely within the spherical region $\|\mathbf{x}\| \leq \epsilon$ (see Fig. 7.5). Next we consider a moving surface $V(\mathbf{x},t) = c$ such that it remains at all times within the region $W(\mathbf{x}) = c$, from which it follows that $V(\mathbf{x},t) = c$ must also remain within the region $\|\mathbf{x}\| \leq \epsilon$. If any motion $\mathbf{x}$ defined by Eq. (7.89) falls within the surface $V(\mathbf{x},t) = c$, then it must remain within that surface if the motion is to be stable. Otherwise at any point of intersection with the surface $\dot{V}(\mathbf{x},t)$ would be positive, and this contradicts the conditions of the theorem. Hence any motion initiated within the spherical region $\|\mathbf{x}\| \leq \eta$, lying entirely within the surface $V(\mathbf{x},t_0) = c$, will remain inside the region $\|\mathbf{x}\| \leq \epsilon$ at all times.

The preceding theorems furnish only sufficient conditions for stability or instability and give no indications of the existence of Liapunov functions. The question remains whether the fact that the trivial solution $\mathbf{x} = \mathbf{0}$ of system (7.89) is known to be stable can be construed to mean that a Liapunov function exists. This problem, known as the *inverse* or *converse problem*, has been investigated by I. G. Malkin and J. L. Massera, who modified the theorems to render the existence of a Liapunov function both necessary and sufficient for stability or instability. Their contributions and other pertinent developments are discussed, for example, by Cesari (Ref. 2, pp. 111–113) and Hahn (Ref. 5, pp. 60–82).

FIGURE 7.5

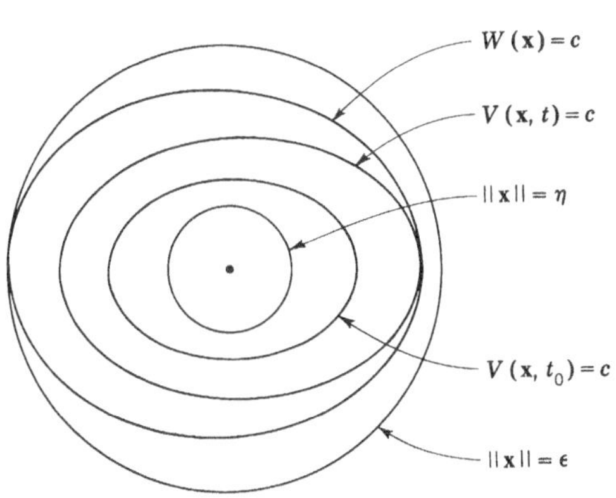

$W(\mathbf{x}) = c$

$V(\mathbf{x}, t) = c$

$\|\mathbf{x}\| = \eta$

$V(\mathbf{x}, t_0) = c$

$\|\mathbf{x}\| = \epsilon$

When the vector function $\mathbf{X}(\mathbf{x},t)$ is periodic in time, the conditions of Theorems 7.8.2 and 7.8.3 can be relaxed somewhat. Indeed it is shown by Krasovskii (Ref. 7, secs. 14 and 15) that in this case $\dot{V}(\mathbf{x},t)$ need only be semidefinite, rather than definite, provided the set of points for which $\dot{V} = 0$ contains no nontrivial half-trajectory. The function $V(\mathbf{x},t)$ is also periodic in this case and with the same period as the vector $\mathbf{X}(\mathbf{x},t)$.

SUGGESTED REFERENCES

1. Bellman, R.: *Stability Theory of Differential Equations*, McGraw-Hill Book Company, New York, 1953.
2. Cesari, L.: *Asymptotic Behavior and Stability Problems in Ordinary Differential Equations*, Springer-Verlag OHG, Berlin, 1963.
3. Coddington, E. A., and M. Levinson: *Theory of Ordinary Differential Equations*, McGraw-Hill Book Company, New York, 1955.
4. Gelfand, I. M., and V. B. Lidskii: "On the Structure of the Regions of Stability of Linear Canonical Systems of Differential Equations with Periodic Coefficients," *Amer. Math. Soc. Trans.*, ser. 2, vol. 8, pp. 143–181, 1958.
5. Hahn, W.: *Theory and Applications of Liapunov's Direct Method*, Prentice-Hall, Inc., Englewood Cliffs, N.J., 1963.
6. Hale, J. K.: *Oscillations in Nonlinear Systems*, McGraw-Hill Book Company, New York, 1963.
7. Krasovskii, N. N.: *Stability of Motion*, Stanford University Press, Stanford, Calif., 1963.
8. Malkin, I. G.: *Theory of Stability of Motion*, U.S. Atomic Energy Commission Translation, AEC-tr-3352, Department of Commerce, Washington, D.C., 1958.
9. McLachlan, N. W.: *Theory and Applications of Mathieu Functions*, Oxford University Press, New York, 1947.
10. Nemytskii, V. V., and V. V. Stepanov: *Qualitative Theory of Differential Equations*, Princeton University Press, Princeton, N.J., 1960.
11. Whittaker, E. T., and G. N. Watson: *A Course in Modern Analysis*, Cambridge University Press, London, 1927.

chapter eight

Analytical Solutions by Perturbation Techniques

From our preceding discussions the conclusion emerges that, except for particular cases, closed-form solutions of the differential equations governing the behavior of dynamical systems are beyond our reach. More success can be anticipated if we confine ourselves to investigating the quality of the motion in the neighborhood of known motions.

Under certain circumstances, it is possible to obtain approximate solutions of the differential equations of the system in the form of power series. This may be the case when the differential equations are nearly linear and autonomous, i.e., when the terms rendering the system nonlinear or nonautonomous are relatively small. Such terms, known as *perturbations*, represent second-order effects identified by a small parameter multiplying the nonlinear or nonautonomous terms. For such systems we may be able to obtain a solution in the form of a power series in the small parameter, in particular if the system is nearly harmonic.

The techniques for treating nearly linear systems are generally referred to as *perturbation methods*. Although Poisson used a perturbation technique to investigate problems of celestial mechanics, modern perturbation theory is attributed to Poincaré. Early attempts by astronomers to use perturbation techniques were plagued by the appearance of terms increasing with time. Such terms, destroying the convergence of the solution, are referred to as *secular terms*. A large variety of perturbation methods suppressing the secular terms have been developed, including a method by Poincaré.

Since the subject of perturbation techniques is quite extensive, for our limited objectives we must confine ourselves to a number of methods which appear suitable for our dynamical problems. An additional perturbation technique, based on the Hamilton-Jacobi theory, will be presented in Sec. 9.10.

8.1 THE FUNDAMENTAL PERTURBATION TECHNIQUE

We shall be interested in developing mathematical techniques for obtaining analytic solutions of the differential equations associated with *weakly nonlinear* autonomous dynamical systems or *weakly nonautonomous* systems or both. Weakly nonlinear systems are sometimes referred to as *quasi-linear* systems. The implication of the terminology weakly nonlinear is that the differential equations governing the system behavior can be separated into one part containing linear terms and a second part, relatively small compared to the first, containing the nonlinear terms. The definition can be extended to the other types of systems, e.g., weakly nonautonomous. The small terms rendering the system nonautonomous or nonlinear will be referred to as *perturbations*.

To introduce the basic idea involved in the treatment of quasi-linear systems, we consider the nonhomogeneous differential equation

$$F(u) = v, \tag{8.1}$$

where $F(u)$ is a form in $u = u(t)$ of a general nature, containing linear and nonlinear terms in u and its derivatives, and $v = v(t)$ is the nonhomogeneous part of the equation. The form $F(u)$ is such that no closed-form solution of Eq. (8.1) is feasible or even possible.

Next we consider the linear nonhomogeneous differential equation

$$L(u) = v, \tag{8.2}$$

where L represents a linear homogeneous differential operator with constant coefficients. The linear homogeneous differential expression $L(u)$ is closely related to $F(u)$ and is such that Eq. (8.2) possesses a closed-form solution. By the definition of linear operators, if u_1 and u_2 are any two functions and c_1 and c_2 any two arbitrary constants, then the relation

$$L(c_1 u_1 + c_2 u_2) = c_1 L(u_1) + c_2 L(u_2) \tag{8.3}$$

must be satisfied. Adding and subtracting $L(u)$ in the left side of Eq. (8.1), we obtain

$$L(u) + [F(u) - L(u)] = v, \tag{8.4}$$

which can be written in the form

$$L(u) = v + N(u), \tag{8.5}$$

where the notation

$$N(u) = L(u) - F(u) \tag{8.6}$$

has been introduced. The forms $F(u)$ to be considered in our applications will be almost linear homogeneous differential expressions differing from the expression $L(u)$ by second-order terms. Hence the form $N(u)$ can be regarded as a perturbation of the linear nonhomogeneous differential equation (8.2).

It will prove convenient to introduce a parameter ϵ and consider the equation

$$L(u) = v + \epsilon N(u), \tag{8.7}$$

which is evidently related to both Eqs. (8.2) and (8.5). For $\epsilon = 0$ Eq. (8.7) reduces to Eq. (8.2), the solution of which we know, whereas for $\epsilon = 1$ it coincides with Eq. (8.5), the solution of which we wish to obtain. The role of the parameter ϵ is to separate the small terms in Eq. (8.5). The presence of such a parameter associated with the small terms enables us to effect the transition between the known and the desired solutions.

We shall be particularly interested in the case in which the parameter ϵ is a small quantity, $\epsilon \ll 1$, so that the term $\epsilon N(u)$ in Eq. (8.7) is relatively small compared with the remaining terms. In this case it seems natural to assume a solution of Eq. (8.7) in the form of a power series in ϵ

$$u = u_0 + \epsilon u_1 + \epsilon^2 u_2 + \cdots, \tag{8.8}$$

where the functions u_i ($i = 0, 1, 2, \ldots$) do not depend on ϵ. The term u_0, representing the solution of Eq. (8.2), is referred to as the *zero-order approximation* of Eq. (8.7) or the *generating solution*. Introducing Eq. (8.8) into (8.7), we obtain

$$L(u_0 + \epsilon u_1 + \epsilon^2 u_2 + \cdots) = v + \epsilon N(u_0 + \epsilon u_1 + \epsilon^2 u_2 + \cdots). \tag{8.9}$$

But L is a linear operator for which the relation

$$L(u_0 + \epsilon u_1 + \epsilon^2 u_2 + \cdots) = L(u_0) + \epsilon L(u_1) + \epsilon^2 L(u_2) + \cdots \tag{8.10}$$

must hold. Furthermore, it will be assumed that the function $N(u)$ can be expanded in a power series in ϵ of the form

$$N(u_0 + \epsilon u_1 + \epsilon^2 u_2 + \cdots) = N_0(u_0) + \epsilon N_1(u_0, u_1) + \epsilon^2 N_2(u_0, u_1, u_2) + \cdots. \tag{8.11}$$

Introducing Eqs. (8.10) and (8.11) into (8.9) and equating the coefficients of like powers of ϵ from both sides of the resulting equation, we obtain the infinite sequence of linear differential equations

$$\begin{aligned} L(u_0) &= v, \\ L(u_1) &= N_0(u_0), \\ L(u_2) &= N_1(u_0,u_1), \end{aligned} \qquad (8.12)$$

$$\cdots \cdots \cdots \cdots,$$

which can be solved recursively by noting that the right side of the equation for u_n contains only variables through u_{n-1} ($n = 1, 2, 3, \ldots$).

Solution (8.8), representing the solution of Eq. (8.7) in terms of a power series in the small parameter ϵ, is referred to as a *formal solution*. It gives rise to successive approximations for the solution of Eq. (8.7), and there is some question about the convergence of such a solution. A proof of the convergence of the power series solution (8.8) for sufficiently small ϵ is given in Ref. 7, sec. 9.4. In our case, the formal solution of Eq. (8.5) is obtained by setting $\epsilon = 1$ which, of course, stipulates that $N(u)$ itself is relatively small.

As an illustration of the procedure, let us consider the second-order system

$$\ddot{x} + \omega^2 x = \epsilon f(x,\dot{x}), \qquad (8.13)$$

where we recognize that the linear operator L for this particular system has the form

$$L = \frac{d^2}{dt^2} + \omega^2. \qquad (8.14)$$

Moreover, Eq. (8.13) is homogeneous, $v = 0$, and $N = f$ is an analytic function of x and $\dot{x}$. Assuming a solution of Eq. (8.13) in the form

$$x = x_0 + \epsilon x_1 + \epsilon^2 x_2 + \cdots, \qquad (8.15)$$

we can write the following power series expansion for f about the point $(x_0,\dot{x}_0)$

$$\begin{aligned} f(x,\dot{x}) = f(x_0,\dot{x}_0) &+ \epsilon\left[x_1 \frac{\partial f(x_0,\dot{x}_0)}{\partial x} + \dot{x}_1 \frac{\partial f(x_0,\dot{x}_0)}{\partial \dot{x}}\right] \\ &+ \epsilon^2\left[x_2 \frac{\partial f(x_0,\dot{x}_0)}{\partial x} + \dot{x}_2 \frac{\partial f(x_0,\dot{x}_0)}{\partial \dot{x}} + \frac{1}{2!} x_1^2 \frac{\partial^2 f(x_0,\dot{x}_0)}{\partial x^2} \right. \\ &\left. + \frac{2}{2!} x_1 \dot{x}_1 \frac{\partial^2 f(x_0,\dot{x}_0)}{\partial x\, \partial \dot{x}} + \frac{1}{2!} \dot{x}_1^2 \frac{\partial^2 f(x_0,\dot{x}_0)}{\partial \dot{x}^2}\right] + \cdots, \end{aligned} \qquad (8.16)$$

which leads to the system of equations

$$\begin{aligned} \ddot{x}_0 + \omega^2 x_0 &= 0, \\ \ddot{x}_1 + \omega^2 x_1 &= f(x_0,\dot{x}_0), \\ \ddot{x}_2 + \omega^2 x_2 &= x_1 \frac{\partial f(x_0,\dot{x}_0)}{\partial x} + \dot{x}_1 \frac{\partial f(x_0,\dot{x}_0)}{\partial \dot{x}}, \end{aligned} \qquad (8.17)$$

$$\cdots \cdots \cdots \cdots \cdots \cdots \cdots.$$

Analytical Solutions by Perturbation Techniques 297

A similar procedure can be used for weakly linear or nonlinear non-autonomous systems. Although the general ideas remain the same, there are many variations of the perturbation method to suit particular systems. Some of these versions will be discussed in this chapter.

Example 8.1

Consider the van der Pol's oscillator of Sec. 5.5, assume that $\mu = \epsilon$ is a small parameter, and derive the first four differential equations corresponding to the set (8.17).

Let us write the van der Pol's equation (5.82) in the form

$$\ddot{x} + x = \epsilon \dot{x}(1 - x^2) = \epsilon f(x,\dot{x}), \tag{a}$$

and use Eq. (8.16) to obtain the expansion

$$\begin{aligned}
f(x,\dot{x}) &= \dot{x}(1 - x^2) \\
&= \dot{x}_0(1 - x_0^2) + \epsilon[\dot{x}_1(1 - x_0^2) - 2\dot{x}_0 x_0 x_1] \\
&\quad + \epsilon^2[\dot{x}_2(1 - x_0^2) - 2\dot{x}_1 x_0 x_1 - \dot{x}_0(x_1^2 + 2x_0 x_2)] + \cdots,
\end{aligned} \tag{b}$$

which leads immediately to the desired equations

$$\begin{aligned}
\ddot{x}_0 + x_0 &= 0, \\
\ddot{x}_1 + x_1 &= \dot{x}_0(1 - x_0^2), \\
\ddot{x}_2 + x_2 &= \dot{x}_1(1 - x_0^2) - 2\dot{x}_0 x_0 x_1, \\
\ddot{x}_3 + x_3 &= \dot{x}_2(1 - x_0^2) - 2\dot{x}_1 x_0 x_1 - \dot{x}_0(x_1^2 + 2x_0 x_2),
\end{aligned} \tag{c}$$

8.2 SECULAR TERMS

Whether a system possesses a periodic solution or not has been shown to be of considerable importance in the study of dynamics. Although many dynamical systems are known to possess periodic solutions, a simple application of the perturbation technique, which for practical reasons must limit the solution to the first several terms, may produce an apparently divergent solution. Terms that grow indefinitely with time are referred to as *secular terms*. In problems of celestial mechanics the secular terms present no significant difficulties because the periods involved are quite long, thus allowing for corrections to be made at relatively large intervals of time. With the advent of low-period systems, an example of which is the van der Pol's oscillator, the problem of secular terms became quite serious, and the simple perturbation technique was no longer adequate; the technique had to be modified to accommodate the new situations.

To illustrate the problem of secular terms we shall consider a simple mass-spring system as discussed in Sec. 1.6. In contrast, however, in the system considered here the spring exhibits a nonlinear response in the sense that the spring restoring force consists of one term proportional to the spring elongation plus a smaller term proportional to the third power of the elongation. Such a spring is said to be a "hard" spring (see Prob. 5.6), and the differential equation of the corresponding *quasi-harmonic* system has the form

$$\ddot{x} + \omega^2(x + \epsilon x^3) = 0, \quad \epsilon \ll 1, \tag{8.18}$$

where $\omega = \sqrt{k/m}$ is the natural frequency of the associated harmonic system, obtained by discarding the nonlinear term. The symbol m represents the mass, whereas k is the slope to the force-displacement curve at $x = 0$ and coincides with the spring constant for the linearized system.

Let the solution of Eq. (8.18) have the form (8.15), so that the perturbation procedure discussed in Sec. 8.1 leads to the system

$$\begin{aligned}
\ddot{x}_0 + \omega^2 x_0 &= 0, \\
\ddot{x}_1 + \omega^2 x_1 &= -\omega^2 x_0^3, \\
\ddot{x}_2 + \omega^2 x_2 &= -3\omega^2 x_0^2 x_1, \\
&\cdots\cdots\cdots\cdots\cdots,
\end{aligned} \tag{8.19}$$

which can be solved in sequence.

The solution of the first of Eqs. (8.19) is simply

$$x_0 = A \cos(\omega t + \varphi). \tag{8.20}$$

Introducing this solution into the second of Eqs. (8.19), we obtain

$$\ddot{x}_1 + \omega^2 x_1 = -\tfrac{3}{4}\omega^2 A^3 \cos(\omega t + \varphi) - \tfrac{1}{4}\omega^2 A^3 \cos 3(\omega t + \varphi), \tag{8.21}$$

which, in turn, has the solution

$$x_1 = -\tfrac{3}{8}\omega t A^3 \sin(\omega t + \varphi) + \tfrac{1}{32}A^3 \cos 3(\omega t + \varphi). \tag{8.22}$$

It is not difficult to see that the first term on the right side of Eq. (8.22) is a secular term, becoming infinitely large as $t \to \infty$.

There are reasons to expect a periodic solution for the system (8.18), however. Indeed, the system is conservative and of the type studied in Sec. 5.3. The only singular point is at the origin of the phase plane, $x = \dot{x} = 0$, and the singularity is a center, as no complex values for x are physically acceptable. Referring to Sec. 5.3, we can write the potential energy expression per unit mass

$$V(x) = -\omega^2 \int_x^0 (x + \epsilon x^3)\, dx = \omega^2 \left(\frac{x^2}{2} + \epsilon \frac{x^4}{4}\right), \tag{8.23}$$

and, introducing that expression into Eq. (5.58), we obtain

$$\tfrac{1}{2}\dot{x}^2 + \omega^2(\tfrac{1}{2}x^2 + \tfrac{1}{4}\epsilon x^4) = E = \text{const}, \tag{8.24}$$

representing the fact that the system's total energy E per unit mass is conserved. For any given value of E the motion is bounded, so that no secular terms are possible. In fact, from Sec. 5.3 we conclude that the motion must be periodic.

There remains the question of explaining the appearance of secular terms in the perturbation solution. The answer lies in the fact that secular terms need not always be interpreted as implying a divergent solution. Perhaps if a sufficiently large number of terms are taken in Eq. (8.15), the perturbation solution may assume the appearance of a periodic function after all. To illustrate this idea, we consider a series expansion of the periodic function

$$\sin(\omega + \epsilon)t = \sin \omega t \cos \epsilon t + \cos \omega t \sin \epsilon t$$

$$= \left(1 - \frac{1}{2!}\epsilon^2 t^2 + \frac{1}{4!}\epsilon^4 t^4 - \cdots\right) \sin \omega t$$

$$+ \left(\epsilon t - \frac{1}{3!}\epsilon^3 t^3 + \frac{1}{5!}\epsilon^5 t^5 - \cdots\right) \cos \omega t. \quad (8.25)$$

If we retained only the first few terms in the series for $\sin \epsilon t$ and $\cos \epsilon t$, we would not be able to conclude that the expansion represents a periodic function.

This discussion points out that a simple application of the perturbation technique, whereby we altered only the *amplitude* of the solution as shown by Eq. (8.15), may not always be satisfactory. Indeed, in this case, the nonlinear term ϵx^3 affects not only the amplitude but *also the frequency of oscillation* of the solution, as might be hinted by expansion (8.25). Due to computational difficulties, however, the need of a perturbation technique able to produce a periodic solution by retaining only a limited number of terms in expansion (8.15) is evident. If the expansion is limited to terms through the nth power in ϵ, the solution is said to be an *approximation of order n*. For reasons of convergence, a low-order approximation can be used only if ϵ is reasonably small. A number of perturbation techniques designed for the express purpose of seeking periodic solutions of nonlinear systems by suppressing the secular terms will be presented in the following few sections.

8.3 LINDSTEDT'S METHOD

Lindstedt's method seeks a periodic solution of a differential equation in the form

$$x = x_0 + \epsilon x_1 + \epsilon^2 x_2 + \cdots = \sum_{n=0}^{\infty} \epsilon^n x_n \quad (8.26)$$

by requiring that each x_n ($n = 0, 1, 2, \ldots$) be periodic. In contrast with the procedure of Sec. 8.2, however, Lindstedt's method does take into consideration the fact that the nonlinear terms may affect the frequency of the periodic solution. The frequency possesses a certain degree of arbitrariness, which is removed by insisting that the solution (8.26) be periodic.

We shall be concerned with second-order dynamical systems of the form

$$\ddot{x} + \omega^2 x = \epsilon f(x, \dot{x}), \tag{8.27}$$

where $f(x, \dot{x})$ is a nonlinear function of x and $\dot{x}$. For $\epsilon = 0$ the period of oscillation is $2\pi/\omega$. The nonlinear term can be expected to alter that period by an amount of order ϵ, so that if system (8.27) possesses a periodic solution, the period may be $T = 2\pi/\omega + O(\epsilon)$. This is equivalent to the statement that the frequency of the periodic oscillation depends on ϵ. Hence, we shall assume that the frequency has the form

$$\Omega = \omega + \epsilon \omega_1 + \epsilon^2 \omega_2 + \cdots, \tag{8.28}$$

where the frequency $\Omega = 2\pi/T$ is as yet unknown. Instead of working with unknown periods, it will prove more convenient to change the independent variable from t to τ to render the period of oscillation in terms of the new variable equal to 2π. We therefore introduce the substitution $\tau = \Omega t$, $d/dt = \Omega d/d\tau$, which reduces Eq. (8.27) to

$$\Omega^2 x'' + \omega^2 x = \epsilon f(x, \Omega x'), \tag{8.29}$$

where primes denote differentiations with respect to τ. Next we expand the function $f(x, \Omega x')$ in a power series in ϵ. Recalling Eqs. (8.26) and (8.28), we can write

$$f(x, \Omega x') = f(x_0, \omega x_0') + \epsilon \left[x_1 \frac{\partial f(x_0, \omega x_0')}{\partial x} + x_1' \frac{\partial f(x_0, \omega x_0')}{\partial x'} + \omega_1 \frac{\partial f(x_0, \omega x_0')}{\partial \Omega} \right]$$
$$+ \epsilon^2 (\cdots) + \cdots. \tag{8.30}$$

Introducing Eqs. (8.26), (8.28), and (8.30) into Eq. (8.29), we obtain the system of equations

$$\omega^2 x_0'' + \omega^2 x_0 = 0,$$
$$\omega^2 x_1'' + \omega^2 x_1 = f(x_0, \omega x_0') - 2\omega \omega_1 x_0'',$$
$$\omega^2 x_2'' + \omega^2 x_2 = x_1 \frac{\partial f(x_0, \omega x_0')}{\partial x} + x_1' \frac{\partial f(x_0, \omega x_0')}{\partial x'} + \omega_1 \frac{\partial f(x_0, \omega x_0')}{\partial \Omega} \tag{8.31}$$
$$- (2\omega \omega_2 + \omega_1^2) x_0'' - 2\omega \omega_1 x_1'',$$
$$\ldots \ldots \ldots \ldots \ldots \ldots \ldots \ldots \ldots \ldots \ldots \ldots$$

Equations (8.31) are solved in sequence, as in Sec. 8.2. Here, however, we have the additional task of determining the quantities ω_{n+1}, which is

Analytical Solutions by Perturbation Techniques

accomplished by requiring that each x_n be periodic

$$x_n(\tau + 2\pi) = x_n(\tau), \qquad n = 0, 1, 2, \ldots . \tag{8.32}$$

As an illustration of the procedure, let us consider the quasi-harmonic system of Sec. 8.2. In terms of our present notation $f(x,\dot{x}) = f(x) = -\omega^2 x^3$, so that the corresponding system of equations becomes

$$\begin{aligned}
x_0'' + x_0 &= 0, \\
x_1'' + x_1 &= -x_0^3 - 2\frac{\omega_1}{\omega} x_0'', \\
x_2'' + x_2 &= -3x_0^2 x_1 - \left[2\frac{\omega_2}{\omega} + \left(\frac{\omega_1}{\omega}\right)^2\right] x_0'' - 2\frac{\omega_1}{\omega} x_1'', \\
&\ldots \ldots \ldots \ldots \ldots \ldots \ldots \ldots \ldots \ldots \ldots ,
\end{aligned} \tag{8.33}$$

which are subject to the periodicity conditions (8.32). It will also be assumed that

$$x_n'(0) = 0, \qquad n = 0, 1, 2, \ldots , \tag{8.34}$$

which can be done by including a phase angle in τ. This is permissible because the system is autonomous.

In view of the initial conditions (8.34), the solution of the first of Eqs. (8.33) is

$$x_0 = A \cos \tau, \tag{8.35}$$

so that the second of Eqs. (8.33) becomes

$$x_1'' + x_1 = A\left(2\frac{\omega_1}{\omega} - \tfrac{3}{4}A^2\right) \cos \tau - \tfrac{1}{4}A^3 \cos 3\tau. \tag{8.36}$$

The first term on the right side of Eq. (8.36) is likely to induce secular terms, thus rendering the solution x_1 nonperiodic. To suppress this possibility, we must set

$$\frac{\omega_1}{\omega} = \tfrac{3}{8}A^2, \tag{8.37}$$

so that, in view of Eqs. (8.34), the solution of Eq. (8.36) reduces to

$$x_1 = \tfrac{1}{32}A^3 \cos 3\tau. \tag{8.38}$$

It is not difficult to show that the periodicity condition, applied to the third of Eqs. (8.33), yields

$$\frac{\omega_2}{\omega} = -\tfrac{15}{256}A^4, \tag{8.39}$$

with the solution of the resulting differential equation assuming the form

$$x_2 = -\tfrac{21}{1024}A^5 \cos 3\tau + \tfrac{1}{1024}A^5 \cos 5\tau. \tag{8.40}$$

Introducing Eqs. (8.35), (8.38), and (8.40) into Eq. (8.26) and denoting the phase angle mentioned above by φ, so that $\tau = \Omega t + \varphi$, we obtain the second-order approximation solution

$$x(t) = A \cos (\Omega t + \varphi) + \epsilon \frac{A^3}{32}\left(1 - \epsilon \frac{21A^2}{32}\right) \cos 3(\Omega t + \varphi)$$

$$+ \epsilon^2 \frac{A^5}{1024} \cos 5(\Omega t + \varphi), \qquad (8.41)$$

in which
$$\Omega \approx \omega\left(1 + \epsilon \frac{3A^2}{8} - \epsilon^2 \frac{15A^4}{256}\right), \qquad (8.42)$$

so that the effect of the spring nonlinearity is reflected in both the amplitude and frequency of the periodic motion.

8.4 THE KRYLOV-BOGOLIUBOV-MITROPOLSKY (KBM) METHOD

A method concerned with the existence of periodic solutions of a quasi-harmonic system has been developed by Krylov and Bogoliubov and put on a more sound mathematical foundation by Bogoliubov and Mitropolsky. Here we follow closely the exposition given in the treatise by the latter two (Ref. 2, chap. 1). The KBM method also builds into the solution a certain degree of arbitrariness, enabling us to produce a periodic solution while removing the arbitrariness. In this respect it reminds us of Lindstedt's method, although the approach is substantially different.

We shall be concerned with the quasi-harmonic autonomous system

$$\ddot{x} + \omega^2 x = \epsilon f(x, \dot{x}), \qquad (8.43)$$

where ϵ is a small parameter. When $\epsilon = 0$, Eq. (8.43) reduces to the differential equation of a harmonic oscillator having the solution

$$x = a \cos \psi, \qquad \psi = \omega t + \varphi, \qquad (8.44)$$

in which the amplitude a, the natural frequency ω, and the phase angle φ are all constant. When $\epsilon \neq 0$ but is small, the right side of Eq. (8.43) may be regarded as a perturbation causing both the amplitude and frequency to vary slowly.

The KBM method suggests a general solution of Eq. (8.43) in the form

$$x = a \cos \psi + \epsilon u_1(a,\psi) + \epsilon^2 u_2(a,\psi) + \cdots, \qquad (8.45)$$

where $u_i(a,\psi)$ ($i = 1, 2, \ldots$) are periodic functions of ψ with period 2π and the quantities a and ψ are functions of time satisfying the equations

$$\begin{aligned}\dot{a} &= \epsilon A_1(a) + \epsilon^2 A_2(a) + \cdots, \\ \dot{\psi} &= \omega + \epsilon B_1(a) + \epsilon^2 B_2(a) + \cdots.\end{aligned} \qquad (8.46)$$

Analytical Solutions by Perturbation Techniques

The functions $u_i(a,\psi)$, $A_i(a)$, and $B_i(a)$ ($i = 1, 2, \ldots$) are determined so that Eq. (8.45), in conjunction with expressions (8.46), represents a solution of Eq. (8.43). The procedure also leads to a system of equations which lends itself to a recursive solution. The arbitrariness built into the solution is removed by insisting that the solution be free of secular terms, which imposes on the functions $u_i(a,\psi)$ the following conditions:

$$\int_0^{2\pi} u_i(a,\psi) \cos \psi \, d\psi = 0,$$
$$\int_0^{2\pi} u_i(a,\psi) \sin \psi \, d\psi = 0, \qquad i = 1, 2, \ldots . \qquad (8.47)$$

The effect of these conditions is to suppress the fundamental harmonic from the functions $u_i(a,\psi)$, leaving a as the full amplitude of the first harmonic in solution (8.45).

The formal solution of Eq. (8.43) is obtained by using Eq. (8.45) to evaluate $\ddot{x} + \omega^2 x$ and $\epsilon f(x,\dot{x})$ and equating the coefficients of like powers of ϵ in both these expressions. To this end, let us differentiate Eq. (8.45) with respect to time and write

$$\dot{x} = \dot{a}\left(\cos \psi + \epsilon \frac{\partial u_1}{\partial a} + \epsilon^2 \frac{\partial u_2}{\partial a} + \cdots\right)$$
$$+ \dot{\psi}\left(-a \sin \psi + \epsilon \frac{\partial u_1}{\partial \psi} + \epsilon^2 \frac{\partial u_2}{\partial \psi} + \cdots\right)$$
$$= -a\omega \sin \psi + \epsilon\left(A_1 \cos \psi - aB_1 \sin \psi + \omega \frac{\partial u_1}{\partial \psi}\right)$$
$$+ \epsilon^2\left(A_2 \cos \psi + A_1 \frac{\partial u_1}{\partial a} - aB_2 \sin \psi + B_1 \frac{\partial u_1}{\partial \psi} + \omega \frac{\partial u_2}{\partial \psi}\right)$$
$$+ \epsilon^3(\cdots) + \cdots, \qquad (8.48)$$

where use has been made of Eqs. (8.46). One more differentiation yields

$$\ddot{x} = \ddot{a}\left(\cos \psi + \epsilon \frac{\partial u_1}{\partial a} + \epsilon^2 \frac{\partial u_2}{\partial a} + \cdots\right)$$
$$+ \ddot{\psi}\left(-a \sin \psi + \epsilon \frac{\partial u_1}{\partial \psi} + \epsilon^2 \frac{\partial u_2}{\partial \psi} + \cdots\right)$$
$$+ \dot{a}^2\left(\epsilon \frac{\partial^2 u_1}{\partial a^2} + \epsilon^2 \frac{\partial^2 u_2}{\partial a^2} + \cdots\right)$$
$$+ 2\dot{a}\dot{\psi}\left(-\sin \psi + \epsilon \frac{\partial^2 u_1}{\partial a \, \partial \psi} + \epsilon^2 \frac{\partial^2 u_2}{\partial a \, \partial \psi} + \cdots\right)$$
$$+ \dot{\psi}^2\left(-a \cos \psi + \epsilon \frac{\partial^2 u_1}{\partial \psi^2} + \epsilon^2 \frac{\partial^2 u_2}{\partial \psi^2} + \cdots\right), \qquad (8.49)$$

in which

$$\ddot{a} = \left(\epsilon \frac{dA_1}{da} + \epsilon^2 \frac{dA_2}{da} + \cdots\right)(\epsilon A_1 + \epsilon^2 A_2 + \cdots) = \epsilon^2 A_1 \frac{dA_1}{da} + \epsilon^3 \cdots,$$

$$\ddot{\psi} = \left(\epsilon \frac{dB_1}{da} + \epsilon^2 \frac{dB_2}{da} + \cdots\right)(\epsilon A_1 + \epsilon^2 A_2 + \cdots) = \epsilon^2 A_1 \frac{dB_1}{da} + \epsilon^3 \cdots,$$

$$\dot{a}^2 = (\epsilon A_1 + \epsilon^2 A_2 + \cdots)^2 = \epsilon^2 A_1^2 + \epsilon^3 \cdots, \qquad (8.50)$$

$$\dot{a}\dot{\psi} = (\epsilon A_1 + \epsilon^2 A_2 + \cdots)(\omega + \epsilon B_1 + \epsilon^2 B_2 + \cdots)$$
$$= \epsilon \omega A_1 + \epsilon^2(\omega A_2 + A_1 B_1) + \epsilon^3 \cdots,$$

$$\dot{\psi}^2 = (\omega + \epsilon B_1 + \epsilon^2 B_2 + \cdots)^2$$
$$= \omega^2 + 2\epsilon \omega B_1 + \epsilon^2(B_1^2 + 2\omega B_2) + \epsilon^3 \cdots.$$

A combination of Eqs. (8.45), (8.49), and (8.50) enables us to write

$$\ddot{x} + \omega^2 x = \epsilon\left[-2\omega A_1 \sin\psi - 2\omega a B_1 \cos\psi + \omega^2\left(\frac{\partial^2 u_1}{\partial \psi^2} + u_1\right)\right]$$
$$+ \epsilon^2\left[\left(A_1 \frac{dA_1}{da} - aB_1^2 - 2\omega a B_2\right)\cos\psi\right.$$
$$- \left(aA_1 \frac{dB_1}{da} + 2\omega A_2 + 2A_1 B_1\right)\sin\psi + 2\omega A_1 \frac{\partial^2 u_1}{\partial a\, \partial \psi}$$
$$\left. + 2\omega B_1 \frac{\partial^2 u_1}{\partial \psi^2} + \omega^2\left(\frac{\partial^2 u_2}{\partial \psi^2} + u_2\right)\right] + \epsilon^3 \cdots. \qquad (8.51)$$

To expand the function $f(x,\dot{x})$ in a power series in ϵ, it will prove convenient to introduce the notation

$$x_0 = a\cos\psi, \qquad \dot{x}_0 = -a\omega \sin\psi, \qquad (8.52)$$

where x_0 is recognized as the zero-order approximation solution, or the generating solution, in which a and $\dot{\psi}$ are constant. Recalling Eqs. (8.45) and (8.48), we can write

$$\epsilon f(x,\dot{x}) = \epsilon f(x_0,\dot{x}_0) + \epsilon^2\left[u_1 \frac{\partial f(x_0,\dot{x}_0)}{\partial x} + \left(A_1 \cos\psi - aB_1 \sin\psi + \omega \frac{\partial u_1}{\partial \psi}\right)\right.$$
$$\left. \times \frac{\partial f(x_0,\dot{x}_0)}{\partial \dot{x}}\right] + \epsilon^3 \cdots. \qquad (8.53)$$

Equating like powers of ϵ in Eqs. (8.51) and (8.53), we obtain the set of differential equations

$$\omega^2\left(\frac{\partial^2 u_k}{\partial \psi^2} + u_k\right) = f_{k-1}(a,\psi) + 2\omega A_k \sin\psi + 2\omega a B_k \cos\psi,$$
$$k = 1, 2, \ldots, \qquad (8.54)$$

Analytical Solutions by Perturbation Techniques

where, for simplicity, we have adopted the notation

$$f_0(a,\psi) = f(x_0,\dot{x}_0),$$

$$f_1(a,\psi) = u_1 \frac{\partial f(x_0,\dot{x}_0)}{\partial x} + \left(A_1 \cos\psi - aB_1 \sin\psi + \omega \frac{\partial u_1}{\partial \psi}\right) \frac{\partial f(x_0,\dot{x}_0)}{\partial \dot{x}}$$

$$+ \left(aB_1{}^2 - A_1 \frac{dA_1}{da}\right) \cos\psi + A_1\left(a\frac{dB_1}{da} + 2B_1\right) \sin\psi \quad (8.55)$$

$$- 2\omega\left(A_1 \frac{\partial^2 u_1}{\partial a\, \partial \psi} + B_1 \frac{\partial^2 u_1}{\partial \psi^2}\right),$$

. . . .

The functions $f_k(a,\psi)$ are periodic in the variable ψ with period 2π, and the explicit expression of each of these functions can be determined provided all the functions $A_i(a)$, $B_i(a)$, and $u_i(a,\psi)$ ($i = 1, 2, \ldots, k$) are known. It follows that Eqs. (8.54) can be solved recursively, thus completing the solution (8.45) and (8.46) to any desired degree of accuracy.

To solve the first of Eqs. (8.54) we represent $f_0(a,\psi)$ and $u_1(a,\psi)$ by the Fourier series

$$f_0(a,\psi) = g_0(a) + \sum_{n=1}^{\infty} [g_n(a) \cos n\psi + h_n(a) \sin n\psi],$$

$$u_1(a,\psi) = v_0(a) + \sum_{n=2}^{\infty} [v_n(a) \cos n\psi + w_n(a) \sin n\psi], \quad (8.56)$$

where, by virtue of conditions (8.47), we have set $v_1(a) = w_1(a) = 0$. Introducing Eqs. (8.56) into the first of Eqs. (8.54), we obtain

$$\omega^2 v_0(a) + \sum_{n=2}^{\infty} \omega^2(1 - n^2)[v_n(a) \cos n\psi + w_n(a) \sin n\psi]$$

$$= g_0(a) + [g_1(a) + 2\omega a B_1(a)] \cos\psi + [h_1(a) + 2\omega A_1(a)] \sin\psi$$

$$+ \sum_{n=2}^{\infty} [g_n(a) \cos n\psi + h_n(a) \sin n\psi]. \quad (8.57)$$

Equating the coefficients of identical harmonics, we obtain the relations

$$g_1(a) + 2\omega a B_1(a) = 0, \quad h_1(a) + 2\omega A_1(a) = 0, \quad (8.58)$$

yielding the functions $A_1(a)$ and $B_1(a)$, as well as

$$v_0(a) = \frac{g_0(a)}{\omega^2},$$

$$v_n(a) = \frac{g_n(a)}{\omega^2(1 - n^2)}, \quad w_n(a) = \frac{h_n(a)}{\omega^2(1 - n^2)}, \quad n = 2, 3, \ldots, \quad (8.59)$$

which determine the function $u_1(a,\psi)$ as

$$u_1(a,\psi) = \frac{1}{\omega^2} g_0(a) + \sum_{n=2}^{\infty} \frac{1}{\omega^2(1-n^2)} [g_n(a) \cos n\psi + h_n(a) \sin n\psi], \quad (8.60)$$

completing the first-order approximation.

Having determined $A_1(a)$, $B_1(a)$, and $u_1(a,\psi)$, we can obtain an explicit expression for $f_1(a,\psi)$ from the second of Eqs. (8.55) and represent it by the Fourier series

$$f_1(a,\psi) = g_0^{(1)}(a) + \sum_{n=1}^{\infty} [g_n^{(1)}(a) \cos n\psi + h_n^{(1)}(a) \sin n\psi]. \quad (8.61)$$

Repeating the procedure, the second of Eqs. (8.54) leads to the relations

$$g_1^{(1)}(a) + 2\omega a B_2(a) = 0, \qquad h_1^{(1)}(a) + 2\omega A_2(a) = 0, \quad (8.62)$$

and the series

$$u_2(a,\psi) = \frac{1}{\omega^2} g_0^{(1)}(a) + \sum_{n=2}^{\infty} \frac{1}{\omega^2(1-n^2)} [g_n^{(1)}(a) \cos n\psi + h_n^{(1)}(a) \sin n\psi]. \quad (8.63)$$

The higher-order approximations can be obtained recursively in the same manner.

For higher-order values of k the evaluation of the function $f_k(a,\psi)$ becomes increasingly laborious, so that we are often forced to confine ourselves to lower-order approximations, which raises the question of convergence of the solution. Bogoliubov and Mitropolsky (Ref. 2, p. 41) point out that if Eqs. (8.46) determine the functions a and ψ with an accuracy of order ϵ^k, then the formal solution (8.45) of corresponding order satisfies Eq. (8.43) with an accuracy of order ϵ^{k+1}. It is further shown by Bogoliubov and Mitropolsky (Ref. 2, p. 48) that a consistent first-order approximation can actually be obtained by representing the solution by the simplified expression

$$x = a \cos \psi, \quad (8.64)$$

in which

$$\dot{a} = \epsilon A_1(a), \qquad \dot{\psi} = \omega + \epsilon B_1(a). \quad (8.65)$$

Similarly, for the second-order approximation we can take

$$x = a \cos \psi + \epsilon u_1(a,\psi), \quad (8.66)$$

where

$$\dot{a} = \epsilon A_1(a) + \epsilon^2 A_2(a), \qquad \dot{\psi} = \omega + \epsilon B_1(a) + \epsilon^2 B_2(a). \quad (8.67)$$

When the quasi-linear system is conservative, a large degree of simplification of the procedure is obtained. In this case Eq. (8.53) reduces to

$$\epsilon f(x,\dot{x}) = \epsilon f(x) = \epsilon f(x_0) + \epsilon^2 u_1 \frac{\partial f(x_0)}{\partial x} + \epsilon^3 \cdots , \quad (8.68)$$

and, since $x_0 = a \cos \psi$, the Fourier series for $f(x)$ will contain no sine terms. In particular, the first of Eqs. (8.56) becomes

$$f_0(a,\psi) = f(a \cos \psi) = \sum_{n=0}^{\infty} g_n(a) \cos n\psi, \quad (8.69)$$

whereas all the coefficients of the sine terms are zero, $h_n(a) = 0$. From Eqs. (8.58) it follows that

$$A_1(a) = 0, \quad B_1(a) = -\frac{1}{2\omega a} g_1(a). \quad (8.70)$$

Using the above results, the first-order approximation takes the form (8.64), in which

$$\dot{a} = 0, \quad \dot{\psi} = \omega - \frac{\epsilon}{2\omega a} g_1(a), \quad (8.71)$$

indicating that the amplitude of the solution is constant, $a = \text{const.}$

For the second-order approximation we must have $u_1(a,\psi)$, which, by Eq. (8.60), assumes the form

$$u_1(a,\psi) = \frac{1}{\omega^2} \sum_{\substack{n=0 \\ n \neq 1}}^{\infty} \frac{1}{1-n^2} g_n(a) \cos n\psi, \quad (8.72)$$

so that the Fourier series representation of $f_1(a,\psi)$, Eq. (8.61), also contains no sine terms

$$f_1(a,\psi) = \sum_{n=0}^{\infty} g_n^{(1)}(a) \cos n\psi, \quad (8.73)$$

with the implication that

$$A_2(a) = 0, \quad B_2(a) = -\frac{1}{2\omega a} g_1^{(1)}(a). \quad (8.74)$$

Hence the second-order approximation is obtained by introducing Eq. (8.72) into (8.66) with

$$\dot{a} = 0, \quad \dot{\psi} = \omega - \epsilon \frac{1}{2\omega a} g_1(a) - \epsilon^2 \frac{1}{2\omega a} g_1^{(1)}(a), \quad (8.75)$$

so that $a = \text{const.}$ It turns out that for quasi-harmonic conservative systems all $A_n(a)$ are zero, indicating that the amplitude a is constant.

Example 8.2

To illustrate the KBM method, we shall obtain the second-order approximation of the quasi-harmonic system involving the hard spring. The system is conservative and defined by the differential equation

$$\ddot{x} + \omega^2 x = -\epsilon\omega^2 x^3, \qquad (a)$$

so that
$$f(x,\dot{x}) = f(x) = -\omega^2 x^3. \qquad (b)$$

The second-order approximation is given by Eqs. (8.66) and (8.67), and since for this conservative case $A_1(a) = A_2(a) = 0$, it remains for us to determine the functions $B_1(a)$, $B_2(a)$, and $u_1(a,\psi)$.

From the first of Eqs. (8.55) and Eq. (b), we conclude that

$$f_0(a,\psi) = f(a \cos \psi) = -\omega^2(\tfrac{3}{4}a^3 \cos \psi + \tfrac{1}{4}a^3 \cos 3\psi), \qquad (c)$$

which is the first of the Fourier series in (8.56). As expected, the series contains no sine terms. Equation (c) indicates that from all the coefficients of the series only two survive

$$g_1(a) = -\tfrac{3}{4}\omega^2 a^3, \qquad g_3(a) = -\tfrac{1}{4}\omega^2 a^3. \qquad (d)$$

From the second of Eqs. (8.70) we obtain

$$B_1(a) = -\frac{1}{2\omega a} g_1(a) = \tfrac{3}{8}\omega a^2, \qquad (e)$$

whereas series (8.72) reduces to the single term

$$u_1(a,\psi) = \tfrac{1}{32}a^3 \cos 3\psi. \qquad (f)$$

To obtain the function $f_1(a,\psi)$, we use the second of Eqs. (8.55), in conjunction with the results just obtained, and write

$$f_1(a,\psi) = -3\omega^2 x_0^2 u_1 + aB_1^2 \cos \psi - 2\omega B_1 \frac{\partial^2 u_1}{\partial \psi^2}$$
$$= \tfrac{15}{128}\omega^2 a^5 \cos \psi + \tfrac{21}{128}\omega^2 a^5 \cos 3\psi - \tfrac{3}{128}\omega^2 a^5 \cos 5\psi. \qquad (g)$$

Comparing Eqs. (g) and (8.73), we conclude that the Fourier series for $f_1(a,\psi)$ contains only three terms with the coefficients

$$g_1^{(1)}(a) = \tfrac{15}{128}\omega^2 a^5, \qquad g_3^{(1)}(a) = \tfrac{21}{128}\omega^2 a^5, \qquad g_5^{(1)}(a) = -\tfrac{3}{128}\omega^2 a^5. \qquad (h)$$

The second of Eqs. (8.74) yields

$$B_2(a) = -\frac{1}{2\omega a} g_1^{(1)}(a) = -\tfrac{15}{256}\omega a^4. \qquad (i)$$

Introducing Eq. (f) into Eq. (8.66), we obtain

$$x = a \cos \psi + \epsilon \tfrac{1}{32} a^3 \cos 3\psi, \qquad (j)$$

Analytical Solutions by Perturbation Techniques

whereas from Eqs. (8.75), (*e*), and (*i*) we conclude that

$$\dot{a} = 0, \qquad \dot{\psi} = \omega + \epsilon^3 \tfrac{3}{8}\omega a^2 - \epsilon^2 \tfrac{15}{256}\omega a^4, \tag{k}$$

which completes the solution. It is worth noting from the second of Eqs. (*k*) that the second-order approximation to the fundamental frequency $\dot{\psi}$ is the same as the one obtained by Lindstedt's method, as can be seen from Eq. (8.42).

8.5 A PERTURBATION TECHNIQUE BASED ON HILL'S DETERMINANTS

In Chap. 7 we discussed methods of solution of nonautonomous systems of differential equations, in particular, systems containing coefficients in the form of periodic functions of time. Even when the systems are linear, the solution of nonautonomous systems presents a problem of extreme difficulty. However, in the case of a linear system for which the terms involving the time-dependent coefficients are relatively small the situation is considerably more hopeful. The perturbation techniques developed in this chapter can also be applied to nearly autonomous linear equations, such as Hill's or Mathieu's equations. To this end, it will be recalled that in the case of Mathieu's equation the periodic solutions represented boundary curves in the parameter plane separating the stable and unstable solutions. Instead of seeking a solution by one of the perturbation techniques presented in this chapter, however, we shall return to Sec. 7.7 and derive the equations of the boundary curves by means of an approach based on Hill's determinants.

Let us consider Mathieu's equation in the form introduced in Sec. 7.7, namely,

$$\ddot{x} + (\delta + 2\epsilon \cos 2t)x = 0, \tag{8.76}$$

and concern ourselves with the nature of the solution as the system parameters δ and ϵ vary. In particular, we shall consider the behavior of the solution in the neighborhood of the δ axis in the parameter plane (δ, ϵ), that is, for small values of ϵ.

In Sec. 7.7 we established that Eq. (8.76) admits periodic solutions of periods $T = \pi$ and $2T = 2\pi$. It was also shown in that section that on the boundary curves defining the periodic solutions of period $2T$ the system parameters must satisfy the determinantal equation

$$\begin{vmatrix} \delta - 1 \pm \epsilon & \epsilon & 0 & 0 & \cdots \\ \epsilon & \delta - 3^2 & \epsilon & 0 & \cdots \\ 0 & \epsilon & \delta - 5^2 & \epsilon & \cdots \\ \cdots & \cdots & \cdots & \cdots & \cdots \end{vmatrix} = 0, \tag{8.77}$$

whereas for the periodic solutions of period T they must satisfy

$$\begin{vmatrix} \delta - 2^2 & \epsilon & 0 & 0 & \cdots \\ \epsilon & \delta - 4^2 & \epsilon & 0 & \cdots \\ 0 & \epsilon & \delta - 6^2 & \epsilon & \cdots \\ \cdots & \cdots & \cdots & \cdots & \cdots \end{vmatrix} = 0 \qquad (8.78)$$

and

$$\begin{vmatrix} \delta & \epsilon & 0 & 0 & 0 & \cdots \\ 2\epsilon & \delta - 2^2 & \epsilon & 0 & 0 & \cdots \\ 0 & \epsilon & \delta - 4^2 & \epsilon & 0 & \cdots \\ 0 & 0 & \epsilon & \delta - 6^2 & \epsilon & \cdots \\ \cdots & \cdots & \cdots & \cdots & \cdots & \cdots \end{vmatrix} = 0. \qquad (8.79)$$

The intersection of the boundary curves with the δ axis can be obtained by simply setting $\epsilon = 0$ in Eqs. (8.77) to (8.79), which yields $\delta = n^2$ ($n = 0$, 1, 2, . . .). This can be easily verified by the Strutt diagram shown in Fig. 7.3. At this point, however, we are interested in obtaining analytical expressions for the boundary curves in the vicinity of the δ axis. For each value of n there are two branches emerging from the points $\delta = n^2$, $\epsilon = 0$ with the exception of $n = 0$, for which there is only one branch. As the diagram is symmetric with respect to the δ axis, we discuss only positive values of ϵ. Ignoring the case $n = 0$, it can be stated that every two branches terminating at the points $\delta = n^2$, $\epsilon = 0$ ($n = 1, 2, 3, \ldots$) enclose regions of instability because they represent boundary curves defining periodic solutions of equal periods (see Sec. 7.7). Each of these regions will be designated according to the value of n.

The regions of instability are determined by means of an approximate method based on the determinants (8.77) to (8.79). The procedure is amazingly simple. Instead of working with infinite determinants, by virtue of the assumption that ϵ is a small quantity, we derive an approximation for the boundary curves by using finite determinants, obtained by considering the principal minor determinants associated with expressions (8.77) to (8.79). A crude approximation of the instability regions of low order can be obtained by using 2 × 2 determinants or, in one case, even a single element. As the number of rows and columns of the determinant is increased, additional regions can be determined while at the same time improving the approximation of the lower-order regions.

As an illustration, we shall derive the first several instability regions of Mathieu's equation. To obtain the boundary curve emerging from $\delta = 0$ (here as well as in the subsequent discussion it is understood that all these

Analytical Solutions by Perturbation Techniques

curves begin at $\epsilon = 0$), we consider the equation corresponding to the 2×2 minor determinant in Eq. (8.79)

$$\begin{vmatrix} \delta & \epsilon \\ 2\epsilon & \delta - 2^2 \end{vmatrix} = \delta(\delta - 4) - 2\epsilon^2 = 0,$$

which yields the boundary curve

$$\delta = -\tfrac{1}{2}\epsilon^2. \tag{8.80}$$

On this curve the periodic motion is of period T.

From the point $\delta = 1$ two boundary curves emerge. They correspond to periods $2T$, and their equations are obtained from the first element of the determinant in (8.77)

$$\delta = 1 \pm \epsilon, \tag{8.81}$$

which represent straight lines.

The curves bounding the instability region terminating in the point $\delta = 4$ are obtained from the determinantal equations

$$\begin{vmatrix} \delta - 2^2 & \epsilon \\ \epsilon & \delta - 4^2 \end{vmatrix} = (\delta - 4)(\delta - 16) - \epsilon^2 = 0$$

and

$$\begin{vmatrix} \delta & \epsilon \\ 2\epsilon & \delta - 2^2 \end{vmatrix} = \delta(\delta - 4) - 2\epsilon^2 = 0,$$

which are both expanded about $\delta = 4$. The first equation yields

$$\delta = 4 - \tfrac{1}{12}\epsilon^2, \tag{8.82}$$

whereas the second one leads to

$$\delta = 4 + \tfrac{1}{2}\epsilon^2. \tag{8.83}$$

On both these curves the motion is periodic with period T.

The determinantal equation

$$\begin{vmatrix} \delta - 1 \pm \epsilon & \epsilon \\ \epsilon & \delta - 3^2 \end{vmatrix} = (\delta - 1 \pm \epsilon)(\delta - 9) - \epsilon^2 = 0$$

enables us not only to determine the instability region associated with $\delta = 9$ but also to obtain a better approximation for the region terminating at $\delta = 1$. In the neighborhood of $\delta = 9$ the boundary curves have the equations

$$\delta = 9 + \frac{\epsilon^2}{8 \pm \epsilon} \approx 9 + \tfrac{1}{8}\epsilon^2(1 \pm \tfrac{1}{8}\epsilon), \tag{8.84}$$

whereas the improved approximation for the instability region associated with $\delta = 1$ is provided by the boundary curves

$$\delta = 1 \pm \epsilon - \tfrac{1}{8}\epsilon^2, \tag{8.85}$$

which differ from the curves (8.81) by a quantity of second order in ϵ. Both Eqs. (8.84) and (8.85) represent curves on which the periodic motion has period $2T$.

The higher-order instability regions can be obtained in a similar fashion. We shall not pursue this problem any further, but before abandoning the subject, it may prove of interest to obtain a better approximation for the instability region corresponding to $\delta = 4$. The 3×3 principal minor determinant associated with Eq. (8.78) leads to the equation

$$\delta = 4 - \frac{32\epsilon^2}{384 - 2\epsilon^2} \approx 4 - \tfrac{1}{12}\epsilon^2(1 + \tfrac{1}{192}\epsilon^2), \tag{8.86}$$

whereas the 3×3 determinant of (8.79) yields

$$\delta = 4 + \frac{20\epsilon^2}{48 + 3\epsilon^2} \approx 4 + \tfrac{5}{12}\epsilon^2(1 - \tfrac{1}{16}\epsilon^2). \tag{8.87}$$

Incidentally, the same determinant provides a better approximation for the boundary curve beginning at $\delta = 0$. The equation of this curve is

$$\delta = -\frac{32\epsilon^2}{64 - 3\epsilon^2} \approx -\tfrac{1}{2}\epsilon^2(1 + \tfrac{3}{64}\epsilon^2). \tag{8.88}$$

The instability regions obtained above are plotted in the parameter plane shown in Fig. 8.1. The first approximations are shown in dotted lines and the second approximations in solid lines. Figure 8.1 represents an enlargement of the neighborhood of the segment $0 < \delta < 9$ of Fig. 7.3.

Of special interest is the "width" of the instability regions. From Eq. (8.85) we observe that the width of the first region is of order ϵ. Equations (8.82) and (8.83) indicate that the width of the second region is proportional to ϵ^2, whereas from Eq. (8.84) we conclude that the width of the third region is of order ϵ^3. Hence for small ϵ the first region is appreciably wider when

FIGURE 8.1

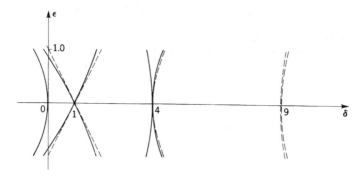

Analytical Solutions by Perturbation Techniques 313

compared with the remaining ones, and it therefore must be considered as the most significant one. The first region is sometimes referred to as the *principal instability region*. The remaining regions become progressively narrower and less important.

8.6 PERIODIC SOLUTIONS OF NONAUTONOMOUS SYSTEMS. DUFFING'S EQUATION

Let us consider the nearly harmonic nonautonomous system

$$\ddot{x} + \omega^2 x = \epsilon f(x, \dot{x}, \Omega t), \qquad (8.89)$$

where ϵ is a small positive number and $f(x, \dot{x}, \Omega t)$ is a periodic function of time with period $T = 2\pi/\Omega$. We wish to explore the existence of periodic solutions of Eq. (8.89) with the same period T. Unlike autonomous systems, in nonautonomous systems the time scale cannot be shifted. Hence, although in this case the frequency of the periodic motion is known, the phase angle of the fundamental harmonic is no longer arbitrary and cannot be assumed zero, as for autonomous systems.

We shall find it convenient to change the time scale so that the period of oscillation becomes 2π. To this end, we introduce a new time $\tau + \varphi = \Omega t$, $d/dt = \Omega d/d\tau$, where φ is the unknown phase angle, and rewrite Eq. (8.89) in the form

$$\Omega^2 x'' + \omega^2 x = \epsilon f(x, \Omega x', \tau + \varphi). \qquad (8.90)$$

As in Sec. 8.3, primes denote differentiations with respect to τ. The phase angle φ is determined by insisting that the initial condition have the convenient form

$$x'(0) = 0. \qquad (8.91)$$

We seek a solution of Eq. (8.90), subject to condition (8.91), in the form of power series in ϵ for $x(\tau)$ as well as for φ

$$x(\tau) = x_0(\tau) + \epsilon x_1(\tau) + \epsilon^2 x_2(\tau) + \cdots, \qquad (8.92)$$

$$\varphi = \varphi_0 + \epsilon \varphi_1 + \epsilon^2 \varphi_2 + \cdots. \qquad (8.93)$$

Substituting Eqs. (8.92) and (8.93) in the function $f(x, \Omega x', \tau + \varphi)$ and expanding the series about the zero-order approximation, we obtain

$$f(x, \Omega x', \tau + \varphi) = f(x_0, \Omega x_0', \tau + \varphi_0)$$
$$+ \epsilon \left[x_1 \frac{\partial f(x_0, \Omega x_0', \tau + \varphi_0)}{\partial x} + x_0' \frac{\partial f(x_0, \Omega x_0', \tau + \varphi_0)}{\partial x'} \right.$$
$$\left. + \varphi_1 \frac{\partial f(x_0, \Omega x_0', \tau + \varphi_0)}{\partial \varphi} \right] + \epsilon^2(\cdots) + \cdots.$$
$$(8.94)$$

Introducing Eqs. (8.92) and (8.93) into Eq. (8.90), considering Eq. (8.94), and equating coefficients of equal powers of ϵ, we arrive at

$$\Omega^2 x_0'' + \omega^2 x_0 = 0,$$
$$\Omega^2 x_1'' + \omega^2 x_1 = f(x_0, \Omega x_0', \tau + \varphi_0),$$
$$\Omega^2 x_2'' + \omega^2 x_2 = x_1 \frac{\partial f(x_0, \Omega x_0', \tau + \varphi_0)}{\partial x} + x_1' \frac{\partial f(x_0, \Omega x_0', \tau + \varphi_0)}{\partial x'}$$
$$+ \varphi_1 \frac{\partial f(x_0, \Omega x_0', \tau + \varphi_0)}{\partial \varphi},$$
$$\cdots\cdots\cdots\cdots\cdots\cdots\cdots\cdots\cdots\cdots\cdots\cdots,$$
(8.95)

which can be solved recursively. The quantities $x_n(\tau)$ ($n = 0, 1, 2, \ldots$) must satisfy the periodicity and the initial conditions

$$x_n(\tau + 2\pi) = x_n(\tau), \quad n = 0, 1, 2, \ldots \quad (8.96)$$

and

$$x_n'(0) = 0, \quad n = 0, 1, 2, \ldots, \quad (8.97)$$

respectively.

To make the discussion more meaningful, it is advisable at this point to abandon the generalities and select a particular function $f(x,\dot{x},\Omega t)$. Hence, we shall consider the function $f(x,\dot{x},\Omega t) = -\omega^2(\alpha x + \beta x^3) + F \cos \Omega t$, so that Eq. (8.89) becomes

$$\ddot{x} + \omega^2 x = \epsilon(-\omega^2 \alpha x - \omega^2 \beta x^3 + F \cos \Omega t), \quad \epsilon \ll 1. \quad (8.98)$$

It is not difficult to recognize that Eq. (8.98) represents the differential equation of motion of a mass–nonlinear spring system subjected to a harmonic force. When β is positive, the spring is hard, and when β is negative, the spring is soft (see Probs. 5.5 and 5.6). For this system Eqs. (8.95) assume the specific form

$$\Omega^2 x_0'' + \omega^2 x_0 = 0,$$
$$\Omega^2 x_1'' + \omega^2 x_1 = -\omega^2(\alpha x_0 + \beta x_0^3) + F \cos(\tau + \varphi_0),$$
$$\Omega^2 x_2'' + \omega^2 x_2 = -\omega^2(\alpha x_1 + 3\beta x_0^2 x_1) - F\varphi_1 \sin(\tau + \varphi_0),$$
$$\cdots\cdots\cdots\cdots\cdots\cdots\cdots\cdots\cdots\cdots\cdots\cdots,$$
(8.99)

whereas the periodicity conditions (8.96) and initial conditions (8.97) remain unchanged.

The solution of the first of Eqs. (8.99) is simply

$$x_0(\tau) = A_0 \cos \frac{\omega}{\Omega} \tau, \quad (8.100)$$

where A_0 is constant. It can be shown with ease that solution (8.100) satisfies conditions (8.96) and (8.97) with $n = 0$ only if

$$\omega = \Omega. \quad (8.101)$$

Analytical Solutions by Perturbation Techniques 315

In further discussion of this case we assume that Eq. (8.101) holds.
Introducing Eq. (8.100) into the second of Eqs. (8.99), we obtain

$$x_1'' + x_1 = -\frac{F}{\omega^2}\sin\varphi_0 \sin\tau - \left(\alpha A_0 + \tfrac{3}{4}\beta A_0^3 - \frac{F}{\omega^2}\cos\varphi_0\right)\cos\tau$$
$$-\tfrac{1}{4}\beta A_0^3 \cos 3\tau. \quad (8.102)$$

The periodicity condition, Eq. (8.96) with $n = 1$, must be invoked to suppress secular terms, which requires that the coefficients of $\cos\tau$ and $\sin\tau$ vanish. Hence, we must have

$$\alpha A_0 + \tfrac{3}{4}\beta A_0^3 - \frac{F}{\omega^2} = 0, \qquad \varphi_0 = 0 \quad (8.103)$$

or
$$\alpha A_0 + \tfrac{3}{4}\beta A_0^3 + \frac{F}{\omega^2} = 0, \qquad \varphi_0 = \pi. \quad (8.104)$$

It follows that if the zero-order response satisfies Eqs. (8.103), that response is in phase with the driving force, whereas if the response satisfies Eqs. (8.104), it is 180° out of phase with the force. But since Eqs. (8.104) do not yield any information which cannot be extracted from Eqs. (8.103), further discussion will concentrate on Eqs. (8.103), because a 180° out-of-phase response is the equivalent of an in-phase response with negative amplitude.

In view of Eqs. (8.103) and Eq. (8.97) with $n = 1$, the solution of Eq. (8.102) becomes

$$x_1(\tau) = A_1 \cos\tau + \tfrac{1}{32}\beta A_0^3 \cos 3\tau, \quad (8.105)$$

where A_1 is yet to be determined.

A substitution of Eqs. (8.100) and (8.105) into the third of Eqs. (8.99) yields

$$x_2'' + x_2 = -F\varphi_1 \sin\tau - (\alpha A_1 + \tfrac{9}{4}\beta A_0^2 A_1 + \tfrac{3}{128}\beta^2 A_0^5)\cos\tau$$
$$- \tfrac{1}{4}\beta A_0^2(3A_1 + \tfrac{1}{8}\alpha A_0 + \tfrac{3}{16}\beta A_0^3)\cos 3\tau - \tfrac{33}{128}\beta^2 A_0^5 \cos 5\tau. \quad (8.106)$$

The periodicity condition for $n = 2$ requires that

$$A_1 = -\frac{3\beta^2 A_0^5}{32(4\alpha + 9\beta A_0^2)}, \qquad \varphi_1 = 0, \quad (8.107)$$

so that the solution of Eq. (8.106) assumes the form

$$x_2(\tau) = A_2 \cos\tau + \tfrac{1}{32}\beta A_0^2(3A_1 + \tfrac{1}{8}\alpha A_0 + \tfrac{3}{16}\beta A_0^3)\cos 3\tau$$
$$+ \tfrac{3}{3,072}\beta^2 A_0^5 \cos 5\tau, \quad (8.108)$$

where A_2 is obtained from the next approximation.

The procedure for deriving the higher-order approximations is evident. We shall stop here and write the second-order approximation to the solution of Eq. (8.98)

$$x(t) = (A_0 + \epsilon A_1 + \epsilon^2 A_2)\cos \Omega t + \frac{\epsilon}{32}\beta A_0^2$$

$$\times [A_0 + \epsilon(3A_1 + \tfrac{1}{8}\alpha A_0 + \tfrac{3}{16}\beta A_0^3)]\cos 3\Omega t + \frac{3\epsilon^2}{3072}\beta^2 A_0^5 \cos 5\Omega t. \tag{8.109}$$

We may note that to the first approximation the phase angle is zero, $\varphi_0 = \varphi_1 = 0$. This is no coincidence, and, in fact, all φ_n ($n = 0, 1, 2, \ldots$) are zero. The explanation is that Eq. (8.98) does not contain $\dot{x}$. When the mass–nonlinear spring system in question possesses viscous damping, the quantities φ_n are not zero and the response is out of phase with the force.

An interesting aspect of the solution unveils itself if we let

$$\omega_0^2 = (1 + \epsilon\alpha)\omega^2 \tag{8.110}$$

in the first of Eqs. (8.103). We note from Eq. (8.98) that ω_0 can be interpreted as the natural frequency of the associated linear system, obtained by setting $\beta = 0$. Recalling that ϵ is a small quantity and introducing Eq. (8.110) into the first of Eqs. (8.103), we arrive at

$$\omega^2 = \omega_0^2 + \tfrac{3}{4}\omega_0^2 \epsilon\beta A_0^2 - \frac{\epsilon F}{A_0}. \tag{8.111}$$

Assuming that ω_0 and $\epsilon\beta$ are known, Eq. (8.111) can be used to plot A_0 versus ω with ϵF as a parameter. In the free vibration case, $F = 0$, and for small $\epsilon\beta$ Eq. (8.111) yields approximately a parabola intersecting the ω axis at $\omega = \omega_0$. Figure 8.2 shows plots of A_0 versus ω for $F = 0$ and $F \neq 0$, corresponding to a positive value of $\epsilon\beta$. We see from that figure that for $F \neq 0$ the plot A_0 versus ω has two branches, one above the positive portion of the curve $F = 0$ and one between the negative portion of $F = 0$ and the ω axis. The branch corresponding to negative A_0 represents the 180° out-of-phase response. For comparison, the linear case corresponding to $\beta = 0$ is also shown in Fig. 8.2. Instead of A_0 versus ω it is customary to plot the curves $|A_0|$ versus ω, obtained by folding the lower half of the plane (A_0, ω) about the ω axis. Plots of $|A_0|$ versus ω for two different values of $\epsilon\beta$, one positive and one negative, and with ϵF as a parameter are shown in Fig. 8.3a and b. The response curves correspond to the hard and soft springs, respectively. We see from these figures that for a hard spring the curves are bent to the right compared to the linear spring, whereas for the soft spring they are bent to the left.

Consider again Fig. 8.3a and denote by T the point at which a vertical axis is tangent to a given response curve. For values of ω on the left of

Analytical Solutions by Perturbation Techniques

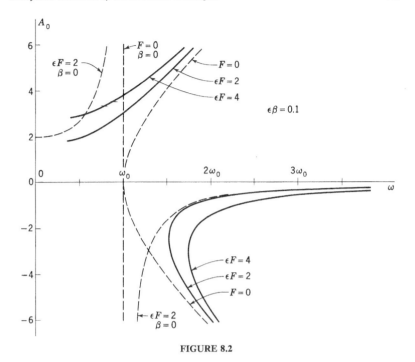

FIGURE 8.2

point T the curve in question has only one value, so that Eq. (8.111) has one real root and two complex roots. For ω corresponding to point T there are three real roots, two of which are coincident with T, whereas for values of ω on the right of point T there are three distinct real roots. As ω increases from a small value, the amplitude of the response increases, but there is no finite value of ω which renders the amplitude infinitely large. Hence, resonance is not possible for the mass–hard spring system, by contrast with the linear system, for which resonance occurs at $\omega = \omega_0$.

Equation (8.98) assumes that the system is undamped. Any physical system, however, possesses a certain amount of damping, no matter how small. Damping alters the response curve $|A_0|$ versus ω to a certain extent, as shown in Fig. 8.4. Unlike the undamped system, in the presence of small viscous damping the amplitude does not increase indefinitely with the driving frequency, nor is the response in phase with the driving force. This introduces the possibility of discontinuities in the response. Indeed, as the driving frequency is increased, the amplitude increases until point 1 is reached, at which point the tangent to the curve is infinite and the amplitude "jumps" suddenly to point 2 on the lower limb of the response curve, from which point it decreases as the frequency continues to increase. On the other hand, if the

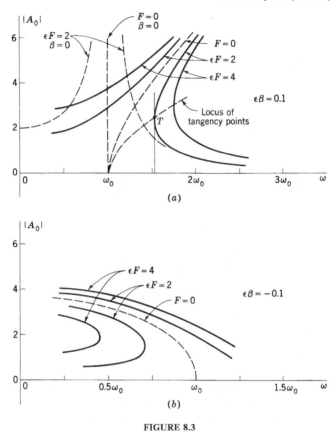

FIGURE 8.3

frequency is decreased from a relatively large value, the amplitude increases continuously until point 3, where the tangent becomes infinite, and a jump to point 4 on the upper limb takes place, from which point the amplitude decreases with a continuous decrease in the driving frequency. The portion of the curve between points 1 and 3 is never traversed and is to be regarded as unstable. Whether the system traverses the arc between 4 and 1 or between 2 and 3 depends on the limb on which the system found itself just before entering one of the two arcs. Clearly the jump takes place after one of the two arcs is traversed. We note that whereas the jump from 3 to 4 could also be expected in an undamped system, the jump from 1 to 2 has no counterpart in the undamped system. The jump phenomenon also occurs in a soft spring, $\beta < 0$, but the jumps in amplitude take place in the reverse direction. The solution of Duffing's equation with damping is left as an exercise to the reader (see Prob. 8.5).

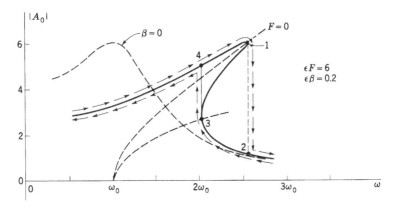

FIGURE 8.4

Returning to the undamped system, Eq. (8.98), and examining solution (8.109), we conclude that the solution consists of a series of harmonic components such that the period is $T = 2\pi/\Omega$. For small ϵ the solution is nearly harmonic, with frequency Ω. The existence of the periodic solution with period $2\pi/\Omega$ requires that ω_0 have a value close to Ω, as indicated by Eqs. (8.101) and (8.110). The question remains whether system (8.98) admits solutions with harmonic components of frequencies smaller than Ω. In the sequel we propose to explore this question.

Let us consider the differential equation

$$\ddot{x} + \omega^2 x = -\epsilon\omega^2\alpha x - \epsilon\beta_0 x^3 + F\cos\Omega t, \quad \epsilon \ll 1, \quad (8.112)$$

where F is not necessarily small. Otherwise Eq. (8.112) is the same as (8.98) if we agree that $\beta_0 = \omega^2\beta$. We are interested in a periodic solution of Eq. (8.112) for the case in which $\Omega = 3\omega$. Letting this solution have the form

$$x(t) = x_0(t) + \epsilon x_1(t) + \epsilon^2 x_2(t) + \cdots, \quad (8.113)$$

we obtain the set of equations

$$\ddot{x}_0 + \frac{\Omega^2}{9}x_0 = F\cos\Omega t,$$

$$\ddot{x}_1 + \frac{\Omega^2}{9}x_1 = -\frac{\Omega^2}{9}\alpha x_0 - \beta_0 x_0^3, \quad (8.114)$$

$$\ddot{x}_2 + \frac{\Omega^2}{9}x_2 = -\frac{\Omega^2}{9}\alpha x_1 - 3\beta_0 x_0^2 x_1,$$

$$\ldots\ldots\ldots\ldots\ldots\ldots\ldots\ldots\ldots,$$

which must be solved recursively.

The solution of the first of Eqs. (8.114) is simply

$$x_0(t) = A_0 \cos \frac{\Omega}{3} t - \frac{9F}{8\Omega^2} \cos \Omega t, \qquad (8.115)$$

so that, after using certain trigonometric relations, the second of Eqs. (8.114) becomes

$$\ddot{x}_1 + \frac{\Omega^2}{9} x_1 = -A_0 \left[\frac{\Omega^2}{9} \alpha + \tfrac{3}{4}\beta_0 A_0^2 - \tfrac{3}{2}\beta_0 A_0 \frac{9F}{8\Omega^2} + \tfrac{3}{2}\beta_0 \left(\frac{9F}{8\Omega^2}\right)^2 \right] \cos \frac{\Omega}{3} t$$

$$+ \left[\tfrac{1}{8}\alpha F - \tfrac{1}{4}\beta_0 A_0^3 + \tfrac{3}{2}\beta_0 A_0^2 \frac{9F}{8\Omega^2} + \tfrac{3}{4}\beta_0 \left(\frac{9F}{8\Omega^2}\right)^3 \right] \cos \Omega t$$

$$+ \tfrac{3}{4}\beta_0 A_0 \frac{9F}{8\Omega^2} \left(A_0 - \frac{9F}{8\Omega^2} \right) \cos \frac{5\Omega}{3} t - \tfrac{3}{4}\beta_0 A_0^2 \left(\frac{9F}{8\Omega^2}\right)^2 \cos \frac{7\Omega}{3} t$$

$$+ \tfrac{1}{4}\beta_0 \left(\frac{9F}{8\Omega^2}\right)^3 \cos 3\Omega t. \qquad (8.116)$$

To eliminate secular terms in the solution of (8.116), the coefficient of $\cos(\Omega t/3)$ must vanish, which yields the quadratic equation in A_0

$$A_0^2 - \frac{9F}{8\Omega^2} A_0 + 2\left(\frac{9F}{8\Omega^2}\right)^2 + \frac{4\Omega^2 \alpha}{27\beta_0} = 0, \qquad (8.117)$$

having the roots

$$A_0 = \frac{1}{2} \frac{9F}{8\Omega^2} \pm \frac{1}{2} \left[\left(\frac{9F}{8\Omega^2}\right)^2 - 8\left(\frac{9F}{8\Omega^2}\right)^2 - \frac{16\Omega^2 \alpha}{27\beta_0} \right]^{\frac{1}{2}}. \qquad (8.118)$$

Since A_0 is by definition a real quantity, we must have

$$-7\left(\frac{9F}{8\Omega^2}\right)^2 - \frac{16\Omega^2 \alpha}{27\beta_0} \geq 0. \qquad (8.119)$$

But, letting $\omega = \Omega/3$ in Eq. (8.110), we obtain the relation

$$\Omega^2 = \frac{9}{\epsilon\alpha} \left(\omega_0^2 - \frac{\Omega^2}{9} \right), \qquad (8.120)$$

so that inequality (8.119) reduces to

$$\Omega^2 \geq 9 \left[\omega_0^2 + \tfrac{21}{16}\epsilon\beta_0 \left(\frac{9F}{8\Omega^2}\right)^2 \right], \qquad (8.121)$$

and we note that $-9F/8\Omega^2$ is the amplitude of the zero-order component of the solution having the frequency Ω. In conclusion if Ω satisfies inequality (8.121), then Eq. (8.112) admits a solution with a harmonic component of frequency $\Omega/3$.

Oscillations with frequencies which are a fraction of the driving frequency are referred to as *subharmonic oscillations*. Hence Duffing's equation, with

Analytical Solutions by Perturbation Techniques 321

no damping, admits a subharmonic oscillation with frequency $\Omega/3$. The subharmonic is said to be of order 3.

If an undamped linear oscillator is subjected to two harmonic forcing functions of different frequencies, then the response is the sum of two harmonic components with the same frequencies as the excitation. In a nonlinear system, however, other frequencies may be excited. To demonstrate this, let us consider the system

$$\ddot{x} + \omega_0^2 x + \epsilon\beta_0 x^3 = F_1 \cos \Omega_1 t + F_2 \cos \Omega_2 t. \tag{8.122}$$

Assuming the solution of Eq. (8.122) in the form (8.113), we obtain the recursive set of equations

$$\begin{aligned}
\ddot{x}_0 + \omega_0^2 x_0 &= F_1 \cos \Omega_1 t + F_2 \cos \Omega_2 t, \\
\ddot{x}_1 + \omega_0^2 x_1 &= -\beta_0 x_0^3, \\
\ddot{x}_2 + \omega_0^2 x_2 &= -3\beta_0 x_0^2 x_1, \\
&\cdots\cdots\cdots\cdots\cdots\cdots.
\end{aligned} \tag{8.123}$$

The forced oscillation solution of the first of Eqs. (8.123) is simply

$$x_0(t) = G_1 \cos \Omega_1 t + G_2 \cos \Omega_2 t, \tag{8.124}$$

where
$$G_1 = \frac{F_1}{\omega_0^2 - \Omega_1^2}, \qquad G_2 = \frac{F_2}{\omega_0^2 - \Omega_2^2}. \tag{8.125}$$

Introducing Eq. (8.124) into the second of Eqs. (8.123), we obtain

$$\begin{aligned}
\ddot{x}_1 + \omega_0^2 x_1 = {}& H_1 \cos \Omega_1 t + H_2 \cos \Omega_2 t \\
& + H_3[\cos(2\Omega_1 + \Omega_2)t + \cos(2\Omega_1 - \Omega_2)t] \\
& + H_4[\cos(\Omega_1 + 2\Omega_2)t + \cos(\Omega_1 - 2\Omega_2)t] \\
& + H_5 \cos 3\Omega_1 t + H_6 \cos 3\Omega_2 t,
\end{aligned} \tag{8.126}$$

in which

$$H_1 = -\frac{3\beta_0 G_1(G_1^2 + 2G_2^2)}{4}, \qquad H_2 = -\frac{3\beta_0 G_2(G_2^2 + 2G_1^2)}{4},$$

$$H_3 = -\frac{3\beta_0 G_1^2 G_2}{4}, \qquad H_4 = -\frac{3\beta_0 G_1 G_2^2}{4}, \tag{8.127}$$

$$H_5 = -\frac{\beta_0 G_1^3}{4}, \qquad H_6 = -\frac{\beta_0 G_2^3}{4}.$$

It is clear from Eq. (8.126) that the solution $x_1(t)$ has harmonic components of frequencies Ω_1, Ω_2, $2\Omega_1 \pm \Omega_2$, $\Omega_1 \pm 2\Omega_2$, $3\Omega_1$, and $3\Omega_2$. Hence, by contrast with the linear system, the response of the nonlinear system (8.122) consists not only of harmonic components of frequencies Ω_1 and Ω_2 but also of higher harmonics of frequencies $3\Omega_1$ and $3\Omega_2$ as well as the socalled *combination tones* with frequencies $2\Omega_1 \pm \Omega_2$ and $\Omega_1 \pm 2\Omega_2$. The terms with frequencies other than Ω_1 and Ω_2 are smaller in magnitude than the terms with frequencies Ω_1 and Ω_2, as they appear only in the first-order component x_1 and not in the zero-order solution x_0. In addition to these frequencies, other higher harmonics and combination tones appear in x_2, but their magnitude will be even smaller than the ones in x_1. It should be stressed that this set of higher harmonics and combination tones is peculiar to system (8.122) and reflects the fact that the nonlinearity is introduced by the term $\epsilon \beta_0 x^3$. For other nonlinear systems entirely different harmonics and combination tones can be expected.

8.7 THE METHOD OF AVERAGING

The basic idea behind the method of averaging is to derive an approximate solution of a nonautonomous system by considering in its place the associated averaged system, which is autonomous. For the procedure to be valid, the system must possess a certain form to be discussed later. We shall follow here the exposition given by Bogoliubov and Mitropolsky (Ref. 2, chaps. 5 and 6), which contains not only a description of the method but also the mathematical justification.

We shall be concerned with a simultaneous set of nonlinear nonautonomous differential equations

$$\ddot{x}_k + \omega_k^2 x_k = \epsilon X_k(x_k, \dot{x}_k, t), \qquad k = 1, 2, \ldots, n, \qquad (8.128)$$

where ϵ is a small parameter. It is shown in Ref. 2 that through a suitable change of variables Eqs. (8.128) can be reduced to a set of first-order differential equations in which the terms in the small parameter ϵ are, as in Eqs. (8.128), on the right side of the equations. Such equations are said to be in the *standard form* and can be written

$$\dot{x}_k = \epsilon X_k(x_1, x_2, \ldots, x_m, t), \qquad k = 1, 2, \ldots, m, \qquad (8.129)$$

where the functions X_k, which are different from the functions X_k in Eqs. (8.128), can be represented by the series

$$X_k(x_1, x_2, \ldots, x_m, t) = \sum_\nu e^{i\nu t} X_{k\nu}(x_1, x_2, \ldots, x_m), \qquad k = 1, 2, \ldots, m, \qquad (8.130)$$

in which the frequencies ν are constant.

Equations (8.129) can be written in a compact form by means of the vector notation

$$\dot{\mathbf{x}} = \epsilon \mathbf{X}(\mathbf{x}, t), \qquad (8.131)$$

Analytical Solutions by Perturbation Techniques 323

where x and **X** are *m*-vectors, the latter being of the type

$$\mathbf{X}(\mathbf{x},t) = \sum_{v} e^{ivt}\mathbf{X}_v(\mathbf{x}). \tag{8.132}$$

For any component k of a vector $\mathbf{F}(\mathbf{x},t)$ we have the relation

$$\frac{d}{dt}F_k(x_1,x_2,\ldots,x_m,t) = \sum_{j=1}^{m}\frac{\partial F_k}{\partial x_j}\frac{dx_j}{dt} + \frac{\partial F_k}{\partial t}$$

$$= \left(\sum_{j=1}^{m}\frac{dx_j}{dt}\frac{\partial}{\partial x_j}\right)F_k + \frac{\partial F_k}{\partial t}, \tag{8.133}$$

where the quantity in parentheses on the right side of (8.133) can be regarded as a scalar operator. With this in mind, we can write

$$\frac{d\mathbf{F}}{dt} = (\dot{\mathbf{x}} \cdot \nabla)\mathbf{F} + \frac{\partial \mathbf{F}}{\partial t}, \tag{8.134}$$

where ∇ is an *m*-dimensional del operator.

Let us assume that $\mathbf{F}(\mathbf{x},t)$ is of the form

$$\mathbf{F}(\mathbf{x},t) = \sum_{v} e^{ivt}\mathbf{F}_v(\mathbf{x}) \tag{8.135}$$

and introduce the *averaging operator*

$$\underset{t}{M}[\mathbf{F}(\mathbf{x},t)] = \mathbf{F}_0(\mathbf{x}), \tag{8.136}$$

which evidently represents an average over the explicit time variable, with the value of **x** regarded as constant. In addition, we define the *integrating operator*

$$\tilde{\mathbf{F}}(\mathbf{x},t) = \sum_{v \neq 0} \frac{e^{ivt}}{iv}\mathbf{F}_v(\mathbf{x}), \tag{8.137}$$

where the operator is equivalent to the indefinite integral of $\mathbf{F}(\mathbf{x},t)$ with respect to time, in which the nonharmonic component in time has been ignored. Moreover

$$\tilde{\tilde{\mathbf{F}}}(\mathbf{x},t) = \sum_{v \neq 0} \frac{e^{ivt}}{(iv)^2}\mathbf{F}_v(\mathbf{x}). \tag{8.138}$$

From relations (8.135) to (8.138), we conclude that

$$\frac{\partial \tilde{\tilde{\mathbf{F}}}}{\partial t} = \tilde{\mathbf{F}}, \quad \frac{\partial \tilde{\mathbf{F}}}{\partial t} = \mathbf{F} - \underset{t}{M}[\mathbf{F}]. \tag{8.139}$$

Let us now return to the system (8.131) and (8.132). Because $\dot{\mathbf{x}}$ is proportional to a vector multiplied by the small parameter ϵ, it is reasonable to expect that **x** varies slowly. Hence it will be assumed that **x** can be represented as a superposition of a smoothly varying term **ξ** and some small oscillatory terms. As a first approximation, it may be possible to ignore the oscillatory terms and write

$$\dot{\mathbf{x}} = \epsilon \mathbf{X}(\mathbf{x},t) \approx \epsilon \mathbf{X}(\boldsymbol{\xi},t) = \epsilon \sum_{\nu} \mathbf{X}_{\nu}(\boldsymbol{\xi})e^{i\nu t} \tag{8.140}$$

or
$$\dot{\mathbf{x}} = \epsilon \mathbf{X}_0(\boldsymbol{\xi}) + \text{small harmonic terms.} \tag{8.141}$$

Assuming that the small harmonic terms cause small oscillations about the slowly changing **ξ** without affecting its value, the first approximation takes the form

$$\dot{\boldsymbol{\xi}} = \epsilon \mathbf{X}_0(\boldsymbol{\xi}) = \epsilon \, M_t \, [\mathbf{X}(\boldsymbol{\xi},t)], \tag{8.142}$$

which represents a set of autonomous differential equations.

In conclusion, the equation of the first approximation, Eq. (8.142), is obtained by averaging the exact differential equation (8.131) with respect to time, where the vector **ξ**, replacing the vector **x**, is treated as constant during the averaging process. This procedure is referred to as the *principle of averaging*.

It should be noted that the principle does not require that $\mathbf{X}(\boldsymbol{\xi},t)$ be of the form (8.132). It does require, however, the existence of an average value

$$\mathbf{X}_0(\boldsymbol{\xi}) = \lim_{T \to \infty} \frac{1}{T} \int_0^T \mathbf{X}(\boldsymbol{\xi},t) \, dt. \tag{8.143}$$

For the second approximation we must consider the small harmonic terms in Eq. (8.141). These terms produce small oscillations $\epsilon(e^{i\nu t}/i\nu)\mathbf{X}_\nu(\boldsymbol{\xi})$ about the vector **ξ** so that, in view of Eq. (8.137), we have

$$\mathbf{x} = \boldsymbol{\xi} + \epsilon \sum_{\nu \neq 0} \frac{e^{i\nu t}}{i\nu} \mathbf{X}_\nu(\boldsymbol{\xi}) = \boldsymbol{\xi} + \epsilon \tilde{\mathbf{X}}(\boldsymbol{\xi},t). \tag{8.144}$$

Introducing Eq. (8.144) into (8.131), we obtain

$$\dot{\mathbf{x}} = \epsilon \mathbf{X}(\boldsymbol{\xi} + \epsilon \tilde{\mathbf{X}}, t) = \epsilon \, M_t \, [\mathbf{X}(\boldsymbol{\xi} + \epsilon \tilde{\mathbf{X}}, t)] + \text{small harmonic terms.} \tag{8.145}$$

Invoking the same argument used in connection with the first approximation and recalling the notation of Eq. (8.134), the equation of the second approximation becomes

$$\dot{\boldsymbol{\xi}} = \epsilon \, M_t \, [\mathbf{X}(\boldsymbol{\xi} + \epsilon \tilde{\mathbf{X}}, t)] = \epsilon \, M_t \, [\mathbf{X}(\boldsymbol{\xi},t) + \epsilon(\tilde{\mathbf{X}} \cdot \nabla)\mathbf{X}(\boldsymbol{\xi},t)]. \tag{8.146}$$

The same pattern is followed for the higher approximations. In practice we confine ourselves to the first or second approximation.

Analytical Solutions by Perturbation Techniques 325

The above results, obtained by means of physical reasoning, can be derived by using a more rigorous mathematical approach. The interested reader should consult Ref. 2, pp. 392–399.

Example 8.3

Consider a viscously damped simple pendulum whose point of support executes in the vertical direction a harmonic motion with an amplitude A which is very small relative to the pendulum length L, $A \ll L$, and with a frequency ω such that $\omega > \omega_0 L/A$, where $\omega_0 = \sqrt{g/L}$ is the natural frequency of the harmonic oscillator associated with the small motion of the simple pendulum with the support fixed. We shall be interested in investigating the stability of the pendulum in the upright vertical position (which was shown to be unstable for the simple pendulum) by means of the averaging method.

Denoting by θ the angular displacement from the equilibrium position in the downward vertical position, the differential equation of motion can be written (see Example 7.2)

$$\ddot{\theta} + 2\zeta\omega_0\dot{\theta} + \omega_0^2\left[1 - \frac{A}{L}\left(\frac{\omega}{\omega_0}\right)^2 \sin \omega t\right] \sin \theta = 0, \qquad (a)$$

where ζ is the damping coefficient. It will be assumed that $\zeta < 1$, so that the pendulum is underdamped.

Let us render the time variable dimensionless by introducing the new variable $\tau = \omega t$, so that Eq. (a) reduces to

$$\theta'' + 2\zeta\frac{\omega_0}{\omega}\theta' + \left(\frac{\omega_0}{\omega}\right)^2\left[1 - \frac{A}{L}\left(\frac{\omega}{\omega_0}\right)^2 \sin \tau\right] \sin \theta = 0, \qquad (b)$$

in which primes denote differentiations with respect to τ. Equation (b) can be further simplified by using the substitutions

$$\frac{A}{L} = \epsilon, \qquad \frac{\omega_0}{\omega} = k\epsilon, \qquad \zeta\frac{\omega_0}{\omega} = \alpha\epsilon, \qquad (c)$$

where ϵ is a small parameter. Because $\omega_0/\omega < A/L = \epsilon$, we note that $\alpha < 1$ and $k < 1$. Introducing expressions (c) into (b), we obtain

$$\theta'' + 2\epsilon\alpha\theta' + (k^2\epsilon^2 - \epsilon \sin \tau) \sin \theta = 0. \qquad (d)$$

First we reduce Eq. (d) to the standard form. We let

$$\theta = \varphi - \epsilon \sin \tau \sin \varphi, \qquad \theta' = \epsilon\Omega - \epsilon \cos \tau \sin \varphi, \qquad (e)$$

differentiate the first of Eqs. (e) with respect to τ, and compare with the second of (e) so that

$$\theta' = \varphi' - \epsilon \cos \tau \sin \varphi - \epsilon\varphi' \sin \tau \cos \varphi = \epsilon\Omega - \epsilon \cos \tau \sin \varphi, \qquad (f)$$

from which we conclude that
$$\epsilon\Omega = \varphi'(1 - \epsilon \sin \tau \cos \varphi). \tag{g}$$
Furthermore, differentiating the second of Eqs. (e) with respect to τ and using Eq. (d), we have
$$\begin{aligned}\theta'' &= \epsilon\Omega' + \epsilon \sin \tau \sin \varphi - \epsilon\varphi' \cos \tau \cos \varphi \\ &= (\epsilon \sin \tau - k^2\epsilon^2) \sin \theta - 2\epsilon\alpha\theta',\end{aligned} \tag{h}$$
or, using Eqs. (e) as well as Eq. (g),
$$\begin{aligned}\Omega' = &-\sin \tau \sin \varphi + \varphi' \cos \tau \cos \varphi + \sin \tau \sin (\varphi - \epsilon \sin \tau \sin \varphi) \\ &- k^2\epsilon \sin (\varphi - \epsilon \sin \tau \sin \varphi) - 2\alpha(\epsilon\Omega - \epsilon \cos \tau \sin \varphi).\end{aligned} \tag{i}$$
Recalling that ϵ is a small parameter, we can show that Eqs. (g) and (i) reduce to the standard form
$$\begin{aligned}\varphi' &= \epsilon\Omega + \epsilon^2 \cdots, \\ \Omega' &= \epsilon(\Omega \cos \tau \cos \varphi - \sin^2 \tau \sin \varphi \cos \varphi \\ &\quad - k^2 \sin \varphi - 2\alpha\Omega + 2\alpha \cos \tau \sin \varphi) + \epsilon^2 \cdots.\end{aligned} \tag{j}$$

Equations (j) can be treated by the method of averaging. In view of the fact that
$$\underset{\tau}{M}[\cos \tau] = 0, \qquad \underset{\tau}{M}[\sin^2 \tau] = \tfrac{1}{2}, \tag{k}$$
the equations of the first approximation become
$$\begin{aligned}\varphi' &= \epsilon\Omega, \\ \Omega' &= -\epsilon(\tfrac{1}{2} \sin \varphi \cos \varphi + k^2 \sin \varphi + 2\alpha\Omega),\end{aligned} \tag{l}$$
which are autonomous.

In the neighborhood of $\varphi = \pi$ the variational equations become
$$\begin{aligned}\varphi' &= \epsilon\Omega, \\ \Omega' &= -\epsilon(\tfrac{1}{2} - k^2)\varphi - 2\epsilon\alpha\Omega,\end{aligned} \tag{m}$$
leading to the characteristic equation
$$\lambda^2 + 2\epsilon\alpha\lambda + \epsilon^2(\tfrac{1}{2} - k^2) = 0. \tag{n}$$
The equilibrium position $\varphi = \pi$ is stable if neither of the eigenvalues λ_1 and λ_2 is real and positive or complex with positive real part. Solving Eq. (n) for λ_1 and λ_2, we conclude that for stability we must have
$$(\epsilon\alpha)^2 - \epsilon^2(\tfrac{1}{2} - k^2) < 0, \tag{o}$$
or, in terms of the system parameters,
$$\omega > \sqrt{2}\,(1 + \zeta^2)^{\frac{1}{2}} \frac{L}{A} \omega_0, \tag{p}$$

which indicates that the upright vertical position is stable provided the frequency with which the point of support oscillates is sufficiently large for inequality (p) to be satisfied.

It should be noted that for $\zeta = 0$ and small θ Eq. (d) can be reduced to Mathieu's equation by a shift in the nondimensional time τ by an amount $\pi/2$, as can be seen from Example 7.2.

PROBLEMS

8.1 Consider the differential equation

$$\ddot{x} + \omega^2 x - \epsilon x^2 = 0, \qquad \epsilon \ll 1,$$

and use Lindstedt's method to obtain a periodic solution. Limit the solution to the second-order approximation.

8.2 Consider the van der Pol equation

$$\ddot{x} + \epsilon \dot{x}(x^2 - 1) + x = 0, \qquad \epsilon \ll 1,$$

and obtain a first-order approximation periodic solution by Lindstedt's method.

8.3 Solve Prob. 8.1 by the KBM method.

8.4 Consider the equation

$$\ddot{x} + \omega^2 x = \epsilon[-\omega^2 \alpha x + \dot{x}(1 - x^2) + F \cos \Omega t], \qquad \epsilon \ll 1,$$

and use the method of Sec. 8.6 to obtain a periodic solution with period $2\pi/\Omega$. Obtain the first approximation only.

8.5 Solve Duffing's equation with small damping

$$\ddot{x} + \omega^2 x = -\epsilon(\omega^2 \alpha x + c\dot{x} + \omega^2 \beta x^3 - F \cos \Omega t), \qquad \epsilon \ll 1,$$

and obtain a first-order approximation periodic solution with period $2\pi/\Omega$. Choose appropriate values for ϵ, c, β, F, and ω_0, where ω_0 is related to ω by Eq. (8.110), and plot the response curve for the soft spring, $\beta < 0$.

8.6 Consider the equation

$$\ddot{x} + \omega^2 x = -\epsilon(\omega^2 \alpha x - \beta_0 x^2) + F \cos \Omega t, \qquad \epsilon \ll 1,$$

and obtain a subharmonic solution.

SUGGESTED REFERENCES

1. Bellman, R.: *Perturbation Techniques in Mathematics, Physics, and Engineering*, Holt, Rinehart and Winston, Inc., New York, 1964.
2. Bogoliubov, N. N., and Y. A. Mitropolsky: *Asymptotic Methods in the Theory of Non-Linear Oscillations* (trans. from Russian), Hindustan Publishing, Delhi, India, 1961.
3. Cesari, L.: *Asymptotic Behavior and Stability Problems in Ordinary Differential Equations*, Springer-Verlag OHG, Berlin, 1963.
4. Cunningham, W. J.: *Introduction to Nonlinear Analysis*, McGraw-Hill Book Company, New York, 1958.
5. Hale, J. K.: *Oscillations in Nonlinear Systems*, McGraw-Hill Book Company, New York, 1963.
6. Hayashi, C.: *Nonlinear Oscillations in Physical Systems*, McGraw-Hill Book Company, New York, 1964.
7. Minorsky, N.: *Nonlinear Oscillations*, D. Van Nostrand Company, Inc., Princeton, N.J., 1962.
8. Stoker, J. J.: *Nonlinear Vibrations*, Interscience Publishers, New York, 1950.
9. Struble, R. A.: *Nonlinear Differential Equations*, McGraw-Hill Book Company, New York, 1962.

chapter nine

Transformation Theory. The Hamilton-Jacobi Equation

Lagrange's equations of motion provide an elegant and concise formulation of the problems of mechanics. Another formulation, built on the Lagrangian mechanics, is the Hamiltonian mechanics, due primarily to Hamilton, as the name indicates. Whereas the Lagrangian formulation is in terms of n second-order differential equations, the Hamiltonian formulation is in terms of $2n$ first-order differential equations. This fact alone does not offer any advantage in the direct integration of the equations of motion, but the Hamiltonian formulation offers a number of other advantages. Among them is the deeper insight it provides into the behavior of dynamical systems, as illustrated by the geometrical description of the motion presented in Chap. 5. The equal status of the canonical variables allows a higher level of abstraction. Moreover, the Hamiltonian formulation proves superior in the indirect integration of the equations of motion by means of suitable transformations, since Hamilton's canonical equations are in a particularly desirable form for this purpose.

We recall from Lagrangian mechanics that when a coordinate is ignorable, the conjugate momentum is conserved, thus providing us with a motion integral. Hence, an indirect method of integrating the equations of motion is to seek coordinate transformations producing ignorable coordinates. A similar situation exists in the Hamiltonian mechanics, but the advantages of the latter are that (1) the equal status of the canonical variables provides for a larger choice of transformations, (2) the Hamiltonian is a function of the variables alone and not of their derivatives, and (3) the procedure involved in solution by means of transformations is more systematic. The restriction placed on the transformations is that the structure of Hamilton's equations be preserved.

The group of transformations of the Hamiltonian mechanics is referred to as contact transformations or as canonical transformations. The transformation theory, attributed to Jacobi, has attained a high degree of sophistication. The purpose of the transformation is to obtain a set of canonical variables which so simplify the Hamiltonian function that the equations of motion in terms of the new variables are directly integrable. The outstanding feature of this method is that the contact transformation is characterized by a single function, namely, the generating function. If a generating function can be found, the entire dynamical problem reduces to one of differentiations and eliminations.

It turns out that in order to derive the generating function it is necessary to solve a single partial differential equation, known as the *Hamilton-Jacobi equation*. This is the main drawback of the method, as there are no general ways of solving partial differential equations. Under certain circumstances, however, the equation can be solved by the separation of variables. In most cases, systems that can be solved by the separation of variables, referred to as separable systems, can also be solved by other means, so that from a practical viewpoint there appears to be little gain. Nevertheless, the theory does provide a deeper understanding of the structure of mechanics, and it possesses an extreme and satisfying beauty. Moreover, the theory provides the stepping-stones for both statistical mechanics and quantum mechanics.

In this chapter we extend the discussion of variational principles, first introduced in Chap. 2, by presenting the principle of least action. Then the subject of contact transformations is covered in some detail, which leads naturally to the concepts of integral invariants, Lagrange and Poisson brackets, and infinitesimal contact transformations. After the basic ingredients have been introduced, a discussion of the Hamilton-Jacobi equation follows, including the cases in which the generating function is time-dependent and time-independent. The latter case is intimately connected with the subject of separable systems. Subsequently, the discussion centers around periodic solutions of a certain type, and to this end the concepts of angle and action variables are introduced. Finally, a perturbation method based on the Hamilton-Jacobi theory is presented.

9.1 THE PRINCIPLE OF LEAST ACTION

It will prove of interest to consider a more general variational principle, known as the *principle of least action*. This, in turn, necessitates the introduction of a new and more general type of variation of the path of the system, namely, one in which changes in time are also permitted. Figure 9.1 illustrates the configuration space in which A represents the original path and B is the varied path. Point P on the unvaried path A represents the point with coordinates q_i through which the system passes at time t. At the same time t the system passes through point Q with coordinates $q_i + \delta q_i$ on the varied path B. At the later time $t + \Delta t$ the system passes through point $q_i + \Delta q_i$ on the same path B. Hence, the variations in the system coordinates can be written

$$\Delta q_i = \delta q_i + \dot{q}_i \Delta t, \tag{9.1}$$

where the new variation, designated as a Δ variation as opposed to a δ variation, is such that the Δ operation and the differentiation with respect to time are no longer interchangeable, as they are in the case of the δ operation. We recall from Sec. 2.2 that the virtual displacements δq_i are contemporaneous and consistent with the system constraints but that in general the varied path

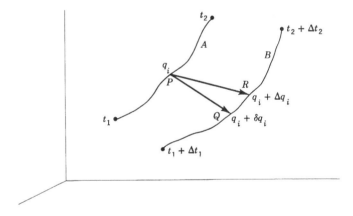

FIGURE 9.1

is not a possible path. This is certainly the case when there are time-dependent constraints. On the other hand, the variations Δq_i do allow a change in time, which gives us the freedom to prescribe the speed with which the system moves along the varied path. We shall consider only those varied paths for which $\dot{q}_i$ are such that if the Hamiltonian H is conserved for the original path, it is conserved for the varied path also. Moreover, the variations Δq_i are assumed to be zero at the end points, which implies that changes in time are possible at these points also.

For any arbitrary function $f = f(q_1, q_2, \ldots, q_n, \dot{q}_1, \dot{q}_2, \ldots, \dot{q}_n, t)$ we obtain the Δ variation

$$\Delta f = \sum_{i=1}^{n} \left(\frac{\partial f}{\partial q_i} \Delta q_i + \frac{\partial f}{\partial \dot{q}_i} \Delta \dot{q}_i \right) + \frac{\partial f}{\partial t} \Delta t$$

$$= \sum_{i=1}^{n} \frac{\partial f}{\partial q_i} (\delta q_i + \dot{q}_i \Delta t) + \sum_{i=1}^{n} \frac{\partial f}{\partial \dot{q}_i} (\delta \dot{q}_i + \ddot{q}_i \Delta t) + \frac{\partial f}{\partial t} \Delta t$$

$$= \sum_{i=1}^{n} \left(\frac{\partial f}{\partial q_i} \delta q_i + \frac{\partial f}{\partial \dot{q}_i} \delta \dot{q}_i \right) + \Delta t \left[\sum_{i=1}^{n} \left(\frac{\partial f}{\partial q_i} \dot{q}_i + \frac{\partial f}{\partial \dot{q}_i} \ddot{q}_i \right) + \frac{\partial f}{\partial t} \right]$$

$$= \delta f + \Delta t \frac{df}{dt}, \qquad (9.2)$$

so that the operation Δ is related to the operation δ by

$$\Delta = \delta + \Delta t \frac{d}{dt}. \qquad (9.3)$$

Next let us define the *action integral* A as

$$A = \int_{t_1}^{t_2} \sum_{i=1}^{n} p_i \dot{q}_i \, dt = \int_{t_1}^{t_2} \sum_{i=1}^{n} \frac{\partial L}{\partial \dot{q}_i} \dot{q}_i \, dt, \qquad (9.4)$$

where the motion takes place along an actual path in the configuration space. For a nonnatural system (see Sec. 2.8), Eq. (9.4) is equivalent to

$$A = \int_{t_1}^{t_2} (2T_2 + T_1) \, dt, \qquad (9.5)$$

whereas for a natural system it reduces to

$$A = \int_{t_1}^{t_2} 2T \, dt. \qquad (9.6)$$

Recalling the definition of the Hamiltonian, namely, $H = \sum_{i=1}^{n} p_i \dot{q}_i - L$, and considering only systems for which H is constant, we see that the action integral can be written

$$A = \int_{t_1}^{t_2} (L + H) \, dt = \int_{t_1}^{t_2} L \, dt + H(t_2 - t_1). \qquad (9.7)$$

Restricting the varied paths to those for which H remains constant, the Δ variation of A assumes the form

$$\Delta A = \Delta \int_{t_1}^{t_2} L \, dt + H(\Delta t_2 - \Delta t_1) = \Delta \int_{t_1}^{t_2} L \, dt + H \Delta t \Big|_{t_1}^{t_2}, \qquad (9.8)$$

where the limits t_1 and t_2 are not fixed but subject to Δ variations. Concentrating on the variation of the integral in (9.8), we obtain

$$\Delta \int_{t_1}^{t_2} L \, dt = \delta \int_{t_1}^{t_2} L \, dt + L \, \Delta t \Big|_{t_1}^{t_2}. \tag{9.9}$$

Moreover, in view of Lagrange's equations, Eqs. (2.91), we have

$$\delta \int_{t_1}^{t_2} L \, dt = \int_{t_1}^{t_2} \sum_{i=1}^{n} \left(\frac{\partial L}{\partial q_i} \delta q_i + \frac{\partial L}{\partial \dot{q}_i} \delta \dot{q}_i \right) dt$$

$$= \int_{t_1}^{t_2} \sum_{i=1}^{n} \left[\frac{d}{dt} \left(\frac{\partial L}{\partial \dot{q}_i} \right) \delta q_i + \frac{\partial L}{\partial \dot{q}_i} \delta \dot{q}_i \right] dt = \sum_{i=1}^{n} \int_{t_1}^{t_2} \frac{d}{dt} \left(\frac{\partial L}{\partial \dot{q}_i} \delta q_i \right) dt. \tag{9.10}$$

By virtue of the fact that Δq_i ($i = 1, 2, \ldots, n$) vanish at t_1 and t_2, introduction of relations (9.1) into (9.10) leads to

$$\delta \int_{t_1}^{t_2} L \, dt = \sum_{i=1}^{n} \int_{t_1}^{t_2} \frac{d}{dt} \left(\frac{\partial L}{\partial \dot{q}_i} (\Delta q_i - \dot{q}_i \Delta t) \right) dt$$

$$= \sum_{i=1}^{n} \frac{\partial L}{\partial \dot{q}_i} \Delta q_i \Big|_{t_1}^{t_2} - \sum_{i=1}^{n} \frac{\partial L}{\partial \dot{q}_i} \dot{q}_i \Delta t \Big|_{t_1}^{t_2} = - \sum_{i=1}^{n} \frac{\partial L}{\partial \dot{q}_i} \dot{q}_i \Delta t \Big|_{t_1}^{t_2}. \tag{9.11}$$

Combining Eqs. (9.8), (9.9), and (9.11) and recalling again the definition of H, we arrive at

$$\Delta A = \left(H + L - \sum_{i=1}^{n} \frac{\partial L}{\partial \dot{q}_i} \dot{q}_i \right) \Delta t \Big|_{t_1}^{t_2} = 0. \tag{9.12}$$

Equation (9.12) is known as the *principle of least action*, which can be stated as follows: *If the initial and final configurations of a system are given, and if the system possesses a Jacobi integral in the form of the Hamiltonian H, then the action integral is stationary for the actual path when compared with neighboring paths satisfying the given conditions.* Actually it would be more appropriate to call it the *principle of stationary action*, but it is more widely known by the first name. The principle was originated by Maupertuis and set on a firm foundation by Euler and Lagrange.

The principle of least action gives necessary and sufficient conditions for the action A to have a stationary value for the actual motion of the system, in a way similar to Hamilton's principle, introduced in Sec. 2.7. The conditions rendering the action a minimum for the actual motion are discussed by Whittaker (Ref. 7, sec. 103). Our interest in the principle is not so much for the purpose of deriving the system equations of motion as for background to the Hamilton-Jacobi theory, to be presented later in this chapter.

Next let us consider the integral

$$S = \int_{t_1}^{t_2} L(q_1, q_2, \ldots, q_n, \dot{q}_1, \dot{q}_2, \ldots, \dot{q}_n, t) \, dt, \tag{9.13}$$

where the integration is along an actual path of the system. If the path of integration and the Lagrangian L are known, then S becomes a function of the end points

$$S = S[q_i(t_1), t_1, q_i(t_2), t_2], \tag{9.14}$$

which is called the *Hamilton principal function*. When the Hamiltonian H is constant, introduction of Eq. (9.14) into (9.7) yields

$$S[q_i(t_1), t_1, q_i(t_2), t_2] = A[q_i(t_1), q_i(t_2)] - H(t_2 - t_1). \tag{9.15}$$

Assuming that the initial conditions at $t = t_1$ are known and letting $t_2 = t$, we obtain

$$S(q_i, t) = A(q_i) - Ht + \text{const}, \tag{9.16}$$

so that for a conservative system Hamilton's principal function can be separated into a function of the coordinates alone and a linear function of time.

If the action integral is expressed in the form (9.6), the principle of least action becomes

$$\Delta \int_{t_1}^{t_2} T \, dt = 0. \tag{9.17}$$

In the special case in which the system kinetic energy is constant, e.g., for the moment-free rigid body of Sec. 4.10, Eq. (9.17) reduces to

$$\Delta(t_2 - t_1) = 0, \tag{9.18}$$

which indicates that of all the paths a conservative system can take between two given points the actual path is the one for which the transit time has a stationary value. On physical grounds it can be concluded that the stationary value is actually a minimum.

Equation (9.6) can be expressed in the form of a line integral, and if this is done, the principle, Eq. (9.17), reduces to the problem of finding the path for which the integral is rendered stationary. In this form the principle is known as the *Jacobi principle*, and it can be shown to reduce to the problem of finding a *geodesic* in a given Riemannian space, namely, the shortest path between two points in that space. (For a discussion of the Jacobi principle, see Ref. 4, pp. 132–140.)

9.2 CONTACT TRANSFORMATIONS

From Sec. 9.1 we can write the Hamilton principal function associated with an n-degree-of-freedom dynamical system in the form

$$S[q(t_1), t_1, q(t_2), t_2] = \int_{t_1}^{t_2} L(q, \dot{q}, t) \, dt, \tag{9.19}$$

where q and $\dot{q}$ represent n-vectors with components consisting of the generalized coordinates q_i and the generalized velocities $\dot{q}_i$, respectively. The principal function S can be regarded as a function of n components $q_{i1} = q_i(t_1)$, n components $q_{i2} = q_i(t_2)$, as well as t_1 and t_2, for a total of $2n + 2$ arguments.

It will prove of interest to examine how S varies as the $2n + 2$ arguments vary. First we concentrate on the variations in the terminal coordinates q_{i1} and q_{i2}. From Eqs. (9.10) and (9.19), we obtain

$$\delta S = \sum_{i=1}^{n} \frac{\partial L}{\partial \dot{q}_i} \delta q_i \Big|_{t_1}^{t_2} = \sum_{i=1}^{n} (p_{i2}\, \delta q_{i2} - p_{i1}\, \delta q_{i1}), \tag{9.20}$$

from which we conclude that

$$\frac{\partial S}{\partial q_{i1}} = -p_{i1}, \qquad \frac{\partial S}{\partial q_{i2}} = p_{i2}, \qquad i = 1, 2, \ldots, n, \tag{9.21}$$

where p_{i1} and p_{i2} represent the momentum components at times t_1 and t_2, respectively. Now we consider the variation in t_2. Referring to Eq. (9.19), denoting by L_2 the value of the system Lagrangian at time t_2, and considering Eqs. (9.21), we can write

$$L_2 = \frac{\partial S}{\partial t_2} + \sum_{i=1}^{n} \frac{\partial S}{\partial q_{i2}} \frac{\partial q_{i2}}{\partial t_2} = \frac{\partial S}{\partial t_2} + \sum_{i=1}^{n} p_{i2} \dot{q}_{i2}. \tag{9.22}$$

In view of the definition of the Hamiltonian, Eq. (9.22) becomes

$$\frac{\partial S}{\partial t_2} = -\left(\sum_{i=1}^{n} p_{i2} \dot{q}_{i2} - L_2\right) = -H_2, \tag{9.23}$$

where H_2 is the value of the Hamiltonian at $t = t_2$. Similarly

$$-L_1 = \frac{\partial S}{\partial t_1} + \sum_{i=1}^{n} \frac{\partial S}{\partial q_{i1}} \frac{\partial q_{i1}}{\partial t_1} = \frac{\partial S}{\partial t_1} - \sum_{i=1}^{n} p_{i1} \dot{q}_{i1}, \tag{9.24}$$

from which it follows that

$$\frac{\partial S}{\partial t_1} = H_1, \tag{9.25}$$

in which H_1 represents the Hamiltonian at $t = t_1$. Hence, allowing variations in all the $2n + 2$ arguments, we obtain the expression

$$dS = \sum_{i=1}^{n} p_{i2}\, dq_{i2} - \sum_{i=1}^{n} p_{i1}\, dq_{i1} - H_2\, dt_2 + H_1\, dt_1. \tag{9.26}$$

Formula (9.26) has far-reaching implications, but before discussing them we shall examine some of the properties of the principal function.

From Eqs. (9.21) we conclude that if the principal function S and the initial conditions q_{i1}, p_{i1} at $t = t_1$ are known, then it is possible to use these equations and solve for the motion q_{i2}, p_{i2} at any other time $t = t_2$. Hence, Eqs. (9.21) can be regarded as the integrals of Hamilton's equations, furnishing the solution of these equations in the phase space.

We recall from Sec. 2.10 that if the Lagrangian L does not depend explicitly on time, then the system possesses a Jacobi integral in the form of the Hamiltonian, $H_1 = H_2 = H$. In this case Eq. (9.26) reduces to

$$dS = \sum_{i=1}^{n} p_{i2}\, dq_{i2} - \sum_{i=1}^{n} p_{i1}\, dq_{i1} - H\, d(t_2 - t_1), \qquad (9.27)$$

from which we conclude that the time enters into S only in the form of the difference $t_2 - t_1$.

Since the vector components p_{i2} are independent of the vector components q_{i1}, there cannot be any relation connecting them, which carries the implication that the Jacobian of the transformation must be different from zero

$$\frac{\partial(p_{i2})}{\partial(q_{j1})} = \frac{\partial(p_{12}, p_{22}, \ldots, p_{n2})}{\partial(q_{11}, q_{21}, \ldots, q_{n1})} = \begin{vmatrix} \frac{\partial p_{12}}{\partial q_{11}} & \frac{\partial p_{12}}{\partial q_{21}} & \cdots & \frac{\partial p_{12}}{\partial q_{n1}} \\ \frac{\partial p_{22}}{\partial q_{11}} & \frac{\partial p_{22}}{\partial q_{21}} & \cdots & \frac{\partial p_{22}}{\partial q_{n1}} \\ \cdots & \cdots & \cdots & \cdots \\ \frac{\partial p_{n2}}{\partial q_{11}} & \frac{\partial p_{n2}}{\partial q_{21}} & \cdots & \frac{\partial p_{n2}}{\partial q_{n1}} \end{vmatrix} \neq 0. \qquad (9.28)$$

In view of Eqs. (9.21), however, Eq. (9.28) can be expressed compactly as

$$\left| \frac{\partial^2 S}{\partial q_{j1}\, \partial q_{i2}} \right| \neq 0. \qquad (9.29)$$

Next let us consider the transformation from the coordinates and momenta at time t_1 to those at time t_2. Regarding the times t_1 and t_2 as fixed, the transformation can be written in the functional form

$$\begin{aligned} q_{i2} &= q_{i2}(q_{11}, q_{21}, \ldots, q_{n1}, p_{11}, p_{21}, \ldots, p_{n1}), \\ p_{i2} &= p_{i2}(q_{11}, q_{21}, \ldots, q_{n1}, p_{11}, p_{21}, \ldots, p_{n1}), \end{aligned} \quad i = 1, 2, \ldots, n. \qquad (9.30)$$

We shall write the corresponding Jacobian of the transformation symbolically as

$$\frac{\partial(q_{i2}, p_{i2})}{\partial(q_{j1}, p_{j1})} = \frac{\partial(q_{12}, q_{22}, \ldots, q_{n2}, p_{12}, p_{22}, \ldots, p_{n2})}{\partial(q_{11}, q_{21}, \ldots, q_{n1}, p_{11}, p_{21}, \ldots, p_{n1})}. \qquad (9.31)$$

Transformation Theory. The Hamilton-Jacobi Equation

The value of the Jacobian does not change if we write it as follows

$$\frac{\partial(q_{i2},p_{i2})}{\partial(q_{j1},p_{j1})} = \frac{\dfrac{\partial(q_{i2},p_{i2})}{\partial(q_{j1},q_{i2})}}{\dfrac{\partial(q_{j1},p_{j1})}{\partial(q_{j1},q_{i2})}}. \tag{9.32}$$

But

$$\frac{\partial(q_{i2},p_{i2})}{\partial(q_{j1},q_{i2})} = \begin{vmatrix} \dfrac{\partial q_{12}}{\partial q_{11}} & \dfrac{\partial q_{12}}{\partial q_{21}} & \cdots & \dfrac{\partial q_{12}}{\partial q_{12}} & \dfrac{\partial q_{12}}{\partial q_{22}} & \cdots \\[4pt] \dfrac{\partial q_{22}}{\partial q_{11}} & \dfrac{\partial q_{22}}{\partial q_{21}} & \cdots & \dfrac{\partial q_{22}}{\partial q_{12}} & \dfrac{\partial q_{22}}{\partial q_{22}} & \cdots \\[2pt] \cdots & \cdots & \cdots & \cdots & \cdots & \cdots \\[2pt] \dfrac{\partial p_{12}}{\partial q_{11}} & \dfrac{\partial p_{12}}{\partial q_{21}} & \cdots & \dfrac{\partial p_{12}}{\partial q_{12}} & \dfrac{\partial p_{12}}{\partial q_{22}} & \cdots \\[4pt] \dfrac{\partial p_{22}}{\partial q_{11}} & \dfrac{\partial p_{22}}{\partial q_{21}} & \cdots & \dfrac{\partial p_{22}}{\partial q_{12}} & \dfrac{\partial p_{22}}{\partial q_{22}} & \cdots \\[2pt] \cdots & \cdots & \cdots & \cdots & \cdots & \cdots \end{vmatrix}$$

$$= (-1)^n \begin{vmatrix} \dfrac{\partial p_{12}}{\partial q_{11}} & \dfrac{\partial p_{12}}{\partial q_{21}} & \cdots & \dfrac{\partial p_{12}}{\partial q_{n1}} \\[4pt] \dfrac{\partial p_{22}}{\partial q_{11}} & \dfrac{\partial p_{22}}{\partial q_{21}} & \cdots & \dfrac{\partial p_{22}}{\partial q_{n1}} \\[2pt] \cdots & \cdots & \cdots & \cdots \\[2pt] \dfrac{\partial p_{n2}}{\partial q_{11}} & \dfrac{\partial p_{n2}}{\partial q_{21}} & \cdots & \dfrac{\partial p_{n2}}{\partial q_{n1}} \end{vmatrix} = (-1)^n \frac{\partial(p_{i2})}{\partial(q_{j1})}$$

$$= (-1)^n \left| \frac{\partial^2 S}{\partial q_{j1}\, \partial q_{i2}} \right|. \tag{9.33}$$

Similarly, it can be shown that

$$\frac{\partial(q_{j1},p_{j1})}{\partial(q_{j1},q_{i2})} = \frac{\partial(p_{i1})}{\partial(q_{j2})} = (-1)^n \left| \frac{\partial^2 S}{\partial q_{j2}\, \partial q_{i1}} \right|. \tag{9.34}$$

Introducing Eqs. (9.33) and (9.34) into (9.32) and recalling that the value of a determinant does not change if the rows and columns are interchanged, we conclude that

$$\frac{\partial(q_{i2},p_{i2})}{\partial(q_{j1},p_{j1})} = \frac{(-1)^n \left| \dfrac{\partial^2 S}{\partial q_{j1}\, \partial q_{i2}} \right|}{(-1)^n \left| \dfrac{\partial^2 S}{\partial q_{j2}\, \partial q_{i1}} \right|} = 1. \tag{9.35}$$

Hence, we conclude that *the Jacobian determinant of the transformation from q_{i1}, p_{i1} to q_{i2}, p_{i2} defined by the integrals of Hamilton's equations is equal to unity*. This statement is known as *Liouville's theorem*.

Next we shall regard the time t_1 as the initial time and, for simplicity, take it as equal to zero, $t_1 = t_0 = 0$. On the other hand, we consider the time t_2 as any arbitrary time, $t_2 = t$. Hence, the quantities q_{i1}, p_{i1} represent the initial conditions q_{i0}, p_{i0}, whereas the quantities q_{i2}, p_{i2} represent the solution q_i, p_i of Hamilton's equations. The relation between the two sets of canonical variables can be written in the form

$$q_i(t) = q_i(q_{10}, q_{20}, \ldots, q_{n0}, p_{10}, p_{20}, \ldots, p_{n0}, t),$$
$$p_i(t) = p_i(q_{10}, q_{20}, \ldots, q_{n0}, p_{10}, p_{20}, \ldots, p_{n0}, t), \quad i = 1, 2, \ldots, n, \quad (9.36)$$

which may be regarded as a mapping of the phase space onto itself. Moreover, Eq. (9.26) reduces to

$$dS = \sum_{i=1}^{n} p_i \, dq_i - \sum_{i=1}^{n} p_{i0} \, dq_{i0} - H \, dt, \qquad (9.37)$$

where $S = S(q, q_0, t)$ is the principal function. Eliminating the quantities p_{i0} from Eqs. (9.36), the momenta p_i can be expressed in terms of q_{i0}, q_i, and t, so that the Hamiltonian can be written in the general form $H = H(q, q_0, t)$.

The Jacobian of the transformation (9.36) is equal to unity and, clearly, cannot vanish. Hence, we can solve Eqs. (9.36) for the quantities q_{i0}, p_{i0} in terms of q_i, p_i, and t. The result can be expressed in the compact form

$$q_{i0} = q_{i0}(q, p, t), \qquad p_{i0} = p_{i0}(q, p, t), \qquad i = 1, 2, \ldots, n. \qquad (9.38)$$

Expressions (9.38) can be regarded as the inverse transformation associated with transformation (9.36).

Transformation (9.36) is such that, for a fixed value of time, the expression

$$\sum_{i=1}^{n} p_i \, dq_i - \sum_{i=1}^{n} p_{i0} \, dq_{i0} \qquad (9.39)$$

forms a perfect differential, which is a property shared by all transformations satisfying Hamilton's equations. A transformation possessing this property is referred to as a *contact transformation*. If expression (9.39) vanishes identically, the transformation is said to be a *homogeneous contact transformation*.

The motion of a dynamical system can be interpreted as being defined by a continuous series of contact transformations in the phase space. To gain perspective, it should be pointed out that transformation (9.36) represents the same solution as the one first introduced in Sec. 5.1 in the form of Eqs. (5.6). Hence, the concepts presented there are equally valid here.

Transformation Theory. The Hamilton-Jacobi Equation

The contact transformations constitute a class of transformations possessing what is referred to as the *group property*. The properties characterizing the group are as follows: (1) the identity transformation belongs to this class; (2) two successive transformations are commutative, and the result is also a contact transformation; (3) two contact transformations satisfy the associative law; and (4) the inverse of each transformation is also a contact transformation.

It is possible to obtain the motion of a dynamical system from a single function, namely, the principal function S. To show this, we write

$$dS = \sum_{i=1}^{n} \frac{\partial S}{\partial q_i} dq_i + \sum_{i=1}^{n} \frac{\partial S}{\partial q_{i0}} dq_{i0} + \frac{\partial S}{\partial t} dt, \quad (9.40)$$

introduce the result in Eq. (9.37), compare coefficients, and obtain the formulas defining the contact transformation in question in the form

$$p_i = \frac{\partial S}{\partial q_i}, \quad p_{i0} = -\frac{\partial S}{\partial q_{i0}}, \quad i = 1, 2, \ldots, n,$$

$$H = -\frac{\partial S}{\partial t}. \quad (9.41)$$

We note that the same equations could be derived from Eqs. (9.21) and (9.23). The problem of producing the function S still remains.

9.3 FURTHER EXTENSIONS OF THE CONCEPT OF CONTACT TRANSFORMATIONS

In Chaps. 5 to 7 we had an opportunity to ascertain some of the advantages of Hamiltonian mechanics as opposed to Lagrangian mechanics. In particular, we mentioned the advantages in the phase space representation of the solutions of the equations of motion of the system, as well as in the stability analysis of these solutions in the neighborhood of equilibrium points. However, closed-form solutions of the equations of motion are just as elusive in the Hamiltonian formulation as in the Lagrangian one, since no direct methods for the integration of these equations are available. An indirect way of solving the equations of motion of the system is by means of coordinate transformations. Indeed a proper choice of coordinates can uncover the so-called ignorable coordinates, leading to motion integrals. Hence, it is natural to seek coordinate transformations capable of producing ignorable coordinates. Not all transformations are acceptable, however, but only those which preserve the structure of the equations of motion. Whereas solutions by coordinate transformations are equally applicable to the Lagrangian mechanics, the Hamiltonian mechanics proves superior in this respect. Since in the Hamiltonian mechanics both the generalized coordinates and momenta are regarded as independent, we have at our disposal twice the number of variables as in the Lagrangian formulation, which provides us with a larger choice of transformations. Moreover, in the Hamiltonian formulation there are established methods for uncovering ignorable coordinates, as opposed to the Lagrangian formulation, where no such methods are available.

From Sec. 2.10 we recall that if in the two-body central force problem the motion is described in terms of the cartesian coordinates $q_1 = x$, $q_2 = y$, no ignorable coordinate is available. On the other hand, the use of the polar coordinates $Q_1 = r$ and $Q_2 = \theta$ reveals that θ is an ignorable coordinate and the associated momentum $P_2 = P_\theta$ is conserved. The relation between the two sets of generalized coordinates is

$$r = (x^2 + y^2)^{\frac{1}{2}}, \qquad \theta = \tan^{-1}\frac{y}{x}, \qquad (9.42)$$

which can be regarded as a coordinate transformation of the form

$$Q_i = Q_i(q_1, q_2, \ldots, q_n), \qquad i = 1, 2, \ldots, n, \qquad (9.43)$$

representing a mapping of the Lagrangian n-dimensional configuration space q onto the n-dimensional configuration space Q. In our particular case $n = 2$. Note that in writing Eqs. (9.43) we have adopted the widely accepted practice of denoting the new coordinates by capital letters. A transformation of the type (9.43) is referred to as a *point transformation*, because to a definite point in the q-space there corresponds a definite point in the Q-space. Since the structure of Lagrange's equations of motion does not depend on the generalized coordinates, it is clear that the equations are invariant with respect to a point transformation.

Next we wish to consider a more general type of transformation, in which not only the coordinates but also the momenta are transformed. This transformation can be written in the general form

$$Q_i = Q_i(q,p,t), \qquad P_i = P_i(q,p,t), \qquad i = 1, 2, \ldots, n, \qquad (9.44)$$

where q and p denote n-vectors with components $q_1, q_2, \ldots, q_n$ and $p_1, p_2, \ldots, p_n$, respectively. The old coordinates q_i, p_i satisfy the Hamiltonian equations

$$\dot{q}_i = \frac{\partial H}{\partial p_i}, \qquad \dot{p}_i = -\frac{\partial H}{\partial q_i}, \qquad i = 1, 2, \ldots, n, \qquad (9.45)$$

in which $H = H(q,p,t)$ is the Hamiltonian of the system. We shall be interested only in those transformations which preserve the structure of the Hamiltonian mechanics. This is equivalent to requiring that there exist a function $K = K(Q,P,t)$, where Q and P are new n-vectors with components

Q_i and P_i, respectively, and that these components satisfy the set of Hamiltonian equations

$$\dot{Q}_i = \frac{\partial K}{\partial P_i}, \qquad \dot{P}_i = -\frac{\partial K}{\partial Q_i}, \qquad i = 1, 2, \ldots, n, \qquad (9.46)$$

in which K is the Hamiltonian corresponding to the new coordinates.

The satisfaction of Eqs. (9.45) implies that the old coordinates satisfy the Hamilton principle, Eq. (2.80). Since the Hamiltonian and the Lagrangian are related by $H = \sum_{i=1}^{n} p_i \dot{q}_i - L$, Hamilton's principle can also be written in the form

$$\delta \int_{t_1}^{t_2} \left[\sum_{i=1}^{n} p_i \dot{q}_i - H(q,p,t) \right] dt = 0. \qquad (9.47)$$

In a similar way, the new coordinates must satisfy

$$\delta \int_{t_1}^{t_2} \left[\sum_{i=1}^{n} P_i \dot{Q}_i - K(Q,P,t) \right] dt = 0. \qquad (9.48)$$

Now let us consider an arbitrary function $F = F(q,Q,t)$. Assuming that the variations in the variables q_i and Q_i are zero at the end points t_1 and t_2, we have

$$\delta \int_{t_1}^{t_2} \frac{\partial F}{dt} dt = \delta\{F[q(t_2),Q(t_2),t_2] - F[q(t_1),Q(t_1),t_1]\} = 0. \qquad (9.49)$$

It follows that the integrands in Eqs. (9.47) and (9.48) are not necessarily equal to one another but are related by

$$\sum_{i=1}^{n} p_i \, dq_i - \sum_{i=1}^{n} P_i \, dQ_i - (H - K)\, dt = dF. \qquad (9.50)$$

Comparing Eq. (9.50) with Eq. (9.37), we conclude that Eqs. (9.44) represent a contact transformation. Because the form of the canonical equations is invariant with respect to the group of contact transformations, the transformation is often referred to as a *canonical transformation*. We shall use the terminologies interchangeably. Note that, whereas Eqs. (9.36) represent a transformation between variables at two different times, Eqs. (9.44) represent a transformation not necessarily involving the time. In fact, some of the most important applications of transformation (9.44) do not involve the time explicitly. We note that the point transformation is a special case of a canonical transformation.

Transformation (9.44) is fully defined by the function F. Indeed, writing

$$dF = \sum_{i=1}^{n} \frac{\partial F}{\partial q_i} dq_i + \sum_{i=1}^{n} \frac{\partial F}{\partial Q_i} dQ_i + \frac{\partial F}{\partial t} dt \qquad (9.51)$$

and comparing coefficients in Eqs. (9.50) and (9.51), we obtain the relations

$$p_i = \frac{\partial F}{\partial q_i}, \qquad i = 1, 2, \ldots, n, \qquad (9.52)$$

$$P_i = -\frac{\partial F}{\partial Q_i}, \qquad i = 1, 2, \ldots, n, \qquad (9.53)$$

and
$$K = H + \frac{\partial F}{\partial t}. \qquad (9.54)$$

If a function F can be found, Eqs. (9.52) to (9.54) can be used to obtain Q_i, P_i ($i = 1, 2, \ldots, n$), and K, thus determining the transformation completely. For this reason F is called the *generating function* of the transformation. In the event F is not an explicit function of time, the transformed Hamiltonian K is equal to the original Hamiltonian H. The remarkable feature of this procedure is that if the *generating function F is given, the solution of the entire dynamical problem consists of differentiations and eliminations.*

It turns out that the indicated functional dependence of F is not the only possible one. Indeed for a valid transformation it is necessary only that F be a function of $2n$ independent variables and time such that n of the variables are old and n of them are new. Hence, the generating function can assume one of the four forms

$$F_1(q,Q,t), \qquad F_2(q,P,t), \qquad F_3(p,Q,t), \qquad F_4(p,P,t). \qquad (9.55)$$

Of course, the first of functions (9.55) is precisely the function F already explored.

In view of Eqs. (9.53), the generating function F_2 can be written in the form of the Legendre transformation

$$F_2(q,P,t) = F(q,Q,t) + \sum_{i=1}^{n} Q_i P_i. \qquad (9.56)$$

Introduction of Eq. (9.56) into Eq. (9.50) yields

$$\sum_{i=1}^{n} p_i \, dq_i + \sum_{i=1}^{n} Q_i \, dP_i - (H - K) \, dt = dF_2, \qquad (9.57)$$

leading to

$$p_i = \frac{\partial F_2}{\partial q_i}, \qquad i = 1, 2, \ldots, n, \qquad (9.58)$$

$$Q_i = \frac{\partial F_2}{\partial P_i}, \qquad i = 1, 2, \ldots, n, \qquad (9.59)$$

and
$$K = H + \frac{\partial F_2}{\partial t}. \qquad (9.60)$$

The equations corresponding to the generating functions F_3 and F_4 can be obtained in a similar manner by using other types of Legendre transformations.

The purpose of the transformations is to produce new ignorable canonical variables, which is the case if some or all of the new canonical variables Q_i and P_i are absent from the new Hamiltonian function K.

It is not difficult to show that Liouville's theorem holds also for the general contact transformation. In this case Eq. (9.35) reduces to the form

$$\frac{\partial(Q,P)}{\partial(q,p)} = \frac{\partial(Q_1, Q_2, \ldots, Q_n, P_1, P_2, \ldots, P_n)}{\partial(q_1, q_2, \ldots, q_n, p_1, p_2, \ldots, p_n)} = 1, \qquad (9.61)$$

which will be referred to as the *extended Liouville theorem*.

The canonical transformations possess the group property, as defined in Sec. 9.2. We note that in the case of the identity transformation the generating function is not an explicit function of time. For two successive transformations, the generating function of the resultant transformation is the sum of the generating functions of the individual transformations, and this resultant depends only on the initial and final coordinates and not on the intermediate ones. Moreover, with regard to the inverse transformation, in transforming from the new to the old canonical coordinates the generating function simply changes sign.

As an illustration of a canonical transformation, let us consider the generating function

$$F_2(q,P,t) = \sum_{i=1}^{n} q_i P_i, \qquad (9.62)$$

which is of a special type since F_2 is not an explicit function of time. Introducing Eq. (9.62) into Eqs. (9.58) to (9.60), we obtain

$$p_i = \frac{\partial F_2}{\partial q_i} = P_i, \qquad Q_i = \frac{\partial F_2}{\partial P_i} = q_i, \qquad i = 1, 2, \cdots, n,$$
$$K = H, \qquad (9.63)$$

which is recognized as the *identity transformation*.

To justify the statement that point transformations are special cases of canonical transformations, we consider a generating function F_2 of the more general form

$$F_2(q,P,t) = \sum_{i=1}^{n} f_i(q)P_i, \qquad (9.64)$$

so that the first $2n$ of Eqs. (9.63) must be replaced by

$$p_k = \frac{\partial F_2}{\partial q_k} = \sum_{i=1}^{n} \frac{\partial f_i(q)}{\partial q_k} P_i,$$
$$Q_k = \frac{\partial F_2}{\partial P_k} = f_k(q), \qquad (9.65)$$

with the last of Eqs. (9.63) remaining unaffected. The second half of Eqs. (9.65) is clearly recognized as the equivalent of Eqs. (9.43). A special case of a point transformation is the orthogonal transformation, for which the functions f_k are defined by

$$f_k(q) = \sum_{j=1}^{n} a_{kj} q_j, \qquad (9.66)$$

where the constants a_{jk} satisfy the relations

$$\sum_{j=1}^{n} a_{ij} a_{kj} = \delta_{ik}, \qquad (9.67)$$

in which δ_{ik} is the Kronecker delta. Introducing Eqs. (9.66) into the first half of Eqs. (9.65), we obtain

$$p_k = \sum_{i=1}^{n} \sum_{j=1}^{n} a_{ij} \frac{\partial q_j}{\partial q_k} P_i = \sum_{i=1}^{n} a_{ik} P_i. \qquad (9.68)$$

Multiplying Eqs. (9.68) through by a_{jk}, summing up over k, and recalling Eqs. (9.67), we arrive at

$$P_j = \sum_{k=1}^{n} a_{jk} p_k. \qquad (9.69)$$

which indicates that the new conjugate momenta also transform orthogonally.

Next let us consider the transformation

$$F(q,Q,t) = \sum_{i=1}^{n} q_i Q_i. \qquad (9.70)$$

Since F does not depend explicitly on t, Eq. (9.54) reduces to $K = H$, whereas Eqs. (9.52) and (9.53) become

$$p_i = Q_i, \qquad P_i = -q_i, \qquad (9.71)$$

so that the transformation leads to a change in the status of the coordinates and momenta in the new variables as opposed to the old ones, with exception of the sign difference. This clearly indicates that in the Hamiltonian mechanics the coordinates and momenta attain equal status, and it is no longer meaningful to make that distinction.

Before obtaining the solution of a system of canonical equations by means of a contact transformation, we must have the generating function. The first reaction is to conclude that there is no particular advantage in using this method of solution as opposed to having to solve the Hamiltonian equations. Such a conclusion would be premature, however, because there are indeed established ways of producing a certain generating function F, and no guessing game is involved. It turns out that such an F must satisfy a partial differential equation, known as the Hamilton-Jacobi equation, which is derived by insisting that the new canonical variables be ignorable. We shall discuss this subject later.

Example 9.1

The Hamiltonian associated with a simple linear oscillator has the form

$$H = \frac{1}{2}\left(\frac{p^2}{m} + m\omega^2 q^2\right), \qquad \omega = \sqrt{\frac{k}{m}}, \qquad (a)$$

where the symbols m, k, and ω are as defined in Sec. 1.6. It is not difficult to derive the associated Hamilton's equations and solve them. The solution is known to represent simple harmonic motion. We shall derive the same result by means of a contact transformation.

Let us consider the contact transformation generated by the function

$$F = \tfrac{1}{2} m\omega q^2 \cot Q. \qquad (b)$$

Introducing Eq. (b) into Eqs. (9.52) and (9.53), we obtain

$$p = \frac{\partial F}{\partial q} = m\omega q \cot Q,$$

$$P = -\frac{\partial F}{\partial Q} = \frac{1}{2}\frac{m\omega q^2}{\sin^2 Q}, \qquad (c)$$

leading to

$$q = \sqrt{\frac{2P}{m\omega}} \sin Q, \qquad p = \sqrt{2m\omega P} \cos Q. \qquad (d)$$

On the other hand, Eq. (9.54) yields

$$K = H = \omega P. \qquad (e)$$

The first of Hamilton's canonical equations, Eqs. (9.46), is

$$\dot{Q} = \frac{\partial K}{\partial P} = \omega = \text{const}, \qquad (f)$$

whereas the second of Eqs. (9.46) is identically satisfied by $P = \text{const}$. The solution of Eq. (f) is simply

$$Q = \omega t + \varphi, \qquad (g)$$

where φ is a phase angle depending on the initial conditions. Introducing Eqs. (e) and (g) into the first of Eqs. (d), we obtain the solution

$$q = \sqrt{\frac{2H}{m\omega^2}} \sin(\omega t + \varphi). \qquad (h)$$

As expected, the amplitude of the simple harmonic motion depends on the energy imparted initially to the system, as reflected by H.

The selection of the generating function F, Eq. (b), may seem a little artificial at this point, but later in this chapter we shall examine more systematic ways of deriving generating functions.

9.4 INTEGRAL INVARIANTS

Contact transformations possess the property of leaving the form of Hamilton's equations of motion invariant. It turns out that certain definite integrals also possess the invariance property, for which reason they are called *integral invariants*.

Let us again consider Eq. (9.26) and write it in the form

$$dS = \left(\sum_{i=1}^{n} p_i\, dq_i - H\, dt\right)_{t=t_2} - \left(\sum_{i=1}^{n} p_i\, dq_i - H\, dt\right)_{t=t_1}. \qquad (9.72)$$

Next let us denote by γ_1 a simple closed curve in the configuration space (see Fig. 9.2), so that through each point of γ_1 there passes a curve C representing a solution of the system equations of motion. We denote by γ_2 another such curve, so that γ_1 and γ_2 are nowhere tangent to C. There is a one-to-one correspondence between the points $P_1(q_{i1},t_1)$ and $P_2(q_{i2},t_2)$ of γ_1 and γ_2, respectively. Hence, to a point $P_1'(q_{i1} + dq_{i1}, t_1 + dt_1)$ on γ_1 there corresponds a point $P_2'(q_{i2} + dq_{i2}, t_2 + dt_2)$ on γ_2. The difference between the values of the integral of the Lagrangian taken along C and C', respectively, defines the increment dS of the principal function $S(q_{i1},q_{i2},t_1,t_2)$

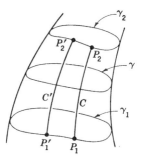

FIGURE 9.2

corresponding to the displacements P_1P_1', P_2P_2', as can be concluded from Eq. (9.72). Hence, the differential in the first pair of parentheses on the right side of Eq. (9.72) refers to the displacement P_2P_2', and a similar interpretation can be provided for the second parentheses.

Now let point P_1 traverse γ_1 until it returns to the original position, with the implication that P_2 likewise goes once around the closed curve γ_2. Because both P_1 and P_2 return to their original position, the net change in S is zero, so that an integration of Eq. (9.72) yields

$$\oint_{\gamma_2} \left(\sum_{i=1}^{n} p_i \, dq_i - H \, dt \right) = \oint_{\gamma_1} \left(\sum_{i=1}^{n} p_i \, dq_i - H \, dt \right). \tag{9.73}$$

Since the curves γ_1 and γ_2 are arbitrary, we can denote by γ a simple closed curve which is intersected only once by any curve C representing a motion integral, where γ and C are nowhere tangent to one another. Hence, Eq. (9.73) is equivalent to the statement

$$I = \oint_{\gamma} \left(\sum_{i=1}^{n} p_i \, dq_i - H \, dt \right) = \text{const.} \tag{9.74}$$

An integral whose value does not change when taken over a given domain, such as curve γ, is said to be an *integral invariant* of the system. If the domain of integration is closed, the integral is said to be *relative*, and if it is open, the integral is *absolute*. In view of these definitions, integral I is a relative integral invariant because curve γ is closed.

If we confine ourselves to surfaces of constant t, then integral I reduces to

$$I = I_1 = \oint_{\gamma} \sum_{i=1}^{n} p_i \, dq_i = \text{const}, \tag{9.75}$$

which is said to be an integral invariant of the *first order*. Equation (9.75) is sometimes referred to as *Poincaré's relative integral invariant*. The integral is also known as the *circulation integral*, and it represents the mathematical expression of *Helmholtz's circulation theorem*, according to which *vortices cannot be created or destroyed in an ideal fluid.*

The curve γ may be regarded as the boundary of a two-dimensional simply connected subspace S_2 of the configuration space. The subspace S_2 is identified completely by not more than two parameters. Denoting these parameters by u and v, the subspace S_2 can be represented parametrically by

$$q_i = q_i(u,v). \tag{9.76}$$

The displacement dq_i along γ has the expression

$$dq_i = \frac{\partial q_i}{\partial u} du + \frac{\partial q_i}{\partial v} dv, \tag{9.77}$$

so that integral (9.75) can be written as

$$I = \oint_\gamma \sum_{i=1}^n \left(p_i \frac{\partial q_i}{\partial u} du + p_i \frac{\partial q_i}{\partial v} dv \right). \tag{9.78}$$

Expression (9.78) represents a line integral which can be transformed into a surface integral by means of Stokes' theorem. Denoting by σ_2 the region in the uv plane which is bounded by the image of γ in that plane and assuming that the orientation of γ is properly defined, Stokes' theorem leads to

$$I = \iint_{\sigma_2} \sum_{i=1}^n \left[\frac{\partial}{\partial v} \left(p_i \frac{\partial q_i}{\partial u} \right) - \frac{\partial}{\partial u} \left(p_i \frac{\partial q_i}{\partial v} \right) \right] du\, dv. \tag{9.79}$$

Integral (9.79) can be written in a more familiar form. To this end, we recognize that the expression

$$\frac{\partial q_i}{\partial u} \frac{\partial p_i}{\partial v} - \frac{\partial q_i}{\partial v} \frac{\partial p_i}{\partial u} = \frac{\partial(q_i,p_i)}{\partial(u,v)} \tag{9.80}$$

represents the Jacobian of the transformation between any pair of canonical variables q_i, p_i and the parameters u, v. In view of Eqs. (9.80), it is easy to see that integral (9.79) can be written as

$$I = \iint_{\sigma_2} \sum_{i=1}^n \frac{\partial(q_i,p_i)}{\partial(u,v)} du\, dv = \text{const.} \tag{9.81}$$

A further transformation in the phase space is possible by noticing that Eqs. (9.76) and the values of p_i on S_2 define a simply connected two-dimensional subspace τ_2 of the phase space, where u and v are the parameters of τ_2.

Since the determinant $\partial(q_i,p_i)/\partial(u,v)$ represents the Jacobian of the transformation between q_i, p_i and u, v, it follows that (9.81) can be written as

$$I_2 = \iint_{\tau_2} \sum_{i=1}^{n} dq_i \, dp_i = \text{const}, \tag{9.82}$$

which represents an *integral invariant of second order*. Because the subspace τ_2 of the phase space is not closed, the integral invariant is absolute.

It turns out that there exist integral invariants of higher order. For example, it can be shown that

$$I_4 = \iiiint_{\tau_4} \sum_{i=1}^{n} \sum_{j=1}^{n} dq_i \, dq_j \, dp_i \, dp_j = \text{const} \tag{9.83}$$

is an *integral invariant of fourth order*, where τ_4 is a four-dimensional subspace of the phase space. Similarly, higher-order integral invariants exist through order $m = 2n$. This last one is of particular importance, so that we consider the integral

$$I_{2n}(t_1) = \int \cdots \int dq_{11} \, dq_{21} \cdots dq_{n1} \, dp_{11} \, dp_{21} \cdots dp_{n1}, \tag{9.84}$$

where q_{i1} and p_{i1} represent the coordinates and momenta at $t = t_1$, respectively. Next we consider

$$I_{2n}(t_2) = \int \cdots \int dq_{21} \, dq_{22} \cdots dq_{n2} \, dp_{21} \, dp_{22} \cdots dp_{n2}, \tag{9.85}$$

where q_{i2} and p_{i2} are the coordinates and momenta at $t = t_2$. But q_{i2}, p_{i2} are related to q_{i1}, p_{i1} by the contact transformation (9.30) with the Jacobian given by (9.31). By (9.35), the Jacobian is equal to unity, from which it follows that $I_{2n}(t_2) = I_{2n}(t_1)$. Since t_1 and t_2 are arbitrary, we must have

$$I_{2n} = \int \cdots \int dq_1 \, dq_2 \cdots dq_n \, dp_1 \, dp_2 \cdots dp_n = \text{const} \tag{9.86}$$

independently of time, where the integration takes place over any integrable volume of the phase space. Expression (9.86) indicates that *the volume of the phase space is invariant under contact transformations*. This statement is known as *Liouville's theorem*.

If the motion is visualized as a fluid flow in the phase space (see Sec. 5.1), this is equivalent to the statement that *the fluid is incompressible*.

9.5 THE LAGRANGE AND POISSON BRACKETS

The conditions for a transformation to be a contact transformation can be expressed in terms of well-known differential expressions called the *Lagrange brackets*. The Lagrange bracket is generally denoted by $[u,v]$, and its definition is

$$[u,v] = \sum_{i=1}^{n}\left(\frac{\partial q_i}{\partial u}\frac{\partial p_i}{\partial v} - \frac{\partial q_i}{\partial v}\frac{\partial p_i}{\partial u}\right) = \sum_{i=1}^{n}\frac{\partial(q_i,p_i)}{\partial(u,v)}. \quad (9.87)$$

Clearly, the Lagrange bracket is antisymmetric for

$$[u,v] = -[v,u]. \quad (9.88)$$

With definition (9.87) in mind, we can write Eq. (9.81) in the form

$$I = \iint_{\sigma_2} [u,v]\, du\, dv = \text{const.} \quad (9.89)$$

The invariance of the integral (9.89) is independent of the choice of parameters u and v. We recall that the invariance of the integral (9.89) is a direct result of the invariance of the circulation integral, Eq. (9.75). Since the invariance of the circulation integral reflects a property of contact transformations, it follows that *the Lagrange bracket $[u,v]$ is invariant under contact transformations*. Hence, the Lagrange bracket must be regarded as a *differential invariant*.

It turns out that Lagrange brackets are not the only differential invariants. Indeed there is another class of differential invariants which are intimately related to the Lagrange brackets and are known as the *Poisson brackets*. The Poisson bracket is defined by the expression

$$(u,v) = \sum_{i=1}^{n}\left(\frac{\partial u}{\partial q_i}\frac{\partial v}{\partial p_i} - \frac{\partial u}{\partial p_i}\frac{\partial v}{\partial q_i}\right). \quad (9.90)$$

It is evident that Poisson's bracket is antisymmetric, as it satisfies

$$(u,v) = -(v,u). \quad (9.91)$$

Next let us suppose that there are $2n$ independent functions $u_1, u_2, \ldots, u_{2n}$ of the variables q_i, p_i. Conversely, $q_1, q_2, \ldots, q_n, p_1, p_2, \ldots, p_n$ may be regarded as functions of $u_1, u_2, \ldots, u_{2n}$. From definitions (9.87) and (9.90), we suspect that there may be some connection between Lagrange's brackets $[u_r,u_s]$ and Poisson's brackets (u_r,u_s). Indeed, let us consider

$$\sum_{r=1}^{2n}(u_r,u_i)[u_r,u_j] = \sum_{r=1}^{2n}\sum_{k=1}^{n}\sum_{l=1}^{n}\left(\frac{\partial u_r}{\partial q_k}\frac{\partial u_i}{\partial p_k} - \frac{\partial u_r}{\partial p_k}\frac{\partial u_i}{\partial q_k}\right)\left(\frac{\partial q_l}{\partial u_r}\frac{\partial p_l}{\partial u_j} - \frac{\partial q_l}{\partial u_j}\frac{\partial p_l}{\partial u_r}\right). \quad (9.92)$$

But

$$\sum_{r=1}^{2n}\frac{\partial u_r}{\partial q_k}\frac{\partial q_l}{\partial u_r} = \sum_{r=1}^{2n}\frac{\partial u_r}{\partial p_k}\frac{\partial p_l}{\partial u_r} = \delta_{kl}, \quad (9.93)$$

Transformation Theory. The Hamilton-Jacobi Equation

where δ_{kl} is the Kronecker delta. Moreover

$$\sum_{r=1}^{2n} \frac{\partial u_r}{\partial q_k} \frac{\partial p_l}{\partial u_r} = \sum_{r=1}^{2n} \frac{\partial u_r}{\partial p_k} \frac{\partial q_l}{\partial u_r} = 0. \tag{9.94}$$

Hence, Eq. (9.92) reduces to

$$\sum_{r=1}^{2n} (u_r, u_i)[u_r, u_j] = \sum_{k=1}^{n} \left(\frac{\partial u_i}{\partial p_k} \frac{\partial p_k}{\partial u_j} + \frac{\partial u_i}{\partial q_k} \frac{\partial q_k}{\partial u_j} \right) = \delta_{ij}. \tag{9.95}$$

If we regard the Poisson bracket (u_r, u_i) as the element P_{ri} of a $2n \times 2n$ matrix $[P]$ and the Lagrange bracket $[u_r, u_j]$ as the element L_{rj} of the $2n \times 2n$ matrix $[L]$, then Eq. (9.95) can be written in the matrix form

$$[P]^T[L] = [1_{2n}], \tag{9.96}$$

where $[1_{2n}]$ is the unit matrix of order $2n$. Since the determinant of the product of two matrices is equal to the product of the determinants of the matrices, it follows from Eq. (9.96) that the determinants of the matrices $[P]$ and $[L]$ are the reciprocals of one another. It is clear that one type of bracket determines the other, so that if Lagrange brackets are invariant under contact transformations, Poisson brackets are also invariant.

The Lagrange and Poisson brackets can be used to test whether a given transformation is a contact transformation or not. To show this, we consider transformation (9.44). If transformation (9.44) is to be a contact transformation, then the new variables Q_i, P_i must be such that Eq. (9.50) is satisfied. Since the old variables q_i, p_i may be regarded as functions of the new variables, the necessary and sufficient condition that the transformation be a contact transformation is that for a fixed value of time the expression

$$\sum_{r=1}^{n} \left[\left(\sum_{i=1}^{n} p_i \frac{\partial q_i}{\partial Q_r} - P_r \right) dQ_r + \left(\sum_{i=1}^{n} p_i \frac{\partial q_i}{\partial P_r} \right) dP_r \right] \tag{9.97}$$

be a perfect differential. The conditions for this are

$$\frac{\partial}{\partial Q_s} \left(\sum_{i=1}^{n} p_i \frac{\partial q_i}{\partial Q_r} - P_r \right) = \frac{\partial}{\partial Q_r} \left(\sum_{i=1}^{n} p_i \frac{\partial q_i}{\partial Q_s} - P_s \right),$$

$$\frac{\partial}{\partial P_s} \left(\sum_{i=1}^{n} p_i \frac{\partial q_i}{\partial P_r} \right) = \frac{\partial}{\partial P_r} \left(\sum_{i=1}^{n} p_i \frac{\partial q_i}{\partial P_s} \right), \tag{9.98}$$

$$\frac{\partial}{\partial P_s} \left(\sum_{i=1}^{n} p_i \frac{\partial q_i}{\partial Q_r} - P_r \right) = \frac{\partial}{\partial Q_r} \left(\sum_{i=1}^{n} p_i \frac{\partial q_i}{\partial P_s} \right).$$

But Q_i and P_i are independent variables, so that conditions (9.98) can easily be shown to reduce to

$$[Q_r, Q_s] = 0, \qquad [P_r, P_s] = 0, \qquad [Q_r, P_s] = \delta_{rs}. \tag{9.99}$$

The necessary and sufficient condition that Eqs. (9.44) represent a contact transformation can also be reduced to the requirement that for a fixed time the expression

$$\sum_{r=1}^{n} \left[\left(\sum_{i=1}^{n} P_i \frac{\partial Q_i}{\partial q_r} - p_r \right) dq_r + \left(\sum_{i=1}^{n} P_i \frac{\partial Q_i}{\partial p_r} \right) dp_r \right] \tag{9.100}$$

be a perfect differential. In a manner similar to the one used to derive conditions (9.99), it can be shown that the requirement that expression (9.100) be a perfect differential is equivalent to the conditions

$$(Q_r, Q_s) = 0, \qquad (P_r, P_s) = 0, \qquad (Q_r, P_s) = \delta_{rs}. \tag{9.101}$$

9.6 INFINITESIMAL CONTACT TRANSFORMATIONS

It would be beneficial if we could assess the effect of a contact transformation on the canonical variables. When the changes in the variables are finite, this effect may prove to be intractable. A discussion of the effect is possible in the case of *infinitesimal contact transformations,* namely, when the new variables Q_i, P_i differ from the original variables q_i, p_i by small quantities whose squares and products are negligible. Hence, let us consider the transformation

$$Q_i = q_i + \delta q_i, \qquad P_i = p_i + \delta p_i, \qquad i = 1, 2, \ldots, n, \tag{9.102}$$

where $\delta q_i, \delta p_i$ represent infinitesimal changes in coordinates and momenta. Introducing expressions (9.102) into Eq. (9.50), we obtain

$$\sum_{i=1}^{n} dp_i \, \delta q_i - \sum_{i=1}^{n} dq_i \, \delta p_i - [H(q,p,t) - K(Q,P,t)] \, dt \\ = d\left[F(q,Q,t) + \sum_{i=1}^{n} p_i \, \delta q_i \right]. \tag{9.103}$$

But the left side of Eq. (9.103) involves only differentials of $q_i, p_i,$ and t, so that the right side of (9.103) must be a function of these variables. Hence, let us denote

$$F(q,Q,t) + \sum_{i=1}^{n} p_i \, \delta q_i = \epsilon G(q,p,t), \tag{9.104}$$

where ϵ is a small parameter, from which it follows that

$$K(Q,P,t) = H(q,p,t) + \epsilon \frac{\partial G(q,p,t)}{\partial t} \tag{9.105}$$

and that

$$\delta q_i = \epsilon \frac{\partial G(q,p,t)}{\partial p_i}, \qquad \delta p_i = -\epsilon \frac{\partial G(q,p,t)}{\partial q_i}. \tag{9.106}$$

Equations (9.106) furnish us with the infinitesimal changes in the canonical variables in terms of a single function $G(q,p,t)$. For this reason, $G(q,p,t)$ is said to be a *generator* of the infinitesimal contact transformation. It can be chosen as an arbitrary function of the variables q_i, p_i, and t, but in general it is taken to represent some dynamical quantity. Several possibilities are examined in the sequel.

Let us consider the case in which G is the system Hamiltonian, $G = H$, and ϵ is a small time interval, $\epsilon = \delta t$. Dividing Eqs. (9.106) by δt and letting $\delta t \to 0$, we obtain

$$\dot{q}_i = \frac{\partial H}{\partial p_i}, \qquad \dot{p}_i = -\frac{\partial H}{\partial q_i}, \tag{9.107}$$

which are recognized as the Hamilton canonical equations. It follows that the motion in the small time interval in question can be described by an infinitesimal contact transformation generated by the Hamiltonian. But contact transformations possess the group property, so that *the motion of the dynamical system can be regarded as a succession of infinitesimal contact transformations*. Hence, the motion of the system may be interpreted as a continuous mapping of the phase space onto itself. If q_0, p_0 are the initial values of the canonical variables q, p at time $t = t_0$, then the equations expressing the solution q, p in terms of q_0, p_0, t represent a contact transformation from q_0, p_0 to q, p, with t playing the role of a parameter.

If the arguments q_i, p_i of any function $f = f(q_1, q_2, \ldots, q_n, p_1, p_2, \ldots, p_n)$ are subjected to transformation (9.102), then the increment of the function can be written in the form

$$\delta f = f(q_i + \delta q_i, p_i + \delta p_i) - f(q_i, p_i) = \sum_{i=1}^{n} \left(\frac{\partial f}{\partial q_i} \delta q_i + \frac{\partial f}{\partial p_i} \delta p_i \right), \tag{9.108}$$

so that insertion of Eqs. (9.106) into (9.108) yields

$$\delta f = \epsilon \sum_{i=1}^{n} \left(\frac{\partial f}{\partial q_i} \frac{\partial G}{\partial p_i} - \frac{\partial f}{\partial p_i} \frac{\partial G}{\partial q_i} \right) = \epsilon(f, G). \tag{9.109}$$

Hence, the Poisson bracket (f, G) symbolically represents the infinite group of infinitesimal contact transformations of the $2n$ variables q_i, p_i.

A case of special interest is the one in which the function f is the Hamiltonian H, in which case Eq. (9.109) reduces to

$$\delta H = \epsilon(H,G). \tag{9.110}$$

In particular, if the generator of the infinitesimal contact transformation is a constant of the motion, $G = G(q,p) = \text{const}$, then the Poisson bracket (H,G) vanishes and so does the increment of the Hamiltonian. It follows that *if the generator of the infinitesimal contact transformation is a constant of the motion, then the transformation leaves the Hamiltonian invariant.* As an illustration of this statement, we consider the generating function

$$G(q,p) = p_s = \text{const}, \tag{9.111}$$

where the momentum p_s represents a constant of the motion. Introducing Eq. (9.111) into (9.110) and recalling that in this case H must be independent of the coordinate q_s, we obtain

$$\delta H = \epsilon \sum_{i=1}^{n} \frac{\partial H}{\partial q_i} \delta_{is} = \epsilon \frac{\partial H}{\partial q_s} = 0. \tag{9.112}$$

It follows that the Hamiltonian is invariant under an infinitesimal contact transformation involving an ignorable coordinate alone.

Next let us consider a generating function in the form of the angular momentum of a system of particles about the z axis. In cartesian coordinates, this can be written as

$$G = \sum_{j=1}^{n} (x_j p_{y_j} - y_j p_{x_j}). \tag{9.113}$$

Insertion of expression (9.113) into Eqs. (9.106) leads to

$$\delta x_i = \epsilon \frac{\partial G}{\partial p_{x_i}} = -\epsilon y_i, \quad \delta p_{x_i} = -\epsilon \frac{\partial G}{\partial x_i} = -\epsilon p_{y_i},$$
$$\delta y_i = \epsilon \frac{\partial G}{\partial p_{y_i}} = \epsilon x_i, \quad \delta p_{y_i} = -\epsilon \frac{\partial G}{\partial y_i} = \epsilon p_{x_i}, \tag{9.114}$$
$$\delta z_i = 0, \quad \delta p_{z_i} = 0.$$

Hence, Eqs. (9.102), written in cartesian coordinates, become

$$\begin{aligned} X_i &= x_i - \epsilon y_i, & P_{x_i} &= p_{x_i} - \epsilon p_{y_i}, \\ Y_i &= y_i + \epsilon x_i, & P_{y_i} &= p_{y_i} + \epsilon p_{x_i}, \\ Z_i &= z_i, & P_{z_i} &= p_{z_i}. \end{aligned} \tag{9.115}$$

Equations (9.115) can be easily interpreted as an infinitesimal rotation of the system about axis z, where ϵ represents a small angle of rotation $\delta\theta$. We note that the angular momentum components transform in the same way as the coordinates.

9.7 THE HAMILTON-JACOBI EQUATION

We have shown (Sec. 9.3) that a contact transformation from the old variables q_i, p_i to the new variables Q_i, P_i can be defined by the new Hamiltonian

$$K(Q,P,t) = H(q,p,t) + \frac{\partial F(q,Q,t)}{\partial t} \tag{9.116}$$

and that the new variables satisfy the canonical equations of motion

$$\dot{Q}_i = \frac{\partial K}{\partial P_i}, \qquad \dot{P}_i = -\frac{\partial K}{\partial Q_i}, \qquad i = 1, 2, \ldots, n. \tag{9.117}$$

Moreover, the variables q_i, p_i are related to the variables Q_i, P_i by means of the expressions

$$p_i = \frac{\partial F}{\partial q_i}, \qquad P_i = -\frac{\partial F}{\partial Q_i}, \qquad i = 1, 2, \ldots, n. \tag{9.118}$$

The interest lies in a transformation rendering the equations of motion into a form which is easy to integrate. Clearly if K is identically zero, then from Eqs. (9.117) we conclude that Q_i and P_i are constants of the motion

$$Q_i = \alpha_i = \text{const}, \qquad P_i = \beta_i = \text{const}, \qquad i = 1, 2, \ldots, n. \tag{9.119}$$

The $2n$ constants α_i, β_i are sufficient to describe the initial conditions and, in fact, may be regarded as being related to them. Moreover, for $K = 0$, the generating function must satisfy the differential equation

$$H(q,p,t) + \frac{\partial F}{\partial t} = 0. \tag{9.120}$$

In view of the first half of Eqs. (9.119), the generating function can be written

$$F = F(q_1, q_2, \ldots, q_n, \alpha_1, \alpha_2, \ldots, \alpha_n, t), \tag{9.121}$$

and Eqs. (9.118) become

$$p_i = \frac{\partial F}{\partial q_i}, \qquad \beta_i = -\frac{\partial F}{\partial \alpha_i}, \qquad i = 1, 2, \ldots, n. \tag{9.122}$$

Inserting the first half of Eqs. (9.122) into (9.120), we obtain

$$H\left(q_1, q_2, \ldots, q_n, \frac{\partial F}{\partial q_1}, \frac{\partial F}{\partial q_2}, \ldots, \frac{\partial F}{\partial q_n}, t\right) + \frac{\partial F}{\partial t} = 0. \tag{9.123}$$

Equation (9.123) represents a first-order partial differential equation in the $n + 1$ independent variables q_i, t, and is known as *the Hamilton-Jacobi equation*. Its solution requires $n + 1$ constants of integration. However, since F does not appear in the equation explicitly but only in the form of partial derivatives with respect to q_i and t, it follows that one of the $n + 1$ constants of integration is additive and may be ignored. The other n constants may be taken as the α_i, which justifies writing the complete solution of Eq. (9.123) in the form (9.121). If we can solve Eq. (9.123) for F, we can use Eqs. (9.122) and write the solution of the dynamical problem in terms of α_i, β_i, and t as

$$q_i = q_i(\alpha_1, \alpha_2, \ldots, \alpha_n, \beta_1, \beta_2, \ldots, \beta_n, t),$$
$$p_i = p_i(\alpha_1, \alpha_2, \ldots, \alpha_n, \beta_1, \beta_2, \ldots, \beta_n, t), \quad i = 1, 2, \ldots, n. \quad (9.124)$$

This requires that the Jacobian determinant $|\partial^2 F/\partial q_i \, \partial \alpha_j|$ be different from zero.

It will prove instructive to examine the nature of the generating function F. Since in our case Q_i ($i = 1, 2, \ldots, n$) are constant and in addition $K = 0$, Eq. (9.50) reduces to

$$dF = \sum_{i=1}^{n} p_i \, dq_i - H \, dt. \quad (9.125)$$

Recalling the relation between the Lagrangian L and the Hamiltonian H, Eq. (9.125) can be integrated, with the result

$$F = \int_{t_1}^{t} L \, dt, \quad (9.126)$$

where t_1 is the initial time. In view of definition (9.14), if we set $t_2 = t$ and $q_i(t_1) = \alpha_i$, we conclude that *the generating function F is simply the Hamilton principal function S*

$$S = S(q_1, q_2, \ldots, q_n, \alpha_1, \alpha_2, \ldots, \alpha_n, t). \quad (9.127)$$

In the sequel we shall use the symbol S instead of F when referring to the generating function. Hence, the canonical variables at any time are obtained from their initial values by means of a contact transformation generated by the principal function S. This conclusion, although interesting, is of no practical value because to obtain S by integrating the Lagrangian L with respect to time we must know the solution, which is the object of the analysis in the first place.

A case of particular interest is the one in which the Hamiltonian does not contain t explicitly, $H = H(q, p)$. In this case, we have a motion integral in the form of the Jacobi integral (see Sec. 2.10)

$$H = H(q, p) = h = \text{const}, \quad (9.128)$$

and the Hamilton-Jacobi equation reduces to

$$H\left(q_i, \frac{\partial S}{\partial q_i}\right) + \frac{\partial S}{\partial t} = 0. \quad (9.129)$$

The solution of Eq. (9.129) can be written in the form

$$S(q_i,\alpha_i,t) = W(q_i,\alpha_i) - ht, \tag{9.130}$$

where the function W must satisfy the differential equation

$$H\left(q_i,\frac{\partial W}{\partial q_i}\right) = h, \tag{9.131}$$

which is independent of time. We shall refer to Eq. (9.131) as the *modified Hamilton-Jacobi equation*.

The function W can be used to obtain the canonical variables q_i, p_i. Indeed, the first half of Eqs. (9.122) become simply

$$p_i = \frac{\partial W}{\partial q_i}, \qquad i = 1, 2, \ldots, n. \tag{9.132}$$

On the other hand, the second half of Eqs. (9.122) cannot be used directly because the constants α_i are not all independent. If they were, the function S would depend on $n + 1$ constants, namely, n constants α_i and h. Hence, h must be a function of α_i, $h = h(\alpha_i)$, where the explicit form of h depends on the nature of the constants α_i.

It is possible to choose one of the constants α_i, say α_1, to coincide with h

$$h = \alpha_1. \tag{9.133}$$

Then the second half of Eqs. (9.122) becomes

$$\beta_1 = -\frac{\partial S}{\partial \alpha_1} = t - \frac{\partial W}{\partial \alpha_1}, \tag{9.134}$$

$$\beta_i = -\frac{\partial S}{\partial \alpha_i} = -\frac{\partial W}{\partial \alpha_i}, \qquad i = 2, 3, \ldots, n. \tag{9.135}$$

Equation (9.134) implies that the time and the Hamiltonian are conjugate canonical variables. Whereas Eqs. (9.135) yield the path in the q-space without reference to time, Eq. (9.134) gives the relation between the position on the path and the time. Equations (9.132) complete the solution.

We note that by choosing $h = \alpha_1$, the symmetry of the relations for α_i is destroyed. To preserve the symmetry, let us choose

$$h = \sum_{i=1}^{n} \alpha_i, \tag{9.136}$$

so that Eqs. (9.134) and (9.135) are replaced by

$$\beta_i = t - \frac{\partial W}{\partial \alpha_i}, \qquad i = 1, 2, \ldots, n. \tag{9.137}$$

By contrast with Eqs. (9.135), Eqs. (9.137) give the motion of the system as functions of time.

Similar Hamilton-Jacobi theories can be formulated by means of other functions generating contact transformations, such as the generating functions F_2, F_3, and F_4 given by (9.55). The function F_2 turns out to be of particular interest, and we shall explore it in further detail. To this end, we recall from Sec. 9.3 that F_2 has the expression

$$F_2(q,P,t) = F(q,Q,t) + \sum_{i=1}^{n} Q_i P_i \qquad (9.138)$$

and that the contact transformation generated by F_2 is defined by

$$p_i = \frac{\partial F_2}{\partial q_i}, \qquad Q_i = \frac{\partial F_2}{\partial P_i}, \qquad i = 1, 2, \ldots, n. \qquad (9.139)$$

If the transformation is to produce new constant canonical variables Q_i, P_i, the new Hamiltonian must be zero, $K = 0$, in which case Eq. (9.60) reduces to

$$H\left(q_i, \frac{\partial F_2}{\partial q_i}, t\right) + \frac{\partial F_2}{\partial t} = 0. \qquad (9.140)$$

It is easy to see that this formulation is not substantially different from the one in terms of the generating function F, since, for constant Q_i and P_i, the functions F_2 and F differ only by an additive constant.

If H does not contain t explicitly, we can write

$$F_2(q,P,t) = W'(q,P) - ht, \qquad (9.141)$$

where h is the value of the constant Hamiltonian and

$$W'(q,P) = W(q,\alpha) + \sum_{i=1}^{n} Q_i P_i. \qquad (9.142)$$

This enables us to express the modified Hamilton-Jacobi equation in the form

$$H\left(q_i, \frac{\partial W'}{\partial q_i}\right) = h, \qquad (9.143)$$

so that the function F_2 also generates a transformation leading to a system for which the Hamiltonian is zero and the canonical variables are constants of the motion.

In the special case in which the Hamiltonian does not depend on t explicitly it is possible to use another simple transformation leading to the same modified Hamilton-Jacobi equation (9.143). This is the contact transformation generated by the time-independent function $W'(q,P)$ alone, and the properties of this transformation are different from the properties of the

Transformation Theory. The Hamilton-Jacobi Equation 359

transformation generated by F_2. In particular, we seek that generating function $W'(q,P)$ for which the new Hamiltonian K is equal to the old Hamiltonian H and for which all the new conjugate variables P_i are constant, although the new variables Q_i may not all be constant. Using such $W'(q,P)$ as the generating function, the transformation equations are

$$p_i = \frac{\partial W'}{\partial q_i}, \qquad Q_i = \frac{\partial W'}{\partial P_i}, \qquad i = 1, 2, \ldots, n. \tag{9.144}$$

The new variables satisfy the canonical equations of motion

$$\dot{Q}_i = \frac{\partial K}{\partial P_i}, \qquad \dot{P}_i = -\frac{\partial K}{\partial Q_i}, \qquad i = 1, 2, \ldots, n, \tag{9.145}$$

where the new Hamiltonian K is constant

$$K = H(q_i, p_i) = h. \tag{9.146}$$

Introducing the first half of Eqs. (9.144) into (9.146), the modified Hamilton-Jacobi equation becomes

$$H\left(q_i, \frac{\partial W'}{\partial q_i}\right) = h, \tag{9.147}$$

which is the same as the form (9.143).

The solution of Eq. (9.147) requires n constants of integration. Of course, we recall that h must depend on these constants. We denote the constants by α_i ($i = 1, 2, \ldots, n$) and choose them to coincide with the new variables P_i, so that the second half of Eqs. (9.145) can be written

$$\dot{P}_i = -\frac{\partial K}{\partial Q_i} = 0, \qquad P_i = \alpha_i = \text{const}, \qquad i = 1, 2, \ldots, n. \tag{9.148}$$

If h is taken to coincide with α_1, then

$$K = \alpha_1 \tag{9.149}$$

and the first half of Eqs. (9.145) becomes

$$\dot{Q}_i = \frac{\partial K}{\partial P_i} = \delta_{1i}, \qquad i = 1, 2, \ldots, n, \tag{9.150}$$

where δ_{1i} is the Kronecker delta, in which one of the indices is 1. Equations (9.150) have the solutions

$$Q_1 = t + \beta_1 = \frac{\partial W'}{\partial \alpha_1}, \tag{9.151}$$

$$Q_i = \beta_i = \frac{\partial W'}{\partial \alpha_i}, \qquad i = 2, 3, \ldots, n, \tag{9.152}$$

where β_i are constants. The only new canonical variable which is not a constant of the motion is Q_1, which is equal to the time t plus the constant β_1. This is another example where the time and the Hamiltonian are conjugate canonical variables. Equations (9.152) yield the path in the q-space, and Eq. (9.151) relates the position on that path with the time t. The first half of Eqs. (9.144) completes the solution by furnishing the conjugate variables p_i. The function W' generating the contact transformation (9.144) is sometimes referred to as *Hamilton's characteristic function*.

It is also possible to choose h as a function of all α_i, thus preserving the symmetry of the first half of Eqs. (9.145). We shall return to this subject later in this chapter.

The Hamilton-Jacobi method requires the solution of a single partial differential equation instead of $2n$ ordinary differential equations, namely, Hamilton's equations. This may not always represent an improvement, as there are no general methods for the solution of partial differential equations. There are cases, however, when the Hamilton-Jacobi equation can be separated into a sequence of ordinary differential equations whose solutions can be obtained by quadratures. Such systems, referred to as *separable*, are discussed in Sec. 9.8. We note that when the Hamiltonian does not depend explicitly on time, the system is separable in part, as the modified Hamilton-Jacobi equation is independent of time.

Example 9.2

Let us consider the harmonic oscillator of Example 9.1. The system Hamiltonian has the expression

$$H = \frac{1}{2}\left(\frac{p^2}{m} + m\omega^2 q^2\right), \qquad \omega = \sqrt{\frac{k}{m}}, \tag{a}$$

which does not contain the time explicitly. Hence, we shall be interested in working with the modified Hamilton-Jacobi equation (9.131), which in our case assumes the form

$$\frac{1}{2}\left[m\omega^2 q^2 + \frac{1}{m}\left(\frac{\partial W}{\partial q}\right)^2\right] = h = \alpha. \tag{b}$$

Solving Eq. (b) for $\partial W/\partial q$, we obtain

$$\frac{\partial W}{\partial q} = m^{\frac{1}{2}}(2\alpha - m\omega^2 q^2)^{\frac{1}{2}}, \tag{c}$$

yielding

$$W = m^{\frac{1}{2}} \int (2\alpha - m\omega^2 q^2)^{\frac{1}{2}} \, dq, \tag{d}$$

where the additive constant has been ignored. Using Eq. (9.134), we arrive at

$$\beta = t - \frac{\partial W}{\partial \alpha} = t - \frac{1}{\omega}\int \frac{dq}{(2\alpha/m\omega^2 - q^2)^{\frac{1}{2}}} = t - \frac{1}{\omega}\sin^{-1}\left(\frac{m\omega^2}{2\alpha}\right)^{\frac{1}{2}} q. \quad (e)$$

Solving Eq. (e) for q and recalling that $\alpha = h$, we have

$$q = \left(\frac{2h}{m\omega^2}\right)^{\frac{1}{2}} \sin \omega(t - \beta). \quad (f)$$

Moreover, Eq. (9.132) leads to

$$p = \frac{\partial W}{\partial q} = m\omega\left(\frac{2\alpha}{m\omega^2} - q^2\right)^{\frac{1}{2}} = (2hm)^{\frac{1}{2}} \cos \omega(t - \beta). \quad (g)$$

Letting $\varphi = -\omega\beta$ in Eq. (h) of Example 9.1, it is easy to see that the present solution is identical to the one obtained there.

9.8 SEPARABLE SYSTEMS

Let us confine ourselves to systems for which the Hamiltonian does not depend on time explicitly. Moreover, we shall assume that the principal function can be written

$$S(q_1, q_2, \ldots, q_n, \alpha_1, \alpha_2, \ldots, \alpha_n, t) = \sum_{i=1}^{n} W_i(q_i, \alpha_1, \alpha_2, \ldots, \alpha_n) - \alpha_1 t. \quad (9.153)$$

Then, if the modified Hamilton-Jacobi equation reduces to n differential equations of the form

$$H_i\left(q_i, \frac{\partial W}{\partial q_i}, \alpha_1, \alpha_2, \ldots, \alpha_i\right) = \alpha_{i+1}, \quad i = 1, 2, \ldots, n, \quad (9.154)$$

where each of these equations depends on the variable q_i alone, the system is said to be *separable* and the method of solution is referred to as the *separation of variables*. Note that the constant on the right side in the last of Eqs. (9.154) is to be replaced by α_n. Although in certain special cases it may be possible to use Stäckel's theorem (see Ref. 5, sec. 18.3) to test whether a system is separable or not, it is often simpler to assume a principal function in the form (9.153) and examine the resulting Hamilton-Jacobi equation.

Equations (9.154) represent a set of n ordinary differential equations of the type

$$\frac{\partial W_i}{\partial q_i} = f_i(q_i, \alpha_1, \alpha_2, \ldots, \alpha_n), \quad i = 1, 2, \ldots, n, \quad (9.155)$$

which can always be integrated. Not all the constants α_i enter into each individual equation. Generally these equations can be solved in sequence such that the solution for q_1 brings in a constant, the solution for q_2 adds another constant, and with each equation solved one more constant is introduced. To illustrate the procedure, let us write Eq. (9.131) as

$$H\left(q_i, \frac{\partial W}{\partial q_i}\right) = \alpha_1. \tag{9.156}$$

The assumption of separability implies that Eq. (9.156) can be segregated into one part in q_1 alone and another part containing the remaining variables. Hence, we must have

$$H_1\left(q_1, \frac{\partial W_1}{\partial q_1}, \alpha_1\right) = H_1^*\left(q_2, \ldots, q_n, \frac{\partial W_2}{\partial q_2}, \ldots, \frac{\partial W_n}{\partial q_n}, \alpha_1\right). \tag{9.157}$$

Since the left side of Eq. (9.157) is an expression in q_1 alone and the right side is an expression in the remaining q_i, the two expressions can be equal to one another if and only if they are both equal to a constant, say α_2. This argument leads to

$$H_1\left(q_1, \frac{\partial W_1}{\partial q_1}, \alpha_1\right) = \alpha_2, \tag{9.158}$$

$$H_1^*\left(q_2, \ldots, q_n, \frac{\partial W_2}{\partial q_2}, \ldots, \frac{\partial W_n}{\partial q_n}, \alpha_1\right) = \alpha_2. \tag{9.159}$$

Equation (9.158) is an ordinary differential equation whose solution has the general form $W_1 = W_1(q_1, \alpha_1, \alpha_2)$. On the other hand, expression (9.159) can be separated further, leading to

$$H_2\left(q_2, \frac{\partial W_2}{\partial q_2}, \alpha_1, \alpha_2\right) = H_2^*\left(q_3, \ldots, q_n, \frac{\partial W_3}{\partial q_3}, \ldots, \frac{\partial W_n}{\partial q_n}, \alpha_1, \alpha_2\right) = \alpha_3. \tag{9.160}$$

Whereas the equation

$$H_2\left(q_2, \frac{\partial W_2}{\partial q_2}, \alpha_1, \alpha_2\right) = \alpha_3 \tag{9.161}$$

is soluble and indeed yields the result $W_2 = W_2(q_2, \alpha_1, \alpha_2, \alpha_3)$, expression H_2^* lends itself to further separation. This procedure, when repeated n times, yields the general solution (9.153). We observe that the last separation leads to

$$H_{n-1}\left(q_{n-1}, \frac{\partial W_{n-1}}{\partial q_{n-1}}, \alpha_1, \alpha_2, \ldots, \alpha_{n-1}\right) = \alpha_n,$$

$$H_{n-1}^* = H_n\left(q_n, \frac{\partial W_n}{\partial q_n}, \alpha_1, \alpha_2, \ldots, \alpha_n\right) = \alpha_n. \tag{9.162}$$

Transformation Theory. The Hamilton-Jacobi Equation

It should be noted that separability depends on the proper choice of coordinates and does not necessarily represent an inherent physical characteristic of the system. A system may become separable as a result of a point transformation, but no general procedure for producing the right point transformation exists.

A case of particular interest is the one in which ignorable coordinates are present. Let us assume that all the generalized coordinates q_i are ignorable with the exception of q_1. Again we seek a solution of the modified Hamilton-Jacobi equation in the form

$$W = \sum_{i=1}^{n} W_i(q_i, \alpha_1, \alpha_2, \ldots, \alpha_n). \tag{9.163}$$

Because the canonical variables conjugate to the ignorable coordinates are constant, we can write

$$\frac{\partial W_i}{\partial q_i} = p_i = \alpha_i, \quad i = 2, 3, \ldots, n, \tag{9.164}$$

so that the modified Hamilton-Jacobi equation (9.156) reduces to

$$H\left(q_1, \frac{\partial W_1}{\partial q_1}, \alpha_2, \alpha_3, \ldots, \alpha_n\right) = \alpha_1. \tag{9.165}$$

The solutions of Eqs. (9.164) are simply

$$W_i = \alpha_i q_i, \quad i = 2, 3, \ldots, n, \tag{9.166}$$

whereas Eq. (9.165) yields the function W_1. Hence, solution (9.148) assumes the particular form

$$W = W_1(q_1, \alpha_1, \alpha_2, \ldots, \alpha_n) + \sum_{i=2}^{n} \alpha_i q_i. \tag{9.167}$$

In the sequel we consider several illustrations of the separation of variables.

Example 9.3

Let us consider the motion of a free particle in a central force field. Using the polar coordinates r and θ, the Hamiltonian can be shown to have the expression

$$H = \frac{1}{2m}\left(p_r^2 + \frac{p_\theta^2}{r^2}\right) + V(r), \tag{a}$$

in which p_r and p_θ are the momenta conjugate to r and θ, respectively, and $V(r)$ is the potential energy. Expression (a) is independent of time, and, moreover, the coordinate θ is ignorable, so that we consider the characteristic function

$$W = W_1(r) + \alpha_2 \theta, \tag{b}$$

where, from Eqs. (9.164), we have

$$\alpha_2 = \frac{\partial W}{\partial \theta} = p_\theta = \text{const.} \tag{c}$$

In view of Eqs. (a) to (c), Eq. (9.165) becomes simply

$$\frac{1}{2m}\left[\left(\frac{\partial W_1}{\partial r}\right)^2 + \frac{\alpha_2^2}{r^2}\right] + V = \alpha_1, \tag{d}$$

which depends only on r. Equation (d) is equivalent to

$$\frac{\partial W_1}{\partial r} = \left[2m(\alpha_1 - V) - \frac{\alpha_2^2}{r^2}\right]^{\frac{1}{2}}, \tag{e}$$

which has the solution

$$W_1 = \int_{r_0}^{r}\left[2m(\alpha_1 - V) - \frac{\alpha_2^2}{r^2}\right]^{\frac{1}{2}} dr. \tag{f}$$

Equations (9.134) and (9.135), in conjunction with expressions (b) and (f), yield

$$\beta_1 = t - \frac{\partial W}{\partial \alpha_1} = t - \int_{r_0}^{r} \frac{m\, dr}{[2m(\alpha_1 - V) - \alpha_2^2/r^2]^{\frac{1}{2}}} \tag{g}$$

and

$$\beta_2 = -\frac{\partial W}{\partial \alpha_2} = \int_{r_0}^{r} \frac{\alpha_2\, dr}{r^2[2m(\alpha_1 - V) - \alpha_2^2/r^2]^{\frac{1}{2}}} - \theta. \tag{h}$$

Comparing Eq. (h) with Eq. (1.97), we conclude that $\beta_2 = -\theta_0$. Moreover, α_1 represents the total energy E, and α_2 is the conserved momentum conjugate to the coordinate θ.

Example 9.4

Another interesting illustration of the Hamilton-Jacobi method is provided by the spinning top of Sec. 4.11. In this case the Hamiltonian has the expression

$$H = \frac{1}{2}\left[\frac{p_\theta^2}{A} + \frac{(p_\phi - p_\psi \cos\theta)^2}{A \sin^2\theta}\right] + \frac{p_\psi^2}{2C^2} + Mgl\cos\theta, \tag{a}$$

so that both ϕ and ψ are ignorable. Hence, the characteristic function can be written

$$W = W_1(\theta) + \alpha_2\phi + \alpha_3\psi, \tag{b}$$

where

$$\alpha_2 = \frac{\partial W}{\partial \phi} = p_\phi, \qquad \alpha_3 = \frac{\partial W}{\partial \psi} = p_\psi \tag{c}$$

are the constant momenta corresponding to the ignorable coordinates. The modified Hamilton-Jacobi equation reduces to

$$\frac{1}{2A}\left(\frac{\partial W_1}{\partial \theta}\right)^2 + \frac{(\alpha_2 - \alpha_3 \cos\theta)^2}{2A \sin^2\theta} + \frac{\alpha_3^2}{2C} + Mgl\cos\theta = \alpha_1, \qquad (d)$$

which has the solution

$$W_1 = \int_{\theta_0}^{\theta} \left[2A\alpha_1 - \frac{(\alpha_2 - \alpha_3\cos\theta)^2}{\sin^2\theta} - \frac{A\alpha_3^2}{C} - 2AMgl\cos\theta\right]^{\frac{1}{2}} d\theta$$

$$= \int_{\theta_0}^{\theta} f(\theta)\, d\theta. \qquad (e)$$

Using Eqs. (9.134) and (9.135), as well as Eqs. (b) and (e), we obtain

$$\beta_1 = t - \frac{\partial W}{\partial \alpha_1} = t - \int_{\theta_0}^{\theta} \frac{A\, d\theta}{f(\theta)}, \qquad (f)$$

$$\beta_2 = -\frac{\partial W}{\partial \alpha_2} = \int_{\theta_0}^{\theta} \frac{(\alpha_2 - \alpha_3\cos\theta)\, d\theta}{f(\theta)\sin^2\theta} - \phi, \qquad (g)$$

$$\beta_3 = -\frac{\partial W}{\partial \alpha_3} = -\int_{\theta_0}^{\theta} \frac{(\alpha_2 - \alpha_3\cos\theta)\cos\theta\, d\theta}{f(\theta)\sin^2\theta} + \frac{A\alpha_3}{C}\int_{\theta_0}^{\theta} \frac{d\theta}{f(\theta)} - \psi. \qquad (h)$$

9.9 ACTION AND ANGLE VARIABLES

Systems possessing periodic solutions are of particular interest in the field of dynamics. We recall from Chap. 5 that periodic solutions correspond to closed trajectories in the phase space. In such cases the motion of the system is generally referred to as *oscillation* or, less frequently but particularly in dynamical astronomy, as *libration*. Periodic solutions are also possible in cases in which not all the canonical variables return to their original position after one complete period. This motion, known as *rotation*, can be visualized as a trajectory in the phase space in which p varies periodically with q. The terminology can perhaps best be illustrated by means of a simple pendulum. If the initial motion imparted to the bob is not great enough to carry it past the upright vertical position, the subsequent motion is oscillatory. On the other hand, if the motion can be made to go past that vertical position, rotational motion is achieved. Since the system is conservative, the momentum at two given angles differing from each other by an integer multiple of 2π will be the same, so that although mathematically the two positions represent different points in the phase space, physically they represent the same configuration. In many problems involving periodic solutions the interest lies primarily in the frequencies of motion and not in the shape of the trajectories in the phase space. The Hamilton-Jacobi theory turns out to be ideally suited for treating such problems. The procedure for obtaining the frequencies of periodic systems by means of the Hamilton-Jacobi theory is referred to as *Delauney's method*.

We shall be concerned with multi-degree-of-freedom systems exhibiting periodic motion and for which the modified Hamilton-Jacobi equation for the generating function W' is separable. Since the variables are separable, the periodic motions are independent of one another. The periods T_i associated with the pairs of canonical variables q_i, p_i ($i = 1, 2, \ldots, n$) are not necessarily equal. Hence the motion of the system may not be simply periodic.

For a single-degree-of-freedom system possessing a constant Hamiltonian and periodic motion the integral

$$J = \oint p \, dq, \tag{9.168}$$

where the integration is extended over a complete period, is a constant of the motion. From Sec. 9.1 we conclude that Eq. (9.168) represents an action integral having units of angular momentum. By analogy with the single-degree-of-freedom system, we choose a set of variables J_i corresponding to the pairs q_i, p_i in the form

$$J_i = \oint p_i \, dq_i, \qquad i = 1, 2, \ldots, n, \tag{9.169}$$

where the integration is extended over one complete period of oscillation, or rotation, of the variable q_i. The variables J_i are called *action variables*, for obvious reasons. By our assumption that the system is separable, Eqs. (9.169) can be written as

$$J_i = \oint \frac{\partial W'_i(q_i, \alpha_1, \alpha_2, \ldots, \alpha_n)}{\partial q_i} \, dq_i, \qquad i = 1, 2, \ldots, n. \tag{9.170}$$

It follows from Eqs. (9.170) that the action variables J_i are independent functions of the constants α_i, so that they can replace them as the constants of integration associated with the solution of the Hamilton-Jacobi equation.

In view of the preceding discussion, we confine ourselves to a separable system for which the Hamiltonian does not contain the time explicitly and consider the time-independent generating function $W'(q,P)$. We shall assume that W' is such that the canonical variables P_i are constant and let these constants coincide with the action variables J_i

$$P_i = J_i, \qquad i = 1, 2, \ldots, n. \tag{9.171}$$

Referring to the second half of Eqs. (9.148), we see that J_i are replacing α_i as constants of the motion. Moreover, it is customary to denote the conjugate variables by w_i, $Q_i = w_i$, and refer to them as *angle variables*. With this notation in mind, Eqs. (9.144) can be rewritten in the form

$$p_i = \frac{\partial W'}{\partial q_i}, \qquad w_i = \frac{\partial W'}{\partial J_i}, \qquad i = 1, 2, \ldots, n. \tag{9.172}$$

Since the Hamiltonian is independent of time, it is a constant of the motion equal to the Jacobi integral h. In addition, the variables J_i are constant, so that the conjugate variables w_i must be ignorable. It follows that

$$H(q_i, p_i) = K(J_i) = h, \tag{9.173}$$

and, consistent with this, the new canonical equations are simply

$$\dot{w}_i = \frac{\partial K}{\partial J_i} = \nu_i, \qquad \dot{J}_i = -\frac{\partial K}{\partial w_i} = 0, \qquad i = 1, 2, \ldots, n, \tag{9.174}$$

in which ν_i are a set of constants, by virtue of the fact that K and J_i are constant. From the first half of Eqs. (9.174) we conclude that the conjugate variables must be linear functions of time

$$w_i = \nu_i t + \delta_i, \qquad i = 1, 2, \ldots, n, \tag{9.175}$$

where δ_i are arbitrary constants of integration, which are the counterparts of the constants β_i of Sec. 9.7. Hence, the time-independent generating function W' produces a set of coordinates increasing linearly with time.

The second half of Eqs. (9.172), together with Eqs. (9.175), can be used to solve for the coordinates q_i in terms of the constants ν_i, δ_i, and the time t. Since during the motion of the system the action variables J_i remain constant, it follows that changes in the angle variables w_i must result from changes in the coordinates q_i. Let us evaluate the change in an angle variable w_i corresponding to a complete period of q_k, during which the remaining coordinates q_j, $j \neq k$, are constant. Denoting this change by Δw_i and using the second half of Eqs. (9.172), we have

$$\Delta w_i = \oint dw_i = \oint \frac{\partial w_i}{\partial q_k} dq_k = \oint \frac{\partial^2 W'}{\partial q_k \, \partial J_i} dq_k. \tag{9.176}$$

But J_i are regarded as mere parameters during the integration process, so that it is permissible to take the derivatives with respect to J_i outside the integral sign. Using Eqs. (9.170), this leads to the result

$$\Delta w_i = \frac{\partial}{\partial J_i} \oint \frac{\partial W'}{\partial q_k} dq_k = \frac{\partial J_k}{\partial J_i} = \delta_{ik}. \tag{9.177}$$

The implication of Eq. (9.177) is that w_i undergoes no change as q_k completes one period except when $k = i$, in which case w_k changes by 1. This result can be interpreted geometrically in terms of the q-space and w-space, as we shall see later.

Denoting by T_i the period associated with q_i, Eqs. (9.175) and (9.177) yield

$$\Delta w_i = \nu_i T_i = 1, \tag{9.178}$$

so that ν_i is the reciprocal of the period and can be identified as the frequency corresponding to q_i. Hence, the procedure enables us to obtain the frequencies of the system without actually carrying out the solution of the differential equations of motion. To this end, we derive the action variables by means of Eqs. (9.169), express the Hamiltonian in terms of these variables, and use the first half of Eqs. (9.174) to obtain the frequencies of the system.

The implication of the preceding discussion is that as the separation coordinate q_i completes one cycle, the corresponding angle variable w_i changes by unity. Hence, for oscillatory motion, q_i can be regarded as a periodic function of the angle variable with the fundamental period equal to 1 and can be represented by the Fourier series

$$q_k = \sum_{r=-\infty}^{\infty} c_{k,r} e^{2\pi i r w_k} = \sum_{r=-\infty}^{\infty} c_{k,r} e^{2\pi i r \nu_k t} e^{2\pi i r \delta_k}, \quad i = \sqrt{-1}, \tag{9.179}$$

where the r's are integers. The coefficients $c_{k,r}$ are obtained by means of the formula

$$c_{k,r} = \int_0^1 q_k e^{-2\pi i r w_k} \, dw_k. \tag{9.180}$$

In the case of rotation, the terms $w_k q_{0k}$ must be subtracted from the left side of Eqs. (9.179), where q_{0k} is the increase in q_k corresponding to a change in w_k by 1. Since rotational coordinates ordinarily represent angles, this problem can be avoided by limiting consideration to only those values of q_k ranging from 0 to 2π.

In general, when the q_k's represent a set of coordinates not necessarily separable, as is often the case when the coordinates are cartesian, they can be regarded as periodic functions of *all* the angle variables w_i with the fundamental period equal to 1. In this case q_k can be expressed as a multiple Fourier series of the type

$$\begin{aligned} q_k &= \sum_{r_1=-\infty}^{\infty} \sum_{r_2=-\infty}^{\infty} \cdots \sum_{r_n=-\infty}^{\infty} c_{k,r_1,r_2,\ldots,r_n} \exp\left(2\pi i \sum_j r_j w_j\right) \\ &= \sum_{r_1=-\infty}^{\infty} \sum_{r_2=-\infty}^{\infty} \cdots \sum_{r_n=-\infty}^{\infty} c_{k,r_1,r_2,\ldots,r_n} \exp\left(2\pi i \sum_j r_j \nu_j t\right) \\ &\quad \times \exp\left(2\pi i \sum_j r_j \delta_j\right), \end{aligned} \tag{9.181}$$

where every one of the indices $r_1, r_2, \ldots, r_n$ takes all integral values, positive, negative, and zero.

In general the q_k's are not periodic functions of time. They would be periodic, with period T, if the quantities $\nu_1 T, \nu_2 T, \ldots, \nu_n T$ were all integers. This is possible only if the fractions $\nu_2/\nu_1, \nu_3/\nu_1, \ldots, \nu_n/\nu_1$ are all rational. It is not difficult to show that if $T = r_j T_j$ ($j = 1, 2, \ldots, n$), where T_j is the reciprocal of ν_j, the system returns to its original state after a time interval T. In this case the frequencies ν_j are said to be *commensurable*, and the motion is simply periodic, describing a closed curve in the configuration space. If the frequencies are not commensurable, the system does not return to the original state in a finite time T. Such motion is called *multiply* or *conditionally periodic*, and it describes a Lissajous curve in the q-space which never closes. For large values of T, however, the motion returns to a state reasonably close to the original one.

It will prove of interest to discuss the motion geometrically in terms of the q-space and the w-space, which can be regarded as mappings of one onto another. If the q_i's are separation coordinates, the configuration space defined by these coordinates represents a limited region of the n-dimensional q-space. Since coordinates q_i vary periodically between certain bounds, if we envision the q_i's as defining a Euclidean n-space, then the entire motion is confined to the interior of an n-dimensional rectangular prism in that space. The prism is parallel to axes q_i, and its size depends on the initial conditions. In view of Eq. (9.177) and its implication, we conclude that the w-space has a periodic structure. If the w-space is also a Euclidean n-space, then it consists of an infinity of unit cubes such that two congruent points in any two cubes correspond to the same point in the q-space. Hence it is possible to represent the motion in the entire w-space by an equivalent motion in the single cube, obtained by transferring the segment of the motion in any cube to the congruent segment of the typical cube. Equations (9.175) indicate that the motion in the w-space takes place along straight lines traversed with constant velocity, by contrast with the motion in the q-space, which may take the form of a complicated Lissajous curve. The motion in the w-space can best be visualized by means of a two-dimensional projection, as shown in Fig. 9.3. Assuming that the lower left corner of the cube represents the initial point $w_i = \delta_i$, corresponding to $t = 0$, we see that as the straight line reaches the right boundary of the unit cube it jumps instantaneously back to the left boundary, from which it continues along a line parallel to the first. The jump represents no physical phenomenon but is a direct result of transferring congruent segments of the motion to the typical cube. When the line reaches the upper boundary, it drops to the lower boundary and continues from there. Depending on the nature of the frequencies ν_j, the line may or may not return to the initial point. To explore this question in more detail,

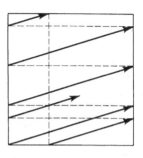

FIGURE 9.3

let us consider the relation

$$r_1\nu_1 + r_2\nu_2 + \cdots + r_n\nu_n = 0, \tag{9.182}$$

where the integers r_j are not all zero. If there are m independent relations of the type (9.182), where $0 < m < n - 1$, then the motion is confined to an $(n - m)$-dimensional plane of the w-space or correspondingly to an $(n - m)$-dimensional subspace of the q-space. In this case, the system is said to be *m-fold degenerate*. If $m = n - 1$, then after a finite time the line returns to the original point and the motion is simply periodic. Such motion, in which the path is closed, is referred to as *completely degenerate*. It is not difficult to show that the existence of $n - 1$ independent relations of the type (9.182) is equivalent to the fractions $\nu_2/\nu_1, \nu_3/\nu_1, \ldots, \nu_n/\nu_1$ being rational. A special case of a completely degenerate system is the one in which all the system frequencies ν_j ($j = 1, 2, \ldots, n$) are equal. If, on the other hand, no relation of the type (9.182) exists, then the path is dense in the w-space; i.e., the motion *enters the neighborhood of any point of that space* (see Ref. 1, app. I). The same statement can be made with regard to the q-space.

It turns out that there is some connection between the degeneracy and separability of a system. Indeed, if the system is nondegenerate, the motion densely fills the permissible region of the configuration space. This region is defined by the separation variables q_i, as pointed out earlier. Clearly the separation of variables in a nondegenerate system must be unique, which means that the Hamilton-Jacobi equation cannot be separated in two different sets of variables. Hence, if the motion is separable in more than one set of variables, we must conclude that the system is degenerate.

Example 9.5

Let us consider the harmonic oscillator of Example 9.2 and find its frequency of oscillation. In that example we derived the result

$$p = \frac{\partial W'}{\partial q} = m^{\frac{1}{2}}(2\alpha - m\omega^2 q^2)^{\frac{1}{2}}, \tag{a}$$

so that, introducing Eq. (a) into (9.169), we obtain the action variable

$$J = \oint p\, dq = \oint \frac{\partial W'}{\partial q}\, dq$$

$$= m^{\frac{1}{2}} \oint (2\alpha - m\omega^2 q^2)^{\frac{1}{2}}\, dq = \frac{2\pi\alpha}{\omega}. \tag{b}$$

Since $K = H = \alpha$, Eq. (b) yields the new Hamiltonian

$$K = \frac{J\omega}{2\pi}, \tag{c}$$

so that, using the first of Eqs. (9.174), we arrive at the frequency

$$\nu = \frac{\partial K}{\partial J} = \frac{\omega}{2\pi}, \tag{d}$$

which, of course, could easily be anticipated.

Example 9.6

The procedure can be applied to obtain the period of motion of a body in a closed orbit. To this end, we consider again the motion of a free particle in a central force field of Example 9.3 and assume that the orbit is elliptical. Denoting by r_p and r_a the minimum and maximum values of the radial distance r, we can write

$$J_r = 2\int_{r_p}^{r_a} p_r\, dr, \qquad J_\theta = 2\int_0^\pi p_\theta\, d\theta. \tag{a}$$

Letting the mass m in Example 9.3 be equal to unity, we obtain the results

$$p_r = \frac{\partial W'}{\partial r} = \left[2(\alpha_1 - V) - \frac{\alpha_2^2}{r^2}\right]^{\frac{1}{2}}, \qquad p_\theta = \frac{\partial W'}{\partial \theta} = \alpha_2, \tag{b}$$

so that the first of Eqs. (a) yields

$$J_r = 2\int_{r_p}^{r_a} \left[2(\alpha_1 - V) - \frac{\alpha_2^2}{r^2}\right]^{\frac{1}{2}} dr. \tag{c}$$

In the special case in which the potential energy has the form $V = -\mu/r$ (see Sec. 1.9), we have

$$\left.\begin{array}{c}r_a\\r_p\end{array}\right\} = \frac{k^2}{\mu}\frac{1 \pm e}{1 - e^2} = -\frac{\mu}{2E}\left[1 \pm \left(1 + \frac{2Ek^2}{\mu^2}\right)^{\frac{1}{2}}\right]. \tag{d}$$

Inserting these values in (c) and integrating, we obtain

$$J_r = -2\pi\alpha_2 + \frac{\sqrt{2}\pi\mu}{(-E)^{\frac{1}{2}}}. \tag{e}$$

On the other hand, from the second of Eqs. (a) we have

$$J_\theta = 2\pi\alpha_2, \qquad (f)$$

so that, combining Eqs. (e) and (f), we arrive at

$$J_r + J_\theta = \frac{\sqrt{2}\pi\mu}{(-E)^{\frac{1}{2}}}. \qquad (g)$$

Since the system Hamiltonian is equal to the total energy E, we can write

$$K = E = -\frac{2\pi^2\mu^2}{(J_r + J_\theta)^2}. \qquad (h)$$

Finally, use of the first half of Eqs. (9.174) as well as of Eq. (g) yields the frequency

$$\nu = \nu_r = \nu_\theta = \frac{4\pi^2\mu^2}{(J_r + J_\theta)^3} = \frac{\mu^{\frac{1}{2}}}{2\pi}\left(-\frac{2E}{\mu}\right)^{\frac{3}{2}} = \frac{\mu^{\frac{1}{2}}}{2\pi a^{\frac{3}{2}}} \qquad (i)$$

where $a = -\mu/2E$ is the semimajor axis of the ellipse. It is easy to see that the reciprocal of Eq. (i) coincides with the period given by Eq. (1.125). Clearly the system is completely degenerate, a result which does not surprise us because the orbit is closed.

9.10 PERTURBATION THEORY

Dynamical systems lending themselves to closed-form solutions are very rare indeed. Systems which admit closed-form solutions represent for the most part simple idealized cases. There is, however, a large class of problems for which analytical solutions can be produced by methods of successive approximations. These problems correspond to systems which differ only slightly from systems possessing closed-form solutions. We shall refer to such systems as *perturbed* or *real systems*. The solution of perturbed systems forms the subject of perturbation theory, and, in fact, Chap. 8 was entirely devoted to problems of this breed. Whereas in Chap. 8 the theory was based primarily on differential equations of second order, in this section we discuss a perturbation theory based on the Hamilton-Jacobi theory. The exposition follows closely that of Born (Ref. 1, sec. 41).

We shall be concerned with the case in which the Hamiltonian of the real system can be written in the form

$$H = H_0 + \epsilon H_1 + \epsilon^2 H_2 + \cdots, \qquad (9.183)$$

where ϵ is a small parameter. For $\epsilon = 0$ the problem reduces to that of an unperturbed system described by the Hamiltonian H_0, where the unperturbed problem is assumed to be soluble by the separation of variables.

We have considerable interest in systems admitting multiple periodic solutions. Hence, our object is to produce a generating function S yielding a contact transformation defined by

$$p_i = \frac{\partial S}{\partial q_i}, \quad w_i = \frac{\partial S}{\partial J_i}, \quad i = 1, 2, \ldots, n, \quad (9.184)$$

transforming the coordinates and momenta q_i, p_i into the angle and action variables w_i, J_i (see Sec. 9.9).

Further consideration will be given to nondegenerate unperturbed systems, namely, systems for which there exists no relation of the form

$$r_1 \nu_1^0 + r_2 \nu_2^0 + \cdots + r_n \nu_n^0 = 0, \quad (9.185)$$

where r_i are integers and ν_i^0 are the frequencies of the unperturbed system.

Next we denote by w_i^0, J_i^0 the angle and action variables associated with the unperturbed system and describe the Hamiltonian in terms of these variables. Since for $\epsilon = 0$ the Hamiltonian must reduce to H_0, a function of the action variables alone, we must have

$$H = H_0(J^0) + \epsilon H_1(w^0, J^0) + \epsilon^2 H_2(w^0, J^0) + \cdots. \quad (9.186)$$

However, w_i^0 and J_i^0 no longer represent angle and action variables, respectively. Indeed, in the presence of perturbations, w_i^0 are not linear functions of time, and J_i^0 are not constant.

To derive the angle and action variables of the perturbed system, we seek a generating function $S(w^0, J)$ defining the contact transformation

$$J_i^0 = \frac{\partial S}{\partial w_i^0}, \quad w_i = \frac{\partial S}{\partial J_i}, \quad i = 1, 2, \ldots, n, \quad (9.187)$$

which transforms the unperturbed variables w_i^0, J_i^0 into the perturbed variables w_i, J_i. The transformation is to be such that the position coordinates of the system are periodic functions of the new angle variables w_i with fundamental period equal to 1, in the same way they were periodic functions of the old variables w_i^0. Moreover, the Hamiltonian H is to be transformed into a function W depending on the J_i's alone, and the function $S - \sum_i w_i J_i$ is to be periodic in the variables w_i with period 1 (see Ref. 1, p. 250). It follows that the new and old angle variables differ by a periodic function with unit period and that the function $S - \sum_i w_i^0 J_i$ is periodic in the variables w_i^0 with unit period. Because for $\epsilon = 0$ the real system must reduce to the unperturbed system, the contact transformation must reduce to the identity transformation when ϵ vanishes. Hence, we consider the time-independent generating function

$$S(w^0, J) = \sum_{i=1}^{n} w_i^0 J_i + \epsilon S_1(w^0, J) + \epsilon^2 S_2(w^0, J) + \cdots, \quad (9.188)$$

where the functions $S_1, S_2, \ldots$ are periodic in the w_i^0. Introducing expression (9.188) into Eqs. (9.187), we obtain

$$J_i^0 = \frac{\partial S}{\partial w_i^0} = J_i + \epsilon \frac{\partial S_1}{\partial w_i^0} + \epsilon^2 \frac{\partial S_2}{\partial w_i^0} + \cdots \qquad (9.189)$$

and

$$w_i = \frac{\partial S}{\partial J_i} = w_i^0 + \epsilon \frac{\partial S_1}{\partial J_i} + \epsilon^2 \frac{\partial S_2}{\partial J_i} + \cdots. \qquad (9.190)$$

By virtue of the fact that the new and old angle variables differ by a periodic function of period equal to unity, it follows from Eqs. (9.190) that $\partial S_1/\partial J_i$, $\partial S_2/\partial J_i, \ldots$ are periodic functions of the w_i^0 with period 1. Without loss of generality, we shall take them to have zero mean values.

Substituting series (9.188) into the Hamilton-Jacobi equation for the perturbed motion, we obtain

$$H_0\left(\frac{\partial S}{\partial w^0}\right) + \epsilon H_1\left(w^0, \frac{\partial S}{\partial w^0}\right) + \epsilon^2 H_2\left(w^0, \frac{\partial S}{\partial w^0}\right) + \cdots = K(J), \quad (9.191)$$

where $K(J)$ is the Hamiltonian expressed in terms of the new action variables only. The function $K(J)$ can also be expanded in a power series in ϵ in the form

$$K(J) = K_0(J) + \epsilon K_1(J) + \epsilon^2 K_2(J) + \cdots. \qquad (9.192)$$

Inserting Eqs. (9.189) into (9.191), we arrive at

$$K(J) = H_0(J) + \epsilon \left[\sum_{i=1}^{n} \frac{\partial H_0}{\partial J_i} \frac{\partial S_1}{\partial w_i^0} + H_1(w^0, J)\right]$$
$$+ \epsilon^2 \left[\sum_{i=1}^{n} \frac{\partial H_0}{\partial J_i} \frac{\partial S_2}{\partial w_i^0} + \sum_{i=1}^{n} \frac{\partial H_1}{\partial J_i} \frac{\partial S_1}{\partial w_i^0} + \sum_{i=1}^{n}\sum_{j=1}^{n} \frac{1}{2!} \frac{\partial^2 H_0}{\partial J_i \partial J_j} \frac{\partial S_1}{\partial w_i^0} \frac{\partial S_1}{\partial w_j^0} \right.$$
$$\left. + H_2(w^0, J)\right] + \cdots, \qquad (9.193)$$

in which $H_0(J), H_1(w^0,J), H_2(w^0,J), \ldots$ are simply $H_0(J^0), H_1(w^0,J^0)$, $H_2(w^0,J^0), \ldots$ with J_i^0 replaced by J_i. Comparing coefficients of like powers of ϵ in Eqs. (9.192) and (9.193), we obtain the set of differential equations

$$K_0(J) = H_0(J), \qquad (9.194)$$

$$K_1(J) = \sum_{i=1}^{n} \frac{\partial H_0}{\partial J_i} \frac{\partial S_1}{\partial w_i^0} + H_1(w^0, J), \qquad (9.195)$$

$$K_2(J) = \sum_{i=1}^{n} \frac{\partial H_0}{\partial J_i} \frac{\partial S_2}{\partial w_i^0} + \sum_{i=1}^{n} \frac{\partial H_1}{\partial J_i} \frac{\partial S_1}{\partial w_i^0}$$
$$+ \sum_{i=1}^{n}\sum_{j=1}^{n} \frac{1}{2!} \frac{\partial^2 H_0}{\partial J_i \partial J_j} \frac{\partial S_1}{\partial w_i^0} \frac{\partial S_1}{\partial w_j^0} + H_2(w^0, J), \qquad (9.196)$$

. . . .

Equations (9.194), (9.195), ... furnish the various orders of approximation to the Hamiltonian. For example, K_0 represents the zero-order approximation and is obtained by simply replacing J_i^0 by J_i in the Hamiltonian of the unperturbed motion. Equation (9.195) for the first-order approximation can be used to determine the functions K_1 and S_1. The first expression on the right side of (9.195) has zero mean value over the unit cube of the w^0-space. Averaging over one period of the w_i^0, it follows that

$$K_1(J) = \overline{H_1(w^0, J)}, \tag{9.197}$$

and we note that the average over one period of the w_i^0 is equal to the time average over the unperturbed trajectory of the system. But H_1 can be written as the sum of its average value and another term which is oscillatory and has zero mean

$$H_1 = \overline{H}_1 + \tilde{H}_1, \tag{9.198}$$

so that, making use of the first half of Eqs. (9.174), Eq. (9.195) becomes

$$\sum_{i=1}^{n} \frac{\partial H_0}{\partial J_i} \frac{\partial S_1}{\partial w_i^0} + \tilde{H}_1 = \sum_{i=1}^{n} \nu_i^0 \frac{\partial S_1}{\partial w_i^0} + \tilde{H}_1 = 0. \tag{9.199}$$

Equation (9.199), when solved for S_1, leads to the new action variables to the first order of approximation.

Turning our attention to Eq. (9.196), we observe that the first sum on the right side averages to zero and the remaining two sums can be regarded as known. Hence, the average of Eq. (9.196) over the unit cube of the w^0-space can be written in the form

$$K_2(J) = \sum_{i=1}^{n} \overline{\frac{\partial H_1}{\partial J_i} \frac{\partial S_1}{\partial w_i^0}} + \sum_{i=1}^{n} \sum_{j=1}^{n} \frac{1}{2!} \overline{\frac{\partial^2 H_0}{\partial J_i \partial J_j} \frac{\partial S_1}{\partial w_i^0} \frac{\partial S_1}{\partial w_j^0}} + \overline{H}_2. \tag{9.200}$$

Again segregating the average value and the oscillatory part with zero mean, Eqs. (9.196) and (9.200) lead to

$$\sum_{i=1}^{n} \nu_i^0 \frac{\partial S_2}{\partial w_i^0} + \sum_{i=1}^{n} \widetilde{\frac{\partial H_1}{\partial J_i} \frac{\partial S_1}{\partial w_i^0}} + \sum_{i=1}^{n} \sum_{j=1}^{n} \widetilde{\frac{1}{2!} \frac{\partial^2 H_0}{\partial J_i \partial J_j} \frac{\partial S_1}{\partial w_i^0} \frac{\partial S_1}{\partial w_j^0}} + \tilde{H}_2 = 0, \tag{9.201}$$

which can be solved for S_2, thus obtaining the new action variables to the second order of approximation. In practical cases it is seldom necessary to go beyond the second-order approximation, and, in fact, the first-order one often suffices.

Returning to Eq. (9.199) for the first-order approximation and recalling that S_1 and H_1 are periodic functions of the old angle variables, we can expand these functions in multiple Fourier series of the type (9.181) as follows:

$$S_1 = \sum_{r_1=-\infty}^{\infty} \sum_{r_2=-\infty}^{\infty} \cdots \sum_{r_n=-\infty}^{\infty} S_{r_1,r_2,\ldots,r_n}^1 \exp\left(2\pi i \sum_j r_j w_j^0\right), \quad (9.202)$$

$$\tilde{H}_1 = \sum_{r_1=-\infty}^{\infty} \sum_{r_2=-\infty}^{\infty} \cdots \sum_{r_n=-\infty}^{\infty} H_{r_1,r_2,\ldots,r_n}^1 \exp\left(2\pi i \sum_j r_j w_j^0\right), \quad (9.203)$$

where $i = \sqrt{-1}$ and r_k are integers. Inserting Eqs. (9.202) and (9.203) into Eq. (9.199) and equating the coefficients of equal harmonics to zero, we obtain the relation

$$2\pi i \sum_{k=1}^{n} r_k \nu_k{}^0 S_{r_1,r_2,\ldots,r_n}^1 + H_{r_1,r_2,\ldots,r_n}^1 = 0, \quad (9.204)$$

which can be solved for $S_{r_1,r_2,\ldots,r_n}^1$ in terms of $H_{r_1,r_2,\ldots,r_n}^1$, provided no relation of the type (9.185) exists. Otherwise the system is degenerate. Even in the absence of degeneracy, a certain choice of the integers r_k can render the sum $\sum_k r_k \nu_k{}^0$ arbitrarily small, raising questions concerning the convergence of the Fourier series (9.202). In spite of this, the method has proved useful in many problems in celestial mechanics. The perturbation method can be also applied to degenerate systems. The interested reader can find a discussion of this subject in the book by Born (Ref. 1, sec. 43).

Example 9.7

As an illustration of the perturbation method, let us consider the oscillator

$$m\ddot{q} + k(q + \epsilon q^3) = 0, \qquad (a)$$

where ϵ is a small parameter. Clearly the unperturbed system coincides with the harmonic oscillator of Examples 9.2 and 9.5. It is not difficult to show that the system Hamiltonian is

$$H = \frac{1}{2}\frac{p^2}{m} + \tfrac{1}{2}m\omega^2(q^2 + \tfrac{1}{2}\epsilon q^4), \qquad \omega^2 = \frac{k}{m}, \qquad (b)$$

so that

$$H_0 = \frac{1}{2}\frac{p^2}{m} + \tfrac{1}{2}m\omega^2 q^2, \qquad (c)$$

$$H_1 = \tfrac{1}{4}m\omega^2 q^4. \qquad (d)$$

From Example 9.2 we obtain the zero-order approximation

$$q = \left(\frac{2h}{m\omega^2}\right)^{\frac{1}{2}} \sin \omega t, \qquad (e)$$

whereas from Example 9.5 we have the relations

$$H_0 = h = \nu^0 J^0, \qquad \omega = 2\pi\nu^0, \tag{f}$$

so that, introducing the notation $\nu^0 t = w^0$, Eqs. (e) and (f) become

$$q = \left(\frac{J_0}{\pi m \omega}\right)^{\frac{1}{2}} \sin 2\pi w^0, \tag{g}$$

$$H_1 = \frac{1}{m}\left(\frac{J^0}{2\pi}\right)^2 \sin^4 2\pi w^0. \tag{h}$$

Equation (9.195) can be used at this point to obtain K_1 and $\partial S_1/\partial w^0$. To this end, we average expression (h) and replace J^0 by J with the result

$$\overline{H_1(w^0,J)} = \int_0^1 H_1(w^0,J)\,dw^0 = \frac{1}{m}\left(\frac{J}{2\pi}\right)^2 \int_0^1 \sin^4 2\pi w^0\,dw^0 = \frac{3}{8m}\left(\frac{J}{2\pi}\right)^2, \tag{i}$$

from which it follows that

$$\widetilde{H_1(w^0,J)} = \frac{1}{m}\left(\frac{J}{2\pi}\right)^2 (\sin^4 2\pi w^0 - \tfrac{3}{8})$$

$$= -\frac{1}{8m}\left(\frac{J}{2\pi}\right)^2 (4\cos 4\pi w^0 - \cos 8\pi w^0). \tag{j}$$

Introducing Eq. (j) into Eq. (9.199), we arrive at

$$\frac{\partial S_1}{\partial w^0} = -\frac{\tilde{H}_1}{\nu^0} = \frac{1}{8m\nu^0}\left(\frac{J}{2\pi}\right)^2 (4\cos 4\pi w^0 - \cos 8\pi w^0), \tag{k}$$

which can be integrated to yield

$$S_1 = \frac{1}{64\pi m\nu^0}\left(\frac{J}{2\pi}\right)^2 (8\sin 4\pi w^0 - \sin 8\pi w^0). \tag{l}$$

The first-order approximation to the action variable J is obtained by introducing Eq. (k) into Eq. (9.189). The result is

$$J^0 = J + \epsilon \frac{\partial S_1}{\partial w^0} = J + \frac{\epsilon}{8m\nu^0}\left(\frac{J}{2\pi}\right)^2 (4\cos 4\pi w^0 - \cos 8\pi w^0). \tag{m}$$

On the other hand, inserting expression (l) into Eq. (9.190), we derive a first-order approximation to the angle variable w in the form

$$w = w^0 + \epsilon \frac{\partial S_1}{\partial J} = w_0 + \frac{\epsilon}{16m\nu^0}\frac{J}{(2\pi)^3}(8\sin 4\pi w^0 - \sin 8\pi w^0). \tag{n}$$

Equations (m) and (n) can be solved for w^0 and J^0 in terms of w and J, which, when introduced into Eq. (g), yield the first-order approximation

$$q = \left(\frac{J}{\pi m\omega}\right)^{\frac{1}{2}} \sin 2\pi w - \frac{\epsilon}{32}\left(\frac{J}{\pi m\omega}\right)^{\frac{3}{2}}(6\sin 2\pi w + \sin 6\pi w). \tag{o}$$

PROBLEMS

9.1 Prove that two successive contact transformations are commutative and the result is also a contact transformation.

9.2 Use the definition given in Sec. 9.2 to prove that the equations

$$Q = (2q)^{1/2} e^k \cos p, \qquad P = (2q)^{1/2} e^{-k} \sin p$$

define a contact transformation.

9.3 Derive expressions for the contact transformations generated by the functions $F_3(p,Q,t)$ and $F_4(p,P,t)$. Note that $F_3(p,Q,t)$ and $F_4(p,P,t)$ can be related to $F_1(q,Q,t)$ by means of a single and a double Legendre transformation, respectively.

9.4 Prove that the requirement that expression (9.100) be a perfect differential leads to conditions (9.101).

9.5 Prove that the conditions for the transformation

$$Q_i = \sum_{j=1}^{2} a_{ij} q_j, \qquad P_i = \sum_{j=1}^{2} b_{ij} p_j, \qquad i = 1, 2,$$

to be a contact transformation can be written in the matrix form

$$[a][b] = [1],$$

where $[a]$ and $[b]$ are the matrices of the constant coefficients a_{ij} and b_{ij}, respectively.

9.6 Prove that the transformation

$$q_1 = \left(\frac{Q_1}{\omega_1}\right)^{1/2} \cos P_1 + \left(\frac{Q_2}{\omega_2}\right)^{1/2} \cos P_2,$$

$$q_2 = -\left(\frac{Q_1}{\omega_1}\right)^{1/2} \cos P_1 + \left(\frac{Q_2}{\omega_2}\right)^{1/2} \cos P_2,$$

$$p_1 = (\omega_1 Q_1)^{1/2} \sin P_1 + (\omega_2 Q_2)^{1/2} \sin P_2,$$

$$p_2 = -(\omega_1 Q_1)^{1/2} \sin P_1 + (\omega_2 Q_2)^{1/2} \sin P_2$$

is a contact transformation. Let the system Hamiltonian have the expression

$$H = \tfrac{1}{2} p_1^2 + \tfrac{1}{2} p_2^2 + \tfrac{1}{4} \omega_1^2 (q_1 - q_2)^2 + \tfrac{1}{4} \omega_2^2 (q_1 + q_2)^2,$$

derive the new Hamiltonian $K(Q_1,Q_2,P_1,P_2)$, and solve the new Hamiltonian equations of motion. What physical system can be described by this formulation?

9.7 Solve the problem of Example 9.3 by expressing the motion in terms of the spherical coordinates r, θ, and φ. Note that this implies that we do not assume advance knowledge of the fact that the motion is planar.

9.8 A projectile is fired with velocity u at an angle θ with respect to the horizontal. Assuming that the gravitational field is uniform, use the Hamilton-Jacobi method and find the equation of the trajectory and the motion of the projectile as a function of time.

9.9 The potential energy of a three-dimensional anisotropic harmonic oscillator has the expression

$$V = \tfrac{1}{2}(k_1 x_1{}^2 + k_2 x_2{}^2 + k_3 x_3{}^2).$$

Use the method of the separation of variables to obtain the solution of the problem.

9.10 Consider the oscillator of Prob. 9.9, express the problem in terms of angle and action variables, and determine the frequencies of the oscillator without actually solving the problem.

9.11 Consider the pendulum of Prob. 2.5 for the case in which the motion is planar, $\phi = 0$. Assume small motions and determine the frequencies of oscillation by the Hamilton-Jacobi method.

9.12 The formulation of Prob. 9.7 can be regarded as describing the motion of an electron in the field of a nucleus. Assuming that the potential energy has the expression $V = -Ze^2/r$, where $-e$ is the charge of the electron and Ze the charge of the nucleus, express the problem in terms of angle and action variables and determine the frequencies of the periodic motion.

9.13 Use the perturbation method of Sec. 9.10 to obtain a first-order approximation solution for the oscillator

$$m\ddot{q} + k(q + \epsilon q^2) = 0,$$

where ϵ is a small parameter.

9.14 Obtain the second-order approximation solution for the system of Example 9.7.

SUGGESTED REFERENCES

1. Born, M.: *The Mechanics of the Atom* (trans. and rev. of 1925 original German ed.), Frederick Ungar Publishing Co., New York, 1960.
2. Corben, H. C., and P. Stehle: *Classical Mechanics*, John Wiley & Sons, Inc., New York, 1950.
3. Goldstein, H.: *Classical Mechanics*, Addison-Wesley Publishing Company, Inc., Reading, Mass., 1950.
4. Lanczos, C.: *The Variational Principles of Mechanics*, 2d ed., University of Toronto Press, Toronto, 1962.
5. Pars, L. A.: *Analytical Dynamics*, William Heinemann, Ltd., London, 1965.
6. Rund, H.: *The Hamilton-Jacobi Theory in the Calculus of Variations*, D. Van Nostrand Company, Inc., Princeton, N.J., 1966.
7. Whittaker, E. T.: *Analytical Dynamics*, 4th ed., Cambridge University Press, London, 1937.

chapter ten

The Gyroscope: Theory and Applications

A *gyroscope* can be defined as a rotating body possessing one axis of symmetry and whose rotation about the symmetry axis is relatively large compared with the rotation about any other axis. The word *gyroscope* was introduced by Foucault to designate a device capable of proving the movement of the earth. In modern usage, a gyroscope is a system consisting of a symmetric rotor spinning rapidly about its symmetry axis and free to move about one or two perpendicular axes. In a man-made device this freedom of movement is achieved by means of a suitable arrangement of gimbals. The rotor axis maintains a fixed direction in space unless subjected to torques. When acted upon by torques, the spin axis tends to rotate about an axis normal to the torque axis. Although in contrast with what we would intuitively expect, this fascinating behavior is in complete accord with the dynamical laws governing the rotational motion of bodies.

A considerable number of spinning bodies, both natural occurrences in the physical world and those made by man, can be regarded as gyroscopes. It is therefore no surprise that the subject of gyroscopic motion and devices has received so much attention. In Chap. 4 we introduced the theory of gyroscopic motion in discussing the spinning symmetric top. In the last few decades the gyroscope has found an ever increasing number of applications, following improved manufacturing techniques. This in turn calls for an

10.1 OSCILLATIONS OF A SYMMETRIC GYROSCOPE

The gyroscope is a device consisting of a symmetric rotor spinning about its symmetry axis while this axis is free to move about one or two axes by means of an appropriate system of suspension. Complete freedom of movement is achieved by using a system of two gimbals, as shown in Fig. 10.1. The outer gimbal is free to rotate about axis Z of the inertial space X, Y, Z, the angle of rotation being denoted by ϕ. The inner gimbal is free to rotate about axis ξ, which remains in the XY plane at all times; the angle of rotation about ξ is θ. The gyroscope's symmetry axis ζ is normal to ξ, and a set of body axes x, y, z is free to rotate about ζ so that the directions of ζ and z coincide at all times; the angle of rotation about ζ is denoted by ψ. It should

FIGURE 10.1

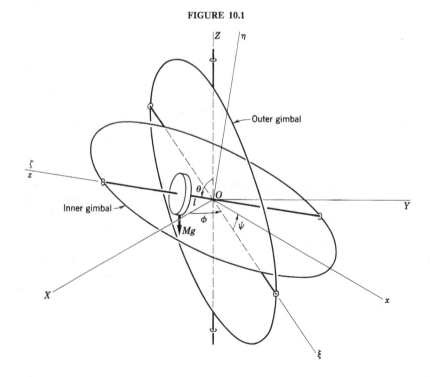

be noted that point O, on the gyroscope's symmetry axis, is always at rest, so that the motion of the gyroscope can be regarded as taking place about the fixed point O. It is easy to see that the symmetric gyroscope of Fig. 10.1 and the symmetric top of Fig. 4.8 are entirely similar systems (with the minor exception that for the gyroscope the angle θ can exceed $\pi/2$), provided the moments of inertia of the gimbals are ignored on the basis that they are small relative to the rotor moments of inertia. It follows that the theory of the symmetric top developed in Sec. 4.11 is perfectly applicable to the symmetric gyroscope, so that in the ensuing discussions we shall rely on results derived there. In pursuing the theory of the symmetric gyroscope a little farther, we shall investigate the behavior of the gyroscope when disturbed from steady precession by a small torque about the ξ axis.

The motion of the gyroscope can be defined by the three Euler's angles ϕ, θ, and ψ. It was shown in Sec. 4.11, however, that two of the coordinates, namely, ϕ and ψ, are cyclic, so that the motion of the symmetric gyroscope can be fully described by a single equation of motion, in conjunction with the associated conserved generalized momenta. From Eqs. (4.109), (4.110), and (4.114) we can write the expressions

$$a = \frac{\beta_\phi}{A} = \dot{\phi} \sin^2 \theta + \frac{C}{A}(\dot{\phi} \cos \theta + \dot{\psi}) \cos \theta = \text{const}, \quad (10.1)$$

$$b = \frac{\beta_\psi}{A} = \frac{C}{A}(\dot{\phi} \cos \theta + \dot{\psi}) = \frac{C}{A} \omega_z = \text{const}, \quad (10.2)$$

whereas Eq. (4.111) leads to the Routhian

$$R = \tfrac{1}{2} A \dot{\theta}^2 - \tfrac{1}{2} A \frac{(a - b \cos \theta)^2}{\sin^2 \theta} - \frac{1}{2} \frac{A^2 b^2}{C} - Mgl \cos \theta. \quad (10.3)$$

Introducing Eq. (10.3) into (4.112), we obtain the differential equation of motion

$$A \ddot{\theta} + A \frac{ab(1 + \cos^2 \theta) - (a^2 + b^2) \cos \theta}{\sin^3 \theta} - Mgl \sin \theta = 0. \quad (10.4)$$

The symmetric gyroscope admits a particular solution representing an equilibrium position. This position is given by the steady precession

$$\theta = \theta_0 = \text{const}, \quad \dot{\phi} = \dot{\phi}_0 = \text{const}, \quad (10.5)$$

where θ_0 satisfies the equation

$$ab(1 + \cos^2 \theta_0) - (a^2 + b^2) \cos \theta_0 = \frac{Mgl}{A} \sin^4 \theta_0. \quad (10.6)$$

On the other hand, from Eq. (4.118), we conclude that $\dot{\phi}_0$ satisfies

$$\dot{\phi}_0 = \frac{a - b \cos \theta_0}{\sin^2 \theta_0}. \quad (10.7)$$

Next let us assume that the system is subjected to a small torque $\epsilon N_\xi(t)$ about axis ξ, where ϵ is a small parameter, so that Eq. (10.4) becomes

$$A\ddot{\theta} + A\frac{ab(1 + \cos^2 \theta) - (a^2 + b^2)\cos \theta}{\sin^3 \theta} - Mgl \sin \theta = \epsilon N_\xi(t). \quad (10.8)$$

Using the perturbation method of Sec. 8.1, a first-order approximation solution of Eq. (10.8) can be obtained in the form

$$\theta = \theta_0 + \epsilon \theta_1. \quad (10.9)$$

Introducing Eq. (10.9) into (10.8), equating the coefficients of ϵ to the first power in both sides of the resulting expression, and considering Eqs. (10.6) and (10.7), we arrive at the first-approximation equation

$$\ddot{\theta}_1 + \omega^2 \theta_1 = \frac{N_\xi}{A}, \quad (10.10)$$

where
$$\omega^2 = b^2 + \dot{\phi}_0^2 \sin^2 \theta_0 - \frac{4Mgl}{A} \cos \theta_0. \quad (10.11)$$

If Eq. (10.10) is to represent the equation of motion of a harmonic oscillator with natural frequency ω, then ω^2 must be a positive quantity. This is the case when the spin is relatively high.

As an interesting sidelight, we see that when $\theta_0 = 0$, the condition that ω^2 be positive again leads to the same stability condition for the sleeping top as that derived by using Eq. (4.135).

The solution of Eq. (10.10) can be written in the form of the convolution integral[1]

$$\theta_1 = \frac{1}{A\omega} \int_0^t N_\xi(\tau) \sin \omega(t - \tau) \, d\tau. \quad (10.12)$$

The first approximation to the precession can be written

$$\dot{\phi} = \dot{\phi}_0 + \epsilon \dot{\phi}_1, \quad (10.13)$$

so that, introducing Eq. (10.13) into (4.118), and considering (10.7), we obtain

$$\dot{\phi}_1 = \frac{b - 2\dot{\phi}_0 \cos \theta_0}{\sin \theta_0} \theta_1 = \frac{b - 2\dot{\phi}_0 \cos \theta_0}{A\omega \sin \theta_0} \int_0^t N_\xi(\tau) \sin \omega(t - \tau) \, d\tau. \quad (10.14)$$

Let $N_\xi(t)$ take the form of an impulsive torque applied at $t = 0$

$$N_\xi(t) = \hat{N}\delta(t), \quad (10.15)$$

---

[1] See, for example, Ref. 3, pp. 283–284.

where $\hat{N}$ is the amplitude of the impulsive torque and $\delta(t)$ is the Dirac delta function. Introducing Eq. (10.15) into solutions (10.12) and (10.14), we can write

$$\theta_1 = \frac{\hat{N}}{A\omega} \sin \omega t \tag{10.16}$$

and
$$\dot{\phi}_1 = \frac{\hat{N}(b - 2\dot{\phi}_0 \cos \theta_0)}{A\omega \sin \theta_0} \sin \omega t, \tag{10.17}$$

so that, assuming that ω^2 is positive, both θ_1 and $\dot{\phi}_1$ are harmonic. Equation (10.17) can be integrated to yield

$$\phi_1 = \frac{\hat{N}(b - 2\dot{\phi}_0 \cos \theta_0)}{A\omega^2 \sin \theta_0} (1 - \cos \omega t), \tag{10.18}$$

where it is recalled that $\phi_1(0) = 0$. Hence the first-approximation solution of the problem can be written

$$\begin{aligned}\theta &= \theta_0 + \epsilon \frac{\hat{N}}{A\omega} \sin \omega t, \\ \phi &= \dot{\phi}_0 t + \epsilon \frac{\hat{N}(b - 2\dot{\phi}_0 \cos \theta_0)}{A\omega^2 \sin \theta_0} (1 - \cos \omega t).\end{aligned} \tag{10.19}$$

Equations (10.19) lend themselves to an easy geometrical interpretation. First let us consider the free gyroscope case, namely, that in which the center of the rotor coincides with the origin O, $l = 0$, so that the torque $Mgl \sin \theta$ about the ξ axis vanishes. Under these circumstances it is possible for the initial motion of the gyroscope to consist of pure spin with the symmetry axis in the fixed position $\theta = \theta_0$, $\phi = 0$, which implies that $\dot{\phi}_0 = 0$ and $\omega = b$. Hence, for this particular case, Eqs. (10.19) reduce to

$$\begin{aligned}\theta &= \theta_0 + \epsilon \frac{\hat{N}}{Ab} \sin bt, \\ \phi &= \epsilon \frac{\hat{N}}{Ab \sin \theta_0} (1 - \cos bt),\end{aligned} \tag{10.20}$$

which indicates that the symmetry axis traces on a unit sphere an ellipse with the origin at $\theta = \theta_0$, $\phi = \epsilon \hat{N}/(Ab \sin \theta_0)$. When $l \neq 0$, the symmetry axis undergoes, in addition to the oscillatory motion, the uniform precession $\dot{\phi}_0$. Hence the path traced by the symmetry axis will no longer be closed but will form loops of the type shown in Fig. 4.10c. Of course, the amplitude and frequency of the oscillatory motion will be different from those of the free gyroscope.

10.2 EFFECT OF GIMBAL INERTIA ON THE MOTION OF A FREE GYROSCOPE

In the analysis of Sec. 10.1 the moments of inertia of the gimbals were assumed to be sufficiently small relative to the rotor moments of inertia to permit their effect on the motion of the gyroscope to be neglected. In particular, in the case of the free gyroscope we established that as a result of an impulsive torque about the ξ axis, the symmetry axis traces an elliptic path on the unit sphere. Laboratory tests, however, reveal that under certain circumstances the motion of the outer gimbal of a free gyroscope is not purely oscillatory but exhibits some drift (Ref. 5). This phenomenon, sometimes referred to as *gimbal walk*, can be shown to be caused by the inertia of the gimbals.

Let us consider the gyroscope of Fig. 10.1 and denote by A', B', and C' the moments of inertia of the inner gimbal about axes ξ, η, and ζ, respectively, and by C'' the moment of inertia of the outer gimbal about axis Z. Next we shall use ϕ, θ, and ψ as generalized coordinates and derive the Lagrange equations of motion for the gyroscope. Noting that the rotor possesses the angular velocity components Ω_ξ, Ω_η, Ω_ζ, as given by Eqs. (4.77), that the inner gimbal rotates with the angular velocities ω_ξ, ω_η, ω_ζ, according to Eqs. (4.76), and that the outer gimbal has the angular velocity $\dot\phi$ about the Z axis, we can write the kinetic energy

$$T = \tfrac{1}{2}[A(\Omega_\xi{}^2 + \Omega_\eta{}^2) + C\Omega_\zeta{}^2 + A'\omega_\xi{}^2 + B'\omega_\eta{}^2 + C'\omega_\zeta{}^2 + C''\dot\phi^2]$$
$$= \tfrac{1}{2}[(A + A')\dot\theta^2 + (A + B')\dot\phi^2 \sin^2\theta + C(\dot\phi\cos\theta + \dot\psi)^2$$
$$+ C'\dot\phi^2 \cos^2\theta + C''\dot\phi^2]. \qquad (10.21)$$

If the center of the gyroscope coincides with the origin O, then $l = 0$, from which it follows that the potential energy is zero, $V = 0$, and the Lagrangian coincides with the kinetic energy, $L = T$. The corresponding Lagrange equations are

$$\frac{d}{dt}\{[(A + B')\sin^2\theta + C'\cos^2\theta + C'']\dot\phi + C(\dot\phi\cos\theta + \dot\psi)\cos\theta\} = N_\phi,$$
$$(A + A')\ddot\theta - (A + B' - C')\dot\phi^2 \sin\theta\cos\theta + C(\dot\phi\cos\theta + \dot\psi)\dot\phi\sin\theta = N_\theta,$$
$$\frac{d}{dt}[C(\dot\phi\cos\theta + \dot\psi)] = N_\psi,$$

$$(10.22)$$

where N_ϕ, N_θ, and N_ψ are the corresponding generalized forces, which in this case turn out to be torques.

We shall be interested in the case in which a small impulsive torque is applied about axis ξ while the rotor spins about its symmetry axis, which is at rest in a position making an angle θ_0 with respect to the Z axis. Letting

$$N_\phi = 0, \qquad N_\theta = \epsilon \hat{N} \delta(t), \qquad N_\psi = 0, \tag{10.23}$$

where ϵ is a small parameter, $\hat{N}$ the amplitude of the impulsive torque, and $\delta(t)$ the Dirac delta function, we conclude from the first and third of Eqs. (10.22) that ϕ and ψ are cyclic. The conservation of the corresponding generalized momenta is expressed by the integrals

$$\beta_\phi = [(A + B')\sin^2\theta + C'\cos^2\theta + C'']\dot\phi + C(\dot\phi\cos\theta + \dot\psi)\cos\theta = \text{const}, \tag{10.24}$$

$$\beta_\psi = C(\dot\phi\cos\theta + \dot\psi) = \text{const}. \tag{10.25}$$

It follows that the gyroscope can be regarded as a single-degree-of-freedom system. From the second of both Eqs. (10.22) and (10.23) as well as from Eqs. (10.24) and (10.25), we obtain the equation describing the system in the form

$$\ddot\theta - \frac{A + B' - C'}{A + A'}\dot\phi^2 \sin\theta\cos\theta + \frac{\beta_\psi}{A + A'}\dot\phi\sin\theta = \epsilon\dot\theta_{\text{in}}\delta(t), \tag{10.26}$$

in which

$$\dot\phi = \frac{\beta_\phi - \beta_\psi\cos\theta}{(A + B')\sin^2\theta + C'\cos^2\theta + C''}. \tag{10.27}$$

Note that in Eq. (10.26) we adopted the notation

$$\dot\theta_{\text{in}} = \frac{\hat{N}}{A + A'}, \tag{10.28}$$

where $\dot\theta_{\text{in}}$ can be interpreted as an initial angular velocity resulting from the impulsive torque.

The solution of Eq. (10.26), in conjunction with Eq. (10.27), will be obtained by means of a perturbation technique. To this end, we limit the solution to the second-order approximation

$$\theta = \theta_0 + \epsilon\theta_1 + \epsilon^2\theta_2, \qquad \dot\phi = \epsilon\dot\phi_1 + \epsilon^2\dot\phi_2, \tag{10.29}$$

where $\theta = \theta_0$ and $\dot\phi = \dot\phi_0 = 0$ represent the equilibrium position prior to the application of the impulsive torque. Note that θ_0 satisfies

$$\beta_\phi - \beta_\psi\cos\theta_0 = 0, \tag{10.30}$$

which is obtained by simply setting $\theta = \theta_0$ and $\dot\phi = 0$ in Eq. (10.27). Introducing Eqs. (10.29) into (10.26), we obtain the equations

$$\ddot\theta_1 + \frac{\beta_\psi}{A + A'}\dot\phi_1\sin\theta_0 = \dot\theta_{\text{in}}\delta(t), \tag{10.31}$$

$$\ddot\theta_2 - \frac{A + B' - C'}{A + A'}\dot\phi_1^2\sin\theta_0\cos\theta_0 + \frac{\beta_\psi}{A + A'}(\dot\phi_2\sin\theta_0 + \theta_1\dot\phi_1\cos\theta_0) = 0,$$

in which

$$\dot{\phi}_1 = \left(\frac{\beta_\psi \sin \theta_0}{I_0}\right)\theta_1, \qquad (10.32)$$

$$\dot{\phi}_2 = \frac{\beta_\psi}{I_0{}^2}\{\theta_1{}^2[\tfrac{1}{2}I_0 - 2(A + B' - C')\sin^2 \theta_0]\cos \theta_0 + \theta_2 I_0 \sin \theta_0\},$$

where the notation

$$I_0 = (A + B')\sin^2 \theta_0 + C'\cos^2 \theta_0 + C'' \qquad (10.33)$$

has been adopted.

Introducing the first of Eqs. (10.32) into the first of Eqs. (10.31), we obtain the first-approximation equation

$$\ddot{\theta}_1 + \omega^2 \theta_1 = \dot{\theta}_{\text{in}}\delta(t), \qquad (10.34)$$

where

$$\omega^2 = \frac{\beta_\psi{}^2 \sin^2 \theta_0}{I_0(A + A')}. \qquad (10.35)$$

The solution of Eq. (10.34) is obtained by means of the convolution integral (see Ref. 3, pp. 283–284) in the form

$$\theta_1 = \frac{\dot{\theta}_{\text{in}}}{\omega}\int_0^t \delta(\tau)\sin \omega(t - \tau)\,d\tau = \frac{\dot{\theta}_{\text{in}}}{\omega}\sin \omega t, \qquad (10.36)$$

so that the first approximation possesses a harmonic component of frequency ω.

A substitution of Eqs. (10.32) and (10.36) into the second of Eqs. (10.31) leads to the second-order approximation equation

$$\ddot{\theta}_2 + \omega^2 \theta_2 = -\frac{3\dot{\theta}_{\text{in}}{}^2 \cos \theta_0}{2I_0 \sin \theta_0}[\tfrac{1}{2}I_0 - (A + B' - C')\sin^2 \theta_0](1 - \cos 2\omega t),$$

$$(10.37)$$

which yields the solution

$$\theta_2 = -\frac{3\dot{\theta}_{\text{in}}{}^2 \cos \theta_0}{2\omega I_0 \sin \theta_0}[\tfrac{1}{2}I_0 - (A + B' - C')\sin^2 \theta_0]$$

$$\times \int_0^t (1 - \cos 2\omega\tau)\sin \omega(t - \tau)\,d\tau$$

$$= -\frac{\dot{\theta}_{\text{in}}{}^2 \cos \theta_0}{2\omega^2 I_0 \sin \theta_0}[\tfrac{1}{2}I_0 - (A + B' - C')\sin^2 \theta_0](3 - 4\cos \omega t + \cos 2\omega t).$$

$$(10.38)$$

Finally, introducing solutions (10.36) and (10.38) into the second of expressions (10.32), we obtain

$$\dot{\phi}_2 = \frac{\beta_\psi \dot{\theta}_{\text{in}}^2 \cos\theta_0}{2I_0^2 \omega^2} \{[\tfrac{1}{2}I_0 - 2(A + B' - C')\sin^2\theta_0](1 - \cos 2\omega t)$$
$$- [\tfrac{1}{2}I_0 - (A + B' - C')\sin^2\theta_0](3 - 4\cos\omega t + \cos 2\omega t)\}$$
$$= -\frac{\beta_\psi \dot{\theta}_{\text{in}}^2 \cos\theta_0}{2I_0 \omega^2}\left[\frac{C' + C''}{I_0} - 4\left(\frac{1}{2} - \frac{A + B' - C'}{I_0}\sin^2\theta_0\right)\cos\omega t\right.$$
$$\left. + \left(1 - 3\frac{A + B' - C'}{I_0}\sin^2\theta_0\right)\cos 2\omega t\right], \qquad (10.39)$$

so that $\dot{\phi}_2$ consists of one constant term and two harmonic terms. The constant term indicates that the outer gimbal possesses a net precession given by the average value

$$\dot{\phi}_{\text{av}} = -\frac{\beta_\psi \dot{\theta}_{\text{in}}^2 \cos\theta_0}{2I_0 \omega^2}\frac{C' + C''}{I_0}, \qquad (10.40)$$

implying that the angle ϕ does not remain constant but regresses with time, confirming the results of the laboratory experiments. This result is identical to the one reported by Plymale and Goodstein (Ref. 5, p. 366). Note that for this precession to occur the initial angle θ_0 must be different from $\pi/2$. Although terms increasing with time, referred to as secular terms (see Sec. 8.2), generally destroy the convergence of the solution, this is not the case here because it is $\dot{\phi}$ and not ϕ which must remain small for the solution to be valid. In certain instruments employing free gyroscopes to establish a fixed direction in space this precession is undesirable as it impairs instrument accuracy.

10.3 EFFECT OF ROTOR SHAFT FLEXIBILITY ON THE FREQUENCY OF OSCILLATION OF A FREE GYROSCOPE

The shaft on which the gyroscope rotor is mounted is generally stiff enough so that the body axes, rigidly attached to the rotor, undergo no translational or rotational displacements induced by the deformation of the shaft. When the rotor shaft is flexible, however, elastic deformations of the shaft affecting the system natural frequencies are possible. For a free gyroscope with stiff bearings, which allow no elastic displacements of the points of support, the elastic translational displacement of the center of the rotor has no effect on the rotational motion of the gimbals. On the other hand, the elastic rotational displacement does have an effect on the motion of the gimbals, as we are about to show.

Let us consider the case in which the rotor elastic shaft undergoes the mode of deformation shown in Fig. 10.2. The body axes x, y, and z, with axis z tangent to the shaft, are obtained from the inner gimbal axes ξ, η, and ζ

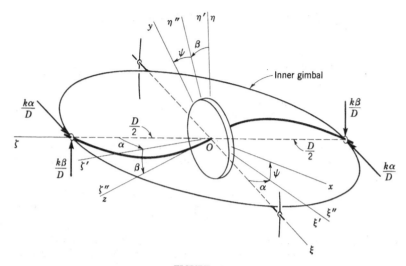

FIGURE 10.2

by means of three rotations. To show this, we follow the motion of a triad which coincides initially with axes ξ, η, and ζ. A rotation α about axis η brings the triad into coincidence with system ξ', η', ζ', a rotation β about ξ' moves the triad into position ξ'', η'', ζ'', and, finally, a rotation ψ about axis ζ'' causes the triad to coincide with the body axes x, y, z. Note that α and β define the elastic angular displacement of the rotor. Due to the symmetry of the rotor about axis ζ'', it will prove expedient to express the angular velocity of the rotor in terms of components about axes ξ'', η'', and ζ'' rather than about the body axes. It is not difficult to show that the angular velocity components of the rotor about axes ξ'', η'', and ζ'' have the expressions

$$\Omega_{\xi''} = \dot\theta \cos\alpha - \dot\phi \cos\theta \sin\alpha + \dot\beta,$$
$$\Omega_{\eta''} = (\dot\phi \sin\theta + \dot\alpha) \cos\beta + (\dot\theta \sin\alpha + \dot\phi \cos\theta \cos\alpha) \sin\beta, \quad (10.41)$$
$$\Omega_{\zeta''} = \dot\psi - (\dot\phi \sin\theta + \dot\alpha) \sin\beta + (\dot\theta \sin\alpha + \dot\phi \cos\theta \cos\alpha) \cos\beta,$$

where $\dot\phi$, $\dot\theta$, and $\dot\psi$ are the precession, nutation, and spin defined in Sec. 4.8.

In general the rotor shaft is quite stiff, so that we shall be interested in deriving the equations of motion for small elastic displacements α and β. In addition, we consider the case in which the inner gimbal is almost normal to the outer gimbal, so that $\theta = \pi/2 + \theta_1$, where θ_1 is a small quantity. Letting

$$\sin\theta \approx 1, \quad \cos\theta \approx -\theta_1, \quad \sin\alpha \approx \alpha, \quad \cos\alpha \approx 1, \quad \sin\beta \approx \beta, \quad \cos\beta \approx 1$$

in Eqs. (10.41), and using the analogy with Eq. (10.21), we can write the kinetic energy in the form

$$T = \tfrac{1}{2}[A(\Omega_{\xi''}^2 + \Omega_{\eta''}^2) + C\Omega_{\zeta''}^2 + A'\omega_{\xi}^2 + B'\omega_{\eta}^2 + C'\omega_{\zeta}^2 + C''\dot{\phi}^2]$$
$$= \tfrac{1}{2}\{A[(\dot{\theta}_1 + \dot{\beta})^2 + (\dot{\phi} + \dot{\alpha})^2] + C[\dot{\psi} - (\dot{\phi} + \dot{\alpha})\beta + (\dot{\theta}_1\alpha - \dot{\phi}\theta_1)]^2$$
$$+ A'\dot{\theta}_1^2 + (B' + C'')\dot{\phi}^2\}, \tag{10.42}$$

where the symbols not defined here have been defined in Sec. 10.2. Denoting by k the resisting torque corresponding to a unit angular displacement α or β, the potential energy is simply

$$V = \tfrac{1}{2}k(\alpha^2 + \beta^2). \tag{10.43}$$

If the system is not subject to any external torques, after ignoring second-order terms, Lagrange's equations of motion reduce to the linear form

$$A(\ddot{\phi} + \ddot{\alpha}) - C(\dot{\psi}\beta + \psi\dot{\beta}) - C\dot{\psi}\dot{\theta}_1 + k\alpha = 0,$$
$$A(\ddot{\theta}_1 + \ddot{\beta}) + C\dot{\psi}(\dot{\phi} + \dot{\alpha}) + k\beta = 0,$$
$$A(\ddot{\phi} + \ddot{\alpha}) - C[\ddot{\psi}(\theta_1 + \beta) + \dot{\psi}(\dot{\theta}_1 + \dot{\beta})] + (B' + C'')\ddot{\phi} = 0, \tag{10.44}$$
$$A(\ddot{\theta}_1 + \ddot{\beta}) + C(\ddot{\psi}\alpha + \dot{\psi}\dot{\alpha}) + A'\ddot{\theta}_1 + C\dot{\psi}\dot{\phi} = 0,$$
$$C\ddot{\psi} = 0.$$

The last of Eqs. (10.44) has the solution

$$C\dot{\psi} = \beta_\psi = \text{const}, \tag{10.45}$$

indicating that for small angular displacements α and β the spin angular momentum is conserved. Moreover, the first four of Eqs. (10.44) can be easily shown to reduce to

$$A(\ddot{\phi} + \ddot{\alpha}) - \beta_\psi(\dot{\theta}_1 + \dot{\beta}) + k\alpha = 0,$$
$$A(\ddot{\theta}_1 + \ddot{\beta}) + \beta_\psi(\dot{\phi} + \dot{\alpha}) + k\beta = 0, \tag{10.46}$$
$$(B' + C'')\ddot{\phi} - k\alpha = 0,$$
$$A'\ddot{\theta}_1 - k\beta = 0.$$

Equations (10.46) are homogeneous and admit a nontrivial solution of the form

$$\phi = \Phi e^{\lambda t}, \qquad \theta_1 = \Theta e^{\lambda t}, \qquad \alpha = a e^{\lambda t}, \qquad \beta = b e^{\lambda t}, \tag{10.47}$$

where Φ, Θ, a, and b are constant amplitudes, provided the determinant of the coefficients vanishes. This leads to the characteristic equation

$$\begin{vmatrix} k + A\lambda^2 & -\lambda\beta_\psi & A\lambda^2 & -\lambda\beta_\psi \\ \lambda\beta_\psi & k + A\lambda^2 & \lambda\beta_\psi & A\lambda^2 \\ -k & 0 & (B' + C'')\lambda^2 & 0 \\ 0 & -k & 0 & A'\lambda^2 \end{vmatrix}$$
$$= \lambda^2[a_0(\lambda^2)^3 + a_1(\lambda^2)^2 + a_2\lambda^2 + a_3] = 0, \tag{10.48}$$

in which
$$a_0 = A^2 A'(B' + C''),$$
$$a_1 = kA[(A + A')(B' + C'') + A'(A + B' + C'')]$$
$$\quad + \beta_\psi^2 A'(B' + C''), \qquad (10.49)$$
$$a_2 = k^2(A + A')(A + B' + C'') + k\beta_\psi^2(A' + B' + C''),$$
$$a_3 = k^2 \beta_\psi^2.$$

It follows from Eq. (10.48) that one of the characteristic values $\lambda^2 = \lambda_0^2$ vanishes, implying that one of the natural frequencies is zero. This is to be expected (see Ref. 3, pp. 112–115) because the potential energy is a positive semidefinite function, whereas the kinetic energy is positive definite by definition. The polynomial in the brackets in Eq. (10.48) leads to a cubic equation in λ^2. An application of the Routh-Hurwitz criterion (Sec. 6.3) leads to

$$\Delta_1 = a_1 = A^2 A'(B' + C'') > 0,$$
$$\Delta_2 = a_1 a_2 - a_3 a_0 = k^2 A[(A + A')(B' + C'') + A'(A + B' + C'')]$$
$$\quad \times [k(A + A')(A + B' + C'') + \beta_\psi^2(A' + B' + C'')]$$
$$\quad + k^2 \beta_\psi^2 A'(B' + C'')[A'(A + B' + C'') + A(B' + C'')]$$
$$\quad + k\beta_\psi^4 A'(B' + C'')(A' + B' + C'') > 0,$$
$$\Delta_3 = k^2 \beta_\psi^2 \Delta_2 > 0,$$

so that all the roots λ_1^2, λ_2^2, and λ_3^2 of the cubic have negative real parts. It can be shown that they possess no imaginary parts, so that $\lambda_j = i\omega_j$ ($j = 1, 2, 3$), where ω_j are the system natural frequencies. Note that Eq. (10.35), for an infinitely stiff rotor shaft, yields in the neighborhood of $\theta_0 = \pi/2$ the nutational frequency

$$\omega = \frac{\beta_\psi}{[(A + A')(A + B' + C'')]^{\frac{1}{2}}}. \qquad (10.50)$$

For a relatively stiff shaft we may expect in our case a natural frequency close to the value given by Eq. (10.50). Indeed an approximation for the lowest natural frequency has the form

$$\omega_1 \approx \left(\frac{a_3}{a_2}\right)^{\frac{1}{2}} = \frac{\beta_\psi}{[(A + A')(A + B' + C'') + \beta_\psi^2 k^{-1}(A' + B' + C'')]^{\frac{1}{2}}}. \qquad (10.51)$$

The estimated value of ω_1 can be improved by assuming a solution of Eq. (10.48) close to the value given by expression (10.51). It is not difficult to show that an improved estimate of ω_1 has the value

$$\omega_1 \approx \left(\frac{a_3}{a_2} + \frac{a_1 a_3^2}{a_2^3}\right)^{\frac{1}{2}}$$

$$\approx \frac{\beta_\psi}{\{(A + A')(A + B' + C'') + \beta_\psi^2 k^{-1} \\ \times [A^2(A + A')^{-1} + (B' + C'')^2(A + B' + C'')^{-1}]\}^{\frac{1}{2}}}. \qquad (10.52)$$

Comparing Eq. (10.50) with either (10.51) or (10.52), we conclude that flexible rotor shafts tend to lower the system natural frequency as opposed to the frequency of completely rigid shafts. Hence, it is conceivable that if the rotor has a slight imbalance, a resonance condition is created when the spin of the rotor $\dot{\psi}$ is equal to the natural frequency ω_1. Recalling Eq. (10.45), we obtain the critical frequency

$$\omega_{\mathrm{cr}} = \left\{\frac{k}{I}\left[1 - \frac{(A + A')(A + B' + C'')}{C^2}\right]\right\}^{\frac{1}{2}}, \qquad (10.53)$$

where I takes the respective value $A' + B' + C''$ or $A^2(A + A')^{-1} + (B' + C'')^2(A + B' + C'')^{-1}$ according to whether ω_1 has the expression (10.51) or (10.52). We note that resonance is possible only if $C^2 > (A + A')(A + B' + C'')$.

Equation (10.48) can be used to obtain the higher natural frequencies ω_2 and ω_3. These higher frequencies can be related to the natural frequencies of a nonspinning rotor mounted on a flexible shaft, where the latter frequencies can be obtained by setting $\beta_\psi = 0$ in the characteristic equation (10.48). Higher frequencies generally imply substantially lower amplitudes of oscillation and are not as important as the lowest one.

10.4 THE GYROCOMPASS

The gyrocompass is designed to determine the direction of true north while avoiding some of the drawbacks of the magnetic compass. A free gyroscope has a tendency to maintain the direction of the spin axis fixed in space, so that a gyroscope on the earth's surface with the spin axis initially pointing north will drift from that direction due to the earth's rotation. To compensate for the earth's rotation, the gyroscope can be made to point north at all times by imparting to it a steady precession through the application of a torque about the node axis. For practical reasons, however, the spin axis does not point exactly north, but the smaller angular deviation is a known quantity.

Let us consider the gyroscope of Fig. 10.3a at a latitude λ on the earth's surface, as shown in Fig. 10.3b. The Z axis in this case is not an inertial axis but coincides with the local vertical at all times. The direction of the gyroscope axis ζ is defined by a rotation ϕ about the Z axis and a rotation θ about the ξ axis. It will be noticed that the angle θ has been redefined in the sense that it is measured from the horizontal plane rather than from the Z axis, as in preceding discussions. The earth's angular velocity Ω is resolved into components $\Omega \cos \lambda$ and $\Omega \sin \lambda$ about axes N and Z, respectively, so that the gyroscope precession consists of the components $\Omega \sin \lambda$ and $\dot{\phi}$. To provide the torque about the axis ξ necessary for producing the precession

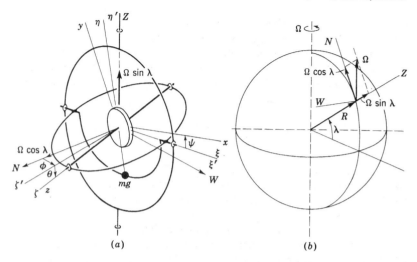

FIGURE 10.3

$\Omega \sin \lambda$, a *pendulous weight* mg is attached at the point $\xi = 0, \eta = -l, \zeta = 0$ of the inner gimbal axes.

In view of the new definitions of the angles ϕ and θ, the angular velocity components of the outer gimbal are

$$\omega_{\xi'} = -\Omega \cos \lambda \sin \phi, \quad \omega_{\eta'} = \dot{\phi} + \Omega \sin \lambda, \quad \omega_{\zeta'} = \Omega \cos \lambda \cos \phi, \quad (10.54)$$

whereas the angular velocity components of the inner gimbal are

$$\begin{aligned}
\omega_{\xi} &= \dot{\theta} - \Omega \cos \lambda \sin \phi, \\
\omega_{\eta} &= (\dot{\phi} + \Omega \sin \lambda) \cos \theta + \Omega \cos \lambda \cos \phi \sin \theta, \\
\omega_{\zeta} &= -(\dot{\phi} + \Omega \sin \lambda) \sin \theta + \Omega \cos \lambda \cos \phi \cos \theta.
\end{aligned} \quad (10.55)$$

The angular velocity components of the rotor about system ξ, η, ζ are

$$\Omega_{\xi} = \omega_{\xi}, \quad \Omega_{\eta} = \omega_{\eta}, \quad \Omega_{\zeta} = \dot{\psi} + \omega_{\zeta}. \quad (10.56)$$

We shall be interested in the moment equations of motion of the system under the assumption that the angles ϕ and θ are small. Note that for small ϕ and θ the Lagrange equations for θ, ϕ, and ψ are equivalent to the torque equations about axes ξ, η, and ζ, respectively. Denoting the moments of inertia of the outer gimbal about axes ξ', η', ζ' by B'', C'', A'', the moments of inertia of the inner gimbal about axes ξ, η, ζ by A', B', C', and the rotor moments of inertia about ξ, η, ζ by A, A, C, the kinetic energy assumes the form

$$T = \tfrac{1}{2}[A(\Omega_\xi^2 + \Omega_\eta^2) + C\Omega_\zeta^2 + A'\omega_\xi^2 + B'\omega_\eta^2 + C'\omega_\zeta^2$$
$$+ A''\omega_{\zeta'}^2 + B''\omega_{\xi'}^2 + C''\omega_{\eta'}^2]$$
$$= \tfrac{1}{2}\{(A + A')(\dot\theta - \Omega \cos \lambda \sin \phi)^2 + (A + B')$$
$$\times [(\dot\phi + \Omega \sin \lambda) \cos \theta + \Omega \cos \lambda \cos \phi \sin \theta]^2$$
$$+ C[\dot\psi - (\dot\phi + \Omega \sin \lambda) \sin \theta + \Omega \cos \lambda \cos \phi \cos \theta]^2$$
$$+ C'[-(\dot\phi + \Omega \sin \lambda) \sin \theta + \Omega \cos \lambda \cos \phi \cos \theta]^2$$
$$+ A''(\Omega \cos \lambda \cos \phi)^2 + B''(\Omega \cos \lambda \sin \phi)^2 + C''(\dot\phi + \Omega \sin \lambda)^2\},$$
(10.57)

whereas the potential energy is simply

$$V = mgl(1 - \cos \theta). \tag{10.58}$$

Performing the appropriate differentiations, it is easy to show that, after ignoring second-order terms in ϕ and θ and their derivatives as well as Ω, the Lagrange equations for ϕ and θ reduce to

$$(A + B' + C'')\ddot\phi + [(2A + A' + B' - C')\Omega \cos \lambda - Cn]\dot\theta$$
$$+ (Cn\Omega \cos \lambda)\phi = 0, \quad (10.59)$$
$$(A + A')\ddot\theta + Cn\dot\phi + mgl\theta + Cn\Omega \sin \lambda = 0,$$

where, from the equation for ψ, we have

$$Cn = C(\dot\psi + \Omega \cos \lambda) = \text{const}, \tag{10.60}$$

which expresses the conservation of the angular momentum about the spin axis.

It is evident that Eqs. (10.59) admit the constant solution

$$\phi_0 = 0, \qquad \theta_0 = -\frac{Cn\Omega \sin \lambda}{mgl}, \tag{10.61}$$

which indicates that an equilibrium position exists in which the gyroscope axis lies in the meridian plane at an angle $(Cn\Omega \sin \lambda)/mgl$ above the horizontal in the Northern Hemisphere and below the horizontal in the Southern Hemisphere. The complete solution of Eqs. (10.59) can be written in the form

$$\phi = \Phi \sin \omega t, \qquad \theta = \theta_0 + \Theta \cos \omega t, \tag{10.62}$$

where the frequency ω must satisfy the characteristic equation

$$\begin{vmatrix} Cn\Omega \cos \lambda - \omega^2(A + B' + C'') & -[(2A + A' + B' - C')\Omega \cos \lambda - Cn]\omega \\ Cn\omega & mgl - \omega^2(A + A') \end{vmatrix} = 0.$$
(10.63)

In general the spin of the rotor is very large, so that Eq. (10.63) yields the two natural frequencies

$$\omega_1 \approx \left(\frac{mgl\Omega \cos \lambda}{Cn}\right)^{\frac{1}{2}}, \qquad \omega_2 \approx \left[\frac{Cn}{(A + A')(A + B' + C'')}\right]^{\frac{1}{2}}. \quad (10.64)$$

The frequency ω_2 is very high and of generally low amplitude difficult to detect. Hence, if disturbed from the equilibrium position, the gyroscope axis will be observed to oscillate about that position with frequency ω_1. In so doing the axis will trace on the unit sphere an elliptical path with the origin given by Eqs. (10.61). In the vicinity of the North Pole, $\lambda \to \pi/2$, the frequency ω_1 becomes very low and the gyrocompass accuracy diminishes.

The oscillation of the gyrocompass is an undesirable effect, as the instrument is expected to indicate the true north. This oscillation can be damped by displacing the pendulous weight a distance e to the east to a position given by $\xi = -e$, $\eta = -l$, $\zeta = 0$, as shown in Fig. 10.4a and b. For small angles θ, the weight produces, in addition to the torque about the ξ axis, a small torque about the η axis

$$N_\eta \approx mge\theta, \quad (10.65)$$

so that Eqs. (10.59) become

$$(A + B' + C'')\ddot{\phi} + [(2A + A' + B' - C')\Omega \cos \lambda - Cn]\dot{\theta}$$
$$+ (Cn\Omega \cos \lambda)\phi - mge\theta = 0, \quad (10.66)$$
$$(A + A')\ddot{\theta} + Cn\dot{\phi} + mgl\theta + Cn\Omega \sin \lambda = 0.$$

FIGURE 10.4

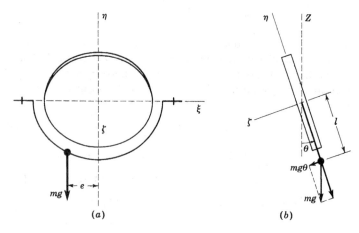

In this case the equilibrium position is simply

$$\phi_0 = -\frac{e}{l}\tan\lambda, \qquad \theta_0 = -\frac{Cn\Omega\sin\lambda}{mgl}, \tag{10.67}$$

so that in the Northern Hemisphere the axis is tilted above the horizontal plane and toward the east.

The general solution of Eqs. (10.66) can be assumed in the form

$$\phi = \phi_0 + \Phi e^{\Gamma t}, \qquad \theta = \theta_0 + \Theta e^{\Gamma t}, \tag{10.68}$$

where Γ must satisfy the characteristic equation

$$\begin{vmatrix} Cn\Omega\cos\lambda + \Gamma^2(A+A'+B'') & [(2A+A'+B'-C')\Omega\cos\lambda - Cn]\Gamma - mge \\ Cn\Gamma & mgl + \Gamma^2(A+A') \end{vmatrix} = 0. \tag{10.69}$$

Again assuming that the spin of the rotor is very large, we obtain the characteristic values

$$\begin{aligned}\left.\begin{array}{c}\Gamma_1 \\ \Gamma_2\end{array}\right\} &= -\frac{mge}{2Cn} \pm i\left(\frac{mgl\Omega\cos\lambda}{Cn}\right)^{\frac{1}{2}}\left[1 - \left(\frac{mge}{2Cn}\right)^2 \frac{Cn}{mgl\Omega\cos\lambda}\right]^{\frac{1}{2}} \\ &= -\gamma\omega_1 \pm i\omega_{1d},\end{aligned} \tag{10.70}$$

where γ is a damping coefficient given by

$$\gamma = \frac{mge}{2Cn}\left(\frac{Cn}{mgl\Omega\cos\lambda}\right)^{\frac{1}{2}} \tag{10.71}$$

and

$$\omega_{1d} = \omega_1(1-\gamma^2)^{\frac{1}{2}} \tag{10.72}$$

is the frequency of the damped oscillation. It is obvious that any disturbance will be damped because both Γ_1 and Γ_2 possess negative real parts.

The analysis above makes no allowance for the fact that the gyrocompass may be mounted in a vehicle moving with a certain velocity relative to the earth's surface. If the vehicle velocity is large in comparison with the linear velocity of a point on the earth's surface, the gyrocompass will register errors because an equivalent angular velocity is introduced as a result of the vehicle motion on a great circle. For this reason the gyrocompass finds its application in slower-moving vehicles, such as ships, rather than aircraft. To show the effect of vehicle velocity, let us consider the case in which the vehicle travels on a great circle in the northwest direction in the Northern Hemisphere, as shown in Fig. 10.5. Denoting the vehicle velocity vector by **v**, the angle between **v** and the meridian direction by μ, and the radius of the great circle by R, we conclude from Fig. 10.5 that there are two additional angular velocity components to account for. As the gyrocompass cannot distinguish between the angular velocity due to the vehicle motion and the one due to the earth's rotation, it will mistake N' for the north. The horizontal correction

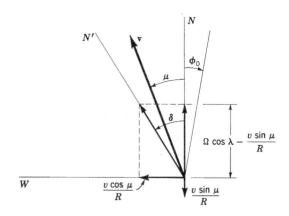

FIGURE 10.5

angle for the damped gyrocompass is $\phi_0 + \delta$ (see Fig. 10.5), where ϕ_0 is obtained from Eqs. (10.67) and δ is the correction due to the vehicle motion

$$\delta = \tan^{-1} \frac{v \cos \mu}{\Omega R \cos \lambda - v \sin \mu} = \tan^{-1} \frac{R\dot{\lambda}}{\Omega R \cos \lambda - v \sin \mu}, \quad (10.73)$$

in which v is the magnitude of the vector $\mathbf{v}$ and $\dot{\lambda}$ represents the rate of change of the latitude. The error δ is relatively small when $\Omega R \gg v$. When v is of the same order of magnitude as the linear velocity $\Omega R \cos \lambda$ of any point on the earth's surface at a latitude λ, the correction becomes significant and instrument reliability ceases to be adequate.

Errors are also introduced by the acceleration of the vehicle. In fact, even for constant velocity the vehicle has an acceleration v^2/R due to the earth's curvature. Vehicle accelerations will be treated in Sec. 10.5.

10.5 THE GYROPENDULUM. SCHULER TUNING

In navigation around the earth it is necessary to know the direction of the local vertical. The simple pendulum can provide the direction of the local vertical only if the point of support is not accelerated. For a vehicle moving around the earth, the direction of the vertical can be obtained by means of a gyropendulum (see Fig. 10.6a). The gyropendulum's outer gimbal axis Z is mounted in the direction of the vehicle's longitudinal axis, so that it coincides with the direction of the vehicle's velocity vector $\mathbf{u}$, where $\mathbf{u}$ makes an angle μ with the north direction, as shown in Fig. 10.6b.

We are interested in deriving the equations of motion for the case in which the vehicle possesses an acceleration $\dot{\mathbf{u}}$. To obtain equations of motion

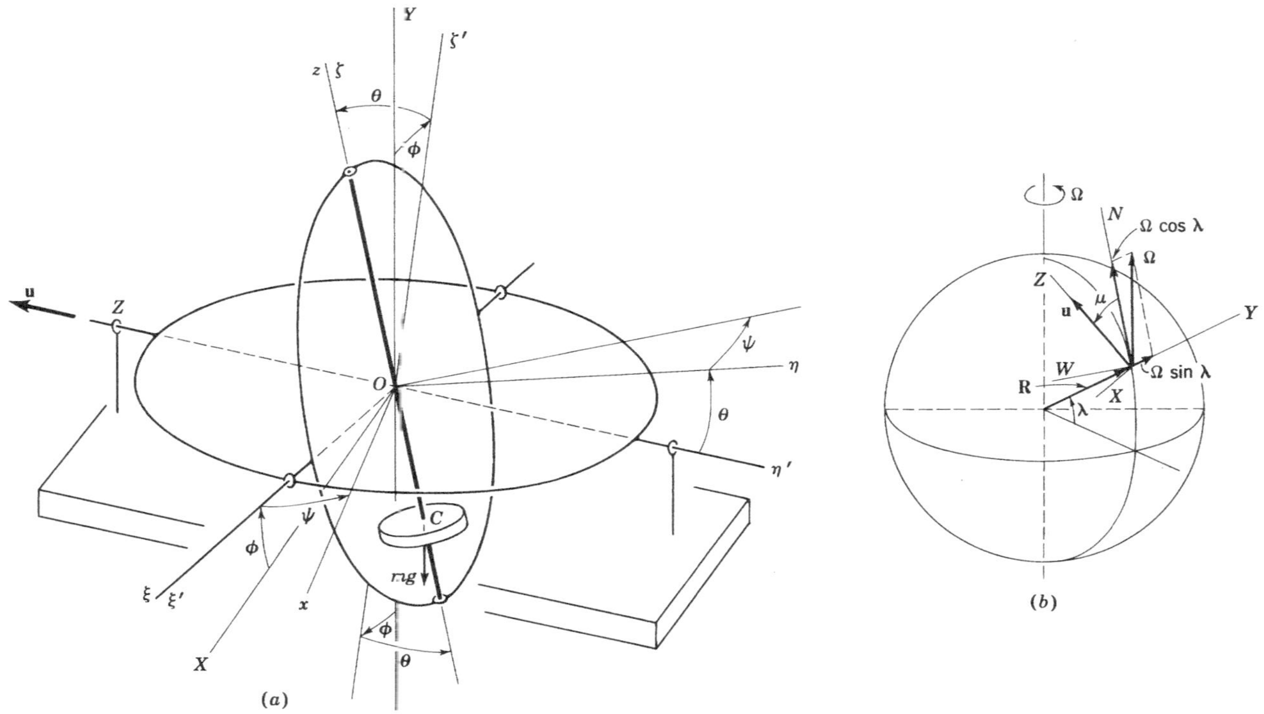

FIGURE 10.6

about three orthogonal axes, we shall use the Lagrange equations in terms of quasi-coordinates formulated in Sec. 4.13. From that section we can write the moment equations of motion about point O (see Fig. 10.6a) in terms of components about axes ξ, η, ζ in the general matrix form

$$\frac{d}{dt}\left\{\frac{\partial T}{\partial \Omega}\right\} + [\omega]\left\{\frac{\partial T}{\partial \Omega}\right\} + [u]\left\{\frac{\partial T}{\partial v}\right\} = \{N_0\}, \qquad (10.74)$$

where T is the kinetic energy and $\{N_0\}$ is the column matrix of the torque components about axes ξ, η, ζ. Moreover, $\{\partial T/\partial \Omega\}$ is a column matrix with elements $\partial T/\partial \Omega_\xi$, $\partial T/\partial \Omega_\eta$, $\partial T/\partial \Omega_\zeta$, where Ω_ξ, Ω_η, Ω_ζ are the angular velocity components of the rotor about axes ξ, η, ζ, and $\{\partial T/\partial v\}$ is a column matrix with elements $\partial T/\partial v_\xi$, $\partial T/\partial v_\eta$, $\partial T/\partial v_\zeta$, where v_ξ, v_η, v_ζ are the velocity components of the rotor center. In addition

$$[\omega] = \begin{bmatrix} 0 & -\omega_\zeta & \omega_\eta \\ \omega_\zeta & 0 & -\omega_\xi \\ -\omega_\eta & \omega_\xi & 0 \end{bmatrix} \qquad (10.75)$$

is the skew-symmetric matrix of the angular velocity components of the inner gimbal axes ξ, η, ζ, and

$$[u] = \begin{bmatrix} 0 & -u_\zeta & u_\eta \\ u_\zeta & 0 & -u_\xi \\ -u_\eta & u_\xi & 0 \end{bmatrix} \qquad (10.76)$$

is the analogous matrix corresponding to the velocity vector **u**. For details concerning the derivation of Eq. (10.74), see Sec. 4.13.

From Fig. 10.6a we conclude that the angular velocity components of axes X, Y, Z are

$$\begin{aligned} \omega_X &= \frac{1}{R}(u - \Omega R \cos \lambda \sin \mu), \\ \omega_Y &= \Omega \sin \lambda, \\ \omega_Z &= \Omega \cos \lambda \cos \mu, \end{aligned} \qquad (10.77)$$

so that the outer gimbal axes have the angular velocity components

$$\begin{aligned} \omega_{\xi'} &= \frac{1}{R}(u - \Omega R \cos \lambda \sin \mu) \cos \phi + \Omega \sin \lambda \sin \phi, \\ \omega_{\eta'} &= -(\dot{\phi} + \Omega \cos \lambda \cos \mu), \\ \omega_{\zeta'} &= -\frac{1}{R}(u - \Omega R \cos \lambda \sin \mu) \sin \phi + \Omega \sin \lambda \cos \phi. \end{aligned} \qquad (10.78)$$

The inner gimbal rotates with an angular velocity $\dot{\theta}$ relative to the outer gimbal, so that its angular velocity components are

$$\omega_\xi = \dot{\theta} + \frac{1}{R}(u - \Omega R \cos \lambda \sin \mu) \cos \phi + \Omega \sin \lambda \sin \phi,$$

$$\omega_\eta = -(\dot{\phi} + \Omega \cos \lambda \cos \mu) \cos \theta$$
$$- \left[\frac{1}{R}(u - \Omega R \cos \lambda \sin \mu) \sin \phi - \Omega \sin \lambda \cos \phi\right] \sin \theta, \quad (10.79)$$

$$\omega_\zeta = (\dot{\phi} + \Omega \cos \lambda \cos \mu) \sin \theta$$
$$- \left[\frac{1}{R}(u - \Omega R \cos \lambda \sin \mu) \sin \phi - \Omega \sin \lambda \cos \phi\right] \cos \theta.$$

Finally, the angular velocity components of the rotor about axes ξ, η, ζ are simply

$$\Omega_\xi = \omega_\xi, \qquad \Omega_\eta = \omega_\eta, \qquad \Omega_\zeta = \dot{\psi} + \omega_\zeta. \quad (10.80)$$

We must distinguish between the velocity **u** of point O, which is the same as the vehicle velocity, and the velocity **v** of the rotor center. It is not difficult to show that the vector **u** has the following components about $\xi, \eta,$ and ζ:

$$u_\xi = -\Omega R \cos \lambda \cos \mu \cos \phi,$$
$$u_\eta = -(u - \Omega R \cos \lambda \sin \mu) \cos \theta + \Omega R \cos \lambda \cos \mu \sin \phi \sin \theta, \quad (10.81)$$
$$u_\zeta = (u - \Omega R \cos \lambda \sin \mu) \sin \theta + \Omega R \cos \lambda \cos \mu \sin \phi \cos \theta.$$

Moreover, the relation between the vectors **u** and **v** can be written in the matrix form

$$\{v\} = \{u\} + [\omega]\{r\}, \quad (10.82)$$

where $[\omega]$ is given by Eq. (10.75) and $\{r\}$ represents the radius vector from the origin O to the rotor center C and has the components $0, 0, -l$. It follows that

$$v_\xi = u_\xi - l\omega_\eta, \qquad v_\eta = u_\eta + l\omega_\xi, \qquad v_\zeta = u_\zeta. \quad (10.83)$$

Denoting the mass of the rotor by m and the principal moments of inertia of the rotor by A, A, C and ignoring the inertia of the gimbals, we can write the kinetic energy expression

$$T = \tfrac{1}{2}m(v_\xi^2 + v_\eta^2 + v_\zeta^2) + \tfrac{1}{2}[A(\Omega_\xi^2 + \Omega_\eta^2) + C\Omega_\zeta^2]$$
$$= \tfrac{1}{2}m[(u_\xi - l\omega_\eta)^2 + (u_\eta + l\omega_\xi)^2 + u_\zeta^2] + \tfrac{1}{2}[A(\Omega_\xi^2 + \Omega_\eta^2) + C\Omega_\zeta^2], \quad (10.84)$$

so that, introducing Eqs. (10.75) and (10.76) in conjunction with (10.84) into Eq. (10.74), we obtain the moment equations

$$ml\dot{u}_\eta + A_0\dot{\omega}_\xi + \omega_\zeta(mlu_\xi - A_0\omega_\eta) + C\Omega_\zeta\omega_\eta - mlu_\zeta\omega_\xi = N_\xi,$$
$$-ml\dot{u}_\xi + A_0\dot{\omega}_\eta + \omega_\zeta(mlu_\eta + A_0\omega_\xi) - C\Omega_\zeta\omega_\xi - mlu_\zeta\omega_\eta = N_\eta, \quad (10.85)$$
$$C\dot{\Omega}_\zeta = N_\zeta,$$

where $A_0 = A + ml^2$ is the moment of inertia of the rotor about any transverse axis through O.

To explore the possibility of using the gyropendulum as an instrument establishing the direction of the local vertical, we must assume that the angles ϕ and θ are small and, moreover, that the spin of the rotor is very large when compared with any other angular velocity. Due to the symmetry of the rotor it follows that $N_\zeta = 0$, so that the third of Eqs. (10.85) yields

$$C\Omega_\zeta = Cn = \text{const.} \tag{10.86}$$

For small ϕ and θ, we have

$$N_\xi = -mgl\theta, \qquad N_\eta = mgl\phi, \tag{10.87}$$

so that, using Eqs. (10.79) to (10.81) as well as Eqs. (10.86) and (10.87) and ignoring higher-order terms, we see that the first two of Eqs. (10.85) reduce to

$$\begin{aligned} Cn(\dot{\phi} + \Omega \cos \lambda \cos \mu - \theta\Omega \sin \lambda) &\approx mgl\theta - ml\dot{u}, \\ Cn\left[\dot{\theta} + \frac{1}{R}(u - \Omega R \cos \lambda \sin \mu) + \phi\Omega \sin \lambda\right] &\approx -mgl\phi, \end{aligned} \tag{10.88}$$

where $\dot{u}$ is the vehicle acceleration, which is in the same direction as u.

Assuming that

$$\frac{mgl}{Cn} = \omega \gg \Omega, \tag{10.89}$$

Eqs. (10.88) can be shown to yield

$$\ddot{\theta} + \omega^2\theta = -\frac{\dot{u}}{R}\left(1 - \frac{\omega^2 R}{g}\right) + \omega\Omega \cos \lambda \cos \mu, \tag{10.90}$$

which for constant $\dot{u}$ has the solution

$$\theta = \alpha \cos(\omega t + \varphi) + \frac{\dot{u}}{R}\left(\frac{R}{g} - \frac{1}{\omega^2}\right) + \frac{\Omega \cos \lambda \cos \mu}{\omega}. \tag{10.91}$$

Introducing Eq. (10.91) into the second of Eqs. (10.88) and ignoring higher-order terms, we obtain

$$\phi = \alpha \sin(\omega t + \varphi) - \frac{1}{R\omega}(u - \Omega R \cos \lambda \sin \mu). \tag{10.92}$$

Solutions (10.91) and (10.92) indicate that the gyropendulum axis will oscillate about the equilibrium position

$$\begin{aligned} \phi_0 &= -\frac{1}{R\omega}(u - \Omega R \cos \lambda \sin \mu), \\ \theta_0 &= \frac{\dot{u}}{R}\left(\frac{R}{g} - \frac{1}{\omega^2}\right) + \frac{\Omega \cos \lambda \cos \mu}{\omega}. \end{aligned} \tag{10.93}$$

In particular, if the system parameters are such that

$$\omega = \sqrt{\frac{g}{R}}, \tag{10.94}$$

then the equilibrium position becomes independent of the acceleration $\dot{u}$. The satisfaction of this condition is referred to as *Schuler tuning* and the period

$$T = 2\pi\sqrt{\frac{R}{g}} = 84.4 \text{ min} \tag{10.95}$$

is called the *Schuler period*. This phenomenon appears in many gyroscopic devices moving on the spherical earth. In fact it also occurs in the gyrocompass of Sec. 10.4. We must remember, however, that solutions (10.91) and (10.92) involve a certain degree of approximation, so that the period may be somewhat different from the Schuler period. Errors ϕ_0 and θ_0 are relatively small, and since they depend only on known quantities, they can easily be compensated for according to Eqs. (10.93).

10.6 RATE AND INTEGRATING GYROSCOPES

For many practical applications it is necessary to measure the angular motion of a given vehicle. The angular motion about an axis normal to a given platform can be recorded by means of a single-gimbal gyroscope whose spin axis is normal to the axis of motion, as shown in Fig. 10.7. Such a gyroscope measures the angular velocity of the platform and is referred to as a *single-axis rate gyroscope*.

Let X, Y, Z be a set of axes attached to the platform, as depicted in Fig. 10.7. The rotor is mounted in a single gimbal, in which it can turn about axis X, so that a rotation about that axis leads to the gimbal axes ξ, η, ζ. The motion of the gimbal relative to the platform is resisted by a torsional spring and dashpot arrangement, so that an angular displacement θ produces a countertorque $k\theta$, and an angular velocity $\dot{\theta}$ gives rise to a resisting torque $c\dot{\theta}$ about axis ξ, where k and c can be regarded as a torsional spring constant and torsional damping coefficient. Axes Y and ξ are referred to as input and output axes, respectively.

If ω_X, ω_Y, and ω_Z denote the angular velocity components of the platform, the gimbal angular velocities become

$$\begin{aligned}\omega_\xi &= \dot{\theta} + \omega_X, \\ \omega_\eta &= \omega_Y \cos\theta + \omega_Z \sin\theta, \\ \omega_\zeta &= -\omega_Y \sin\theta + \omega_Z \cos\theta.\end{aligned} \tag{10.96}$$

The rotor has the angular velocity $\dot{\psi}$ relative to the gimbal, so that the angular

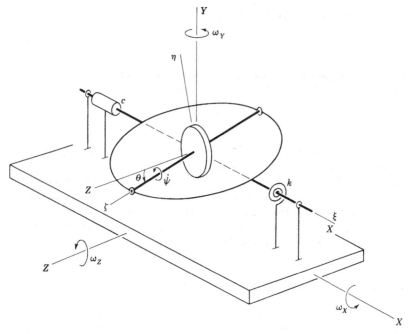

FIGURE 10.7

velocity of the rotor in terms of components about the gimbal axes can be written

$$\Omega_\xi = \omega_\xi, \qquad \Omega_\eta = \omega_\eta, \qquad \Omega_\zeta = \dot{\psi} + \omega_\zeta. \tag{10.97}$$

Because the center of mass and the geometric center O of the rotor and gimbal assembly coincide, the equations for the rotational motion of the system about point O are independent of the equations for the translational motion. The torque equations can be derived by means of the formulation of Sec. 2.12, which takes into account damping forces of the type under consideration. Denoting by A, A, C and A', B', C' the moments of inertia of the rotor and gimbal, respectively, about axes ξ, η, ζ, the kinetic energy takes the form

$$\begin{aligned}T &= \tfrac{1}{2}[A(\Omega_\xi^2 + \Omega_\eta^2) + C\Omega_\zeta^2 + A'\omega_\xi^2 + B'\omega_\eta^2 + C'\omega_\zeta^2] \\&= \tfrac{1}{2}[(A + A')(\dot{\theta} + \omega_X)^2 + (A + B')(\omega_Y \cos\theta + \omega_Z \sin\theta)^2 \\&\quad + C(\dot{\psi} - \omega_Y \sin\theta + \omega_Z \cos\theta)^2 + C'(-\omega_Y \sin\theta + \omega_Z \cos\theta)^2].\end{aligned} \tag{10.98}$$

The potential energy is simply

$$V = \tfrac{1}{2}k\theta^2, \tag{10.99}$$

whereas Rayleigh's dissipation function has the form (see Sec. 2.12)
$$F = \tfrac{1}{2}c\dot{\theta}^2. \tag{10.100}$$
Recalling that $L = T - V$, where L is the Lagrangian, and introducing Eqs. (10.98) to (10.100) into Lagrange's equations for a system with dissipative forces, Eq. (2.149), we obtain

$$(A + A')(\ddot{\theta} + \dot{\omega}_X) - (A + B' - C')(\omega_Y \cos\theta + \omega_Z \sin\theta)$$
$$\times (-\omega_Y \sin\theta + \omega_Z \cos\theta) + C(\dot{\psi} - \omega_Y \sin\theta + \omega_Z \cos\theta)$$
$$\times (\omega_Y \cos\theta + \omega_Z \sin\theta) + c\dot{\theta} + k\theta = 0, \tag{10.101}$$
$$\frac{d}{dt}[C(\dot{\psi} - \omega_Y \sin\theta + \omega_Z \cos\theta)] = 0.$$

The second of Eqs. (10.101) yields the integral
$$C(\dot{\psi} - \omega_Y \sin\theta + \omega_Z \cos\theta) = Cn = \text{const}, \tag{10.102}$$
so that the first of Eqs. (10.101) becomes

$$(A + A')\ddot{\theta} + c\dot{\theta} + k\theta + Cn(\omega_Y \cos\theta + \omega_Z \sin\theta)$$
$$+ (A + B' - C')(\omega_Y \cos\theta + \omega_Z \sin\theta)(\omega_Y \sin\theta - \omega_Z \cos\theta) = -(A + A')\dot{\omega}_X. \tag{10.103}$$

Assuming that the rotor spin is very large compared with the rotation of the platform and that angle θ is small, Eq. (10.103) reduces to
$$(A + A')\ddot{\theta} + c\dot{\theta} + k\theta \approx Cn\omega_Y - (A + A')\dot{\omega}_X, \tag{10.104}$$
which resembles the equation of a damped oscillator.

If the platform undergoes a steady rotation, then $\dot{\omega}_X = 0$. Moreover, assuming zero initial conditions, Eq. (10.104) has the solution

$$\theta = -\frac{Cn\omega_Y}{\omega(A + A')}\int_0^t e^{-\zeta\omega(t-\tau)}\sin\omega_d(t - \tau)\,d\tau$$
$$= -\frac{Cn\omega_Y}{k}\left[1 - \frac{\omega}{\omega_d}e^{-\zeta\omega t}\cos(\omega_d t + \varphi)\right], \tag{10.105}$$

where $\omega = \sqrt{k/(A + A')}$ is the natural frequency of the system, $\zeta = c/2\omega(A + A')$ is a damping coefficient, $\omega_d = \omega(1 - \zeta^2)^{\frac{1}{2}}$ is the damped frequency of oscillation (which stipulates that the system is underdamped, $\zeta < 1$), and φ is the phase angle. Note that the assumption that the initial angle θ is equal to zero implies that the force in the spring is such that when there is no rotation of the platform, the gyro axis is parallel to the platform. It is obvious that as $t \to \infty$ the oscillatory part of the solution is damped out and the instrument records an angular deflection

$$\theta = -\frac{Cn}{k}\omega_Y, \tag{10.106}$$

which is proportional to the rate of rotation about the input axis Y. The displacement θ is measured either electrically or mechanically.

On the other hand, if there is no spring restraining the motion, $k = 0$, and if there is rotation of the platform only about the input axis, then the response becomes

$$\theta = -\frac{Cn}{c}\int_0^t \omega_Y(\tau)\left\{1 - \exp\left[\frac{-c(t-\tau)}{A+A'}\right]\right\}d\tau, \qquad (10.107)$$

which is not restricted to constant ω_Y. The ideal response of the instrument is given by

$$\theta = -\frac{Cn}{c}\int_0^t \omega_Y(\tau)\, d\tau, \qquad (10.108)$$

which is obtained approximately when the second term of the integrand in Eq. (10.107) is small. Equation (10.108) indicates that the angle θ is proportional to the cumulative rotation of the platform about the input axis Y, for which reason the instrument is referred to as an *integrating gyroscope*.

Sometimes a second gimbal is added, in which case the instrument can record angular velocities of the platform about two orthogonal axes. The single-axis gyro, however, is more commonly used.

Inertial navigation is a method which makes use of the dynamical characteristics of simple mass-spring systems and gyroscopes to detect translational and rotational motion. An inertial navigator requires, among other things, a stable platform, whose purpose is to maintain sensing instruments, such as accelerometers and gyroscopes, in a precise known orientation in space. Such a stable platform contains three single-axis gyroscopes which can sense any angular disturbance of the platform about three orthogonal axes and supply this information to corresponding servomotors capable of applying compensating torques designed to maintain the proper orientation in space. (For a detailed discussion of inertial platforms, see Ref. 4, chap. 3.)

PROBLEMS

10.1 Consider the system of Fig. 10.1 and derive the moment equations of motion under the assumption that external torques are applied about axes Z and ξ. Assume the rotor spin to be very large and obtain a first-order solution of the equations of motion by means of a perturbation technique for the case in which a small impulsive torque is applied about the outer gimbal axis Z alone.

10.2 Solve Prob. 10.1 for the case in which the torque applied about axis Z is constant. Comment on the validity of the results.

10.3 Derive the equations of motion for the system of Fig. 10.1 by including the inertia of the gimbals. Use the modified Euler's equations (4.66) and note that the various components possess different velocities.

10.4 Discuss the energy of the gyrocompass for both the undamped and damped cases. Assume that the vehicle in which the gyrocompass is mounted has no motion relative to the rotating earth.

10.5 Derive the equations of motion for the gyrocompass of Fig. 10.3 under the assumption that the vehicle in which the gyrocompass is mounted possesses both heading velocity and acceleration. The vehicle travels on a great circle on the surface of the rotating earth.

10.6 Derive the Lagrangian equations of motion for the gyropendulum of Fig. 10.6.

SUGGESTED REFERENCES

1. Arnold, R. N., and L. Maunder: *Gyrodynamics*, Academic Press, Inc., New York, 1961.
2. Deimel, R. F.: *Mechanics of the Gyroscope*, Dover Publications, Inc., New York, 1950.
3. Meirovitch, L.: *Analytical Methods in Vibrations*, The Macmillan Company, New York, 1967.
4. O'Donnell, C. F. (ed.): *Inertial Navigation Analysis and Design*, McGraw-Hill Book Company, New York, 1964.
5. Plymale, B. T., and R. Goodstein: "Nutation of a Free Gyro Subjected to an Impulse," *J. Appl. Mech.*, September, 1955, pp. 365–366.
6. Thomson, W. T.: *Introduction to Space Dynamics*, John Wiley & Sons, Inc., New York, 1961.

chapter eleven

Problems in Celestial Mechanics

It must be stated at the outset that this excursion into celestial mechanics is a relatively modest one, as the field of celestial mechanics (the common name for the branch of mechanics treating the motion of celestial bodies under their mutual gravitational attraction) is far too extensive to be treated in one chapter. Nevertheless celestial mechanics contains a wealth of fascinating problems ideally suited for illustrating the principles of dynamics. In fact the motion of celestial bodies was a prime factor in the development of classical mechanics. The purpose of this chapter is twofold: (1) some of the problems in celestial mechanics are classical and cannot be ignored in a text on dynamics, and (2) certain techniques and developments of celestial mechanics are equally valuable in modern applications from the field of astrodynamics or space vehicle dynamics.

The first part of this chapter extends the study of particle dynamics of Chap. 1. The many-body problem and in particular the three-body problem receive special attention. The discussion of the many-body problem is based on the assumption that the bodies behave like mass particles. This, in turn, stipulates that the mass distribution of each of the bodies is spherically symmetric or that their dimensions are very small compared with the distances between them. In certain cases these assumptions cannot be justified on the basis of physical reality, and a more detailed knowledge of the gravitational field in the neighborhood of a body of arbitrary mass distribution is required. The lack of inertial symmetry will be shown to be responsible for the precession and nutation of the earth's axis. The subject of gravitational torques is basic to an analysis of the attitude stability of artificial satellites. Finally, the perturbation of the two-body Keplerian orbit is discussed, and expressions for the rate of change of the orbital elements as a function of the perturbing forces are derived. These expressions will be used in Chap. 12 to study the deviation of artificial satellites from Keplerian orbits.

11.1 KEPLER'S EQUATION. ORBIT DETERMINATION

In Sec. 1.9 we discussed the motion of a two-body system in a central force field varying inversely proportional to the square of the distance between the bodies. We showed in that section that the equation of the orbit described by one body relative to another is a conic section. A conic section of particular interest in planetary motion is the ellipse. In this section we shall pursue the study of the elliptical orbit a little farther and seek to derive an expression for the position in the orbit as a function of time.

It will prove convenient to introduce first a number of definitions. The angle θ, measured from the pericentron A as shown in Fig. 11.1, is called the *true anomaly*. The circle of radius a, where a is the semimajor axis of the ellipse, is tangent to the ellipse at the pericentron and apocentron. The vertical through a point R on the ellipse intersects the circle at Q. The angle ACQ, denoted by w, is known as the *eccentric anomaly*. The *mean motion* is defined as

$$n = \frac{2\pi}{P} = \left(\frac{\mu}{a^3}\right)^{\frac{1}{2}}, \tag{11.1}$$

where P is the period obtained from Eq. (1.125) and $\mu = G(m_1 + m_2)$, in which G is the gravitational constant and m_1 and m_2 are the masses of the two bodies. The mean motion is equivalent to an average orbital angular velocity. The quantity

$$M = n(t - T), \tag{11.2}$$

in which T is known as the *time of pericentron passage*, is called the *mean anomaly*, and it represents the angle that would have been described by the radius vector had the motion been uniform with the mean angular velocity n.

Recalling the properties of the ellipse, we conclude from Fig. 11.1 that

$$r \cos \theta = a(\cos w - e), \qquad r \sin \theta = a \sin w - QR. \tag{11.3}$$

Since the ellipse is obtained from the circle by an affine transformation, we

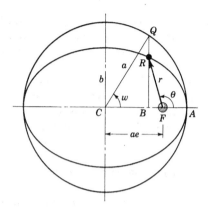

FIGURE 11.1

have the relation

$$\frac{RB}{QB} = \frac{b}{a} = (1 - e^2)^{\frac{1}{2}}, \tag{11.4}$$

where b is the semiminor axis of the ellipse. It follows that

$$QR = QB - RB = a[1 - (1 - e^2)^{\frac{1}{2}}] \sin w, \tag{11.5}$$

so that Eqs. (11.3) assume the form

$$r \cos \theta = a(\cos w - e), \qquad r \sin \theta = a(1 - e^2)^{\frac{1}{2}} \sin w. \tag{11.6}$$

A number of operations leads to the relations

$$r = a(1 - e \cos w) \tag{11.7}$$

and
$$\tan \frac{\theta}{2} = \left(\frac{1+e}{1-e}\right)^{\frac{1}{2}} \tan \frac{w}{2} \tag{11.8}$$

which express the polar coordinates r and θ in terms of the eccentric anomaly w.

A relation between the time t and the true anomaly θ is obtained by introducing Eq. (1.117) with $\theta_0 = 0$ into the second of Eqs. (1.106) and integrating. The result is

$$t = \int_0^\theta \frac{r^2 \, d\alpha}{k} = \frac{k^3}{\mu^2} \int_0^\theta \frac{d\alpha}{(1 + e \cos \alpha)^2}$$
$$= \frac{k^3}{\mu^2} \frac{1}{1 - e^2} \left[-\frac{e \sin \theta}{1 + e \cos \theta} + \frac{2}{(1 - e^2)^{\frac{1}{2}}} \tan^{-1} \frac{(1 - e^2)^{\frac{1}{2}} \tan \frac{1}{2}\theta}{1 + e} \right], \tag{11.9}$$

where α is a dummy variable of integration and k is a constant representing the angular momentum per unit mass. The quantities μ and k are related by Eq. (1.125) defining the period

$$P = \frac{k^3}{\mu^2} \frac{2\pi}{(1-e^2)^{\frac{3}{2}}}, \qquad (11.10)$$

so that using Eqs. (11.1), (11.2), (11.6), and (11.8) to (11.10), we obtain finally

$$n(t - T) = M = w - e \sin w, \qquad (11.11)$$

which is known as *Kepler's equation*.

Kepler's equation allows us to calculate the mean anomaly M if the eccentric anomaly w and the eccentricity e are given. Quite often, however, M and e are known, and the interest lies in calculating w. This presents a slight problem because Eq. (11.11) is transcendental and can be solved only by a numerical procedure. For small eccentricities, however, we can obtain an approximate solution by using an expansion for w in the form of a power series in e, similar to the formal perturbation solution of Sec. 8.1. Indeed, introducing

$$w = w_0 + e w_1 + e^2 w_2 + e^3 w_3 + \cdots \qquad (11.12)$$

into the right side of Eq. (11.11) and equating coefficients of equal powers of e, we obtain

$$\begin{aligned}
w_0 &= M, \\
w_1 &= \sin w_0 = \sin M, \\
w_2 &= w_1 \cos w_0 = \tfrac{1}{2} \sin 2M, \\
w_3 &= w_2 \cos w_0 - \tfrac{1}{2} w_1^2 \sin w_0 = \tfrac{1}{8}(3 \sin 3M - \sin M), \\
&\quad \cdots \cdots \cdots \cdots \cdots \cdots \cdots \cdots \cdots \cdots \cdots,
\end{aligned} \qquad (11.13)$$

so that, to the third order of approximation, the desired solution has the form

$$w = M + e \sin M + \tfrac{1}{2} e^2 \sin 2M + \tfrac{1}{8} e^3 (3 \sin 3M - \sin M). \qquad (11.14)$$

Higher-order approximations can be obtained in a similar fashion, but the procedure becomes increasingly laborious.

In our preceding discussion of a two-body system there was an implicit assumption that the orbital plane is known, and the problem was reduced to the determination of the orbit in that plane. For a complete determination of the orbit, however, we must also establish the orientation of the orbital plane in space as well as the position of the orbit in the plane. Because the motion of a two-body system can be described by three second-order differential equations, a complete determination of this motion requires six constants of integration. These constants may assume a number of forms, such as initial position and velocity or angular observations. The constants can be used to evaluate a corresponding number of orbital elements, from which it follows that to determine an orbit in space six orbital elements are needed.

For the purpose of this discussion, let us consider a two-body system consisting of the sun S and a planet m, and let the origin of an inertial space X', Y', Z' coincide with the center of S (see Fig. 11.2). A coordinate system with the origin at the sun's center is called a *heliocentric reference frame*. The $X'Y'$ plane is taken as the *ecliptic plane*, defined as the plane containing the apparent path of the sun around the earth. Axis X' points toward the *vernal equinox*, also referred to as the *first point of Aries*, which represents the intersection of the earth's equatorial plane by the sun's path going north. Although technically speaking the vernal equinox and the ecliptic plane precess very slowly relative to an inertial space, for many practical purposes both can be assumed to be fixed in that space. The orbit plane of a given body m intersects the $X'Y'$ plane along the line $N'SN$ called the *line of nodes*. If the orbit intersects the $X'Y'$ plane and in so doing rises above that plane, then the corresponding point, such as point N in Fig. 11.2, is said to be an *ascending node*; otherwise it is a *descending node*. The angle Ω, measured in the $X'Y'$ plane from axis X' to axis SN, is called the *longitude of the ascending node*. The orbit plane makes an angle i with respect to the $X'Y'$ plane. The angle i, called the orbit *inclination*, is defined uniquely by the angle between axis Z' and the angular momentum vector $\mathbf{L}$, where the latter is normal to the orbit plane. Finally, we denote by θ_0 the angle between the ascending node and the perihelion, as shown in Fig. 11.2, thus lending geometrical meaning to angle θ_0 in Eq. (1.117). The elements Ω, i, and θ_0 establish the position of the orbit relative to an inertial space. The remaining orbit elements, such as the eccentricity e, the semimajor axis a, and the mean anomaly M, determine the orbit shape and the position of the body in orbit.

FIGURE 11.2

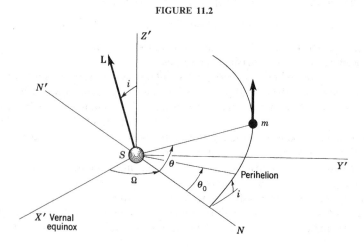

There are various methods for determining the orbit from observations, depending on the nature of the observations, but they lie outside the scope of this text. The interested reader should consult Ref. 6, chaps. 4 and 5.

11.2 THE MANY-BODY PROBLEM

The two-body mathematical model has the advantage of lending itself to a relatively simple mathematical formulation admitting a closed-form solution. There are many physical systems that can be represented adequately by such a model. Since the sun is the dominant factor in the solar system, the sun and any one of the planets can be regarded as a two-body system. Because the gravitational forces fall off rapidly with distance from the center of a planet, the action of one planet upon another must be regarded as a second-order perturbation. Nevertheless, there are certain phenomena which cannot be explained by means of the simple two-body system, so that the motion of a system of n bodies, and in particular the three-body problem, is not without interest. The solar system is truly an n-body system. In this section we shall formulate the many-body problem and examine some of its interesting features, thus preparing the groundwork for a treatment of the three-body problem.

Let us consider a system of n bodies possessing spherical symmetry, so that they can be regarded as point masses. It will be assumed that the only forces acting upon the system are due to the mutual gravitational attraction, which varies inversely with the distance squared. Figure 11.3 shows an

FIGURE 11.3

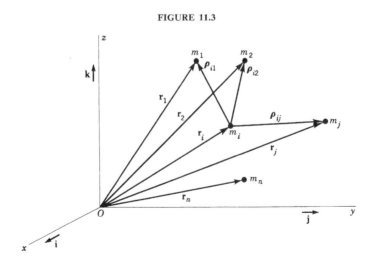

n-body system in which the absolute position of mass m_i is denoted by $\mathbf{r}_i$ and the vector from mass m_i to mass m_j by $\boldsymbol{\rho}_{ij}$. The equation of motion of mass m_i has the form

$$m_i \ddot{\mathbf{r}}_i = G \sum_{j=1}^{n} \delta_{ij}^* \frac{m_i m_j}{\rho_{ij}^3} \boldsymbol{\rho}_{ij}, \qquad i = 1, 2, \ldots, n, \qquad (11.15)$$

where $\delta_{ij}^* = 1 - \delta_{ij}$ can be regarded as a *complementary Kronecker delta*, in which δ_{ij} is the ordinary Kronecker delta.

Summing up over the entire system and recalling that $\boldsymbol{\rho}_{ji} = -\boldsymbol{\rho}_{ij}$, we obtain

$$\sum_{i=1}^{n} m_i \ddot{\mathbf{r}}_i = \mathbf{0}, \qquad (11.16)$$

where the right side has reduced to zero in pairs. Equation (11.16) can be integrated twice with the result

$$\sum_{i=1}^{n} m_i \mathbf{r}_i = \mathbf{c}_1 t + \mathbf{c}_2. \qquad (11.17)$$

Denoting the total mass of the system by M, such that $M = \sum_{i=1}^{n} m_i$, and recalling Eq. (1.52), we conclude that the position vector $\mathbf{R}$ of the system center of mass can be written

$$\mathbf{R} = \frac{1}{M}(\mathbf{c}_1 t + \mathbf{c}_2), \qquad (11.18)$$

which indicates that the center of mass of the system, often referred to as the *barycenter*, is at rest (if $\mathbf{c}_1 = \mathbf{0}$) or moves with uniform velocity. The vectors $\mathbf{c}_1$ and $\mathbf{c}_2$ represent six integrals of the equations of motion.

Next let us write, in sequence, the vector product of $\mathbf{r}_i$ and the equation with the same index i of system (11.15), sum up the results, and obtain

$$\sum_{i=1}^{n} \mathbf{r}_i \times m_i \ddot{\mathbf{r}}_i = G \sum_{i=1}^{n} \sum_{j=1}^{n} \delta_{ij}^* \frac{m_i m_j}{\rho_{ij}^3} (\mathbf{r}_i \times \boldsymbol{\rho}_{ij}). \qquad (11.19)$$

But

$$\mathbf{r}_i \times \boldsymbol{\rho}_{ij} = \mathbf{r}_i \times (\mathbf{r}_j - \mathbf{r}_i) = \mathbf{r}_i \times \mathbf{r}_j,$$

and, similarly,

$$\mathbf{r}_i \times \boldsymbol{\rho}_{ji} = -\mathbf{r}_i \times \mathbf{r}_j.$$

It follows that the terms on the right side of (11.19) cancel out in pairs, and because masses m_i are constant, we can write

$$\frac{d}{dt}\left(\sum_{i=1}^{n} \mathbf{r}_i \times m_i \dot{\mathbf{r}}_i\right) = 0. \qquad (11.20)$$

The expression enclosed by the parentheses is recognized as the angular momentum of the entire system about point O, so that

$$\mathbf{L}_0 = \sum_{i=1}^{n} \mathbf{r}_i \times m_i \dot{\mathbf{r}}_i = \text{const}, \qquad (11.21)$$

which represents the statement of the *conservation of the angular momentum*. The three constant components of $\mathbf{L}_0$ constitute three additional integrals of the motion.

Equation (11.21) implies that both the magnitude and direction of vector $\mathbf{L}_0$ are constant. The constant direction of $\mathbf{L}_0$ defines a plane called the *invariable plane* by Laplace. For the solar system the invariable plane is inclined at about 1°35' with respect to the ecliptic, between the orbital planes of Jupiter and Saturn.

If x, y, z denote an inertial system with the origin at O, Eq. (11.21) can be written in terms of the cartesian components

$$\sum_{i=1}^{n} m_i(y_i \dot{z}_i - z_i \dot{y}_i) = L_x,$$

$$\sum_{i=1}^{n} m_i(z_i \dot{x}_i - x_i \dot{z}_i) = L_y, \qquad (11.22)$$

$$\sum_{i=1}^{n} m_i(x_i \dot{y}_i - y_i \dot{x}_i) = L_z,$$

which are referred to as *integrals of area*.

Taking the scalar product of $\dot{\mathbf{r}}_i$ and the equation with the same index i of system (11.15) for every value of the index and summing up, we obtain

$$\sum_{i=1}^{n} m_i \dot{\mathbf{r}}_i \cdot \ddot{\mathbf{r}}_i = G \sum_{i=1}^{n} \sum_{j=1}^{n} \delta_{ij}^* \frac{m_i m_j}{\rho_{ij}^3} \dot{\mathbf{r}}_i \cdot \boldsymbol{\rho}_{ij}. \qquad (11.23)$$

The right side of Eq. (11.23) can be shown to involve terms of the form

$$\frac{m_i m_j}{\rho_{ij}^3}(\dot{\mathbf{r}}_i \cdot \boldsymbol{\rho}_{ij} + \dot{\mathbf{r}}_j \cdot \boldsymbol{\rho}_{ji}) = -\frac{m_i m_j}{\rho_{ij}^3} \dot{\boldsymbol{\rho}}_{ij} \cdot \boldsymbol{\rho}_{ij},$$

so that Eq. (11.23) can be shown to be equivalent to

$$\frac{d}{dt}\left(\sum_{i=1}^{n} \tfrac{1}{2} m_i \dot{\mathbf{r}}_i \cdot \dot{\mathbf{r}}_i\right) + \frac{d}{dt}\left(-\frac{1}{2} \sum_{i=1}^{n} \sum_{j=1}^{n} G \delta_{ij}^* \frac{m_i m_j}{\rho_{ij}}\right) = 0. \qquad (11.24)$$

But

$$\frac{1}{2} \sum_{i=1}^{n} m_i \dot{\mathbf{r}}_i \cdot \dot{\mathbf{r}}_i = T \qquad (11.25)$$

is the kinetic energy of the system, and, moreover,

$$-\frac{1}{2}\sum_{i=1}^{n}\sum_{j=1}^{n} G\delta_{ij}^{*}\frac{m_i m_j}{\rho_{ij}} = V \qquad (11.26)$$

is the potential energy of the system. It follows from Eqs. (11.24) to (11.26) that

$$T + V = E = \text{const}, \qquad (11.27)$$

where E is recognized as the total energy of the system. Equation (11.27) represents the statement of the *conservation of energy* of the system.

The energy integral, Eq. (11.27), constitutes the tenth constant of integration, and no other integrals are known to exist. The existence of additional integrals has been investigated by Bruns and Poincaré, who showed that the 10 integrals obtained are the only independent integrals of the system expressible in terms of simple known functions. A discussion of the theorems of Bruns and Poincaré is given in Ref. 15, chap. 14. The general solution of the n-body problem is beyond reach. A complete solution requires $6n$ constants of integration, and there are only 10 available. Even for the three-body system, requiring only 18 constants, a closed-form solution of the general problem does not appear feasible. There exist, however, special solutions of the three-body problem obtained by Lagrange, which are discussed in the next section.

11.3 THE THREE-BODY PROBLEM

The general problem of a system consisting of three bodies of spherically symmetric mass distribution and acted upon by their mutual gravitational attraction has preoccupied many brilliant minds since the middle of the eighteenth century. To this day the problem is still receiving considerable attention. The two-body problem can be regarded as solved, but unfortunately the same cannot be said about the three-body problem. Although Sundman (Ref. 14) produced a solution of the general three-body problem as a power series in time, the radius of convergence of the series is too small to provide real insight into the behavior of the system for sufficiently large values of time. Hence, we must rely on numerical integrations by means of digital computers for a solution of the general n-body problem, where $n \geq 3$. Exact solutions of the three-body problem, however, were shown by Lagrange to exist for some special cases, in all of which the motion of the bodies takes place in the same plane.

Let us consider the special case in which the three bodies move in a plane. It will prove convenient to refer the motion to a set of cartesian axes x, y, z

rotating with respect to an inertial space with constant angular velocity ω about axis z. Denoting by x_i, y_i the position of masses m_i ($i = 1, 2, 3$) relative to the moving axes, the system kinetic energy can be written

$$T = \frac{1}{2} \sum_{i=1}^{3} m_i[(\dot{x}_i - \omega y_i)^2 + (\dot{y}_i + \omega x_i)^2]. \tag{11.28}$$

From Eq. (11.26) we have the potential energy

$$V = -\tfrac{1}{2}G \sum_{i=1}^{3} \sum_{j=1}^{3} \delta_{ij}^* \frac{m_i m_j}{\rho_{ij}}, \tag{11.29}$$

where δ_{ij}^* is the complementary Kronecker delta defined in Sec. 11.2 and

$$\rho_{ij} = [(x_j - x_i)^2 + (y_j - y_i)^2]^{\frac{1}{2}}, \qquad i,j = 1, 2, 3. \tag{11.30}$$

We shall be interested in deriving the Hamilton canonical equations for the system. First we note that for this nonnatural system the kinetic energy can be written in the form

$$T = T_2 + T_1 + T_0, \tag{11.31}$$

where
$$T_2 = \frac{1}{2} \sum_{i=1}^{3} m_i(\dot{x}_i^2 + \dot{y}_i^2),$$

$$T_1 = \omega \sum_{i=1}^{3} m_i(x_i \dot{y}_i - y_i \dot{x}_i), \tag{11.32}$$

$$T_0 = \tfrac{1}{2}\omega^2 \sum_{i=1}^{3} m_i(x_i^2 + y_i^2).$$

Recalling the definition of the generalized moments, we have

$$p_{x_i} = \frac{\partial L}{\partial \dot{x}_i} = m_i(\dot{x}_i - \omega y_i),$$
$$\qquad\qquad\qquad\qquad\qquad\qquad i = 1, 2, 3. \tag{11.33}$$
$$p_{y_i} = \frac{\partial L}{\partial \dot{y}_i} = m_i(\dot{y}_i + \omega x_i),$$

Solving Eqs. (11.33) for $\dot{x}_i$ and $\dot{y}_i$ ($i = 1, 2, 3$) and introducing the result into the first of expressions (11.32), we obtain

$$T_2 = \frac{1}{2} \sum_{i=1}^{3} \left[\frac{1}{m_i}(p_{x_i}^2 + p_{y_i}^2) + 2\omega(y_i p_{x_i} - x_i p_{y_i}) + \omega^2(x_i^2 + y_i^2) \right]. \tag{11.34}$$

From Eq. (2.181) we conclude that for this nonnatural case the Hamiltonian is

$$H = T_2 - T_0 + V = \frac{1}{2}\sum_{i=1}^{3}\frac{1}{m_i}(p_{x_i}^2 + p_{y_i}^2) + \omega\sum_{i=1}^{3}(y_i p_{x_i} - x_i p_{y_i}) + V,$$
(11.35)

where V is given by Eq. (11.29).

Noting that the Lagrangian L does not contain the time explicitly, we conclude that the Hamiltonian is constant (see Sec. 2.10)

$$H = h = \text{const.} \tag{11.36}$$

It follows that the system possesses a *Jacobi integral* replacing the energy integral as a constant of the motion.

Introduction of Eq. (11.35) into Eqs. (2.170) yields the Hamilton canonical equations

$$\dot{x}_k = \frac{1}{m_k}p_{x_k} + \omega y_k, \qquad \dot{y}_k = \frac{1}{m_k}p_{y_k} - \omega x_k,$$

$$\dot{p}_{x_k} = \omega p_{y_k} - \frac{\partial V}{\partial x_k}, \qquad \dot{p}_{y_k} = -\omega p_{x_k} - \frac{\partial V}{\partial y_k}, \qquad k = 1, 2, 3, \quad (11.37)$$

which constitute a set of 12 first-order differential equations of the type discussed in Sec. 5.1. Equations (11.37) possess equilibrium positions at points for which the right sides of the equations vanish

$$p_{x_k} + \omega m_k y_k = 0, \qquad p_{y_k} - \omega m_k x_k = 0,$$

$$\omega p_{y_k} - \frac{\partial V}{\partial x_k} = 0, \qquad -\omega p_{x_k} - \frac{\partial V}{\partial y_k} = 0, \qquad k = 1, 2, 3, \quad (11.38)$$

which can be reduced to

$$\omega^2 m_k x_k - \frac{\partial V}{\partial x_k} = 0, \qquad \omega^2 m_k y_k - \frac{\partial V}{\partial y_k} = 0, \qquad k = 1, 2, 3. \quad (11.39)$$

Physically Eqs. (11.39) express the fact that the gravitational forces must balance the centrifugal forces on each of the three bodies. Mathematically they represent the conditions that the function

$$U = V - T_0 = V - \tfrac{1}{2}\omega^2 \sum_{k=1}^{3} m_k(x_k^2 + y_k^2) \tag{11.40}$$

have a stationary value at an equilibrium position, where U may be regarded as a *modified potential energy* function taking into account the centrifugal forces. We recall encountering the function $U = V - T_0$ in our discussion of the stability of canonical systems (Sec. 6.9).

Introduction of Eq. (11.29) into Eqs. (11.39) leads to

$$\omega^2 x_k - G \sum_{i=1}^{3} \delta^*_{ik} \frac{m_i}{\rho_{ik}^3} (x_k - x_i) = 0,$$
$$\omega^2 y_k - G \sum_{i=1}^{3} \delta^*_{ik} \frac{m_i}{\rho_{ik}^3} (y_k - y_i) = 0,$$
$$k = 1, 2, 3, \quad (11.41)$$

from which we conclude that

$$\sum_{k=1}^{3} m_k x_k = 0, \quad \sum_{k=1}^{3} m_k y_k = 0, \quad (11.42)$$

with the not unexpected implication that the center of mass of the system is at rest at the origin of the coordinate system.

If we select the coordinate system so that the positive x axis passes through m_3, we have $x_3 > 0$ and $y_3 = 0$. Letting $k = 3$ in the second of Eqs. (11.41), we obtain

$$\frac{m_1 y_1}{\rho_{13}^3} + \frac{m_2 y_2}{\rho_{23}^3} = 0, \quad (11.43)$$

which, in view of the second of Eqs. (11.42) with $y_3 = 0$, reduces to the form

$$m_1 y_1 \left(\frac{1}{\rho_{13}^3} - \frac{1}{\rho_{23}^3} \right) = 0, \quad (11.44)$$

having two solutions, $\rho_{13} = \rho_{23} = \rho$ and $y_1 = 0$.

Considering first the case in which $\rho_{13} = \rho_{23} = \rho$, we conclude from the first of Eqs. (11.41) with $k = 3$ and the second of Eqs. (11.41) with $k = 1$ that $\omega^2 = GM/\rho^3$, where M is the total mass of the system, and that $\rho_{12} = \rho_{13} = \rho$. Physically this case corresponds to an equilibrium position in which *the three bodies form an equilateral triangle* rotating with respect to an inertial space with constant angular velocity ω about an axis normal to the plane of the motion; in the process the bodies describe coplanar circular orbits.

When $y_1 = 0$, it follows from the second of Eqs. (11.42) that *the three bodies lie along a straight line*, because y_3 is also zero. The straight line is the x axis, which, by definition, rotates with angular velocity ω in the plane of the motion. In this case the first half of Eqs. (11.41), for $x_1 < x_2 < x_3$, becomes

$$\omega^2 x_1 + G \left[\frac{m_2}{(x_2 - x_1)^2} + \frac{m_3}{(x_3 - x_1)^2} \right] = 0,$$
$$\omega^2 x_2 - G \left[\frac{m_1}{(x_1 - x_2)^2} - \frac{m_3}{(x_3 - x_2)^2} \right] = 0, \quad (11.45)$$
$$\omega^2 x_3 - G \left[\frac{m_1}{(x_1 - x_3)^2} + \frac{m_2}{(x_2 - x_3)^2} \right] = 0.$$

To obtain the position of m_2 relative to m_1 and m_3, we let $x_3 - x_2 = \rho$ and arbitrarily choose $x_3 - x_1 = 1$, which implies a mere scale adjustment. Introducing these relations into Eqs. (11.45) and using elementary operations, we obtain the equation

$$(m_1 + m_2)\rho^5 + (3m_1 + 2m_2)\rho^4 + (3m_1 + m_2)\rho^3 - (m_2 + 3m_3)\rho^2 \\ - (2m_2 + 3m_3)\rho - (m_2 + m_3) = 0. \quad (11.46)$$

It turns out that Eq. (11.46) has only one acceptable solution for ρ because, according to our assumption, ρ must be positive. It follows that Eq. (11.46) determines the relative position of the three masses uniquely.

The equilateral-triangle and the straight-line configurations are essentially the *stationary solutions* of Lagrange. In both cases the system rotates with constant angular velocity about the barycenter and, in so doing, describes circles in the plane of the motion relative to an inertial space. To produce this circular motion, the initial velocity of each body must be normal to the radius vector in the plane of the motion and of magnitude proportional to the radial distance from the barycenter. Of course, the respective initial configuration must be either an equilateral triangle or a straight line.

Equations (11.37) admit another special solution in which the bodies form an equilateral triangle in the plane of the motion, but, in contrast with the stationary motion, the distance between the bodies can change with time. This implies that the velocity of each mass relative to the rotating coordinate system x, y, z takes place along the radial directions from the origin O and the magnitudes of these velocities are proportional to the radial distances. Mathematically this can be expressed in the form

$$\dot{x}_k = \alpha(t)x_k, \quad \dot{y}_k = \alpha(t)y_k, \quad k = 1, 2, 3. \quad (11.47)$$

It is not difficult to show that Eqs. (11.47) satisfy the equations of motion (11.37) provided that Eqs. (11.42) hold true. This motion can be produced by projecting the bodies with an initial velocity proportional in magnitude to the respective radial distances from the barycenter and in directions making equal angles with the respective radius vectors from the barycenter. In this case the motions of the three bodies are no longer circular but general conic sections (see Ref. 10, pp. 313–318) with the focus at the system barycenter. Of course, when the conics are ellipses, the motion of the three bodies is periodic.

11.4 THE RESTRICTED THREE-BODY PROBLEM

Of particular interest in the problem of three bodies is the special case in which the mass of one of the bodies is so small that its motion does not affect the motion of the remaining two bodies. The motion of the two massive bodies m_1 and m_2 is obtained from the solution of the two-body problem and can be assumed to be known. The problem is further restricted by considering the case in which m_1 and m_2 move in *coplanar circular orbits* about their barycenter with the constant angular velocity ω whereas the infinitesimal mass m moves under the combined gravitational attraction of both m_1 and m_2. Under these circumstances, the problem is reduced to the investigation of a three-degree-of-freedom system or, if m is further restricted to the plane of motion of m_1 and m_2, to a two-degree-of-freedom system.

Let us consider a coordinate system x, y, z with the origin at O rotating relative to an inertial space with angular velocity ω about axis z. Setting $\ddot{r} = 0$, $r = \rho$, and $\dot{\theta} = \omega$ in the first of Eqs. (1.106), we obtain an expression for the angular velocity

$$\omega^2 = \frac{\mu}{\rho^3} = \frac{G(m_1 + m_2)}{\rho^3}, \qquad (11.48)$$

where ρ is the distance between m_1 and m_2. Without loss of generality we can choose the coordinate system so that bodies m_1 and m_2 are located on the x axis at distances a and b from their barycenter, where the latter coincides with the origin O, as shown in Fig. 11.4. The position of the infinitesimal mass m is given by the coordinates x, y, z, and the radius vectors from m to m_1 and m_2 are denoted by $\boldsymbol{\rho}_1$ and $\boldsymbol{\rho}_2$, respectively. The problem formulation will follow the pattern of Sec. 11.3, but because the motion of m_1 and m_2 is known, we must concern ourselves only with the motion of m. Allowing for motion of m normal to the plane of motion of m_1 and m_2, the kinetic energy of m can be written

$$T = \tfrac{1}{2}m[(\dot{x} - \omega y)^2 + (\dot{y} + \omega x)^2 + \dot{z}^2] = T_2 + T_1 + T_0, \qquad (11.49)$$

FIGURE 11.4

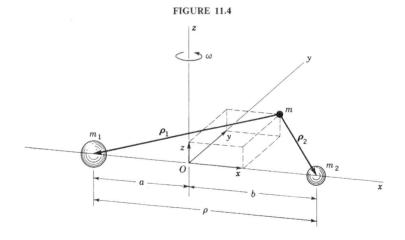

where
$$T_2 = \tfrac{1}{2}m(\dot{x}^2 + \dot{y}^2 + \dot{z}^2),$$
$$T_1 = m\omega(x\dot{y} - y\dot{x}), \quad (11.50)$$
$$T_0 = \tfrac{1}{2}m\omega^2(x^2 + y^2).$$

The potential energy is simply

$$V = -Gm\left(\frac{m_1}{\rho_1} + \frac{m_2}{\rho_2}\right), \quad (11.51)$$

in which ρ_1 and ρ_2 are the magnitudes of the vectors $\boldsymbol{\rho}_1$ and $\boldsymbol{\rho}_2$, respectively. Their expressions are

$$\rho_1 = [(x+a)^2 + y^2 + z^2]^{\frac{1}{2}}, \quad \rho_2 = [(x-b)^2 + y^2 + z^2]^{\frac{1}{2}}. \quad (11.52)$$

Next we express the kinetic energy in terms of generalized momenta rather than velocities and, by analogy with Eqs. (11.35) and (11.36), write the Hamiltonian

$$H = T_2 - T_0 + V = \frac{1}{2}\frac{1}{m}(p_x{}^2 + p_y{}^2 + p_z{}^2) + \omega(yp_x - xp_y) + V$$
$$= h = \text{const}, \quad (11.53)$$

in which

$$p_x = m(\dot{x} - \omega y), \quad p_y = m(\dot{y} + \omega x), \quad p_z = m\dot{z} \quad (11.54)$$

are the generalized momenta. As expected, Eq. (11.53) indicates that the system possesses a Jacobi integral, which is the counterpart of the energy integral for this nonnatural system.

Taking the appropriate derivatives, we obtain the Hamilton canonical equations

$$\dot{x} = \frac{1}{m}p_x + \omega y, \quad \dot{y} = \frac{1}{m}p_y - \omega x, \quad \dot{z} = \frac{1}{m}p_z,$$
$$\dot{p}_x = \omega p_y - \frac{\partial V}{\partial x}, \quad \dot{p}_y = -\omega p_x - \frac{\partial V}{\partial y}, \quad \dot{p}_z = -\frac{\partial V}{\partial z}. \quad (11.55)$$

Equations (11.55) can be reduced to three second-order differential equations of motion

$$\ddot{x} - 2\omega\dot{y} = -\frac{1}{m}\frac{\partial U}{\partial x},$$
$$\ddot{y} + 2\omega\dot{x} = -\frac{1}{m}\frac{\partial U}{\partial y}, \quad (11.56)$$
$$\ddot{z} = -\frac{1}{m}\frac{\partial U}{\partial z},$$

in which U represents a modified potential energy function

$$U = V - T_0 = -Gm\left(\frac{m_1}{\rho_1} + \frac{m_2}{\rho_2}\right) - \tfrac{1}{2}m\omega^2(x^2 + y^2), \quad (11.57)$$

Problems in Celestial Mechanics

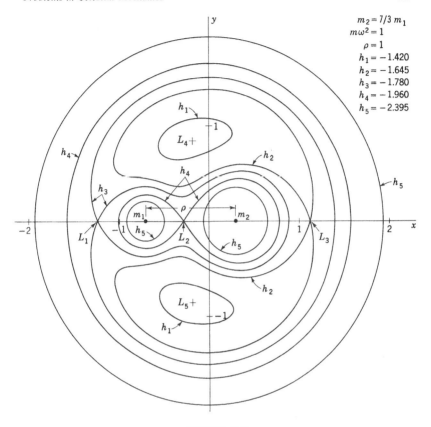

FIGURE 11.5

accounting not only for gravitational forces but also for centrifugal forces.

Denoting by v the total velocity of m relative to the rotating coordinate system, we obtain from Eqs. (11.53), (11.54), and (11.57) the relation

$$v^2 = \frac{2}{m}(h - U), \qquad (11.58)$$

which indicates that motion is possible only in the region $h - U > 0$ of the phase space. The surfaces

$$h = U = \text{const} \qquad (11.59)$$

are called *surfaces of zero relative velocity*. They can be plotted in the three-dimensional space x, y, z. Figure 11.5 shows the intersection of these surfaces and the plane $z = 0$ for given values of h.

When the right sides of Eqs. (11.55) become zero, the mass m has an equilibrium position relative to the rotating system of coordinates. From

Eqs. (11.56), in conjunction with Eqs. (11.52) and (11.57), we conclude that the equilibrium positions satisfy the equations

$$\frac{\partial U}{\partial x} = Gm\left[\frac{m_1(x+a)}{\rho_1{}^3} + \frac{m_2(x-b)}{\rho_2{}^3}\right] - m\omega^2 x = 0,$$

$$\frac{\partial U}{\partial y} = Gm\left(\frac{m_1 y}{\rho_1{}^3} + \frac{m_2 y}{\rho_2{}^3}\right) - m\omega^2 y = 0, \qquad (11.60)$$

$$\frac{\partial U}{\partial z} = Gm\left(\frac{m_1 z}{\rho_1{}^3} + \frac{m_2 z}{\rho_2{}^3}\right) = 0,$$

which, in view of Eq. (11.48), reduce to

$$\frac{m_1(x+a)}{\rho_1{}^3} + \frac{m_2(x-b)}{\rho_2{}^3} - \frac{(m_1+m_2)x}{\rho^3} = 0,$$

$$y\left(\frac{m_1}{\rho_1{}^3} + \frac{m_2}{\rho_2{}^3} - \frac{m_1+m_2}{\rho^3}\right) = 0, \qquad (11.61)$$

$$z = 0.$$

The third of Eqs. (11.61) indicates that all the equilibrium positions are in the xy plane. The second of conditions (11.61) has two solutions, $y = 0$ and $\rho_1 = \rho_2 = \rho$. The first is recognized as the *straight-line configuration*, and the second as the *equilateral-triangle configuration* of Lagrange.

The position of m in the straight-line solution is obtained by letting $y = 0$ in the first of Eqs. (11.61), with the result

$$\frac{m_1(x+a)}{[(x+a)^2]^{\frac{3}{2}}} + \frac{m_2(x-b)}{[(x-b)^2]^{\frac{3}{2}}} - \frac{(m_1+m_2)x}{(a+b)^3} = 0. \qquad (11.62)$$

It turns out that Eq. (11.62) has three solutions corresponding to three positions commonly denoted by L_1, L_2, and L_3. On the other hand, there are two positions, L_4 and L_5, corresponding to the equilateral-triangle solution $\rho_1 = \rho_2 = \rho$. The five positions L_i ($i = 1, 2, \ldots, 5$), shown in Fig. 11.5, are known as *Lagrange libration points*.

The Lagrange solutions of the restricted three-body problem appear at first sight to be of strictly academic interest. There are, however, astrophysical phenomena which can be explained in terms of these solutions, one being the *Gegenschein* (literally counterglow), which is a very faint glow observed at night in a position in the sky exactly opposite to that of the sun. If masses m_1 and m_2 represent the sun and earth, respectively, it can be argued that this glow results from the sun's illuminations of meteoric material temporarily trapped in the equilibrium position L_3, thus providing an example of the straight-line solution. On the other hand, if m_1 and m_2 are the sun and Jupiter, the presence of about a dozen asteroids, called the *Trojans*, oscillating about the positions L_4 and L_5 serves as an illustration of the equilateral-triangle solution.

Another case of particular interest is that in which the two finite bodies m_1 and m_2 perform elliptic orbits in inertial space about their common center of mass. This case is known as the *elliptical restricted three-body problem*. Because the distance ρ between m_1 and m_2 varies periodically, so does the angular velocity ω of the axis through these bodies. This implies that the Lagrangian depends explicitly on time so that no Jacobi integral exists for the system. In this case the equations of motion are nonautonomous. When linearized, they reduce to two simultaneous ordinary differential equations with periodic coefficients of the Hill type. A discussion of this problem is presented by Alfriend and Rand (Ref. 1).

11.5 STABILITY OF MOTION NEAR THE LAGRANGIAN POINTS

In the preceding section we established that the circular restricted three-body problem admits five equilibrium points L_i, known as Lagrangian points. The question remains whether the motion in the neighborhood of these points is stable or not. To test the stability of these points we shall perform an infinitesimal analysis. Having the choice of working with six first-order differential equations, Eqs. (11.55), or with three second-order ones, Eqs. (11.56), we choose the latter, derive the associated set of variational equations, and test the stability of these equations.

Let the motion in the neighborhood of the Lagrangian points be given by

$$x = x_0 + \xi, \qquad y = y_0 + \eta, \qquad z = z_0 + \zeta, \tag{11.63}$$

where x_0, y_0, z_0 represent the equilibrium positions satisfying Eqs. (11.60) and ξ, η, ζ are small perturbations. Introducing Eqs. (11.63) into (11.56) and retaining only the first-order terms in ξ, η, and ζ, we obtain the variational equations

$$\ddot{\xi} - 2\omega\dot{\eta} + \frac{1}{m}\left(\frac{\partial^2 U}{\partial x^2}\xi + \frac{\partial^2 U}{\partial x\,\partial y}\eta + \frac{\partial^2 U}{\partial x\,\partial z}\zeta\right) = 0,$$

$$\ddot{\eta} + 2\omega\dot{\xi} + \frac{1}{m}\left(\frac{\partial^2 U}{\partial x\,\partial y}\xi + \frac{\partial^2 U}{\partial y^2}\eta + \frac{\partial^2 U}{\partial y\,\partial z}\zeta\right) = 0, \tag{11.64}$$

$$\ddot{\zeta} + \frac{1}{m}\left(\frac{\partial^2 U}{\partial x\,\partial z}\xi + \frac{\partial^2 U}{\partial y\,\partial z}\eta + \frac{\partial^2 U}{\partial z^2}\zeta\right) = 0,$$

in which $\partial^2 U/\partial x^2$, $\partial^2 U/(\partial x\,\partial y)$, etc., are evaluated at the equilibrium point under investigation and are to be regarded as constant. Performing the appropriate differentiations in Eq. (11.57), we obtain the expressions

$$\frac{\partial^2 U}{\partial x^2} = Gm\left\{m_1\left[\frac{1}{\rho_1^3} - \frac{3(x+a)^2}{\rho_1^5}\right] + m_2\left[\frac{1}{\rho_2^3} - \frac{3(x-b)^2}{\rho_2^5}\right]\right\} - m\omega^2,$$

$$\frac{\partial^2 U}{\partial x\,\partial y} = -Gm\left[m_1\frac{3(x+a)y}{\rho_1^5} + m_2\frac{3(x-b)y}{\rho_2^5}\right],$$

$$\frac{\partial^2 U}{\partial x\,\partial z} = -Gm\left[m_1\frac{3(x+a)z}{\rho_1^5} + m_2\frac{3(x-b)z}{\rho_2^5}\right],$$

$$\frac{\partial^2 U}{\partial y^2} = Gm\left[m_1\left(\frac{1}{\rho_1^3} - \frac{3y^2}{\rho_1^5}\right) + m_2\left(\frac{1}{\rho_2^3} - \frac{3y^2}{\rho_2^5}\right)\right] - m\omega^2, \quad (11.65)$$

$$\frac{\partial^2 U}{\partial y\,\partial z} = -Gm\left(m_1\frac{3yz}{\rho_1^5} + m_2\frac{3yz}{\rho_2^5}\right),$$

$$\frac{\partial^2 U}{\partial z^2} = Gm\left[m_1\left(\frac{1}{\rho_1^3} - \frac{3z^2}{\rho_1^5}\right) + m_2\left(\frac{1}{\rho_2^3} - \frac{3z^2}{\rho_2^5}\right)\right].$$

Let us first check the stability of the libration point L_3, for which $y = z = 0$. Because in this case $x + a > 0$ and $x - b > 0$, we can set $x + a = \rho_1$ and $x - b = \rho_2$ in Eqs. (11.65) and write

$$\left.\frac{\partial^2 U}{\partial x^2}\right|_{L_3} = -m(2\Omega^2 + \omega^2), \quad \left.\frac{\partial^2 U}{\partial y^2}\right|_{L_3} = m(\Omega^2 - \omega^2),$$

$$\left.\frac{\partial^2 U}{\partial z^2}\right|_{L_3} = m\Omega^2, \quad \left.\frac{\partial^2 U}{\partial x\,\partial y}\right|_{L_3} = \left.\frac{\partial^2 U}{\partial x\,\partial z}\right|_{L_3} = \left.\frac{\partial^2 U}{\partial y\,\partial z}\right|_{L_3} = 0, \quad (11.66)$$

where Ω^2 is defined by

$$\Omega^2 = G\left(\frac{m_1}{\rho_1^3} + \frac{m_2}{\rho_2^3}\right). \quad (11.67)$$

Introducing Eqs. (11.66) into Eqs. (11.64), we obtain the variational equations associated with the equilibrium position L_3

$$\ddot{\xi} - 2\omega\dot{\eta} - (2\Omega^2 + \omega^2)\xi = 0,$$
$$\ddot{\eta} + 2\omega\dot{\xi} + (\Omega^2 - \omega^2)\eta = 0, \quad (11.68)$$
$$\ddot{\zeta} + \Omega^2\zeta = 0.$$

The third of Eqs. (11.68) is uncoupled from the first two, and it leads us to conclude that the mass m undergoes harmonic motion in the z direction with frequency Ω independently of its motion in the xy plane. The first two of Eqs. (11.68), on the other hand, yield the characteristic equation

$$\begin{vmatrix} \lambda^2 - (2\Omega^2 + \omega^2) & -2\omega\lambda \\ 2\omega\lambda & \lambda^2 + \Omega^2 - \omega^2 \end{vmatrix}$$
$$= \lambda^4 - (\Omega^2 - 2\omega^2)\lambda^2 - (2\Omega^2 + \omega^2)(\Omega^2 - \omega^2) = 0, \quad (11.69)$$

which has the solution

$$\lambda^2 = \tfrac{1}{2}(\Omega^2 - 2\omega^2) \pm \tfrac{1}{2}(9\Omega^4 - 8\Omega^2\omega^2)^{\frac{1}{2}}. \quad (11.70)$$

Problems in Celestial Mechanics

To determine the nature of the eigenvalues λ_i ($i = 1, 2, 3, 4$), we turn to Eq. (11.62), write it in the form

$$\frac{x}{G}(\Omega^2 - \omega^2) + \frac{m_1 a}{\rho_1^3} - \frac{m_2 b}{\rho_2^3} = 0,$$

and recall that $m_1 a = m_2 b$. Moreover, for the L_3 equilibrium position, we have $\rho_1 > \rho_2$ and $x > 0$, from which it follows that $\Omega^2 > \omega^2$. Taking the positive sign on the right side of Eq. (11.70), we conclude that in this case $\lambda^2 > 0$. Corresponding to $\lambda^2 > 0$ we obtain two real eigenvalues of equal magnitude but opposite signs. Taking the negative sign on the right side of Eq. (11.70), we obtain $\lambda^2 < 0$, to which correspond two purely imaginary complex conjugate eigenvalues. For instability it is sufficient that at least one of the eigenvalues possess a positive real part, and because corresponding to $\lambda^2 > 0$ there is a real and positive eigenvalue, we conclude that *the motion in the neighborhood of L_3 is unstable*. In a similar fashion, it can be shown that *the motion in the neighborhood of L_1 and L_2 is also unstable*.

Next let us turn our attention to the equilibrium point L_4, defined by $\rho_1 = \rho_2 = \rho$, $x + a = \rho/2$, $x - b = -\rho/2$, $y = (\sqrt{3}/2)\rho$, and $z = 0$. Introducing these values into expressions (11.65), we have

$$\left.\frac{\partial^2 U}{\partial x^2}\right|_{L_4} = -\tfrac{3}{4}m\omega^2, \quad \left.\frac{\partial^2 U}{\partial y^2}\right|_{L_4} = -\tfrac{9}{4}m\omega^2, \quad \left.\frac{\partial^2 U}{\partial z^2}\right|_{L_4} = m\omega^2,$$

$$\left.\frac{\partial^2 U}{\partial x\,\partial y}\right|_{L_4} = -\frac{3\sqrt{3}}{4}m\omega^2(1 - 2R), \quad \left.\frac{\partial^2 U}{\partial x\,\partial z}\right|_{L_4} = \left.\frac{\partial^2 U}{\partial y\,\partial z}\right|_{L_4} = 0,$$

(11.71)

where $R = m_2/(m_1 + m_2)$. The corresponding variational equations become

$$\ddot{\xi} - 2\omega\dot{\eta} - \tfrac{3}{4}\omega^2\xi - \frac{3\sqrt{3}}{4}\omega^2(1 - 2R)\eta = 0,$$

$$\ddot{\eta} + 2\omega\dot{\xi} - \frac{3\sqrt{3}}{4}\omega^2(1 - 2R)\xi - \tfrac{9}{4}\omega^2\eta = 0,$$

(11.72)

$$\ddot{\zeta} + \omega^2\zeta = 0.$$

The third of Eqs. (11.72) again shows that the motion in the z direction is independent of the motion in the xy plane. The motion is harmonic with frequency ω, which is the same as the frequency of rotation of the axis through m_1 and m_2. The first two of Eqs. (11.72) lead to the characteristic equation

$$\begin{vmatrix} \lambda^2 - \tfrac{3}{4}\omega^2 & -2\omega\lambda - \dfrac{3\sqrt{3}}{4}\omega^2(1 - 2R) \\ 2\omega\lambda - \dfrac{3\sqrt{3}}{4}\omega^2(1 - 2R) & \lambda^2 - \tfrac{9}{4}\omega^2 \end{vmatrix}$$

$$= \lambda^4 + \omega^2\lambda^2 + \tfrac{27}{4}\omega^2 R(1 - R) = 0, \quad (11.73)$$

yielding the solution

$$\lambda^2 = -\frac{\omega^2}{2} \pm \frac{\omega^2}{2}[1 - 27R(1 - R)]^{\frac{1}{2}}. \quad (11.74)$$

For stability we must have $\lambda^2 < 0$, which requires that

$$1 - 27R(1 - R) \geq 0. \quad (11.75)$$

It is evident on physical grounds that $R(1 - R) = m_1 m_2/(m_1 + m_2)^2$ is positive, so that the radical in Eq. (11.74) cannot exceed unity. Assuming that $m_2 < m_1$, condition (11.75) is equivalent to the statement that the *motion in the vicinity of the equilibrium point L_4 is stable provided $m_2 < 0.0385(m_1 + m_2)$*.

We must be aware of the fact that, even when the condition $m_2 < 0.0385(m_1 + m_2)$ is satisfied, the motion in the neighborhood of L_4 is *only infinitesimally stable*, as the conclusion was reached on the basis of the linear approximation. From the above stability statement we cannot infer that the equilateral-triangle solutions are stable in the sense of Liapunov. Indeed, no proof has been offered that the motion in the neighborhood of L_4 is Liapunov stable (see Ref. 16, p. 372). An attempt to use the Liapunov direct method with the Hamiltonian as the Liapunov function proves fruitless, as the Hamiltonian, which in our case is the Jacobi integral, turns out not to be positive definite at L_4, so that on the basis of such an analysis we cannot conclude that the motion in the neighborhood of L_4 is bounded.

More light can be shed on the problem by considering nonlinear terms in the expansion of the potential function U about L_4. A step in this direction has been made by Breakwell and Pringle (Ref. 2). Because the motion in the neighborhood of the equilateral libration points exhibits critical behavior, it is also natural to examine how second-order effects, such as perturbing forces from a fourth body, affect its stability. The stability of the equilateral-triangle solutions corresponding to the earth-moon system has received a great deal of attention due to the possibility of stationing artificial satellites at these points. The sun's perturbing effects on the stability of the L_4 and L_5 libration points has been investigated by a number of researchers. We shall not go into the details of this problem but refer the interested reader to the work by Schechter (Ref. 12), which is one of the most complete.

11.6 THE EQUATIONS OF RELATIVE MOTION. DISTURBING FUNCTION

A truly inertial system of coordinates is difficult to establish or even justify, although the concept has its merits on practical grounds. Moreover, it is often desirable to investigate the relative motion rather than the absolute one. In particular, this is so when we wish to choose the origin of a frame of reference at a point other than the center of mass of the system. For example, it is conceivable that on occasion it is advantageous to place the origin at the sun's center or the earth's center. This points to the need for the derivation of a set of equations for the relative motion.

We have shown (Sec. 11.2) that the equations for the absolute motion of a system of n bodies can be written as

$$m_i \ddot{\mathbf{r}}_i = G \sum_{j=1}^{n} \delta_{ij}^* \frac{m_i m_j}{\rho_{ij}^3} \boldsymbol{\rho}_{ij}, \quad i = 1, 2, \ldots, n, \tag{11.76}$$

where $\mathbf{r}_i$ is the absolute position of body m_i in space and $\boldsymbol{\rho}_{ij}$ gives the position of m_j relative to m_i. From Fig. 11.3 we observe that

$$\boldsymbol{\rho}_{ij} = \mathbf{r}_j - \mathbf{r}_i = -\boldsymbol{\rho}_{ji}. \tag{11.77}$$

Let us now divide Eqs. (11.76) by m_i and obtain

$$\ddot{\mathbf{r}}_i = G \sum_{k=1}^{n} \delta_{ik}^* \frac{m_k}{\rho_{ik}^3} \boldsymbol{\rho}_{ik}, \quad i = 1, 2, \ldots, n, \tag{11.78}$$

where the dummy index j has been replaced by k. Similarly

$$\ddot{\mathbf{r}}_j = G \sum_{k=1}^{n} \delta_{jk}^* \frac{m_k}{\rho_{jk}^3} \boldsymbol{\rho}_{jk}, \quad j = 1, 2, \ldots, n. \tag{11.79}$$

Subtracting Eqs. (11.78) from (11.79) and recalling relation (11.77), we arrive after a few elementary operations at

$$\ddot{\boldsymbol{\rho}}_{ij} + G \frac{m_i + m_j}{\rho_{ij}^3} \boldsymbol{\rho}_{ij} - G \sum_{k=1}^{n} \delta_{ik}^* \delta_{jk}^* m_k \left(\frac{\boldsymbol{\rho}_{jk}}{\rho_{jk}^3} - \frac{\boldsymbol{\rho}_{ik}}{\rho_{ik}^3} \right),$$
$$i, j = 1, 2, \ldots, n,$$
$$i \neq j, \tag{11.80}$$

which are the differential equations for the relative motion. But

$$\boldsymbol{\rho}_{jk} = \boldsymbol{\rho}_{ik} - \boldsymbol{\rho}_{ij}, \tag{11.81}$$

so that Eqs. (11.80) can be rewritten in the form

$$\ddot{\boldsymbol{\rho}}_{ij} + G \frac{m_i + m_j}{\rho_{ij}^3} \boldsymbol{\rho}_{ij} = G \sum_{k=1}^{n} \delta_{ik}^* \delta_{jk}^* m_k \left(\frac{\boldsymbol{\rho}_{ik} - \boldsymbol{\rho}_{ij}}{\rho_{jk}^3} - \frac{\boldsymbol{\rho}_{ik}}{\rho_{ik}^3} \right),$$
$$i, j = 1, 2, \ldots, n,$$
$$i \neq j, \tag{11.82}$$

where

$$\rho_{jk} = [(\boldsymbol{\rho}_{ik} - \boldsymbol{\rho}_{ij}) \cdot (\boldsymbol{\rho}_{ik} - \boldsymbol{\rho}_{ij})]^{\frac{1}{2}}. \tag{11.83}$$

Next let us introduce the scalar function

$$R_{ij} = G \sum_{k=1}^{n} \delta_{ik}^* \delta_{jk}^* m_k \left(\frac{1}{\rho_{jk}} - \frac{\boldsymbol{\rho}_{ij} \cdot \boldsymbol{\rho}_{ik}}{\rho_{ik}^3} \right) \qquad (11.84)$$

and also define a del operator by

$$\boldsymbol{\nabla}_{ij} = \frac{\partial}{\partial x_{ij}} \mathbf{i} + \frac{\partial}{\partial y_{ij}} \mathbf{j} + \frac{\partial}{\partial z_{ij}} \mathbf{k}, \qquad (11.85)$$

where x_{ij}, y_{ij}, z_{ij} are the components of the vector $\boldsymbol{\rho}_{ij}$ along a cartesian system of axes defined by the unit vectors $\mathbf{i}, \mathbf{j}, \mathbf{k}$. From Eqs. (11.84) and (11.85) we conclude that

$$\boldsymbol{\nabla}_{ij} R_{ij} = G \sum_{k=1}^{n} \delta_{ik}^* \delta_{jk}^* m_k \left(\frac{\boldsymbol{\rho}_{ik} - \boldsymbol{\rho}_{ij}}{\rho_{jk}^3} - \frac{\boldsymbol{\rho}_{ik}}{\rho_{ik}^3} \right), \qquad (11.86)$$

because $\boldsymbol{\rho}_{ik}$ does not depend on x_{ij}, y_{ij}, and z_{ij}. This enables us to write Eqs. (11.82) in the form

$$\ddot{\boldsymbol{\rho}}_{ij} + G \frac{m_i + m_j}{\rho_{ij}^3} \boldsymbol{\rho}_{ij} = \boldsymbol{\nabla}_{ij} R_{ij}, \qquad \begin{array}{l} i = 1, 2, \ldots, n, \\ i \neq j. \end{array} \qquad (11.87)$$

The functions R_{ij} are referred to as *disturbing* or *perturbative functions*. Such functions are used in the study of planetary motion to describe the disturbing effect of other planets whose motions are generally known. For example, it can be used to describe the influence of the sun on a three-body system consisting of the earth, the moon, and an artificial satellite in the neighborhood of an equilateral-triangle libration point. Although in our derivations the disturbing effect is due to gravitational attraction, the disturbing function may also result from other sources. We shall have an opportunity to return to this subject later in this chapter and again in Chap. 12.

11.7 GRAVITATIONAL POTENTIAL AND TORQUES FOR AN ARBITRARY BODY

In our discussion of the many-body problem we made the standard assumptions that the only forces acting upon the system are due to the mutual gravitational attraction and that the bodies possess spherically symmetric distribution or are sufficiently remote from one another to behave like particles. If a body in a central force gravitational field does not possess spherical symmetry, the center of gravity does not coincide with the center of mass and, in addition to a resultant force, differential-gravity torques are also present. These gravitational torques become significant in many problems of dynamics, as we shall have an opportunity to find out.

Problems in Celestial Mechanics

Let us consider a two-body system consisting of one body m_1 possessing spherical symmetry and another body of total mass m_2 having an arbitrary mass distribution. Because m_1 is symmetric, it can be assumed that its geometric center coincides with the center of force, as shown in Fig. 11.6. The interest lies in specific expressions for the gravitational potential as well as differential-gravity torques acting upon m_2.

The potential energy corresponding to an inverse square force field can be written in the integral form

$$V = -Gm_1 \int_{m_2} \frac{dm_2}{r}, \tag{11.88}$$

where $r = |\mathbf{R} - \boldsymbol{\rho}|$ is the distance between the center of force and the differential element of mass dm_2. The reciprocal of the radial distance can be expanded in the power series

$$\begin{aligned}
r^{-1} &= R^{-1}\left[1 - \frac{2\mathbf{R}\cdot\boldsymbol{\rho}}{R^2} + \left(\frac{\rho}{R}\right)^2\right]^{-\frac{1}{2}} \\
&= R^{-1} + R^{-3}\mathbf{R}\cdot\boldsymbol{\rho} + \tfrac{1}{2}R^{-3}[3R^{-2}(\mathbf{R}\cdot\boldsymbol{\rho})^2 - \rho^2] \\
&\quad + \tfrac{1}{2}R^{-4}[5R^{-3}(\mathbf{R}\cdot\boldsymbol{\rho})^3 - 3R^{-1}(\mathbf{R}\cdot\boldsymbol{\rho})\rho^2] + \tfrac{1}{8}R^{-5}[35R^{-4}(\mathbf{R}\cdot\boldsymbol{\rho})^4 \\
&\quad - 30R^{-2}(\mathbf{R}\cdot\boldsymbol{\rho})^2\rho^2 + 3\rho^4] + \cdots.
\end{aligned} \tag{11.89}$$

Letting $\mathbf{R}\cdot\boldsymbol{\rho} = R\rho\cos\beta$, we can write Eq. (11.89) in the compact formula

$$\frac{1}{r} = \frac{1}{R}\sum_{n=0}^{\infty}\left(\frac{\rho}{R}\right)^n P_n(\cos\beta), \tag{11.90}$$

where $P_n(\cos\beta)$ are Legendre polynomials of argument $\cos\beta$ and degree n.

FIGURE 11.6

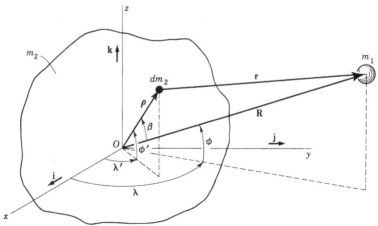

Substituting $\cos \beta = \mu$ in Eq. (11.89) and comparing with Eq. (11.90), we conclude that the first few polynomials have the form

$$P_0(\mu) = 1,$$
$$P_1(\mu) = \mu,$$
$$P_2(\mu) = \tfrac{1}{2}(3\mu^2 - 1),$$
$$P_3(\mu) = \tfrac{1}{2}(5\mu^3 - 3\mu),$$
$$P_4(\mu) = \tfrac{1}{8}(35\mu^4 - 30\mu^2 + 3),$$
$$\dots \dots \dots \dots \dots \dots \dots \dots$$
(11.91)

The Legendre polynomials satisfy the recurrence formula (see Ref. 4, p. 151)

$$nP_n(\mu) = (2n - 1)\mu P_{n-1}(\mu) - (n - 1)P_{n-2}(\mu). \tag{11.92}$$

Introducing Eq. (11.90) into (11.88), we obtain

$$V = -\frac{Gm_1}{R} \sum_{n=0}^{\infty} \int_{m_2} \left(\frac{\rho}{R}\right)^n P_n(\cos \beta)\, dm_2. \tag{11.93}$$

For a more detailed expression for V, we express $\mathbf{R}$ and $\boldsymbol{\rho}$ in terms of the spherical coordinates R, λ, ϕ and ρ, λ', ϕ', respectively, and conclude that

$$\cos \beta = \frac{\mathbf{R} \cdot \boldsymbol{\rho}}{R\rho} = \sin \phi \sin \phi' + \cos \phi \cos \phi' \cos (\lambda - \lambda'). \tag{11.94}$$

Next we introduce the *Laplacian coefficient* $L_n(\phi, \lambda, \phi', \lambda')$, namely, a *surface spherical harmonic* of degree n defined as the Legendre polynomial of degree n and argument $\cos \beta$, where $\cos \beta$ is given by Eq. (11.94). The Laplacian coefficient can be represented by a series of the form (Ref. 4, p. 211)

$$L_n(\lambda, \phi, \lambda', \phi') = P_n(\cos \beta) = P_n[\sin \phi \sin \phi' + \cos \phi \cos \phi' \cos (\lambda - \lambda')]$$
$$= P_n(\sin \phi) P_n(\sin \phi')$$
$$+ 2 \sum_{m=1}^{n} \left[\frac{(n-m)!}{(n+m)!} P_{nm}(\sin \phi) P_{nm}(\sin \phi') \cos m(\lambda - \lambda')\right],$$
(11.95)

in which the quantities

$$P_{nm}(\mu) = \frac{(1-\mu^2)^{m/2}}{2^n} \sum_{i=0}^{k} \frac{(-1)^i (2n - 2i)!\, \mu^{n-m-2i}}{i!\,(n-m-2i)!\,(n-i)!} \tag{11.96}$$

are referred to as associated Legendre functions of nth degree and mth order (Ref. 4, p. 197). The integer k has the values

$$k = \begin{cases} \dfrac{n-m}{2} & n - m \text{ even,} \\[2mm] \dfrac{n-m-1}{2} & n - m \text{ odd.} \end{cases} \tag{11.97}$$

Note that for $m = 0$ the functions $P_{nm}(\mu)$ reduce to the Legendre polynomials (11.91). Introducing Eq. (11.95) into (11.93), we obtain the expression

$$V = -\frac{Gm_1}{R} \sum_{n=0}^{\infty} \int_{m_2} \left(\frac{\rho}{R}\right)^n L_n(\lambda,\phi,\lambda',\phi') \, dm_2$$

$$= -\frac{Gm_1 m_2}{R} - \frac{Gm_1}{R} \sum_{n=1}^{\infty} \left[\frac{P_n(\sin\phi)}{R^n} \int_{m_2} \rho^n P_n(\sin\phi') \, dm_2 \right.$$

$$+ 2 \sum_{m=1}^{n} \frac{(n-m)!}{(n+m)!} P_{nm}(\sin\phi) \cos m\lambda \int_{m_2} \rho^n P_{nm}(\sin\phi') \cos m\lambda' \, dm_2$$

$$\left. + 2 \sum_{m=1}^{n} \frac{(n-m)!}{(n+m)!} P_{nm}(\sin\phi) \sin m\lambda \int_{m_2} \rho^n P_{nm}(\sin\phi') \sin m\lambda' \, dm_2 \right]. \tag{11.98}$$

The form of Eq. (11.98) is convenient for the evaluation of the gravitational potential energy when the shape of m_2 is nearly spherical, in which case the mass differential dm_2 can be written $\gamma \rho^2 \, d\rho \, d\lambda' \, d\phi'$, where γ is the density. It should be recalled that the earth and the moon are nearly spherical. Denoting by a the mean equatorial radius of m_2, Eq. (11.98) can be written

$$V = -\frac{Gm_1 m_2}{R}\left\{1 + \sum_{n=1}^{\infty}\left[\left(\frac{a}{R}\right)^n C_{n0} P_n(\sin\phi)\right.\right.$$

$$\left.\left. + \sum_{m=1}^{n} \left(\frac{a}{R}\right)^n P_{nm}(\sin\phi)(C_{nm} \cos m\lambda + S_{nm} \sin m\lambda)\right]\right\}, \tag{11.99}$$

in which the coefficients

$$C_{n0} = \frac{1}{a^n m_2} \int_{m_2} \rho^n P_{nm}(\sin\phi') \, dm_2,$$

$$C_{nm} = \frac{2}{a^n m_2} \frac{(n-m)!}{(n+m)!} \int_{m_2} \rho^n P_{nm}(\sin\phi') \cos m\lambda' \, dm_2, \tag{11.100}$$

$$S_{nm} = \frac{2}{a^n m_2} \frac{(n-m)!}{(n+m)!} \int_{m_2} \rho^n P_{nm}(\sin\phi') \sin m\lambda' \, dm_2$$

reflect the mass distribution as well as the shape of the body. The integrands in Eqs. (11.100) are given by MacMillan (Ref. 8, p. 372) in terms of the cartesian coordinates x, y, z.

Equation (11.99) represents the solution of $\nabla^2 V = 0$ in terms of spherical coordinates and has essentially the form recommended by The International Astronomical Union of 1961, where $P_{nm}(\sin\phi) \cos m\lambda$ and $P_{nm}(\sin\phi) \sin m\lambda$ are referred to as *tesseral harmonics* of nth degree and mth order. Tesseral harmonics of order zero, $P_{n0} = P_n$, are called *zonal harmonics*. They are Legendre polynomials depending on the latitude ϕ but not on the longitude λ. They are the only ones present in the case of bodies of revolution. Tesseral harmonics for which the degree is equal to the order, $n = m \neq 0$, are referred to as *sectorial harmonics*. The combination of tesseral harmonics

$$Y_n(\lambda,\phi) = \sum_{m=0}^{n} (C_{nm} \cos m\lambda + S_{nm} \sin m\lambda) P_{nm}(\sin \phi) \quad (11.101)$$

represents a *surface spherical harmonic* of degree n, and the quantities $r^{-(n+1)} Y_n$ and $r^n Y_n$ are referred to as *solid spherical harmonics*.

Experiments have been conducted to determine the exact shape of the earth (see Ref. 7) by observing the motion of satellites orbiting the earth. Similar work has been conducted in an effort to determine the shape of the moon.

For a body of arbitrary shape and mass distribution, it proves convenient to express the gravitational potential energy by means of inertial integrals in terms of the cartesian coordinates x, y, z as well as the direction cosines between these coordinates and the radius vector $\mathbf{R}$ to m_1. To this end, we write

$$\boldsymbol{\rho} = x\mathbf{i} + y\mathbf{j} + z\mathbf{k},$$
$$\mathbf{R} = R(l\mathbf{i} + m\mathbf{j} + n\mathbf{k}), \quad (11.102)$$
$$\mathbf{R} \cdot \boldsymbol{\rho} = R(lx + my + nz),$$

where l, m, and n are the corresponding direction cosines. Substituting Eqs. (11.89) and (11.102) into (11.88), we obtain

$$V = -\frac{Gm_1 m_2}{R} - \frac{Gm_1}{R^2} \int_{m_2} (lx + my + nz) \, dm_2$$
$$- \frac{Gm_1}{2R^3} \int_{m_2} [3(lx + my + nz)^2 - (x^2 + y^2 + z^2)] \, dm_2$$
$$- \frac{Gm_1}{2R^4} \int_{m_2} [5(lx + my + nz)^3 - 3(lx + my + nz)(x^2 + y^2 + z^2)] \, dm_2$$
$$- \frac{Gm_1}{8R^5} \int_{m_2} [35(lx + my + nz)^4 - 30(lx + my + nz)^2(x^2 + y^2 + z^2)$$
$$+ 3(x^2 + y^2 + z^2)^2] \, dm_2 + \cdots . \quad (11.103)$$

Next we introduce the *inertial integrals* defined by

$$J_{x^p y^q z^r} = \int_{m_2} x^p y^q z^r \, dm_2, \quad p, q, r \geq 0, \quad (11.104)$$

where p, q, and r are integers. The inertial integrals (11.104) are related to the coefficients C_{n0}, C_{nm}, and S_{nm}, Eqs. (11.100). Expressions (11.104) enable us to write Eq. (11.103) in the form

$$V = -\frac{Gm_1m_2}{R} - \frac{Gm_1m_2}{R^2}(l\bar{x} + m\bar{y} + n\bar{z})$$

$$- \frac{Gm_1}{2R^3}[(3l^2 - 1)J_{xx} + (3m^2 - 1)J_{yy} + (3n^2 - 1)J_{zz}$$

$$+ 6(lmJ_{xy} + lnJ_{xz} + mnJ_{yz})]$$

$$- \frac{Gm_1}{2R^4}[l(5l^2 - 3)J_{xxx} + m(5m^2 - 3)J_{yyy} + n(5n^2 - 3)J_{zzz}$$

$$+ 3m(5l^2 - 1)J_{xxy} + 3n(5l^2 - 1)J_{xxz} + 3l(5m^2 - 1)J_{xyy}$$

$$+ 3n(5m^2 - 1)J_{yyz} + 3l(5n^2 - 1)J_{xzz} + 3m(5n^2 - 1)J_{yzz}$$

$$+ 6lmnJ_{xyz}]$$

$$- \frac{Gm_1}{8R^5}\{(35l^4 - 30l^2 + 3)J_{xxxx} + (35m^4 - 30m^2 + 3)J_{yyyy}$$

$$+ (35n^2 - 30n^2 + 3)J_{zzzz} + 20lm(7l^2 - 3)J_{xxxy}$$

$$+ 20lm(7m^2 - 3)J_{xyyy} + 20ln(7l^2 - 3)J_{xxxz}$$

$$+ 20ln(7n^2 - 3)J_{xzzz} + 20mn(7m^2 - 3)J_{yyyz}$$

$$+ 20mn(7n^2 - 3)J_{yzzz} + 6[35l^2m^2 - 5(l^2 + m^2) + 1]J_{xxyy}$$

$$+ 6[35l^2n^2 - 5(l^2 + n^2) + 1]J_{xxzz}$$

$$+ 6[35m^2n^2 - 5(m^2 + n^2) + 1]J_{yyzz} + 60mn(7l^2 - 1)J_{xxyz}$$

$$+ 60ln(7m^2 - 1)J_{xyyz} + 60lm(7n^2 - 1)J_{xyzz}\} + \cdots, \quad (11.105)$$

where $\bar{x} = J_x$, $\bar{y} = J_y$, $\bar{z} = J_z$ denote the coordinates of the center of mass C relative to axes x, y, z. The second-order inertial integrals can be given the more familiar form

$$J_{xx} = \int_{m_2} x^2 \, dm_2 = \tfrac{1}{2}(I_{yy} + I_{zz} - I_{xx}),$$

$$J_{yy} = \int_{m_2} y^2 \, dm_2 = \tfrac{1}{2}(I_{zz} + I_{xx} - I_{yy}),$$

$$J_{zz} = \int_{m_2} z^2 \, dm_2 = \tfrac{1}{2}(I_{xx} + I_{yy} - I_{zz}),$$

$$J_{xy} = \int_{m_2} xy \, dm_2 = I_{xy},$$

$$J_{xz} = \int_{m_2} xz \, dm_2 = I_{xz},$$

$$J_{yz} = \int_{m_2} yz \, dm_2 = I_{yz},$$

(11.106)

where I_{xx}, I_{yy}, I_{zz} are the mass moments of inertia and I_{xy}, I_{xz}, I_{yz} are the mass products of inertia introduced in Sec. 4.2.

For many practical purposes it is sufficient to retain in V only terms in the inertial integrals through second order. Moreover, if the origin of axes x, y, z is chosen to coincide with the center of mass, then $\bar{x} = \bar{y} = \bar{z} = 0$, so that V reduces to

$$V \approx -\frac{Gm_1m_2}{R} - \frac{Gm_1}{4R^3}[(3l^2 - 1)(I_{yy} + I_{zz} - I_{xx})$$
$$+ (3m^2 - 1)(I_{xx} + I_{zz} - I_{yy}) + (3n^2 - 1)(I_{xx} + I_{yy} - I_{zz})$$
$$+ 12(lmI_{xy} + lnI_{xz} + mnI_{yz})]. \tag{11.107}$$

Letting x, y, z be the principal axes of body m_2 and denoting the moments of inertia about these axes by A, B, C, respectively, Eq. (11.107) can be further reduced to

$$V \approx -\frac{Gm_1m_2}{R} - \frac{Gm_1}{4R^3}[(3l^2 - 1)(B + C - A) + (3m^2 - 1)(A + C - B)$$
$$+ (3n^2 - 1)(A + B - C)]. \tag{11.108}$$

Again retaining the equivalent of terms in inertial integrals through second order only, we can give V a different form by writing

$$V \approx \frac{Gm_1}{R}\left[\int_{m_2} dm_2 + \frac{1}{R}\int_{m_2} \rho \cos\beta \, dm_2 + \frac{1}{2R^2}\int_{m_2} \rho^2(3\cos^2\beta - 1) \, dm_2\right]. \tag{11.109}$$

The second integral in (11.109) vanishes if the origin coincides with the center of mass. Moreover, when the moment of inertia of m_2 about an axis coinciding with **R** is denoted by

$$I = \int_{m_2} \rho^2 \sin^2\beta \, dm_2, \tag{11.110}$$

Eq. (11.109) becomes

$$V = -\frac{Gm_1m_2}{R} - \frac{Gm_1}{2R^3}(A + B + C - 3I), \tag{11.111}$$

which is known as *MacCullagh's formula*.

If all the moments of inertia of m_2 are equal, $A = B = C = I$, or if the radial distance R between m_1 and m_2 is very large, Eq. (11.108) or (11.111) yields

$$V = -\frac{Gm_1m_2}{R}, \tag{11.112}$$

so that in either of these two cases the assumption that the two bodies behave

Problems in Celestial Mechanics

like particles is justified. This implies that the gravitational forces pass through the center of mass of the bodies and there are no torques involved.

The case in which differential-gravity torques are present is not without interest. The torques can be obtained by differentiating the gravitational potential energy V. It is often desirable, however, to work with torque components about three orthogonal axes, such as x, y, z in Fig. 11.6. From that figure we see that the force exerted by m_1 on the differential element of mass dm_2 can be written

$$d\mathbf{F} = \frac{Gm_1\, dm_2}{r^3}\mathbf{r}, \tag{11.113}$$

so that the torque about O is simply

$$\mathbf{N} = \int_{m_2} \boldsymbol{\rho} \times d\mathbf{F} = Gm_1 \int_{m_2} \frac{\boldsymbol{\rho} \times \mathbf{R}}{r^3}\, dm_2. \tag{11.114}$$

For practical reasons we truncate the expressions for the torques by including terms only through second order in ρ. Correspondingly, we have

$$r^{-3} = R^{-3}\left[1 - \frac{2\mathbf{R}\cdot\boldsymbol{\rho}}{R^2} + \left(\frac{\rho}{R}\right)^2\right]^{-\frac{3}{2}} \approx R^{-3}\left[1 + \frac{3(lx + my + nz)}{R}\right]. \tag{11.115}$$

Moreover,

$$\boldsymbol{\rho} \times \mathbf{R} = R[(ny - mz)\mathbf{i} + (lz - nx)\mathbf{j} + (mx - ly)\mathbf{k}]. \tag{11.116}$$

Assuming that the origin O is at the center of mass of m_2, introducing Eqs. (11.115) and (11.116) into (11.114), and performing the integration, we obtain

$$\begin{aligned}\mathbf{N} &\approx \frac{Gm_1}{R^2}\int_{m_2}\left[1 + \frac{3(lx + my + nz)}{R}\right][(ny - mz)\mathbf{i} \\ &\quad + (lz - nx)\mathbf{j} + (mx - ly)\mathbf{k}]\, dm_2 \\ &= \frac{3Gm_1}{R^3}\{[mn(I_{zz} - I_{yy}) + lnI_{xy} - lmI_{xz} + (n^2 - m^2)I_{yz}]\mathbf{i} \\ &\quad + [ln(I_{xx} - I_{zz}) + lmI_{yz} - mnI_{xy} + (l^2 - n^2)I_{xz}]\mathbf{j} \\ &\quad + [lm(I_{yy} - I_{xx}) + mnI_{xz} - lnI_{yz} + (m^2 - l^2)I_{xy}]\mathbf{k}\}. \end{aligned}\tag{11.117}$$

Finally, if x, y, and z are principal axes, the torque components become

$$N_x = \frac{3Gm_1}{R^3}(C - B)mn,$$

$$N_y = \frac{3Gm_1}{R^3}(A - C)ln, \tag{11.118}$$

$$N_z = \frac{3Gm_1}{R^3}(B - A)lm,$$

which confirms the statement that no torques exist if all three moments of inertia are equal to one another. This does not necessarily imply spherical symmetry, as a homogeneous cube, for example, satisfies this condition. It should be stressed, however, that Eqs. (11.118) are limited to second-order inertial integrals only. Since Eqs. (11.118) involve only differences in the moments of inertia, for a nearly symmetric body these terms may be sufficiently small to mean that ignoring higher-order inertial integrals cannot be justified. Even for a homogeneous cube torques are present, although they may be too small to be concerned with. They involve terms in the fourth-order inertial integrals divided by R^5, so that they are very small indeed.

11.8 PRECESSION AND NUTATION OF THE EARTH'S POLAR AXIS

In Sec. 11.1 we stated that the vernal equinox is frequently assumed to be fixed in an inertial space, although in reality it is not. Indeed, as we are about to show, the vernal equinox regresses at a very slow rate. In fact, when it received the name "first point of Aries," it lay in the constellation Aries; it is now in Pisces. This precession of the equinoxes, often referred to as *lunisolar precession*, is due to gravitational torques exerted on the earth by the moon and sun because of the earth's unequal moments of inertia. It is commonly assumed, however, that the earth possesses axial symmetry with principal moments of inertia A, A, C about axes x, y, z, respectively, where z is the polar axis.

Let us assume that the vernal equinox is not fixed in space but precessing with an angular velocity $\dot{\lambda}$ in the ecliptic plane XY (see Fig. 11.7). The radius vector $\mathbf{R}_S$ from the earth's center to the sun's center makes an angle λ_S measured in the ecliptic plane from the vernal equinox. The earth's equatorial plane xy is inclined by an angle ϵ with respect to the ecliptic, so that the sun's coordinates relative to the x, y, z system are

$$x_S = R_S \cos \lambda_S, \quad y_S = R_S \sin \lambda_S \cos \epsilon, \quad z_S = -R_S \sin \lambda_S \sin \epsilon. \quad (11.119)$$

The lunar ascending node makes in the ecliptic plane an angle λ_M with respect to the vernal equinox, and the radius vector $\mathbf{R}_M$ from the earth's center to the moon's center makes an angle λ'_M with respect to the lunar ascending node. The angle λ'_M is measured in the moon's orbital plane, which is inclined by a small angle i (of about 5°9′) with respect to the ecliptic. A series of coordinate transformations yields the moon's coordinates

$$\begin{aligned} x_M &= R_M \cos (\lambda_M + \lambda'_M), \\ y_M &= R_M [\sin (\lambda_M + \lambda'_M) \cos \epsilon + i \sin \lambda'_M \sin \epsilon], \\ z_M &= R_M [-\sin (\lambda_M + \lambda'_M) \sin \epsilon + i \sin \lambda'_M \cos \epsilon], \end{aligned} \quad (11.120)$$

where the fact that angle i is small has been taken into account.

Problems in Celestial Mechanics

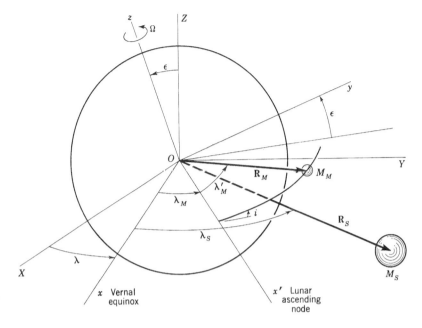

FIGURE 11.7

The angular velocity components of the system x, y, z are

$$\omega_x = \dot{\epsilon}, \qquad \omega_y = \dot{\lambda} \sin \epsilon, \qquad \omega_z = \dot{\lambda} \cos \epsilon, \qquad (11.121)$$

whereas, assuming that Ω is much larger than $\dot{\lambda}$, which is undoubtedly the case, the angular velocity components of the earth about these axes are

$$\Omega_x = \dot{\epsilon}, \qquad \Omega_y = \dot{\lambda} \sin \epsilon, \qquad \Omega_z = \Omega + \dot{\lambda} \cos \epsilon \approx \Omega. \qquad (11.122)$$

Using the modified Euler's equations (4.66), we obtain the moment equations of motion about the earth's center O in the form

$$A\ddot{\epsilon} - A\dot{\lambda}^2 \sin \epsilon \cos \epsilon + C\Omega\dot{\lambda} \sin \epsilon = N_x,$$
$$A(\ddot{\lambda} \sin \epsilon + 2\dot{\lambda}\dot{\epsilon} \cos \epsilon) - C\Omega\dot{\epsilon} = N_y, \qquad (11.123)$$
$$C\dot{\Omega} = N_z,$$

where the torque expressions are in accordance with Eqs. (11.118).

But the direction cosines are simply

$$l_M = \frac{x_M}{R_M}, \qquad m_M = \frac{y_M}{R_M}, \qquad n_M = \frac{z_M}{R_M},$$
$$l_S = \frac{x_S}{R_S}, \qquad m_S = \frac{y_S}{R_S}, \qquad n_S = \frac{z_S}{R_S}, \qquad (11.124)$$

so that introducing Eqs. (11.119), (11.120), and (11.124) into (11.118), we obtain

$$N_x = 3G(C - A)\left(\frac{M_M}{R_M^3} m_M n_M + \frac{M_S}{R_S^3} m_S n_S\right)$$

$$= 3G(C - A)\left\{\frac{M_M}{R_M^3}[-\sin^2(\lambda_M + \lambda_M')\sin\epsilon\cos\epsilon\right.$$

$$\left. + i\sin\lambda_M'\sin(\lambda_M + \lambda_M')(\cos^2\epsilon - \sin^2\epsilon)] - \frac{M_S}{R_S^3}\sin^2\lambda_S\sin\epsilon\cos\epsilon\right\},$$

$$N_y = 3G(A - C)\left(\frac{M_M}{R_M^3} l_M n_M + \frac{M_S}{R_S^3} l_S n_S\right) \quad (11.125)$$

$$= 3G(A - C)\left\{\frac{M_M}{R_M^3}[-\sin(\lambda_M + \lambda_M')\cos(\lambda_M + \lambda_M')\sin\epsilon\right.$$

$$\left. + i\sin\lambda_M'\cos(\lambda_M + \lambda_M')\cos\epsilon] - \frac{M_S}{R_S^3}\sin\lambda_S\cos\lambda_S\sin\epsilon\right\},$$

$$N_z = 0.$$

From the third of Eqs. (11.123) and (11.125) we conclude that the earth's rotation Ω is uniform, which is not unexpected. To solve the remaining two equations, we shall assume that the radial distances R_M and R_S are constant, with the implication that the associated orbits are circular; this, of course, is only approximately true. Moreover, it is known that λ_M varies slowly, completing 2π in about 18.6 years. On the other hand, λ_S and λ_M' complete 2π in one sidereal year and one sidereal month, respectively. These angles can be regarded as linear functions of time. Denoting by ϵ_0 the almost constant value of about 23°27′ between the earth's equatorial plane and the ecliptic, the first two of Eqs. (11.123) together with the first two of Eqs. (11.125) lead to

$$\dot{\lambda} = -\frac{3G(C - A)}{\Omega C}\left\{\frac{M_M}{R_M^3}\left[\sin^2(\lambda_M + \lambda_M')\cos\epsilon_0\right.\right.$$

$$\left.\left. - i\sin\lambda_M'\sin(\lambda_M + \lambda_M')\frac{\cos 2\epsilon_0}{\sin\epsilon_0}\right] + \frac{M_S}{R_S^3}\sin^2\lambda_S\cos\epsilon_0\right\},$$

$$\dot{\epsilon} = -\frac{3G(C - A)}{\Omega C}\left\{\frac{M_M}{R_M^3}[\tfrac{1}{2}\sin 2(\lambda_M + \lambda_M')\sin\epsilon_0\right.$$

$$\left. - i\sin\lambda_M'\cos(\lambda_M + \lambda_M')\cos\epsilon_0] + \frac{M_S}{R_S^3}\tfrac{1}{2}\sin 2\lambda_S\sin\epsilon_0\right\},$$

(11.126)

which represent the equations for the precession and nutation of the earth's axis. The precession has a component producing a net increase in the angle λ

with time, a phenomenon referred to as the *precession of the equinoxes*. In addition, both λ and ϵ possess small oscillatory components.

To show that the precession does indeed have a constant component, we shall average Eqs. (11.126) twice, the first time over the angle $\lambda_M + \lambda'_M$ as well as λ_S and the second time over the angle λ_M. The first essentially represents averages over orbital periods, whereas the second represents an average over the precessional period of the lunar node. The first averaging process yields

$$\bar{\lambda} = -\frac{3G(C-A)}{2\Omega C} \left[\frac{M_M}{R_M^3} \left(\cos \epsilon_0 - i \cos \lambda_M \frac{\cos 2\epsilon_0}{\sin \epsilon_0} \right) + \frac{M_S}{R_S^3} \cos \epsilon_0 \right],$$

$$\bar{\epsilon} = -\frac{3G(C-A)}{2\Omega C} \frac{M_M}{R_M^3} i \sin \lambda_M \cos \epsilon_0,$$
(11.127)

while the second one leads to the anticipated result

$$\bar{\bar{\lambda}} = -\frac{3G(C-A)}{2\Omega C} \left(\frac{M_M}{R_M^3} + \frac{M_S}{R_S^3} \right) \cos \epsilon_0,$$

$$\bar{\bar{\epsilon}} = 0.$$
(11.128)

Because for the earth $C > A$, the first of Eqs. (11.128) indicates that *the vernal equinox is regressing*. On the other hand, the average nutation is zero.

Using the values

$$G = 6.668 \times 10^{-8} \text{ dyne cm}^2\text{gm}^{-2},$$
$$M_M = 7.380 \times 10^{25} \text{ gm}, \qquad M_S = 1.991 \times 10^{33} \text{ gm},$$
$$R_M = 384.405 \times 10^3 \text{ km}, \qquad R_S = 149.6 \times 10^6 \text{ km},$$
$$C = 0.3340 M_E a^2, \qquad A = 0.3329 M_E a^2,$$

where M_E = earth's mass,

a = average radius of earth,

$\Omega = 2\pi$ rad/day,

we obtain the precession period of about 25,200 years. It should be noted that in spite of the fact that the moon's mass is much smaller than the sun's mass, the moon's effect is more than twice as large because it is much closer to the earth.

As indicated above, in addition to the average precession of the earth's axis, there are also oscillatory components of both precession and nutation. Ignoring the annual fluctuations due to the sun and the monthly fluctuations due to the moon, the oscillatory components can be obtained from Eqs. (11.127) if we recognize that the constant terms on the right side of the equations yield $\bar{\bar{\lambda}}$ and $\bar{\bar{\epsilon}} = 0$. With this in mind, it is easy to see that the oscillatory components of precession and nutation, $\bar{\lambda}_{osc}$ and $\bar{\epsilon}_{osc}$, are both due to the moon's effect only. They have different amplitudes but the same period of 18.6 years and are 90° out of phase. Hence, due to this effect alone, the earth's axis describes an ellipse on the celestial sphere. Since the amplitude of the average precession is much larger than the amplitudes of the precessional and nutational oscillations, and since, moreover, the period of the steady precession is much larger than the period of oscillation, the path described by the earth's axis on the celestial sphere has a wavy form similar to the curve in Fig. 4.10a except that the direction of motion is reversed.

A more complete account of this problem can be found in Ref. 11, chap. 22.

11.9 VARIATION OF THE ORBITAL ELEMENTS

The motion of two spherically symmetric bodies, subjected to forces resulting purely from their mutual gravitational attraction varying inversely as the square of the distance between the bodies, consists of conic sections described by each of the two bodies relative to the barycenter as a focus. The motion of the bodies with respect to one another is also a conic section. In the solar system the motion of each planet with respect to the sun can be regarded, in the first approximation, as a Keplerian orbit. We indicated in Sec. 11.1 that the motion of a body in a Keplerian orbit is fully described by six independent orbital elements such as a, e, i, Ω, ω, and T, where ω denotes the angle between the ascending node and the perihelion (angle denoted by θ_0 in Fig. 11.2) and the time of perihelion passage T is substituted for the mean anomaly M. By motion we understand the position and velocity vectors. Conversely, if the position and velocity vectors are known at a given time, the six orbital elements can be determined uniquely.

In a more exact treatment of the two-body problem, however, we must consider forces which, although only of second-order magnitude, may have an effect upon the motion of the system. For example, the motion of a planet around the sun is affected not only by the primary attraction due to the sun but also by the gravitational attraction due to other planets in the solar system. The motion of an artificial satellite in a resisting medium or in a gravitational field differing slightly from the spherically symmetric one commonly assumed in the two-body problem are two other examples of practical interest. We must recognize that both these circumstances are realizable for satellites orbiting the earth and that the second is realizable for moon satellites.

As long as the perturbing forces are absent, the orbital elements in a two-body system remain constant and the motion of one body relative to the other consists of the *undisturbed* Keplerian orbit. If there are small perturbing forces acting upon the system, however, the motion will differ somewhat from the Keplerian orbit. A solution for the perturbed motion may be possible by the method of Sec. 8.1, but it may not be very illuminating. A different approach is based on the recognition that in the presence of perturbing forces the orbital elements are no longer constant but change slowly with time. This is essentially the *method of variation of parameters*, in which the orbital elements play the role of the parameters. In some way it resembles the technique of Sec. 8.4 if it is recognized that the amplitude and frequency of the nonlinear oscillator in that section could be regarded as the time-dependent system parameters. The method described in this section was developed by Lagrange and is particularly suited for the treatment of orbit perturbations. It has been used extensively in planetary motion.

Let us concern ourselves with the motion of a planet m under the gravitational attraction of the sun as the primary force. Also acting upon m are perturbing forces of the type discussed in Sec. 11.7, which can be described by a disturbing function R (see Sec. 11.6) or arising from other sources provided the force can be written in the form ∇R. Denoting by $\mathbf{r}$ the radius vector from the sun to the planet m, Eq. (11.87) becomes

$$\ddot{\mathbf{r}} + \frac{\mu}{r^3}\mathbf{r} = \nabla R, \tag{11.129}$$

where $\mu = G(M + m)$, in which M is the sun's mass. The quantities $\mathbf{r}$ and ∇R can be written in terms of the cartesian coordinates x, y, z in the form

$$\mathbf{r} = x\mathbf{i} + y\mathbf{j} + z\mathbf{k} \tag{11.130}$$

and

$$\nabla R = \frac{\partial R}{\partial x}\mathbf{i} + \frac{\partial R}{\partial y}\mathbf{j} + \frac{\partial R}{\partial z}\mathbf{k}, \tag{11.131}$$

where x, y, z constitute an inertial system of axes and $\mathbf{i}, \mathbf{j}, \mathbf{k}$ are the corresponding unit vectors.

In the absence of perturbing forces, Eq. (11.131) reduces to

$$\ddot{\mathbf{r}} + \frac{\mu}{r^3}\mathbf{r} = \mathbf{0}, \tag{11.132}$$

which has the solution

$$\mathbf{r} = \mathbf{r}(c_1, c_2, \ldots, c_6, t), \qquad \dot{\mathbf{r}} = \frac{\partial \mathbf{r}(c_1, c_2, \ldots, c_6, t)}{\partial t}, \tag{11.133}$$

where c_k ($k = 1, 2, \ldots, 6$) are the constant elements a, e, i, Ω, ω, and T, or any other set of six independent orbital elements determining the motion in the Keplerian orbit. The partial derivative symbol has been used in the second of Eqs. (11.133) to emphasize the fact that the elements c_k are constant.

If perturbing forces are acting upon the system, the elements c_k are, in general, no longer constant but time-dependent. Our object is to obtain expressions relating the rate of change of the orbital elements and the disturbing function R.

Let us denote the position of m at time $t = t_0$ by P. If we suppose for the moment that the disturbing forces acting upon the system are suddenly removed, then mass m follows the Keplerian orbit tangent to the original orbit at P and defined by the constant elements a_0, e_0, i_0, Ω_0, ω_0, and T_0 (see dashed curve in Fig. 11.8). Under the assumption that the position vector $\mathbf{r}$ and the velocity vector $\dot{\mathbf{r}}$ are known at time t_0, we can use Eqs. (11.133) and solve for the elements a_0, e_0, . . . , T_0, thus determining the Keplerian orbit. However, mass m does not follow the path PQ_0 but the slightly different *true path* PQ, according to Eq. (11.129). Because the velocity vector at point P is the same for the true path as for the Keplerian path, we can write

$$\left.\frac{d\mathbf{r}}{dt}\right|_P = \left.\frac{\partial \mathbf{r}}{\partial t}\right|_P. \qquad (11.134)$$

The orbit PQ_0, having simultaneously the same position and velocity as the true orbit, is called the *osculating orbit*, and the constant elements a_0, e_0, . . . , T_0 are referred to as *osculating elements* at time t_0. For a short interval the osculating orbit can be regarded as a good approximation to the true orbit. On the true path PQ the orbital elements are not constant but vary slowly with time because the perturbing forces are supposed to be small. At point Q corresponding to time t_1, however, we can envision a new osculating orbit with the elements a_1, e_1, . . . , T_1. The small differences $a_1 - a_0$, $e_1 - e_0$, . . . , $T_1 - T_0$, corresponding to the time interval $\Delta t = t_1 - t_0$, are the perturbations of the orbital elements due to the perturbing forces. A simple limiting process yields the rate of change of the elements at t_0.

Let us consider the position $\mathbf{r} = \mathbf{r}(c_1, c_2, \ldots, c_6, t)$ of mass m in the true orbit and regard the elements c_k as time-dependent. The velocity in the true orbit is

$$\frac{d\mathbf{r}}{dt} = \frac{\partial \mathbf{r}}{\partial t} + \sum_{k=1}^{6} \frac{\partial \mathbf{r}}{\partial c_k} \frac{dc_k}{dt}. \qquad (11.135)$$

FIGURE 11.8

Problems in Celestial Mechanics

The rates of change $\dot{c}_k$ of the elements c_k satisfy

$$\sum_{k=1}^{6} \frac{\partial \mathbf{r}}{\partial c_k} \dot{c}_k = 0, \tag{11.136}$$

which is merely an expression of the fact that the velocity in the true orbit and the velocity in the osculating orbit are the same. With this in mind, the second derivative of $\mathbf{r}$ with respect to time can be written

$$\ddot{\mathbf{r}} = \frac{\partial^2 \mathbf{r}}{\partial t^2} + \sum_{k=1}^{6} \frac{\partial^2 \mathbf{r}}{\partial t \, \partial c_k} \dot{c}_k = \frac{\partial^2 \mathbf{r}}{\partial t^2} + \sum_{k=1}^{6} \frac{\partial \dot{\mathbf{r}}}{\partial c_k} \dot{c}_k, \tag{11.137}$$

in which

$$\frac{\partial^2 \mathbf{r}}{\partial t \, \partial c_k} = \frac{\partial}{\partial c_k} \left(\frac{\partial \mathbf{r}}{\partial t} \right) = \frac{\partial \dot{\mathbf{r}}}{\partial c_k} \tag{11.138}$$

has been substituted by virtue of the fact that $\partial \mathbf{r}/\partial t = \dot{\mathbf{r}}$. Introducing Eq. (11.137) into (11.129), we obtain

$$\frac{\partial^2 \mathbf{r}}{\partial t^2} + \frac{\mu}{r^3} \mathbf{r} + \sum_{k=1}^{6} \frac{\partial \dot{\mathbf{r}}}{\partial c_k} \dot{c}_k = \nabla R. \tag{11.139}$$

But for the unperturbed system the true and the osculating orbits are one and the same. Hence, Eq. (11.132) is equivalent to

$$\frac{\partial^2 \mathbf{r}}{\partial t^2} + \frac{\mu}{r^3} \mathbf{r} = 0, \tag{11.140}$$

so that Eq. (11.139) becomes

$$\sum_{k=1}^{6} \frac{\partial \dot{\mathbf{r}}}{\partial c_k} \dot{c}_k = \nabla R. \tag{11.141}$$

Equations (11.136) and (11.141) constitute a set of six equations in the unknowns $\dot{c}_k$ ($k = 1, 2, \ldots, 6$) whose solution represents the desired rates of change of the orbital elements c_k in terms of partial derivatives of the disturbing function R. Essentially the equations provide the transformation from the set of variables consisting of the position and velocity of m to a set of variables involving the orbital elements.

The solution of Eqs. (11.136) and (11.141) was first carried out by Lagrange. Although the method of solution is particularly elegant, the details are quite lengthy, so that we give here only the results[1]

---

[1] For the details of the solution see, for example, Ref. 3, pp. 277–289.

$$\frac{da}{dt} = \frac{2}{na}\frac{\partial R}{\partial \sigma},$$

$$\frac{de}{dt} = \frac{1}{na^2 e}\left[(1-e^2)\frac{\partial R}{\partial \sigma} - (1-e^2)^{\frac{1}{2}}\frac{\partial R}{\partial \omega}\right],$$

$$\frac{d\sigma}{dt} = -\frac{1-e^2}{na^2 e}\frac{\partial R}{\partial e} - \frac{2}{na}\frac{\partial R}{\partial a},$$

$$\frac{d\Omega}{dt} = \frac{1}{na^2(1-e^2)^{\frac{1}{2}}\sin i}\frac{\partial R}{\partial i},$$
(11.142)

$$\frac{d\omega}{dt} = \frac{(1-e^2)^{\frac{1}{2}}}{na^2 e}\frac{\partial R}{\partial e} - \frac{\cot i}{na^2(1-e^2)^{\frac{1}{2}}}\frac{\partial R}{\partial i},$$

$$\frac{di}{dt} = \frac{1}{na^2(1-e^2)^{\frac{1}{2}}}\left(\cot i\frac{\partial R}{\partial \omega} - \csc i\frac{\partial R}{\partial \Omega}\right),$$

where $\sigma = -nT$ and n is the mean motion. Equations (11.142) represent one version of *Lagrange's planetary equations*. Of course, to obtain the rate of change of the orbital elements we must derive expressions for partial derivatives of the disturbing function with respect to the orbital elements. This is done in Sec. 11.10.

At times it is more convenient to work with the mean anomaly M rather than σ. The two quantities are related by

$$M = n(t-T) = nt + \sigma = \mu^{\frac{1}{2}}a^{-\frac{3}{2}}t + \sigma, \quad (11.143)$$

and we should be aware that $R = R(a,e,i,\Omega,\omega,M)$ depends on a both explicitly and through M, as shown by Eq. (11.143). From that equation we conclude, after a few algebraic operations, that the first three of Eqs. (11.142) must be replaced by

$$\frac{da}{dt} = \frac{2}{na}\frac{\partial R}{\partial M},$$

$$\frac{de}{dt} = \frac{1}{na^2 e}\left[(1-e^2)\frac{\partial R}{\partial M} - (1-e^2)^{\frac{1}{2}}\frac{\partial R}{\partial \omega}\right], \quad (11.144)$$

$$\frac{dM}{dt} = n - \frac{1-e^2}{na^2 e}\frac{\partial R}{\partial e} - \frac{2}{na}\frac{\partial R}{\partial a},$$

where the derivative $\partial R/\partial a$ is to be taken explicitly with respect to a.

In Chap. 12 we shall discuss perturbations in the orbital elements of an earth satellite caused by the oblateness of the earth as well as by resisting forces due to the earth's atmosphere.

11.10 THE RESOLUTION OF THE DISTURBING FUNCTION

The right side of Eqs. (11.142) contains derivatives with respect to the orbital elements a, e, σ, Ω, ω, and i, which seems to imply that the disturbing function is to be expressed in terms of the orbital elements rather than the position and velocity components. As it turns out, this is not always necessary.

Let us consider the motion of a planet m with respect to the sun and denote by x, y, z a set of inertial coordinates and by $\mathbf{i}$, $\mathbf{j}$, $\mathbf{k}$ the corresponding unit vectors. Let us also introduce a set of radial and transverse orbital axes defined by the unit vectors $\mathbf{u}_r$ and $\mathbf{u}_\theta$, respectively, as shown in Fig. 11.9. The third axis, perpendicular to the orbital plane, coincides with the direction of the angular momentum vector $\mathbf{L}$ and is associated with the unit vector $\mathbf{u}_L$. The elements Ω, i, and ω appear in Fig. 11.9; a and e are defined by Eqs. (11.6) to (11.8) in terms of the radial distance r, the true anomaly θ, and the eccentric anomaly w, whereas σ is defined by Eq. (11.143).

If m_1 denotes the mass of a certain perturbing planet and $\mathbf{r}_1$ and $\boldsymbol{\rho}$ are the radius vectors from the sun and planet m to m_1, respectively, then Eq. (11.84) yields the disturbing function

$$R = Gm_1\left(\frac{1}{\rho} - \frac{\mathbf{r} \cdot \mathbf{r}_1}{r_1^3}\right), \tag{11.145}$$

which is to be used in conjunction with Eq. (11.129) for the motion of m relative to the sun. Equation (11.145) can be expressed in terms of the

FIGURE 11.9

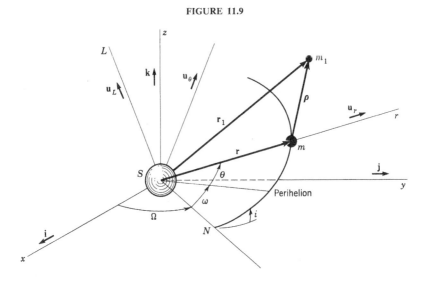

orbital elements, thus enabling us to evaluate the derivatives appearing on the right side of Eqs. (11.142). In this section, however, we shall derive expressions for the partial derivatives of R with respect to the orbital elements for the case in which the disturbing force is given in terms of three mutually orthogonal components, two of which are in the orbital plane. The method of obtaining these expressions is associated with the name of Gauss.

Let us consider the case in which the disturbing function R is defined by

$$\nabla R = \mathbf{F} = F_r \mathbf{u}_r + F_\theta \mathbf{u}_\theta + F_L \mathbf{u}_L, \qquad (11.146)$$

where $\mathbf{F}$ is the disturbing force with components F_r, F_θ, and F_L in the radial, transverse, and L directions, as defined in Fig. 11.9. The corresponding unit vectors $\mathbf{u}_r$, $\mathbf{u}_\theta$, and $\mathbf{u}_L$ can be expressed in terms of inertial components as follows:

$$\begin{aligned}
\mathbf{u}_r &= [\cos \Omega \cos (\omega + \theta) - \sin \Omega \sin (\omega + \theta) \cos i]\mathbf{i} \\
&\quad + [\sin \Omega \cos (\omega + \theta) + \cos \Omega \sin (\omega + \theta) \cos i]\mathbf{j} \\
&\quad + [\sin (\omega + \theta) \sin i]\mathbf{k}, \\
\mathbf{u}_\theta &= -[\cos \Omega \sin (\omega + \theta) + \sin \Omega \cos (\omega + \theta) \cos i]\mathbf{i} \qquad (11.147)\\
&\quad - [\sin \Omega \sin (\omega + \theta) - \cos \Omega \cos (\omega + \theta) \cos i]\mathbf{j} \\
&\quad + [\cos (\omega + \theta) \sin i]\mathbf{k}, \\
\mathbf{u}_L &= (\sin \Omega \sin i)\mathbf{i} - (\cos \Omega \sin i)\mathbf{j} + (\cos i)\mathbf{k}.
\end{aligned}$$

The radius vector $\mathbf{r}$ can be written

$$\mathbf{r} = x\mathbf{i} + y\mathbf{j} + z\mathbf{k} = r\mathbf{u}_r, \qquad (11.148)$$

where, from Eq. (11.7), the radial distance is given by the equation

$$r = a(1 - e \cos w). \qquad (11.149)$$

Moreover, introducing Eq. (11.143) into (11.11), we write Kepler's equation in the form

$$M = nt + \sigma = w - e \sin w. \qquad (11.150)$$

The partial derivatives of R with respect to any orbital element c_k can be written

$$\frac{\partial R}{\partial c_k} = \frac{\partial R}{\partial x}\frac{\partial x}{\partial c_k} + \frac{\partial R}{\partial y}\frac{\partial y}{\partial c_k} + \frac{\partial R}{\partial z}\frac{\partial z}{\partial c_k} = \nabla R \cdot \frac{\partial \mathbf{r}}{\partial c_k}, \qquad k = 1, 2, \ldots, 6, \qquad (11.151)$$

where ∇R has the form (11.146) and $\mathbf{r} = r\mathbf{u}_r$, according to Eq. (11.148). It will be noted that $\mathbf{u}_r$ depends on some but not all the orbital elements. Taking the partial derivatives of $\mathbf{r}$ with respect to the orbital elements and considering Eqs. (11.6) to (11.8) and (11.147) to (11.150), as well as certain results from Sec. 1.9, we obtain

$$\frac{\partial \mathbf{r}}{\partial a} = \frac{r}{a}\mathbf{u}_r, \qquad \frac{\partial \mathbf{r}}{\partial e} = -a \cos \theta\, \mathbf{u}_r + r\left(\frac{a}{r} + \frac{1}{1-e^2}\right) \sin \theta\, \mathbf{u}_\theta,$$

$$\frac{\partial \mathbf{r}}{\partial \sigma} = \frac{ae \sin \theta}{(1-e^2)^{\frac{1}{2}}}\mathbf{u}_r + \frac{a^2}{r}(1-e^2)^{\frac{1}{2}}\mathbf{u}_\theta,$$

$$\frac{\partial \mathbf{r}}{\partial \Omega} = r \cos i\, \mathbf{u}_\theta - r \cos(\omega + \theta) \sin i\, \mathbf{u}_L, \qquad (11.152)$$

$$\frac{\partial \mathbf{r}}{\partial \omega} = r\mathbf{u}_\theta, \qquad \frac{\partial \mathbf{r}}{\partial i} = r \sin(\omega + \theta)\mathbf{u}_L.$$

Introduction of Eqs. (11.146) and (11.152) into (11.151) yields the desired partial derivatives in terms of the components of force radial, transverse, and perpendicular to the orbital plane

$$\frac{\partial R}{\partial a} = F_r \frac{r}{a}, \qquad \frac{\partial R}{\partial e} = -F_r a \cos \theta + F_\theta r\left(\frac{a}{r} + \frac{1}{1-e^2}\right) \sin \theta,$$

$$\frac{\partial R}{\partial \sigma} = F_r \frac{ae \sin \theta}{(1-e^2)^{\frac{1}{2}}} + F_\theta \frac{a^2}{r}(1-e^2)^{\frac{1}{2}},$$

$$\frac{\partial R}{\partial \Omega} = F_\theta r \cos i - F_L r \cos(\omega + \theta) \sin i, \qquad (11.153)$$

$$\frac{\partial R}{\partial \omega} = F_\theta r, \qquad \frac{\partial R}{\partial i} = F_L r \sin(\omega + \theta).$$

In many applications it is convenient to write the disturbing force in terms of tangential and normal components in the orbital plane, as well as a third component perpendicular to the orbital plane, so that the force vector has the expression

$$\mathbf{F} = F_T \mathbf{u}_T + F_N \mathbf{u}_N + F_L \mathbf{u}_L, \qquad (11.154)$$

where $\mathbf{u}_T$, $\mathbf{u}_N$, and $\mathbf{u}_L$ are the corresponding unit vectors. The component F_L is the same as in Eq. (11.146), whereas the components F_T and F_N are related to F_r and F_θ by

$$F_r = F_T \cos \alpha - F_N \sin \alpha, \qquad F_\theta = F_T \sin \alpha + F_N \cos \alpha, \qquad (11.155)$$

in which α is the angle between the radial and tangential directions, as shown in Fig. 11.10. The tangential direction coincides with the direction of the velocity vector $\mathbf{v}$, which can be written

$$\mathbf{v} = v\mathbf{u}_T = \dot{r}\mathbf{u}_r + r\dot{\theta}\mathbf{u}_\theta, \qquad (11.156)$$

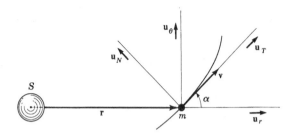

FIGURE 11.10

where v is the magnitude of $\mathbf{v}$. Using results from Sec. 1.9, we obtain

$$\sin \alpha = \frac{r\dot\theta}{v} = \frac{1 + e \cos \theta}{(1 + e^2 + 2e \cos \theta)^{\frac{1}{2}}},$$
$$\cos \alpha = \frac{\dot r}{v} = \frac{e \sin \theta}{(1 + e^2 + 2e \cos \theta)^{\frac{1}{2}}}.$$
(11.157)

Impulsive forces on a space vehicle are conveniently expressed in terms of tangential and normal components, whereas forces due to atmospheric drag on a satellite are tangent to the orbit and act in a direction opposite to the direction of the motion.

PROBLEMS

11.1 Consider the restricted three-body problem and plot the three planar projections of four surfaces of zero relative velocity for the case in which the two massive bodies are the sun and the earth.

11.2 Prove that the libration points L_1 and L_2 are unstable.

11.3 Write the equations for the relative motion corresponding to a three-body system (see Sec. 11.6). Let one of the bodies be small compared with the remaining two.

11.4 Obtain the complete solution of Eqs. (11.127). Note that $\lambda_M = \dot\lambda_M t$, where $\dot\lambda_M$ can be regarded as constant.

11.5 Derive Eqs. (11.152).

SUGGESTED REFERENCES

1. Alfriend, K. T., and R. H. Rand: "Stability of the Triangular Points in the Elliptic Restricted Problem of Three Bodies," *AIAA J.*, vol. 7, no. 6, pp. 1024–1028, 1969.
2. Breakwell, J. V., and R. Pringle, Jr.: "Resonances Affecting Motion Near the Earth-Moon Equilateral Libration Points," *Progr. Astron. Aeron.*, vol. 17, Academic Press, Inc., New York, pp. 55–74, 1966.
3. Brouwer, D., and G. M. Clemence: *Methods of Celestial Mechanics*, Academic Press, Inc., New York, 1961.
4. Byerly, W. E.: *An Elementary Treatise on Fourier's Series and Spherical, Cylindrical, and Ellipsoidal Harmonics*, Ginn and Company, Boston, 1893.
5. Danby, J. M. A.: *Fundamentals of Celestial Mechanics*, The Macmillan Company, New York, 1962.
6. Deutsch, R.: *Orbital Dynamics of Space Vehicles*, Prentice-Hall, Inc., Englewood Cliffs, N.J., 1963.
7. Kozai, Y.: "Tesseral Harmonics of the Gravitational Potential of the Earth as Derived from Satellite Motions," *Astron. J.*, vol. 66, no. 7, pp. 355–358, 1961.
8. MacMillan, W. D.: *The Theory of the Potential*, Dover Publications, Inc., New York, 1930.
9. McCuskey, S. W.: *Introduction to Celestial Mechanics*, Addison-Wesley Publishing Company, Inc., Reading, Mass., 1963.
10. Moulton, F. R.: *Celestial Mechanics*, The Macmillan Company, New York, 1914.
11. Plummer, H. C.: *An Introductory Treatise on Dynamical Astronomy* (republication of the original 1918 ed.), Dover Publications, Inc., New York, 1960.
12. Schechter, H. B.: "Three-Dimensional Nonlinear Stability Analysis of the Sun-Perturbed Earth-Moon Equilateral Points," *AIAA J.*, vol. 6, no. 7, pp. 1223–1228, 1968.
13. Smart, W. M.: *Celestial Mechanics*, John Wiley & Sons, Inc., New York, 1953.
14. Sundman, K. F.: "Mémoire sur le Problème des Trois Corps," *Acta Math.*, vol. 36, pp. 105–192, 1912.
15. Whittaker, E. T.: *Analytical Dynamics*, Cambridge University Press, London, 1937.
16. Wintner, A.: *The Analytical Foundation of Celestial Mechanics*, Princeton University Press, Princeton, N.J., 1947.

chapter twelve

Problems in Spacecraft Dynamics

The development of spacecraft in the last few decades has stirred renewed interest in the field of dynamics. By contrast with the motion of celestial bodies, space vehicles can perform a large variety of motions. Moreover, some vehicles carry with them means of propulsion and hence are capable of altering their own motion. Many of the methods of celestial mechanics are equally applicable to the motion of space vehicles, but, nevertheless, new approaches are necessary. This chapter is not meant to provide an exhaustive treatment of space vehicle dynamics, and it should be regarded as merely an attempt to present certain facets of the problems. Its primary purpose is to illustrate how various methods of analytical dynamics can be applied to physical problems.

Artificial satellites can be regarded as particles or as bodies of finite dimensions, depending whether the interest lies in the satellite's orbital motion or in the rotational motion about the satellite's center of mass. The first part of this chapter is devoted to orbital motion. In particular, the problems associated with orbit modification by means of short impulses are introduced, and the effects of a nonspherical gravitational field as well as atmospheric drag are discussed. A great deal of attention is devoted to the attitude motion of satellites. If the satellite dimensions are small relative to the radial distance from the center of force, the orbital motion of the satellite's

Problems in Spacecraft Dynamics

center of mass is not affected by the rotational motion about the center of mass. Attitude motion provides a rich source of dynamical problems lending themselves to neat mathematical formulation as well as to analytical treatment. In fact the area may be regarded as a modern extension of the problem of lunar libration, which preoccupied Lagrange about two centuries ago. To obtain solutions to the problems of attitude motion, the methods of Chaps. 5 to 9 prove very useful. Finally the subject of variable-mass systems, with emphasis on rocket dynamics, is discussed, and a relatively complete formulation of the problem is provided.

12.1 TRANSFER ORBITS. CHANGES IN THE ORBITAL ELEMENTS DUE TO A SMALL IMPULSE

The motion of a planet in a central force gravitational field, with no perturbing forces present, consists of a two-body Keplerian orbit. The same is equally true for a space vehicle, provided that it behaves like a particle. By contrast, however, space vehicles carrying means of propulsion possess the ability to alter their orbit. Intermediate orbits are generally necessary because of intentional or unintentional failure to place the vehicle in the desired orbit. The first case may arise when it is more desirable to launch the satellite in an intermediate orbit and then use a *transfer orbit* to place the satellite in the final, or target, orbit. In the second case, the vehicle fails to achieve the desired orbit, due to inaccuracies in the launch, and an *orbit modification* is necessary to achieve proper orbit. The transfer orbits fall into two broad categories, the first involving transfer between orbits in a single force field and the second consisting of transfers from one force field to another. For example, transfer within a single force field takes place when both the intermediate and the final orbits are earth orbits. Two force fields are involved when the transfer takes place, for example, between an earth orbit and a moon orbit.

A change in orbit is achieved when the space vehicle engine applies a thrust to the vehicle. This engine thrust can be assumed to last a short time, in which case the thrust can be regarded as impulsive, or it can be continuous over longer periods; a sequence of a given number of short impulses can also be used. If the engine thrust is impulsive, the vehicle can be assumed to undergo instantaneous changes in velocity. Because changes in position take a finite amount of time to develop, it can be assumed that positions do not change during impulse, which results in a substantial simplification in the analysis. Only impulsive thrusts are considered here, and the words impulse and velocity change will be used synonymously.

A two-body Keplerian orbit is planar, and all its orbital elements are constant. When an impulsive thrust is applied to the vehicle, some, but not necessarily all, of the orbital elements change. For example, if the thrust vector is in the orbit plane, no change of the orbital plane will occur. This is the case when the transfer is to take place between *coplanar orbits*; it is the simplest case of orbit transfer. To achieve a transfer between *noncoplanar orbits*, which implies a change in the orbit ascending node and inclination, the thrust must have a component perpendicular to the orbit plane.

To illustrate the problem, we shall calculate the impulse necessary to perform an orbit change involving two noncoplanar intersecting orbits, denoted orbits 1 and 2 in Fig. 12.1. Because the radial distance to the intersection point must be the same and from Sec. 1.9 we can show that the distance has the expression $r = a(1 - e^2)/(1 + e \cos \theta)$, we conclude that the orbital elements a_i, e_i, and θ_i ($i = 1, 2$) pertaining to orbits 1 and 2, respectively, must satisfy the relation

$$\frac{a_1(1 - e_1^2)}{1 + e_1 \cos \theta_1} = \frac{a_2(1 - e_2^2)}{1 + e_2 \cos \theta_2}. \tag{12.1}$$

Moreover, the unit vector $\mathbf{u}_r$ in the radial direction must be the same for both orbits, so that, from the first of Eqs. (11.147), we have

$$\begin{aligned} \cot i_1 \sin i_2 &= \sin (\Omega_2 - \Omega_1) \cot (\omega_2 + \theta_2) + \cos (\Omega_2 - \Omega_1) \cos i_2, \\ \cot i_2 \sin i_1 &= \cos (\Omega_2 - \Omega_1) \cos i_1 - \sin (\Omega_2 - \Omega_1) \cos (\omega_1 + \theta_1). \end{aligned} \tag{12.2}$$

Equations (12.1) and (12.2) indicate that not all the orbital elements of orbit 2

FIGURE 12.1

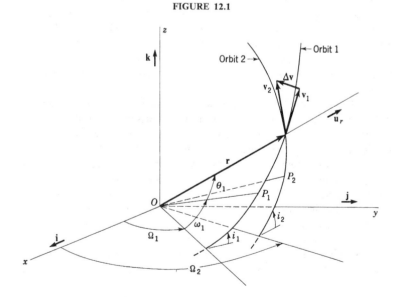

are arbitrary, so that in general a single impulse will not yield any desired terminal orbit. Hence, orbit 2 must be regarded as a transfer orbit. Of course, at the end of the transfer orbit another impulse is necessary to place the vehicle in the target orbit.

Next let us calculate the impulse necessary to achieve the transfer orbit, on the assumption that the departure and transfer orbits are known. Denoting by $\mathbf{v}_1$ and $\mathbf{v}_2$ the velocity vectors in orbits 1 and 2, respectively, at the intersection point and using Eq. (1.126), we obtain the corresponding magnitudes

$$v_j = \left[\mu\left(\frac{2}{r} - \frac{1}{a_j}\right)\right]^{\frac{1}{2}}, \quad j = 1, 2, \qquad (12.3)$$

whereas the directions are given by the unit vectors tangent to the respective orbits

$$\mathbf{u}_{Tj} = \mathbf{u}_r \cos \alpha_j + \mathbf{u}_{\theta j} \sin \alpha_j, \quad j = 1, 2, \qquad (12.4)$$

in which $\sin \alpha_j$ and $\cos \alpha_j$ are defined by Eqs. (11.157) and the unit vectors $\mathbf{u}_{\theta j}$ are obtained from the second of Eqs. (11.147). This enables us to calculate the impulse

$$\Delta \mathbf{v} = \mathbf{v}_2 - \mathbf{v}_1 = v_2 \mathbf{u}_{T2} - v_1 \mathbf{u}_{T1}. \qquad (12.5)$$

For convenience, the unit vector $\mathbf{u}_{T2}$ should be expressed in terms of components tangential, normal, and perpendicular to orbit relating to orbit 1.

In our discussion of transfer orbits the timing of the impulse played no particular role. In many practical applications, however, transfer orbits are used by an orbiting vehicle to intercept another vehicle in a different orbit. Such an orbit is referred to as a *rendezvous transfer orbit*, and for this type of transfer orbit the timing of the impulse is of major importance. The time factor is brought into consideration by virtue of the fact that the two vehicles must reach the same point in space and time.

Although a large variety of transfer orbits could be discussed, no essentially new dynamical principles are involved, so that we shall not pursue the subject any further. The interested reader can find a brief exposition of the subject in the book by Deutsch (Ref. 4, chap. 13) and an extended coverage in the one by Ehricke (Ref. 5, chaps. 3 and 4).

We have indicated how to calculate the impulse necessary to change a given orbit into another given orbit lying in a different plane. Next we pose the converse problem, namely, how a small impulse affects the elements of a given orbit. To answer this question we turn to the derivations of Secs. 11.9 and 11.10.

Assuming that the mass of the vehicle equals unity, the impulse-momentum relation in terms of the components radial, transverse, and perpendicular to the orbit yields the velocity increments

$$\Delta v_r = F_r \Delta t, \qquad \Delta v_\theta = F_\theta \Delta t, \qquad \Delta v_L = F_L \Delta t, \qquad (12.6)$$

where F_r, F_θ, and F_L are the corresponding thrust components and Δt is the thrust duration. Substituting Eqs. (12.6) into Eqs. (11.153) multiplied by Δt, we obtain

$$\frac{\partial R}{\partial a} \Delta t = \Delta v_r \frac{r}{a}, \qquad \frac{\partial R}{\partial e} \Delta t = -\Delta v_r a \cos \theta + \Delta v_\theta r \left(\frac{a}{r} + \frac{1}{1 - e^2}\right) \sin \theta,$$

$$\frac{\partial R}{\partial \sigma} \Delta t = \Delta v_r \frac{ae \sin \theta}{(1 - e^2)^{\frac{1}{2}}} + \Delta v_\theta \frac{a^2}{r} (1 - e^2)^{\frac{1}{2}},$$

$$\frac{\partial R}{\partial \Omega} \Delta t = \Delta v_\theta r \cos i - \Delta v_L r \cos(\omega + \theta) \sin i, \qquad (12.7)$$

$$\frac{\partial R}{\partial \omega} \Delta t = \Delta v_\theta r, \qquad \frac{\partial R}{\partial i} \Delta t = \Delta v_L r \sin(\omega + \theta).$$

Introducing Eqs. (12.7) into Eqs. (11.142) written in incremental form, we arrive at the changes in the orbital elements

$$\Delta a = \frac{2}{n} \left[\Delta v_r \frac{e \sin \theta}{(1 - e^2)^{\frac{1}{2}}} + \Delta v_\theta \frac{a}{r} (1 - e^2)^{\frac{1}{2}} \right],$$

$$\Delta e = \frac{(1 - e^2)^{\frac{1}{2}}}{na} \left[\Delta v_r \sin \theta + \Delta v_\theta \frac{a}{er} \left(1 - e^2 - \frac{r^2}{a^2}\right) \right],$$

$$\Delta \sigma = \frac{1 - e^2}{nae} \left[\Delta v_r \left(\cos \theta - \frac{r}{a} \frac{2e}{1 - e^2}\right) - \Delta v_\theta \left(1 + \frac{r}{a} \frac{1}{1 - e^2}\right) \sin \theta \right],$$

$$\Delta \Omega = \Delta v_L \frac{r \sin(\omega + \theta)}{na^2 (1 - e^2)^{\frac{1}{2}} \sin i}, \qquad (12.8)$$

$$\Delta \omega = -\frac{(1 - e^2)^{\frac{1}{2}}}{nae} \left[\Delta v_r \cos \theta - \Delta v_\theta \left(1 + \frac{r}{a} \frac{1}{1 - e^2}\right) \sin \theta \right.$$

$$\left. + \Delta v_L \frac{r}{a} \frac{e \cot i \sin(\omega + \theta)}{1 - e^2} \right],$$

$$\Delta i = \Delta v_L \frac{r \cos(\omega + \theta)}{na^2 (1 - e^2)^{\frac{1}{2}}}.$$

The physical implication of Eqs. (12.8) is quite simple. Before the impulse is applied, the body follows a Keplerian orbit with the constant elements $a, e, \sigma, \Omega, \omega, i$. Immediately after the impulse the orbit is again Keplerian but with the constant elements $a + \Delta a$, $e + \Delta e$, $\sigma + \Delta \sigma$, $\Omega + \Delta \Omega$, $\omega + \Delta \omega$, $i + \Delta i$, respectively. Equations (12.8) provide a great deal of insight into the problem of orbit change. From these equations it is obvious that to change the orientation of the orbit plane the impulse must have a component perpendicular to that plane. Moreover, the timing of the impulse governs the magnitude of this change. For example, the largest change $\Delta\Omega$ is obtained when the impulse is applied in the position for which $r \sin(\omega + \theta)$ has a maximum. Recalling that $r = a(1 - e^2)/(1 + e \cos \theta)$, we conclude that these positions are given by those values for which the true anomaly θ satisfies the equation

$$\cos(\omega + \theta) + e \cos \omega = 0. \tag{12.9}$$

For small e the positions approach $\omega + \theta = \pi/2, 3\pi/2$. These positions are approximately at right angles with the line of nodes. In a similar manner the largest change Δi takes place when θ satisfies

$$\sin(\omega + \theta) + e \sin \omega = 0, \tag{12.10}$$

which for small e yields the approximate solutions $\omega + \theta = 0, \pi$, so that the positions approach the ascending and descending nodal points, respectively. Of course, an impulse Δv_L also produces a change $\Delta\omega$. On the other hand, a Δv_L impulse alone does not alter the orbit shape or size but only its plane orientation. We can easily visualize this by recognizing that an impulse perpendicular to the orbit has no effect upon the velocities in the plane of the orbit.

12.2 PERTURBATIONS OF A SATELLITE ORBIT IN THE GRAVITATIONAL FIELD OF AN OBLATE PLANET

As a first approximation, the orbit of an artificial satellite can be assumed to be Keplerian, which implies that there are no disturbing forces of any kind acting upon the satellite. A more refined investigation, however, must account for any second-order factors which may perturb the satellite orbit. For satellites close to the earth's surface, the major disturbing effects are the lack of spherical symmetry of the earth and atmospheric drag, whereas for moon satellites only the former exists. We shall investigate the motion of an earth satellite under both these effects by assuming that the earth is an oblate spheroid. Although a more exact orbital theory of an earth satellite should include the simultaneous effects of the earth's oblateness and atmospheric drag, both these effects are of second order compared with the primary central force attraction, so that a reasonable approximation can be expected by treating them separately.

Let us first calculate the perturbing forces on a satellite due to the earth's oblateness. Denoting the moments of inertia about the symmetry axis (coinciding with the polar axis) and about any diametrical axis in the equatorial plane by C and A, respectively, we use Eq. (11.108) and write the gravitational potential energy

$$V = -\frac{GM_e m}{r} - \frac{Gm}{4r^3}[(3l^2 - 1)C + (3m^2 - 1)C + (3n^2 - 1)(2A - C)]$$

$$= -\frac{GM_e m}{r} - \frac{Gm(C - A)}{2r^3}\left(1 - \frac{3z^2}{r^2}\right), \tag{12.11}$$

where M_e denotes the mass of the earth and m the mass of the satellite. Moreover, $r = (x^2 + y^2 + z^2)^{1/2}$ is the radial distance between M_e and m, in which x, y, and z represent the coordinates of m relative to a set of axes with the origin at the earth's center, x coinciding with the vernal equinox and z with the earth's polar axis. We assume that the disturbing function is due to the earth's oblateness alone, so that, ignoring the primary gravitational effect, we can write the disturbing function

$$R = -\frac{1}{m}\left[V - \left(-\frac{GM_e m}{r}\right)\right] = \frac{G(C - A)}{2r^3}\left(1 - \frac{3z^2}{r^2}\right) \tag{12.12}$$

if we recall from Sec. 11.6 that disturbing functions are to be taken for a perturbed body of mass equal to unity. The corresponding disturbing force is simply

$$\nabla R = \mathbf{F} = -\frac{3G(C - A)}{2r^5}\left[\left(1 - \frac{5z^2}{r^2}\right)\mathbf{r} + 2z\mathbf{k}\right]. \tag{12.13}$$

We must call attention to the fact that $\mathbf{F}$ *is not a central force*, as clearly indicated by Eq. (12.13). In this particular case, because the earth is assumed to possess axial symmetry, the only component other than the central force component is parallel to the earth's symmetry axis.

The disturbing force $\mathbf{F}$ can be conveniently expressed in terms of the components radial, transverse, and perpendicular to the orbit, relating to the orbit of m about M_e, in the form

$$F_r = \mathbf{F} \cdot \mathbf{u}_r = -\frac{3G(C - A)}{2r^4}\left[\left(1 - \frac{5z^2}{r^2}\right) + \frac{2z}{r}\mathbf{k} \cdot \mathbf{u}_r\right],$$

$$F_\theta = \mathbf{F} \cdot \mathbf{u}_\theta = -\frac{3G(C - A)}{r^4}\frac{z}{r}\mathbf{k} \cdot \mathbf{u}_\theta, \tag{12.14}$$

$$F_L = \mathbf{F} \cdot \mathbf{u}_L = -\frac{3G(C - A)}{r^4}\frac{z}{r}\mathbf{k} \cdot \mathbf{u}_L,$$

where $\mathbf{u}_r$, $\mathbf{u}_\theta$, $\mathbf{u}_L$ are the associated unit vectors, given by Eqs. (11.147) in terms of the unit vectors $\mathbf{i}$, $\mathbf{j}$, $\mathbf{k}$ along x, y, z, respectively. But $z/r = \mathbf{k} \cdot \mathbf{u}_r$, so that, taking the scalar product of $\mathbf{k}$ and each of Eqs. (11.147) and introducing the results into Eqs. (12.14), we can rewrite these equations as

Problems in Spacecraft Dynamics

$$F_r = -\frac{3G(C-A)}{2r^4}[1 - 3\sin^2(\omega+\theta)\sin^2 i],$$

$$F_\theta = -\frac{3G(C-A)}{2r^4}\sin 2(\omega+\theta)\sin^2 i, \qquad (12.15)$$

$$F_L = -\frac{3G(C-A)}{2r^4}\sin(\omega+\theta)\sin 2i.$$

Equations (12.15), when introduced into Eqs. (11.153), yield the partial derivatives of R with respect to the orbital elements $a, e, \ldots$, which in turn can be substituted into Eqs. (11.142) and (11.144) for the rates of change $a, e, \ldots$. The equations for the rates of change of the orbital elements thus obtained are coupled and do not lend themselves to an easy solution. We can, however, obtain a first-approximation solution of these equations, valid over a relatively short time interval, by assuming that the orbital elements appearing in the right side of the equations correspond to the osculating orbit and hence are constant. As an illustration, let us use the appropriate equation in (11.142), (11.153), and (12.15) and write

$$\frac{d\Omega}{dt} = \frac{1}{na^2(1-e^2)^{\frac{1}{2}}\sin i}\frac{\partial R}{\partial i} = -\frac{3G(C-A)\sin^2(\omega+\theta)\sin 2i}{2na^2(1-e^2)^{\frac{1}{2}}r^3 \sin i}. \qquad (12.16)$$

It will prove convenient to change the independent variable from t to θ, and to do so we use certain results from the two-body problem (see Sec. 11.1), so that

$$\frac{d\Omega}{d\theta} = \frac{d\Omega}{dt}\frac{1}{\dot\theta} = \frac{d\Omega}{dt}\frac{r^2}{k} = \frac{r^2}{na^2(1-e^2)^{\frac{1}{2}}}\frac{d\Omega}{dt}. \qquad (12.17)$$

Combining Eqs. (12.16) and (12.17), we have

$$\frac{d\Omega}{d\theta} = -\frac{3G(C-A)\cos i}{n^2 a^5(1-e^2)^2}(1 + e\cos\theta)\sin^2(\omega+\theta). \qquad (12.18)$$

Under the assumption that the orbital elements on the right side of Eq. (12.18) are constant, we can integrate that equation with respect to θ over a complete period with the result

$$\Delta\Omega_{\text{per}} = -\frac{3G(C-A)\cos i}{2n^2 a^5(1-e^2)^2}. \qquad (12.19)$$

Equation (12.19) implies a secular change in Ω as *the line of nodes regresses* by an amount $|\Delta\Omega_{\text{per}}|$ for each revolution of m about M_e. The magnitude of the regression depends on the orbit inclination i, so that the regression is relatively large for small inclinations and vanishes for $i = \pi/2$. This result can easily be explained physically by the fact that for polar orbits there is no regression of the nodes due to the axial symmetry assumption.

This method of solution can form the basis of an incremental procedure for the determination of a satellite orbit in the gravitational field of an oblate planet. According to this procedure, Eqs. (11.142) are integrated by using on the right side of the equations the elements from the osculating orbit. This enables us to calculate a new osculating orbit at the end of a short time increment Δt. The procedure is repeated by using a new osculating orbit after every time increment.

Equation (12.11) does not describe the gravitational potential energy in sufficient detail, as it is limited to second-order inertial integrals. For a nearly spherical body an expression in terms of the spherical harmonics derived in Sec. 11.7 is more suitable. The motion of an artificial satellite in the gravitational field of the oblate earth has been investigated by Kozai (Ref. 9), among others. The following analysis is essentially his.

Assuming axial symmetry, Eq. (11.99) yields the gravitational potential energy

$$V = -\frac{GM_e m}{r}\left[1 - \sum_{s=2}^{\infty} J_s\left(\frac{1}{r}\right)^s P_s(\sin \phi)\right]$$

$$= -\frac{GM_e m}{r}\left[1 - J_2\left(\frac{1}{r}\right)^2 \tfrac{1}{2}(3 \sin^2 \phi - 1) - J_3\left(\frac{1}{r}\right)^3 \tfrac{1}{2}(5 \sin^3 \phi - 3 \sin \phi)\right.$$

$$\left. - J_4\left(\frac{1}{r}\right)^4 \tfrac{1}{8}(35 \sin^4 \phi - 30 \sin^2 \phi + 3) + \cdots\right], \qquad (12.20)$$

where r is the distance to the satellite, $P_s(\sin \phi)$ are the familiar Legendre polynomials, and J_s are constants related to the inertial integrals. The term in the series which would correspond to $s = 1$ is zero by the assumption that the origin coincides with the earth's center of mass. We recall that ϕ denotes the latitude of the satellite, which is related to the orbital elements i and ω and the true anomaly θ by

$$\sin \phi = \sin i \sin(\omega + \theta). \qquad (12.21)$$

With this in mind, the disturbing function takes the form

$$R = -\frac{1}{m}\left[V - \left(-\frac{GM_e m}{r}\right)\right]$$

$$= \mu\left\{\frac{3}{2}\frac{J_2}{a^3}\left(\frac{a}{r}\right)^3 [\tfrac{1}{3} - \tfrac{1}{2}\sin^2 i + \tfrac{1}{2}\sin^2 i \cos 2(\omega + \theta)]\right.$$

$$+ \frac{3}{2}\frac{J_3}{a^4}\left(\frac{a}{r}\right)^4 [(1 - \tfrac{5}{4}\sin^2 i)\sin(\omega + \theta) + \tfrac{5}{12}\sin^2 i \sin 3(\omega + \theta)]\sin i$$

$$- \frac{35}{8}\frac{J_4}{a^5}\left(\frac{a}{r}\right)^5 [\tfrac{3}{35} - \tfrac{3}{7}\sin^2 i + \tfrac{3}{8}\sin^4 i + (\tfrac{3}{7} - \tfrac{1}{2}\sin^2 i)\sin^2 i \cos 2(\omega + \theta)$$

$$\left. + \tfrac{1}{8}\sin^4 i \cos 4(\omega + \theta)] + \cdots\right\}, \qquad (12.22)$$

where we substituted the approximate value $\mu \approx GM_e$, obtained by ignoring the mass of the satellite, which is generally small compared with the mass of the earth.

It will prove even more convenient to work with the mean anomaly M as an independent variable instead of the true anomaly θ. The relation between the two is

$$\frac{d\theta}{dM} = \frac{d\theta}{dt}\frac{dt}{dM} = \left(\frac{a}{r}\right)^2 (1 - e^2)^{\frac{1}{2}}, \tag{12.23}$$

which is obtained in a way similar to that for Eq. (12.17). From Sec. 11.1 we observe that a/r and θ are periodic functions of M with coefficients in the form of a power series in e. Substitution of a/r and θ into Eq. (12.22) leads us to the conclusion that terms in R not depending on ω or M are secular, terms involving ω but not M are periodic with a long period, and terms depending on M are short-period.

The secular, long-period, and short-period terms in R can be separated by averaging certain quantities with respect to M over one period of motion. Such average values, obtained by Tisserand (Ref. 16, chap. 16), are as follows

$$\overline{\left(\frac{a}{r}\right)^3} = \frac{1}{2\pi}\int_0^{2\pi}\left(\frac{a}{r}\right)^3 dM = (1 - e^2)^{-\frac{3}{2}},$$

$$\overline{\left(\frac{a}{r}\right)^3 \sin 2\theta} = \overline{\left(\frac{a}{r}\right)^3 \cos 2\theta} = 0, \quad \overline{\left(\frac{a}{r}\right)^4 \cos \theta} = e(1 - e^2)^{-\frac{5}{2}},$$

$$\overline{\left(\frac{a}{r}\right)^4 \sin \theta} = \overline{\left(\frac{a}{r}\right)^4 \cos 3\theta} = \overline{\left(\frac{a}{r}\right)^4 \sin 3\theta} = 0, \tag{12.24}$$

$$\overline{\left(\frac{a}{r}\right)^5} = \left(1 + \frac{3e^2}{2}\right)(1 - e^2)^{-\frac{7}{2}}, \quad \overline{\left(\frac{a}{r}\right)^5 \cos 2\theta} = \frac{3e^2}{4}(1 - e^2)^{-\frac{7}{2}},$$

$$\overline{\left(\frac{a}{r}\right)^5 \sin 2\theta} = \overline{\left(\frac{a}{r}\right)^5 \cos 4\theta} = \overline{\left(\frac{a}{r}\right)^5 \sin 4\theta} = 0,$$

leading to the first four components of R

$$R_1 = \frac{3}{2}\frac{\mu J_2}{a^3}(\tfrac{1}{3} - \tfrac{1}{2}\sin^2 i)(1 - e^2)^{-\frac{3}{2}},$$

$$R_2 = -\frac{35}{8}\frac{\mu J_4}{a^5}(\tfrac{3}{35} - \tfrac{3}{7}\sin^2 i + \tfrac{3}{8}\sin^4 i)(1 + \tfrac{3}{2}e^2)(1 - e^2)^{-\frac{7}{2}},$$

$$R_3 = \mu\left[\frac{3}{2}\frac{J_3}{a^4}\sin i(1 - \tfrac{5}{4}\sin^2 i)e(1 - e^2)^{-\frac{5}{2}}\sin \omega \right.$$
$$\left. - \frac{35}{8}\frac{J_4}{a^5}\sin^2 i(\tfrac{9}{28} - \tfrac{3}{8}\sin^2 i)e^2(1 - e^2)^{-\frac{7}{2}}\cos 2\omega\right], \tag{12.25}$$

$$R_4 = \frac{3}{2}\frac{\mu J_2}{a^3}\left(\frac{a}{r}\right)^3\left\{(\tfrac{1}{3} - \tfrac{1}{2}\sin^2 i)\left[1 - \left(\frac{r}{a}\right)^3(1 - e^2)^{-\frac{3}{2}}\right]\right.$$
$$\left. + \tfrac{1}{2}\sin^2 i \cos 2(\omega + \theta)\right\},$$

where R_1 and R_2 are recognized as first- and second-order secular components of R, respectively, R_3 is a long-period and R_4 a short-period component. It is interesting to note at this point that R_1 is responsible for the term $\Delta\Omega_{\text{per}}$, according to Eq. (12.19), whereas R_2 was ignored in that analysis. Moreover, it is easy to see that the short-period component R_4 is accounted for in Eqs. (12.15).

The first-order secular perturbations are obtained by letting the disturbing function equal R_1 alone. Denoting the initial values of the orbital elements minus the periodic perturbations by the subscript zero and introducing $R = R_1$ into the last three of Eqs. (11.142) and Eqs. (11.144), we can calculate the secular perturbations of the first order with the result

$$\bar{a} = a_0, \quad \bar{e} = e_0, \quad \bar{i} = i_0,$$

$$\bar{M} = M_0 + \bar{n}t,$$

$$\bar{\Omega} = \Omega_0 - \frac{3}{2}\frac{J_2\bar{n}t \cos i}{a^2(1-e^2)^2}, \qquad (12.26)$$

$$\bar{\omega} = \omega_0 + \frac{3}{2}\frac{J_2\bar{n}t}{a^2(1-e^2)^2}(2 - \tfrac{5}{2}\sin^2 i),$$

$$\bar{n} = n_0 + \frac{3}{2}\frac{J_2 n_0}{a^2(1-e^2)^{3/2}}(1 - \tfrac{3}{2}\sin^2 i),$$

where the unperturbed mean motion n_0 is related to the unperturbed semimajor axis a_0 by $n_0^2 a_0^3 = \mu$. From Eqs. (12.26) we conclude that only M, Ω, and ω undergo secular perturbations. It will be noticed that by calculating the change of $\bar{\Omega}$ corresponding to a time interval equal to a complete orbital period, we obtain the value $\Delta\bar{\Omega}$, which is the same as the value $\Delta\Omega_{\text{per}}$ given by Eq. (12.19). From the equation for $\bar{\Omega}$ we reach the conclusion that $\bar{\Omega} - \Omega_0$ vanishes for a polar orbit, whereas from the equation for $\bar{\omega}$ we see that the perigee advances or regresses in the orbital plane according to whether $\sin^2 i$ is smaller or larger than 0.8. Hence, the perigee advances for $i < 63°26'$ and regresses for $i > 63°26'$.

The first-order short-period perturbations can be obtained by using the disturbing function equal to R_4. The elements a, e, i, ω, and n on the right sides of Eqs. (11.142) and (11.144) are assumed constant except in the equation for M. In that equation, in the term consisting of n alone, n is to be regarded as a function of time satisfying $n^2 a^3 = \mu$, which can be considered known if a is known. The independent variable is transformed from t to θ, as in Eq. (12.17), which enables us to integrate with respect to θ. As an illustration, the equation for the orbit inclination takes the form

$$\frac{di}{d\theta} = \frac{di}{dt}\frac{1}{\dot{\theta}} = \frac{r^2}{na^2(1-e^2)^{1/2}}\frac{di}{dt} = \frac{\cot i}{n^2 a^2(1-e^2)}\left(\frac{r}{a}\right)^2 \frac{\partial R_4}{\partial \omega}, \qquad (12.27)$$

Problems in Spacecraft Dynamics

which can be integrated with respect to θ, yielding the increment Δi corresponding to a given angle $\Delta \theta$. For the results of the integrations see the paper by Kozai (Ref. 9).

It turns out that the mean values with respect to M of the short-period perturbations are not zero except for the value corresponding to a. The mean values of the short-period changes with respect to M are

$$\overline{\Delta a} = 0,$$

$$\overline{\Delta e} = \frac{3}{2} \frac{J_2 \sin^2 i}{a^2(1-e^2)^2} \frac{1-e^2}{6e} \overline{\cos 2\theta} \cos 2\omega,$$

$$\overline{\Delta M} = -\frac{3}{2} \frac{J_2 \sin^2 i}{a^2(1-e^2)^{3/2}} \left(\frac{1}{8} + \frac{2+e^2}{12e^2} \overline{\cos 2\theta} \right) \sin 2\omega,$$

$$\overline{\Delta \Omega} = -\frac{1}{4} \frac{J_2 \cos i}{a^2(1-e^2)^2} \overline{\cos 2\theta} \sin 2\omega, \qquad (12.28)$$

$$\overline{\Delta \omega} = \frac{3}{2} \frac{J_2}{a^2(1-e^2)^2} \left[\sin^2 i \left(\frac{1}{8} + \frac{1-e^2}{6e^2} \overline{\cos 2\theta} \right) + \tfrac{1}{6} \cos^2 i \, \overline{\cos 2\theta} \right] \sin 2\omega,$$

$$\overline{\Delta i} = -\frac{1}{8} \frac{J_2 \sin 2i}{a^2(1-e^2)^2} \overline{\cos 2\theta} \cos 2\omega,$$

in which

$$\overline{\cos 2\theta} = \frac{e^2[1+2(1-e^2)]^{1/2}}{[1+(1-e^2)^{1/2}]^2}. \qquad (12.29)$$

Details of these calculations as well as a discussion of the long-period perturbations can be found in Kozai (Ref. 9).

12.3 THE EFFECT OF ATMOSPHERIC DRAG ON SATELLITE ORBITS

In this section we consider the case in which the Keplerian elliptic orbit of an earth satellite, governed by the primary two-body central force, is disturbed by a resisting force due to the atmospheric drag. Of course, this implies an orbit with a relatively small perigee height above the earth's surface. The shape of the satellite is important, and, in addition to drag, lift forces may be present. Moreover, if the satellite is spinning, the cross-sectional area which must enter into the drag calculation keeps changing. The analysis can be simplified considerably by assuming a spherical satellite, so that the drag force per unit mass can be approximated by the formula

$$\mathbf{F} = -\frac{1}{2m} C_D A \rho v^2 \mathbf{u}_T, \qquad (12.30)$$

where C_D = aerodynamic drag coefficient,

A = satellite cross-sectional area,

$\rho = \rho(r)$ = air density (assumed to be a function only of the radial distance r from the earth's center),

v = satellite velocity,

$\mathbf{u}_T$ = unit vector tangent to orbit.

The value of the coefficient C_D ranges between 1 and 2 depending on the relative magnitude of the mean free path of the atmospheric molecules. If the mean free path is small compared with the satellite dimensions, the coefficient is close to 1, and if the mean free path is large relative to the satellite dimensions, C_D is close to 2.

From Eqs. (11.155), (11.157), and (12.30) we obtain the radial and transverse forces

$$F_r = -\frac{1}{2m} C_D A \rho v^2 \frac{e \sin \theta}{(1 + e^2 + 2e \cos \theta)^{\frac{1}{2}}},$$

$$F_\theta = -\frac{1}{2m} C_D A \rho v^2 \frac{1 + e \cos \theta}{(1 + e^2 + 2e \cos \theta)^{\frac{1}{2}}},$$

(12.31)

and, of course, the force component perpendicular to the orbit plane is zero. Introducing Eqs. (12.31) in Eqs. (11.153), with $\partial R/\partial \sigma$ replaced by $\partial R/\partial M$, and substituting the result into the last three of Eqs. (11.142) and Eqs. (11.144), we arrive at

$$\frac{da}{dt} = -\frac{A}{m} \frac{C_D \rho v^2 (1 + e^2 + 2e \cos \theta)^{\frac{1}{2}}}{n(1 - e^2)^{\frac{1}{2}}},$$

$$\frac{de}{dt} = -\frac{A}{m} \frac{C_D \rho v^2 (1 - e^2)(e + \cos \theta)}{na(1 + e^2 + 2e \cos \theta)^{\frac{1}{2}}},$$

$$\frac{dM}{dt} = n + \frac{A}{m} \frac{C_D \rho v^2 (1 - e^2)(1 + e^2 + 2e \cos \theta) \sin \theta}{nae(1 + e \cos \theta)(1 + e^2 + 2e \cos \theta)^{\frac{1}{2}}},$$

$$\frac{d\Omega}{dt} = 0,$$

$$\frac{d\omega}{dt} = -\frac{A}{m} \frac{C_D \rho v^2 (1 - e^2) \sin \theta}{nae(1 + e^2 + 2e \cos \theta)^{\frac{1}{2}}},$$

$$\frac{di}{dt} = 0.$$

(12.32)

The fourth and sixth of Eqs. (12.32) indicate that the longitude of the ascending node Ω and the orbit plane inclination i are not affected by drag, so that the orbit remains planar and its orientation in space does not change.

The factor A/m appearing on the right side of Eqs. (12.32) indicates that the drag effect can be reduced by selecting the ratio of cross-sectional area to mass as small as possible, a fact which agrees with intuition. The third and fifth of Eqs. (12.32) contain the term $\sin \theta$, which ensures that the solutions M and ω are periodic. Moreover, because the drag forces are small, the amplitude of oscillation must be relatively small, so that these equations will be ignored in the sequel. The first two of Eqs. (12.32), however, indicate secular changes in a and e; further discussion will be limited to these equations.

The first two of Eqs. (12.32) can be treated more effectively by changing the independent variable from the time t to the eccentric anomaly w. To this end, we recall the relations (see Secs. 1.9 and 11.1)

$$r = \frac{a(1 - e^2)}{1 + e \cos \theta} = a(1 - e \cos w), \qquad r^2 \dot{\theta} = k,$$
$$M = n(t - T) = w - e \sin w, \qquad k^2 = \mu a(1 - e^2) = n^2 a^4 (1 - e^2), \tag{12.33}$$

enabling us to write

$$v^2 = \dot{r}^2 + r^2 \dot{\theta}^2 = \frac{n^2 a^2 (1 + e^2 + 2e \cos \theta)}{1 - e^2}. \tag{12.34}$$

Introducing Eqs. (12.33) and (12.34) into the first two of Eqs. (12.32) and changing the independent variable to w, we arrive at

$$\frac{da}{dw} = -\frac{A}{m} C_D \rho a^2 \frac{(1 + e \cos w)^{3/2}}{(1 - e \cos w)^{1/2}},$$
$$\frac{de}{dw} = -\frac{A}{m} C_D \rho a \frac{(1 - e^2)(1 + e \cos w)^{1/2} \cos w}{(1 - e \cos w)^{1/2}}. \tag{12.35}$$

Assuming, as in Sec. 12.2, that the orbital elements on the right side of Eqs. (12.35) correspond to the osculating orbit, the changes in a and e over one revolution of the satellite in its orbit become

$$\Delta a = -\frac{A}{m} C_D a^2 \int_0^{2\pi} \rho(r) \frac{(1 + e \cos w)^{3/2}}{(1 - e \cos w)^{1/2}} \, dw,$$
$$\Delta e = -\frac{A}{m} C_D a (1 - e^2) \int_0^{2\pi} \rho(r) \frac{(1 + e \cos w)^{1/2} \cos w}{(1 - e \cos w)^{1/2}} \, dw. \tag{12.36}$$

The integrals in Eqs. (12.36) must be evaluated numerically after expressing ρ in terms of the eccentric anomaly w.

Perhaps the behavior of the satellite can be better explained by introducing the perigee and apogee distances

$$r_p = a(1 - e), \qquad r_a = a(1 + e). \tag{12.37}$$

With this in mind, Eqs. (12.36) can be shown to lead to

$$\Delta r_p = -\frac{A}{m} C_D a^2 (1 - e) \int_0^{2\pi} \rho \frac{(1 - \cos w)(1 + e \cos w)^{\frac{1}{2}}}{(1 - e \cos w)^{\frac{1}{2}}} dw,$$

$$\Delta r_a = -\frac{A}{m} C_D a^2 (1 + e) \int_0^{2\pi} \rho \frac{(1 + \cos w)(1 + e \cos w)^{\frac{1}{2}}}{(1 - e \cos w)^{\frac{1}{2}}} dw.$$

(12.38)

If the eccentricity is not very small, it turns out that Δr_a is much larger in magnitude than Δr_p, so that the satellite spirals toward the earth, and in the process both the major axis and the eccentricity decrease. The satellite orbit becomes closer and closer to circular while losing altitude, until the increasing friction due to denser air in the lower fringes of the atmosphere causes the satellite to burn up.

12.4 THE ATTITUDE MOTION OF ORBITING SATELLITES. GENERAL CONSIDERATIONS

In Sec. 11.8 we had an opportunity to examine the motion of the earth about its own center of mass. In particular, we have been able to show that the precession and nutation of the earth's polar axis are a direct result of the gravitational torques exerted on the earth by the moon and the sun. These torques are due to the fact that the earth does not possess inertial symmetry. With the advent of artificial satellites, this classical problem of dynamical astronomy has acquired a very modern flavor. The problem of satellite orientation in space, referred to as *attitude*, is in general considerably more involved, however, for various reasons. Among them we should mention, by way of contrast, that the satellite's moments of inertia may all be different and its rotational motion may not be a simple uniform high spin, as in the case of the earth. The situation may be complicated further if the satellite contains internal moving parts or the orbit is not circular. Moreover, in addition to the forces due to the earth's gravitational field, the forces acting upon an earth satellite may arise from a variety of other sources such as the atmospheric drag, the sun's radiation pressure, the earth's magnetic and electric fields, meteoroid and cosmic-ray bombardments, etc.[1] In general the most significant factor is the earth's gravitational field.

Thanks to a multitude of applications, the problem of attitude motion of satellites has received considerable attention, as witnessed by the abundance of technical papers and reports related to this subject. Because of functional considerations, earth satellites are generally required to point either toward the earth or toward a fixed point in an inertial space. This can be accomplished by active control systems or by passive means, stabilization

---

[1] For a discussion of the various factors affecting the attitude of satellites, see Ref. 15.

in the latter case being provided either by gravitational torques or by a spinning motion initially imparted to the satellite. Accordingly, the two methods of stabilization are referred to as *gravity-gradient* and *spin stabilization*.

For many practical purposes, a satellite can be regarded as a rigid body whose dimensions are small relative to the distance to the center of force. As a result, the attitude motion can be assumed to have no effect upon the orbital motion of the center of mass of the satellite, where the orbital motion of the satellite is governed by the primary inverse square central force law. Of particular interest is the case in which the orbit of a satellite about the spherically symmetric earth is circular. The mathematical model consisting of a rigid body in a circular orbit has the advantage of lending itself to a neat analytical treatment while describing a common physical situation relatively well.

To discuss the attitude motion of a rigid satellite in a circular orbit, it is convenient to introduce an *orbital frame of reference* a_1, a_2, a_3 with the origin O at the center of mass of the body, axis a_1 in the radial direction, axis a_2 tangent to the orbit in the direction of motion, and axis a_3 normal to the orbit plane, as shown in Fig. 12.2. Body axes x_1, x_2, x_3 are obtained from axes a_1, a_2, a_3 by means of three independent rotations to be specified later. The relation between the two sets of axes x_i and a_j can be written in terms of the direction cosines l_{ij} in the compact matrix form

$$\{x\} = [l]\{a\}, \tag{12.39}$$

where the direction cosines depend on the three independent rotations chosen.

The equations of motion about the body axes x_1, x_2, x_3 can be readily written in terms of quasi-coordinates (see Sec. 4.12) as follows

$$\frac{d}{dt}\left\{\frac{\partial T}{\partial \omega}\right\} + [\omega]\left\{\frac{\partial T}{\partial \omega}\right\} = \{N\}, \tag{12.40}$$

FIGURE 12.2

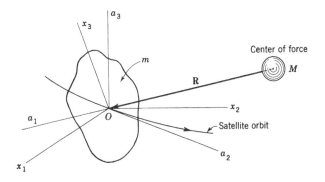

where T = body kinetic energy,

$\left\{\dfrac{\partial}{\partial \omega}\right\}$ = column matrix of partial derivatives with respect to body angular velocities ω_i ($i = 1, 2, 3$),

$[\omega]$ = skew-symmetric matrix associated with ω_i,

$\{N\}$ = torque vector about body axes.

From Eqs. (11.118), we obtain the torque components

$$N_1 = 3\Omega^2(C - B)l_{31}l_{21},$$
$$N_2 = 3\Omega^2(A - C)l_{11}l_{31}, \qquad (12.41)$$
$$N_3 = 3\Omega^2(B - A)l_{21}l_{11},$$

where Ω is the orbital angular velocity and A, B, C are the moments of inertia about the body axes x_1, x_2, x_3, respectively, chosen to coincide with the principal axes. Moreover, we substituted $\Omega^2 = \mu/R^3 \approx GM/R^3$ in Eqs. (12.41), by virtue of the fact that the mass of the satellite is small compared with the earth's mass M.

Let us define the orientation of the body axes x_i relative to the orbital axes a_j in two ways, as depicted in Fig. 12.3a and b. The reason for introducing two sets of rotations will become evident later, when specific equilibrium positions are discussed. Of course, to discuss a particular equilibrium position only one of the two sets of rotations is used. In the first case, shown in Fig. 12.3a, the three rotations are θ_2 about a_2, θ_1 about b_1, and ψ about c_3. In the second case, shown in Fig. 12.3b, the rotations are given by the familiar Euler's angles: ϕ about a_3, θ about b_1, and ψ about c_3. The angular

FIGURE 12.3

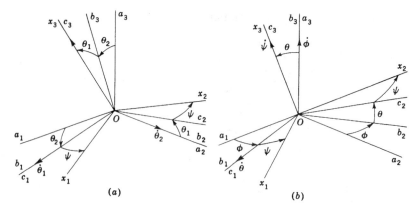

velocity components of the body about axes x_1, x_2, x_3, for the two sets of rotations, are

$$\omega_1 = \Omega l_{13} + \dot{\theta}_2 \cos\theta_1 \sin\psi + \dot{\theta}_1 \cos\psi = \Omega l_{13} + \dot{\phi}\sin\theta\sin\psi + \dot{\theta}\cos\psi,$$
$$\omega_2 = \Omega l_{23} + \dot{\theta}_2 \cos\theta_1 \cos\psi - \dot{\theta}_1 \sin\psi = \Omega l_{23} + \dot{\phi}\sin\theta\cos\psi - \dot{\theta}\sin\psi, \quad (12.42)$$
$$\omega_3 = \Omega l_{33} - \dot{\theta}_2 \sin\theta_1 + \dot{\psi} = \Omega l_{33} + \dot{\phi}\cos\theta + \dot{\psi}.$$

Similarly, the direction cosines between axes x_i and a_j have the expressions

$$l_{11} = \cos\theta_2 \cos\psi + \sin\theta_2 \sin\theta_1 \sin\psi = \cos\phi\cos\psi - \sin\phi\cos\theta\sin\psi,$$
$$l_{12} = \cos\theta_1 \sin\psi = \sin\phi\cos\psi + \cos\phi\cos\theta\sin\psi,$$
$$l_{13} = -\sin\theta_2 \cos\psi + \cos\theta_2 \sin\theta_1 \sin\psi = \sin\theta\sin\psi,$$
$$l_{21} = -\cos\theta_2 \sin\psi + \sin\theta_2 \sin\theta_1 \cos\psi = -\cos\phi\sin\psi - \sin\phi\cos\theta\cos\psi,$$
$$l_{22} = \cos\theta_1 \cos\psi = -\sin\phi\sin\psi + \cos\phi\cos\theta\cos\psi, \quad (12.43)$$
$$l_{23} = \sin\theta_2 \sin\psi + \cos\theta_2 \sin\theta_1 \cos\psi = \sin\theta\cos\psi,$$
$$l_{31} = \sin\theta_2 \cos\theta_1 = \sin\phi\sin\theta,$$
$$l_{32} = -\sin\theta_1 = -\cos\phi\sin\theta,$$
$$l_{33} = \cos\theta_2 \cos\theta_1 = \cos\theta.$$

Equations (12.39) to (12.43) define fully the differential equations for the attitude motion for both cases represented in Fig. 12.3.

At times it is more convenient to work with the Lagrange equations of motions. For this, and perhaps other reasons, we are interested in deriving expressions for the kinetic and potential energies. Using Eqs. (12.42), the rotational kinetic energy can be written

$$T = \tfrac{1}{2}(A\omega_1^2 + B\omega_2^2 + C\omega_3^2) = T_2 + T_1 + T_0, \quad (12.44)$$

where

$$T_2 = \tfrac{1}{2}[A(\dot{\theta}_2 \cos\theta_1 \sin\psi + \dot{\theta}_1 \cos\psi)^2 + B(\dot{\theta}_2 \cos\theta_1 \cos\psi - \dot{\theta}_1 \sin\psi)^2$$
$$\quad + C(\dot{\psi} - \dot{\theta}_2 \sin\theta_1)^2]$$
$$= \tfrac{1}{2}[A(\dot{\phi}\sin\theta\sin\psi + \dot{\theta}\cos\psi)^2 + B(\dot{\phi}\sin\theta\cos\psi - \dot{\theta}\sin\psi)^2$$
$$\quad + C(\dot{\psi} + \dot{\phi}\cos\theta)^2],$$
$$T_1 = \Omega[Al_{13}(\dot{\theta}_2 \cos\theta_1 \sin\psi + \dot{\theta}_1 \cos\psi) \quad (12.45)$$
$$\quad + Bl_{23}(\dot{\theta}_2 \cos\theta_1 \cos\psi - \dot{\theta}_1 \sin\psi) + Cl_{33}(\dot{\psi} - \dot{\theta}_2 \sin\theta_1)]$$
$$= \Omega[Al_{13}(\dot{\phi}\sin\theta\sin\psi + \dot{\theta}\cos\psi)$$
$$\quad + Bl_{23}(\dot{\phi}\sin\theta\cos\psi - \dot{\theta}\sin\psi) + Cl_{33}(\dot{\psi} + \dot{\phi}\cos\theta)],$$
$$T_0 = \tfrac{1}{2}\Omega^2(Al_{13}^2 + Bl_{23}^2 + Cl_{33}^2).$$

Moreover, from Eq. (11.108), we have the potential energy

$$V = -\tfrac{3}{4}\Omega^2[l_{11}^2(B + C - A) + l_{21}^2(A + C - B) + l_{31}^2(A + B - C)]. \tag{12.46}$$

From Eqs. (12.44) to (12.46) we conclude that the Lagrangian $L = T - V$ does not depend explicitly on the time t, and because this is a nonnatural system, it follows that the system possesses a Jacobi integral in the form of the Hamiltonian

$$H = T_2 - T_0 + V = h = \text{const}, \tag{12.47}$$

rather than the total energy E (see Sec. 2.13). If the moments of inertia A, B, C are all different, this is the only motion integral available.

The differential equations describing the rotational motion of a satellite are highly nonlinear, and no closed-form solution of the complete equations is possible. Under these circumstances we must be content with information concerning the behavior of the system in the neighborhood of known motions. In particular, we would like to be able to state whether the system is stable or unstable. In this connection the cases corresponding to gravity-gradient and spin stabilization stand out in importance. These two cases turn out to correspond to positions of equilibrium, so that the investigation reduces to the familiar problem of studying the stability of motion in the neighborhood of an equilibrium point. We recall that such analyses may involve the solution of the eigenvalue problem associated with the variational equations, or the testing of a certain function for sign-definiteness if the Liapunov direct method is used.

12.5 THE ATTITUDE STABILITY OF EARTH-POINTING SATELLITES

Satellites pointing toward the earth at all times while the center of mass is moving in a circular orbit about the earth are of considerable interest because of their many uses. From a dynamical point of view, an earth-pointing satellite may be envisioned as a rigid body rotating about an axis normal to the orbital plane with an angular velocity equal to the orbital angular velocity, namely, the velocity with which the body revolves in its orbit. This is equivalent to the body's being at rest with respect to the orbital frame of reference a_1, a_2, a_3 introduced in Sec. 12.4. Under certain circumstances, this motion can be realized by the simple action of the gravitational torques, for which reason such satellites are referred to as gravity-gradient stabilized. We note that the dynamical system is autonomous.

Mathematically the position in which the satellite is always pointing toward the earth corresponds to an equilibrium point of the system. This represents a desired position about which the body oscillates if disturbed and if the equilibrium is stable. It remains to ascertain whether the position is indeed stable or unstable, where in the latter case the equilibrium represents a position the motion tends away from. To study the stability of motion in the neighborhood of the equilibrium point we shall use the Liapunov direct method.

Let us consider the satellite of Fig. 12.2 and examine the stability in the neighborhood of an equilibrium point for which the body is at rest relative to the orbital frame of reference, so that $\dot{\theta}_2 = \dot{\theta}_1 = \dot{\psi} = 0$ or $\dot{\phi} = \dot{\theta} = \dot{\psi} = 0$, depending on the set of rotations used (see Fig. 12.3). To define the equilibrium points completely, it remains to determine the value of the angular coordinates corresponding to those points. The system being canonical, these positions are given by the equations

$$\frac{\partial H}{\partial q_i} = 0, \qquad i = 1, 2, 3, \tag{12.48}$$

(see Sec. 5.1) in which the generalized coordinates q_i ($i = 1, 2, 3$) are either θ_2, θ_1, ψ or ϕ, θ, ψ. We have shown (Sec. 12.4) that for the type of systems considered the Hamiltonian is constant

$$H = T_2 - T_0 + V = h = \text{const}, \tag{12.49}$$

which represents the Jacobi integral associated with this nonnatural system. But T_2 is quadratic in the generalized velocities, which are zero at the equilibrium points considered, from which it follows that the equilibrium positions relative to the orbital reference frame satisfy

$$\frac{\partial U}{\partial q_i} = 0, \qquad i = 1, 2, 3, \tag{12.50}$$

where, using Eqs. (12.45) and (12.46), we obtain

$$U = V - T_0$$
$$= -\tfrac{1}{2}\Omega^2[A(l_{13}{}^2 - 3l_{11}{}^2) + B(l_{23}{}^2 - 3l_{21}{}^2) + C(l_{33}{}^2 - 3l_{31}{}^2)], \tag{12.51}$$

which is a function depending on the generalized coordinates alone. We pause briefly at this point to look into the meaning of the function U. We notice that if the body has zero angular velocity relative to the orbital frame of reference, then the function U is equal in value to the Jacobi integral h. Hence, the function

$$U = h = \text{const} \tag{12.52}$$

represents a family of *surfaces of zero relative velocity* in the coordinates θ_2, θ_1, ψ or ϕ, θ, ψ. These surfaces are entirely analogous to the ones discussed in Sec. 11.4 in connection with the restricted three-body problem. Introducing Eqs. (12.43) into (12.51) and using (12.50), we obtain the following equations defining the equilibrium positions for the set θ_2, θ_1, ψ

$$\frac{\partial U}{\partial \theta_2} = 4\Omega^2(Al_{11}l_{13} + Bl_{21}l_{23} + Cl_{31}l_{33}) = 0,$$

$$\frac{\partial U}{\partial \theta_1} = -\Omega^2[A \cos \theta_1 \sin \psi (l_{13} \cos \theta_2 - 3l_{11} \sin \theta_2)$$
$$+ B \cos \theta_1 \cos \psi (l_{23} \cos \theta_2 - 3l_{21} \sin \theta_2) \quad (12.53)$$
$$- C \sin \theta_1 (l_{33} \cos \theta_2 - 3l_{31} \sin \theta_2)] = 0,$$

$$\frac{\partial U}{\partial \psi} = -\Omega^2(A - B)(l_{13}l_{23} - 3l_{11}l_{21}) = 0,$$

whereas the equilibrium positions for the set ϕ, θ, ψ must satisfy

$$\frac{\partial U}{\partial \phi} = -3\Omega^2(Al_{11}l_{12} + Bl_{21}l_{22} + Cl_{31}l_{32}) = 0,$$

$$\frac{\partial U}{\partial \theta} = -\Omega^2[A \sin \psi (l_{13} \cos \theta - 3l_{11} \sin \phi \sin \theta)$$
$$+ B \cos \psi (l_{23} \cos \theta - 3l_{21} \sin \phi \sin \theta) \quad (12.54)$$
$$- C(l_{33} \sin \theta + 3l_{31} \sin \phi \cos \theta)] = 0,$$

$$\frac{\partial U}{\partial \psi} = -\Omega^2(A - B)(l_{13}l_{23} - 3l_{11}l_{21}) = 0.$$

The equilibrium positions turn out to be those in which the body principal axes are aligned with the orbital axes. These positions, denoted by E_1, E_2, and E_3, are shown in Fig. 12.4a, b, and c, respectively. We notice that in the first two cases we use the set θ_2, θ_1, ψ, whereas in the last we use ϕ, θ, ψ. The reason is to prevent indeterminacy, which may result when two angular velocity components are collinear (the angular velocity Ω is excepted). In all three equilibrium positions E_i the angular velocities are perpendicular

FIGURE 12.4

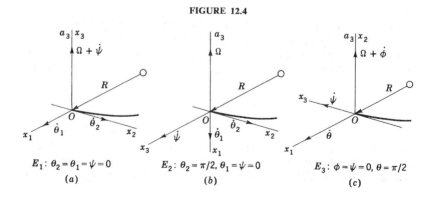

$E_1: \theta_2 = \theta_1 = \psi = 0$ $\quad E_2: \theta_2 = \pi/2, \theta_1 = \psi = 0$ $\quad E_3: \phi = \psi = 0, \theta = \pi/2$
(a) $\qquad\qquad\qquad$ (b) $\qquad\qquad\qquad$ (c)

to one another, so that it suffices to check the stability of only one of them, for example, E_1.

Because the Hamiltonian is constant, its total time derivative $\dot{H}$ is zero, so that if H is sign-definite at an equilibrium point, the Hamiltonian can be regarded as a Liapunov function and the corresponding equilibrium point is stable. But T_2 is positive definite by definition, so that the system is stable if U is positive definite. Hence the problem reduces to that of testing the function U for positive definiteness. Since U depends only on the coordinates and not on the velocities, this represents a substantial reduction in the amount of calculation necessary for testing. To determine whether U is positive definite or not, we shall make use of Sylvester's criterion (see Sec. 6.7), which requires the calculation of the associated Hessian matrix. The elements of this matrix for the coordinates θ_2, θ_1, ψ are

$$\frac{\partial^2 U}{\partial \theta_2^2} = 4\Omega^2[A(l_{13}^2 - l_{11}^2) + B(l_{23}^2 - l_{21}^2) + C(l_{33}^2 - l_{31}^2)],$$

$$\frac{\partial^2 U}{\partial \theta_2 \partial \theta_1} = 4\Omega^2[A \cos\theta_1 \sin\psi(l_{11} \cos\theta_2 + l_{13} \sin\theta_2)$$
$$+ B \cos\theta_1 \cos\psi(l_{21} \cos\theta_2 + l_{23} \sin\theta_2)$$
$$- C \sin\theta_1(l_{31} \cos\theta_2 + l_{33} \sin\theta_2)],$$

$$\frac{\partial^2 U}{\partial \theta_2 \partial \psi} = -\Omega^2(A - B)(l_{11}l_{23} + l_{13}l_{21}), \quad (12.55)$$

$$\frac{\partial^2 U}{\partial \theta_1^2} = -\Omega^2[(A - B) \sin 2\theta_2 \sin\theta_1 \sin 2\psi$$
$$+ (A \sin^2\psi + B \cos^2\psi - C)(\cos^2\theta_2 - 3\sin^2\theta_2)\cos 2\theta_1],$$

$$\frac{\partial^2 U}{\partial \theta_1 \partial \psi} = -\Omega^2(A - B)\cos\theta_1[\cos\psi(l_{13}\cos\theta_2 - 3l_{11}\sin\theta_2)$$
$$+ \sin\psi(l_{23}\cos\theta_2 - 3l_{21}\sin\theta_2)],$$

$$\frac{\partial^2 U}{\partial \psi^2} = -\Omega^2(A - B)[l_{23}^2 - l_{13}^2 - 3(l_{21}^2 - l_{11}^2)].$$

Using Eqs. (12.55), the Hessian matrix corresponding to the equilibrium point E_1, $\theta_2 = \theta_1 = \psi = 0$, takes the form

$$[\mathscr{H}]_{E_1} = \begin{bmatrix} 4\Omega^2(C - A) & 0 & 0 \\ 0 & \Omega^2(C - B) & 0 \\ 0 & 0 & -3\Omega^2(A - B) \end{bmatrix}, \quad (12.56)$$

so that the matrix is positive definite and the equilibrium position E_1 is *stable* if

$$C > B > A, \quad (12.57)$$

with the physical implication that the axis of minimum moment of inertia is aligned with the radial direction and the axis of maximum moment of inertia is normal to the orbital plane. The interpretation of this result is that a small perturbation from the equilibrium position, in which the longer axis of the body coincides with the local vertical, will give rise to counteracting gravity torques seeking to restore the body to its original position. For no damping, the body will oscillate about the equilibrium position. The present analysis gives no clue to the stability of equilibrium in which the axis of minimum moment of inertia is normal to the orbital plane or tangent to the orbit; in such a case the Hessian matrix (12.56) is not positive definite. If there is damping in the system, however, such that $\dot{H} = -2F \le 0$, where F is the Rayleigh's dissipation function, and, moreover, the damping is pervasive, then the configuration given by Eq. (12.57) is *asymptotically stable* and the cases in which the axis of minimum moment of inertia is normal to the local vertical become *unstable* (see Theorems 6.9.5 and 6.9.6). It can be shown that an analysis of the equilibrium points E_2 and E_3 yields no new information.

The same problem can be treated by means of an infinitesimal analysis, which amounts to deriving the system variational equations about the equilibrium positions and solving the corresponding eigenvalue problem. This is the approach used by DeBra and Delp (Ref. 3), whose infinitesimal analysis, examining the equilibrium point E_1 for the undamped case, confirms the conclusion reached here. In addition, however, they obtain another criterion according to which stability is possible in a subdomain of $B > A > C$. This implies that for certain ratios of the moments of inertia there exists a stable equilibrium position in which the axis of intermediate moment of inertia coincides with the local vertical, a result not predicted by the Liapunov direct method.

The same general problem has also been investigated by Breakwell and Pringle (Ref. 2), who studied the oscillatory character of the motion about the equilibrium position corresponding to gravity-gradient stabilization by means of the method of averaging (see Sec. 8.7) in combination with canonical transformations (see Chap. 9). They refer to the resonance oscillation of the satellite in circular orbit as *internal resonance*. In addition, they discuss the case of *external resonance* for satellite motion in an elliptic orbit of small eccentricity.

The problem of an unsymmetrical satellite in a circular orbit rotating relative to the orbital axes, such that one of the axes is nearly normal to the orbital plane, leads to a nonautonomous system of two simultaneous differential equations of motion with periodic coefficients resembling the Mathieu equation. A solution of these equations for certain values of the system parameters can be obtained by means of Floquet's theory in conjunction with numerical integration, as indicated in Sec. 7.1. This is essentially the approach used by Kane and Shippy (Ref. 8) in treating this problem.

Problems in Spacecraft Dynamics

When the body is nearly symmetrical, it is possible to obtain more detailed information concerning the instability of the satellite attitude. In particular, it is possible to plot the instability regions in the parameter space and to identify the resonance frequencies causing the instability. This is done by deriving expressions for the curves dividing regions of stability and instability in planes defined by the system parameters. The approach is based on methods described in Chap. 7. This problem is discussed by Meirovitch and Wallace (Ref. 12).

12.6 THE ATTITUDE STABILITY OF SPINNING SYMMETRICAL SATELLITES

When two of the moments of inertia of a satellite are equal, inequality (12.57) cannot be satisfied, so that no stability, in the sense that the body assumes a unique orientation in space, can be established. In such cases it is perhaps more meaningful to define stability in terms of the orientation of the symmetry axis. Because stability is achieved by giving the satellite a spinning motion about the symmetry axis, this type of stabilization is referred to as *spin stabilization*. The problem of a symmetrical satellite is simpler than the problem of a nonsymmetrical one by virtue of the fact that the coordinate ψ, associated with the rotation about the symmetry axis, is cyclic. On the other hand, because the satellite spins relative to the orbital system of axes, the number of equilibrium configurations is increased. As for the nonsymmetrical case, the center of mass of the satellite is assumed to move in a circular orbit, so that this system is also autonomous.

The motion of the satellite is defined as in Sec. 12.4, but, thanks to symmetry, we need not work with the set of body axes x_1, x_2, x_3, as it is more advantageous to use axes c_1, c_2, c_3. Under these circumstances, we must distinguish between the angular velocities of the system c_1, c_2, c_3

$$\begin{aligned}
\omega_1 &= -\Omega \sin \theta_2 + \dot{\theta}_1 = \dot{\theta}, \\
\omega_2 &= \Omega \cos \theta_2 \sin \theta_1 + \dot{\theta}_2 \cos \theta_1 = (\Omega + \dot{\phi}) \sin \theta, \\
\omega_3 &= \Omega \cos \theta_2 \cos \theta_1 - \dot{\theta}_2 \sin \theta_1 = (\Omega + \dot{\phi}) \cos \theta,
\end{aligned} \quad (12.58)$$

and the angular velocity components of the body about these axes

$$\Omega_1 = \omega_1, \qquad \Omega_2 = \omega_2, \qquad \Omega_3 = \omega_3 + \dot{\psi}. \quad (12.59)$$

The direction cosines between axes c_1, c_2, c_3 and the orbital axes a_1, a_2, a_3 are obtained by setting $\psi = 0$ in Eqs. (12.43).

Denoting the moments of inertia about the symmetry axis and any transverse axis normal to the symmetry axis by C and A, respectively, the body kinetic energy of rotation takes the form

$$\begin{aligned}
T &= \tfrac{1}{2}\{A[(\Omega \sin \theta_2 - \dot{\theta}_1)^2 + (\Omega \cos \theta_2 \sin \theta_1 + \dot{\theta}_2 \cos \theta_1)^2] \\
&\quad + C(\Omega \cos \theta_2 \cos \theta_1 - \dot{\theta}_2 \sin \theta_1 + \dot{\psi})^2\} \\
&= \tfrac{1}{2}\{A[\dot{\theta}^2 + (\Omega + \dot{\phi})^2 \sin^2 \theta] + C[(\Omega + \dot{\phi}) \cos \theta + \dot{\psi}]^2\}, \quad (12.60)
\end{aligned}$$

whereas the potential energy becomes simply

$$V = \tfrac{3}{2}(C - A)\Omega^2 \sin^2 \theta_2 \cos^2 \theta_1 = \tfrac{3}{2}(C - A)\Omega^2 \sin^2 \phi \sin^2 \theta. \quad (12.61)$$

Equations (12.60) and (12.61) reveal first that the Hamiltonian is a Jacobi integral (which is also true for the nonsymmetrical body) and that ψ is a cyclic coordinate. The corresponding conserved angular momentum can be written

$$\begin{aligned}
\beta_\psi &= \frac{\partial L}{\partial \dot\psi} = C(\Omega \cos \theta_2 \cos \theta_1 - \dot\theta_2 \sin \theta_1 + \dot\psi) \\
&= C[(\Omega + \dot\phi) \cos \theta + \dot\psi] = Cn = \text{const}, \quad (12.62)
\end{aligned}$$

where n is the constant angular velocity about the symmetry axis.

Because ψ is cyclic, we can reduce the problem to a two-degree-of-freedom one by using the Routh method for the ignoration of coordinates (see Sec. 2.11). According to this procedure, we replace the Lagrangian $L = T - V$ by the Routhian R, where the latter depends only on the nonignorable coordinates and associated velocities as well as the conserved momentum β_ψ. Using Eq. (2.134), we obtain

$$\begin{aligned}
R &= L - \beta_\psi \dot\psi \\
&= \tfrac{1}{2}A(\dot\theta_1{}^2 + \dot\theta_2{}^2 \cos^2 \theta_1) + A\Omega(-\dot\theta_1 \sin \theta_2 + \dot\theta_2 \cos \theta_2 \sin \theta_1 \cos \theta_1) \\
&\quad + \tfrac{1}{2}A\Omega^2(\sin^2 \theta_2 + \cos^2 \theta_2 \sin^2 \theta_1) - \tfrac{3}{2}(C - A)\Omega^2 \sin^2 \theta_2 \cos^2 \theta_1 \\
&\quad - \frac{1}{2C}\beta_\psi[\beta_\psi - 2C(\Omega \cos \theta_2 \cos \theta_1 - \dot\theta_2 \sin \theta_1)] \\
&= \tfrac{1}{2}A(\dot\theta^2 + \dot\phi^2 \sin^2 \theta) + A\Omega\dot\phi \sin^2 \theta + \tfrac{1}{2}A\Omega^2 \sin^2 \theta \\
&\quad - \tfrac{3}{2}(C - A)\Omega^2 \sin^2 \phi \sin^2 \theta - \frac{1}{2C}\beta_\psi[\beta_\psi - 2C(\Omega + \dot\phi) \cos \theta]. \quad (12.63)
\end{aligned}$$

The Hamiltonian can be written in terms of the nonignorable coordinates and associated velocities by using the Routhian in the following way (see Sec. 2.13)

$$\begin{aligned}
H &= \frac{\partial R}{\partial \dot\theta_2} \dot\theta_2 + \frac{\partial R}{\partial \dot\theta_1} \dot\theta_1 - R \\
&= \tfrac{1}{2}A(\dot\theta_1{}^2 + \dot\theta_2{}^2 \cos^2 \theta_1) - \tfrac{1}{2}A\Omega^2(\sin^2 \theta_2 + \cos^2 \theta_2 \sin^2 \theta_1) \\
&\quad + \tfrac{3}{2}(C - A)\Omega^2 \sin^2 \theta_2 \cos^2 \theta_1 - \beta_\psi \Omega \cos \theta_2 \cos \theta_1 + \frac{1}{2C}\beta_\psi{}^2 \\
&= \frac{\partial R}{\partial \dot\phi}\dot\phi + \frac{\partial R}{\partial \dot\theta}\dot\theta - R \\
&= \tfrac{1}{2}A(\dot\theta^2 + \dot\phi^2 \sin^2 \theta) - \tfrac{1}{2}A\Omega^2 \sin^2 \theta + \tfrac{3}{2}(C - A)\Omega^2 \sin^2 \phi \sin^2 \theta \\
&\quad - \beta_\psi \Omega \cos \theta + \frac{1}{2C}\beta_\psi{}^2 = \text{const}. \quad (12.64)
\end{aligned}$$

Equation (12.64) represents the two versions of the Jacobi integral in which the conservation of the angular momentum about the symmetry axis has been accounted for. In this manner we need deal only with a fourth-order system instead of a sixth-order one.

The Jacobi integral, namely, the Hamiltonian, can be written in the general form

$$H = T_2^* + U = h = \text{const}, \tag{12.65}$$

where T_2^* is a positive definite quadratic function in the velocities $\dot{\theta}_2$ and $\dot{\theta}_1$, or $\dot{\phi}$ and $\dot{\theta}$, and U is a function depending only on the coordinates θ_2 and θ_1, or ϕ and θ, respectively. The function U has the explicit forms

$$\begin{aligned} U &= -\tfrac{1}{2}A\Omega^2(\sin^2\theta_2 + \cos^2\theta_2 \sin^2\theta_1) - \beta_\psi \Omega \cos\theta_2 \cos\theta_1 \\ &\quad + \tfrac{3}{2}(C-A)\Omega^2 \sin^2\theta_2 \cos^2\theta_1 \\ &= -\tfrac{1}{2}A\Omega^2 \sin^2\theta - \beta_\psi \Omega \cos\theta + \tfrac{3}{2}(C-A)\Omega^2 \sin^2\phi \sin^2\theta. \end{aligned} \tag{12.66}$$

As in the preceding section, we recognize that the function

$$U = h = \text{const} \tag{12.67}$$

represents *curves of zero relative velocity* in the coordinates θ_2 and θ_1, or ϕ and θ, which can be conveniently plotted on a unit sphere. We shall return to this subject later in this section.

The system has equilibrium positions relative to the orbital frame of reference at points for which $\dot{\theta}_2 = \dot{\theta}_1 = 0$ and θ_2 and θ_1 satisfy the equations

$$\begin{aligned} \frac{\partial U}{\partial \theta_2} &= [(3C - 4A)\Omega \cos\theta_2 \cos\theta_1 + \beta_\psi]\Omega \sin\theta_2 \cos\theta_1 = 0, \\ \frac{\partial U}{\partial \theta_1} &= -[A\Omega \cos^2\theta_2 \cos\theta_1 - \beta_\psi \cos\theta_2 \\ &\quad + 3(C-A)\Omega \sin^2\theta_2 \cos\theta_1]\Omega \sin\theta_1 = 0, \end{aligned} \tag{12.68}$$

or at points for which $\dot{\phi} = \dot{\theta} = 0$ and ϕ and θ are the solutions of

$$\begin{aligned} \frac{\partial U}{\partial \phi} &= 3(C-A)\Omega^2 \sin\phi \cos\phi \sin^2\theta = 0, \\ \frac{\partial U}{\partial \theta} &= -[A\Omega \cos\theta - \beta_\psi - 3(C-A)\Omega \sin^2\phi \cos\theta]\Omega \sin\theta = 0. \end{aligned} \tag{12.69}$$

Although geometrically the number of equilibrium positions is relatively large, some of the positions are dynamically equivalent, and they lead to the same stability requirements. For this resaon, we shall concentrate on three positions which fully describe the dynamical behavior of the system. These equilibrium points in terms of θ_2 and θ_1 are

E_1: $\quad\quad\quad \theta_2 = \theta_1 = 0,$

E_2: $\quad\quad\quad \theta_2 = 0, \quad \theta_1 = \cos^{-1}\dfrac{\beta_\psi}{A\Omega},$ $\quad\quad\quad$ (12.70)

E_3: $\quad\quad\quad \theta_1 = 0, \quad \theta_2 = \cos^{-1}\left[-\dfrac{\beta_\psi}{(3C - 4A)\Omega}\right],$

whereas in terms of ϕ and θ they are

E_1: $\quad\quad\quad \theta = 0, \quad \phi$ arbitrary,

E_2: $\quad\quad\quad \phi = 0, \quad \theta = \cos^{-1}\dfrac{\beta_\psi}{A\Omega},$ $\quad\quad\quad$ (12.71)

E_3: $\quad\quad\quad \phi = \dfrac{\pi}{2}, \quad \theta = \cos^{-1}\left[-\dfrac{\beta_\psi}{(3C - 4A)\Omega}\right].$

It is not difficult to see that both expressions, (12.70) and (12.71), represent the same equilibrium positions geometrically. Because for $\theta = 0$ the angular velocities $\dot\phi$ and $\dot\psi$ are collinear, we shall use the rotations θ_2 and θ_1 to test the stability of the equilibrium point E_1 and ϕ and θ for the points E_2 and E_3.

Following the pattern of Sec. 12.5, we investigate the system stability at the equilibrium points by means of the Liapunov direct method with the Hamiltonian as the Liapunov function. Because $\dot H$ is zero, we can expect at the most mere stability, as asymptotic stability is not possible. The problem reduces to testing H for sign-definiteness, and, since T_2^* is positive definite by definition, the problem reduces further to testing U for positive definiteness at the equilibrium point. To determine whether U is positive definite, we again use Sylvester's criterion, which necessitates the calculation of the Hessian matrix $[\mathscr{H}]$.

At this point a digression into the nature of the Hessian matrix appears in order. First we note that $[\mathscr{H}]$ evaluated at an equilibrium point can be regarded as the matrix of the coefficients of a quadratic form, namely, the function U in the neighborhood of the equilibrium point in question. The function U can be envisioned as a surface in a three-dimensional Euclidean space defined by $z = U$, θ_2, and θ_1 or $z = U$, ϕ, and θ. When $[\mathscr{H}]$ is positive definite at a given equilibrium point, the function U has a minimum at that point, and, conversely, if $[\mathscr{H}]$ is negative definite, then U has a maximum. If $[\mathscr{H}]$ is sign-variable at an equilibrium point, then U has a saddle point. The elements of $[\mathscr{H}]$ depend on the system parameters, and the matrix $[\mathscr{H}]$ changes as these parameters change. At an equilibrium point for which the determinant of $[\mathscr{H}]$ becomes zero, $|\mathscr{H}| = 0$, there is a bifurcation point, with the implication that the quality of equilibrium undergoes a change at that point. We must remember, however, that the Liapunov function H is defined not only by coordinates but also by velocities. Hence, because T_2^* is positive definite at an equilibrium point for which U has a maximum, H has a saddle point.

The elements of the Hessian $[\mathcal{H}]$, corresponding to the coordinates θ_2 and θ_1, are

$$\frac{\partial^2 U}{\partial \theta_2^2} = [(3C - 4A)\Omega \cos 2\theta_2 + \beta_\psi \cos \theta_2]\Omega \cos \theta_1,$$

$$\frac{\partial^2 U}{\partial \theta_2 \, \partial \theta_1} = -[2(3C - 4A)\Omega \cos \theta_2 \cos \theta_1 + \beta_\psi]\Omega \sin \theta_2 \sin \theta_1, \quad (12.72)$$

$$\frac{\partial^2 U}{\partial \theta_1^2} = -[A \cos^2 \theta_2 - 3(C - A) \sin^2 \theta_2]\Omega^2 \cos 2\theta_1 + \beta_\psi \Omega \cos \theta_2 \cos \theta_1,$$

so that the Hessian matrix for the equilibrium point E_1 becomes

$$[\mathcal{H}]_{E_1} = \begin{bmatrix} (3C - 4A)\Omega^2 + \beta_\psi \Omega & 0 \\ 0 & -A\Omega^2 + \beta_\psi \Omega \end{bmatrix}. \quad (12.73)$$

Hence, the point E_1 is stable if the conditions

$$\beta_\psi > (4A - 3C)\Omega, \qquad \beta_\psi > A\Omega \quad (12.74)$$

are satisfied. Conditions (12.74) indicate that it is possible to stabilize an equilibrium position, which may otherwise be unstable, provided the satellite is imparted a sufficiently large spin about the symmetry axis. The amount of spin necessary depends on the ratio of the moments of inertia. It will prove convenient to introduce the nondimensional parameters

$$b = \frac{\beta_\psi}{C\Omega} = \frac{n}{\Omega}, \qquad r = \frac{C}{A}, \quad (12.75)$$

so that conditions (12.74) become

$$b > \frac{4}{r} - 3, \qquad b > \frac{1}{r}. \quad (12.76)$$

This enables us to envision a parameter plane b versus r with inequalities (12.76) defining a region in that plane in which the equilibrium point E_1 is stable. The parameter plane is shown in Fig. 12.5 and the corresponding stability region designated by E_1. We note that, due to physical considerations, r varies between 0 and 2, approaching 0 for a very thin rod and 2 for a flat disk.

To investigate the remaining equilibrium points, we write the Hessian matrix elements

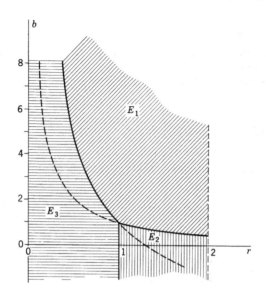

FIGURE 12.5

$$\frac{\partial^2 U}{\partial \phi^2} = 3(C - A)\Omega^2 \cos 2\phi \sin^2 \theta,$$

$$\frac{\partial^2 U}{\partial \phi\, \partial \theta} = \tfrac{3}{2}(C - A)\Omega^2 \sin 2\phi \sin 2\theta, \tag{12.77}$$

$$\frac{\partial^2 U}{\partial \theta^2} = -[A - 3(C - A) \sin^2 \phi]\Omega^2 \cos 2\theta + \beta_\psi \Omega \cos \theta.$$

This enables us to evaluate the Hessian matrix associated with the equilibrium point E_2, defined by the second set of conditions in (12.71),

$$[\mathscr{H}]_{E_2} = \begin{bmatrix} 3(C - A)\Omega^2 \left(1 - \dfrac{\beta_\psi^2}{A^2\Omega^2}\right) & 0 \\ 0 & A\Omega^2 \left(1 - \dfrac{\beta_\psi^2}{A^2\Omega^2}\right) \end{bmatrix}. \tag{12.78}$$

In terms of the parameters b and r, we conclude that the equilibrium point E_2 is stable if

$$b < \frac{1}{r}, \quad r > 1. \tag{12.79}$$

The corresponding region is denoted by E_2 in Fig. 12.5.

The Hessian matrix corresponding to E_3 has the form

$$[\mathcal{H}]_{E_3} = \begin{bmatrix} -3(C-A)\Omega^2\left[1 - \dfrac{\beta_\psi^2}{(3C-4A)^2\Omega^2}\right] & 0 \\ 0 & -(3C-4A)\Omega^2\left[1 - \dfrac{\beta_\psi^2}{(3C-4A)^2\Omega^2}\right] \end{bmatrix}$$
(12.80)

leading to the stability conditions

$$r < 1, \quad b < \frac{4}{r} - 3. \tag{12.81}$$

The stability region is denoted by E_3 in Fig. 12.5.

Let us now return to the curves of zero relative velocity given by Eq. (12.67). Because U depends on two coordinates, θ_2 and θ_1, or ϕ and θ, a geometric representation of these curves is feasible. This representation is not necessarily confined to the neighborhood of equilibrium points. At a stable equilibrium point the Hamiltonian, and hence the function U, has a relative minimum $H_{\min} = U_{\min} = h_0$. For a different value of h, for example, $h = h_1 > h_0$, we can plot a contour $U = h_1 =$ const, and for every point on this contour the symmetry axis of the satellite is at rest relative to the orbital axes a_1, a_2, a_3. For a value h_1 relatively close to h_0, so that the equilibrium point is an isolated minimum, $U = h_1$ represents a closed contour defining the region $U < h_1$ and enclosing the equilibrium point; motion relative to the orbital axes is possible only in that region, and this motion is bounded oscillatory. As the value of h is increased, the region of oscillatory motion expands, provided the contour $U = h$ does not cross any singularity. The maximum region of bounded motion about a minimum point of U is defined by separatrix curves passing through saddle points, where the curves can be thought of as *bounds on the libration* of the satellite about the stable equilibrium point. It should be stressed that the *libration within this region can reach large amplitudes and is according to the complete nonlinear equations of motion*. Hence, it is not limited to motion described by the variational equations, which are valid only in the neighborhood of the equilibrium. In crossing the separatrices the quality of the system's behavior changes, and the motion is no longer bounded about the original equilibrium point, although it may still be bounded about a different equilibrium point.

Curves of zero relative velocity can be plotted on a plane, rather than a unit sphere. They are obtained by projecting the contours on given planes, such as a_1a_2, a_1a_3, and a_2a_3, corresponding to the orbital axes; a point on a contour has the coordinates $a_1 = l_{31}, a_2 = l_{32}, a_3 = l_{33}$. Figure 12.6a and b, obtained using this scheme, shows typical plots illustrating the equilibrium

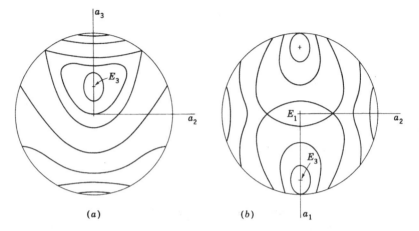

FIGURE 12.6

positions E_1 and E_3 for a given set of values for b and r. These values satisfy the inequalities

$$b < \frac{1}{r}, \quad r < 1, \qquad (12.82)$$

so that U has a maximum at E_1 and a minimum at E_3. Figure 12.6b shows another stable equilibrium point corresponding to the equilibrium position $\phi = 3\pi/2$, $\theta = \cos^{-1}[-\beta_\psi/(3C - 4A)\Omega]$, which is dynamically equivalent to E_3.

The special case of zero spin, $b = 0$, discussed by Auelmann (Ref. 1), corresponds simply to the r axis in Fig. 12.5. The work was extended by Pringle, who used Poincaré's bifurcation theory to obtain the results presented here. In fact, Fig. 12.6 is taken from his paper.[1] Likins (Ref. 10), in pointing out that a Liapunov analysis with the Hamiltonian as the Liapunov function yields the same information as Poincaré's bifurcation theory, derived the same results. In addition, he presented an infinitesimal analysis and a comparison with the Liapunov analysis.

The problem was extended by Meirovitch and Wallace (Ref. 13), who studied the simultaneous effect of aerodynamic and gravitational torques on the attitude stability of spinning symmetrical satellites by means of the Liapunov direct method.

The problem of a symmetrical satellite with the spin axis in the neighborhood of the normal to the orbital plane and with the center of mass moving in an elliptic orbit leads to a nonautonomous system consisting of

---

[1] Ref. 14: R. Pringle, Jr., "Bounds on the Librations of a Symmetrical Satellite," *AIAA J.*, vol. 2, no. 5, pp. 908–912, 1964.

two simultaneous differential equations for θ_1 and θ_2. The equations are nonlinear and possess periodic coefficients. The coordinate ψ is cyclic, and the periodic coefficients are due to the presence of the radial distance R in the expressions for the gravitational torques.

The linearized system has been discussed by Kane and Barba (Ref. 7) by means of the semianalytical method presented in Sec. 7.1. The nonlinear system and the linearized one have been treated by Wallace and Meirovitch (Ref. 17), who determined instability regions in the parameter plane b versus r for various values of the orbit eccentricity.

12.7 VARIABLE-MASS SYSTEMS

By a variable-mass system we understand a *system of changing composition*. To clarify the meaning of this statement, we introduce the concept of *control volume*, which is a definite volume enclosed by a *control surface*. Although the shape of the control volume is assumed to be fixed, the identity of the matter within the control volume may change with time. Newton's second law $\mathbf{F} = d(m\mathbf{v})/dt = m\mathbf{a}$ was formulated for a single particle, and it can be extended to systems of particles, provided the composition of the system does not change. When the system composition does change, it is no longer possible to equate the time derivative of a sum to the sum of time derivatives because the summation at different times involves different sets of particles. In this case the proper procedure is to write the force equation in the form $\mathbf{F} = \dot{\mathbf{p}}$, where the rate of change of $\mathbf{p}$ is obtained by means of a limiting process involving the calculation of $\mathbf{p}$ at two different instants, a time interval Δt apart, dividing the difference by Δt, and letting $\Delta t \to 0$. In so doing, we must make sure that the same total mass is involved, although at one time it is entirely inside the control volume and the other time part of the mass is outside.

Let us consider a system occupying at time t a certain volume in space, namely, the control volume enclosed by the control surface shown in solid line in Fig. 12.7. The same amount of matter occupies at time $t + \Delta t$ the volume enclosed by the dashed line. The difference between the two volumes at time t is denoted by I and at time $t + \Delta t$ by III, with the volume common to both denoted by II. An area differential on the control surface pointing outward in a direction normal to the surface at that point is designated by the vector $d\mathbf{A}$. The control volume is fixed relative to axes x, y, z, which may be moving with respect to the inertial space X, Y, Z. The position of any mass element dm relative to the moving system x, y, z is denoted by $\mathbf{r}$ and relative to the inertial space X, Y, Z by $\mathbf{R}$ so that

$$\mathbf{R} = \mathbf{R}_0 + \mathbf{r}, \qquad (12.83)$$

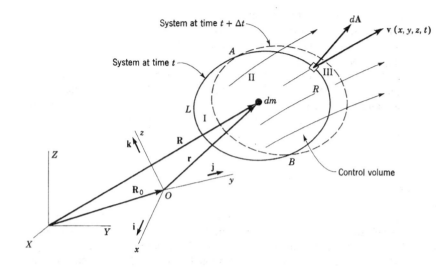

FIGURE 12.7

where $\mathbf{R}_0$ is the position of the origin O of axes x, y, z with respect to the inertial space. The velocity of the mass element dm relative to the control volume, hence with respect to axes x, y, z, is denoted by the vector $\mathbf{v}(x,y,z,t)$.

We shall first derive the equations of motion of the system by assuming that the control volume is at rest, so that axes x, y, z are to be regarded as an inertial system. Letting $dm = \rho\, dv$, where ρ is the density and dv the associated volume element, we can write the linear momentum at time t in the form

$$\mathbf{p} = \int_{cv} \mathbf{v}\, dm = \left(\int_{\mathrm{I}} \mathbf{v}\rho\, dv + \int_{\mathrm{II}} \mathbf{v}\rho\, dv \right)_t, \qquad (12.84)$$

whereas at time $t + \Delta t$ we have

$$\mathbf{p} + \Delta\mathbf{p} = \left(\int_{\mathrm{II}} \mathbf{v}\rho\, dv + \int_{\mathrm{III}} \mathbf{v}\rho\, dv \right)_{t+\Delta t}, \qquad (12.85)$$

so that the momentum rate of change can be written

$$\begin{aligned}\dot{\mathbf{p}} = &\lim_{\Delta t \to 0} \frac{(\int_{\mathrm{II}} \mathbf{v}\rho\, dv)_{t+\Delta t} - (\int_{\mathrm{II}} \mathbf{v}\rho\, dv)_t}{\Delta t} \\ &+ \lim_{\Delta t \to 0} \frac{(\int_{\mathrm{III}} \mathbf{v}\rho\, dv)_{t+\Delta t}}{\Delta t} - \lim_{\Delta t \to 0} \frac{(\int_{\mathrm{I}} \mathbf{v}\rho\, dv)_t}{\Delta t}.\end{aligned} \qquad (12.86)$$

Problems in Spacecraft Dynamics

But as $\Delta t \to 0$, volume II tends to become identified with the control volume, so that the first limiting process on the right side of Eq. (12.86) yields

$$\lim_{\Delta t \to 0} \frac{(\int_{\text{II}} \mathbf{v}\rho \, dv)_{t+\Delta t} - (\int_{\text{II}} \mathbf{v}\rho \, dv)_t}{\Delta t} = \frac{\partial}{\partial t} \int_{cv} \mathbf{v}\rho \, dv. \tag{12.87}$$

Furthermore, as $\Delta t \to 0$, volume III approaches zero, so that in the limit

$$\lim_{\Delta t \to 0} \frac{(\mathbf{v}\rho \, dv|_{\text{III}})_{t+\Delta t}}{\Delta t} = \mathbf{v}(\rho \mathbf{v} \cdot d\mathbf{A})_{\text{III}}, \tag{12.88}$$

where the right side of (12.88) represents the rate of efflux of $\mathbf{p}$ through the area element $d\mathbf{A}$ of the control surface ARB. A similar treatment of the last term on the right side of Eq. (12.86) results in the rate of influx of $\mathbf{p}$ across a differential element of the control surface ALB. But the sum of the two integrals yields the net efflux rate, so that the total force, which is equal to the time rate of change of momentum, becomes

$$\mathbf{F} = \mathbf{F}_S + \mathbf{F}_B = \frac{\partial}{\partial t}\int_{cv} \mathbf{v}\rho \, dv + \int_{cs} \mathbf{v}(\rho \mathbf{v} \cdot d\mathbf{A}), \tag{12.89}$$

where $\mathbf{F}_S$ and $\mathbf{F}_B$ are the resultants of the surface and body forces, respectively.

Next we consider the case in which the control volume translates and rotates relative to the inertial space. It will prove convenient to assume that part of the mass is rigidly attached to the control volume, hence moving together with axes x, y, z, and part of it is moving relative to these axes. In fact axes x, y, z can be regarded as a set of body axes, so that, using Eq. (3.46) in conjunction with Eq. (3.43) for the absolute acceleration $\mathbf{a}$, we obtain the equation of motion

$$\mathbf{F} = \mathbf{F}_S + \mathbf{F}_B - \int_m \mathbf{a} \, dm$$
$$= \int_m [\mathbf{a}_0 + \dot{\mathbf{v}} + 2\boldsymbol{\omega} \times \mathbf{v} + \dot{\boldsymbol{\omega}} \times \mathbf{r} + \boldsymbol{\omega} \times (\boldsymbol{\omega} \times \mathbf{r})] \, dm, \tag{12.90}$$

where m = total mass of system,
$\mathbf{a}_0 = \ddot{\mathbf{R}}_0$ = acceleration of origin,
$\boldsymbol{\omega}$ = angular velocity of axes x, y, z.

Denoting by m_f the mass moving relative to the body axes and recognizing that if the body axes were fixed in an inertial space, only the term $\int_{m_f} \dot{\mathbf{v}} \, dm$ would survive, we can rewrite the force equation (12.90) in the form

$$\mathbf{F}_S + \mathbf{F}_B + \mathbf{F}_C + \mathbf{F}_U + \mathbf{F}_R = m\mathbf{a}_0 + \dot{\boldsymbol{\omega}} \times \int_m \mathbf{r} \, dm + \boldsymbol{\omega} \times \left(\boldsymbol{\omega} \times \int_m \mathbf{r} \, dm\right), \tag{12.91}$$

where
$$\mathbf{F}_C = -2\boldsymbol{\omega} \times \int_{m_f} \mathbf{v}\, dm \tag{12.92}$$

is recognized as the *Coriolis force*,

$$\mathbf{F}_U = -\frac{\partial}{\partial t}\int_{m_f} \mathbf{v}\, dm \tag{12.93}$$

is a *force due to the unsteadiness of the relative motion*, and

$$\mathbf{F}_R = -\int_A \mathbf{v}(\rho\mathbf{v}\cdot d\mathbf{A}) \tag{12.94}$$

will be referred to as a *reactive force*, where the integral is taken over the area A of the control surface. We must note that $\mathbf{F}_C$, $\mathbf{F}_U$, and $\mathbf{F}_R$ are not true forces in the same sense as the actual forces $\mathbf{F}_S$ and $\mathbf{F}_B$. The terms on the right side of Eq. (12.91) may be regarded as being due to the entire mass m moving as a rigid body, where m is the mass of the system at time t. The partial time derivative $\partial/\partial t$ in Eq. (12.93) is to be taken by regarding axes x, y, z as fixed.

The torque equations can be derived in an analogous way. First, regarding axes x, y, z as fixed in an inertial space, we use the force element corresponding to Eq. (12.89) and obtain

$$\mathbf{N} = \mathbf{N}_S + \mathbf{N}_B = \int_F \mathbf{r} \times d\mathbf{F} = \frac{\partial}{\partial t}\int_{m_f} \mathbf{r} \times \mathbf{v}\, dm + \int_A (\mathbf{r} \times \mathbf{v})(\rho\mathbf{v}\cdot d\mathbf{A}), \tag{12.95}$$

where $\mathbf{N}$ is the torque about the origin O, including the torques $\mathbf{N}_S$ and $\mathbf{N}_B$ due to surface and body forces, respectively. The integral over m_f on the right side of Eq. (12.95) can be easily explained by recalling that $\partial/\partial t$ implies a time rate of change with axes x, y, z regarded as fixed.

Similarly, in the event that the body axes x, y, z are translating and rotating relative to the inertial space, the force element from Eq. (12.90) yields the torque about point O

$$\mathbf{N} = \mathbf{N}_S + \mathbf{N}_B = \int_m \mathbf{r} \times [\mathbf{a}_0 + \dot{\mathbf{v}} + 2\boldsymbol{\omega}\times\mathbf{v} + \dot{\boldsymbol{\omega}}\times\mathbf{r} + \boldsymbol{\omega}\times(\boldsymbol{\omega}\times\mathbf{r})]\, dm, \tag{12.96}$$

which can be rewritten in the form

$$\mathbf{N}_S + \mathbf{N}_B + \mathbf{N}_C + \mathbf{N}_U + \mathbf{N}_R = \int_m \mathbf{r} \times [\mathbf{a}_0 + \dot{\boldsymbol{\omega}}\times\mathbf{r} + \boldsymbol{\omega}\times(\boldsymbol{\omega}\times\mathbf{r})]\, dm, \tag{12.97}$$

where the terms on the right side can be interpreted as due to rigid body motion. Moreover,

$$\mathbf{N}_C = -2 \int_{m_f} \mathbf{r} \times (\boldsymbol{\omega} \times \mathbf{v}) \, dm,$$

$$\mathbf{N}_U = -\frac{\partial}{\partial t} \int_{m_f} \mathbf{r} \times \mathbf{v} \, dm, \qquad (12.98)$$

$$\mathbf{N}_R = -\int_A (\mathbf{r} \times \mathbf{v})(\rho \mathbf{v} \cdot d\mathbf{A})$$

are the Coriolis torque, the torque due to the unsteadiness of the relative motion, and the reactive torque, respectively. Again in taking the partial derivative with respect to time in the expression for $\mathbf{N}_U$ we must regard x, y, z as fixed.

The above equations must be supplemented by the continuity equation

$$\int_A \rho \mathbf{v} \cdot d\mathbf{A} = -\frac{\partial}{\partial t} \int_{m_f} dm, \qquad (12.99)$$

which expresses the fact that the net efflux rate of mass across the control surface must equal the rate of mass decrease within the control volume.

Equations (12.91) and (12.97) can be given an interesting physical interpretation by recalling that the system consists of one part with rigid and another part with changing composition. We observe that the right sides of these equations represent the motion of the system as if it were rigid in its entirety. Hence Eqs. (12.91) and (12.97) can be regarded as the equations of motion of a fictitious rigid body of instantaneous mass m, provided that the actual surface and body forces acting upon the system are supplemented by three equivalent forces, namely, the Coriolis force, the force due to the unsteadiness of the relative motion, and the reactive force. This is the substance of a statement referred to as the *principle of solidification for a system of changing composition* (see Ref. 6, p. 13).

12.8 ROCKET DYNAMICS

The formulation of Sec. 12.7 is applicable to a large number of systems, ranging from the lawn sprinkler to the turbine and the rocket engine. In particular, the solidification principle is ideally suited for treating the problems associated with the motion of a rocket. In this case the body axes x, y, z are attached to the rocket casing, and the control volume can be regarded as consisting of the volume occupied by the fuel and oxidizer, the combustion chamber, and the nozzle. Although in a liquid-fuel rocket the unburned fuel and oxidizer can move relative to the rocket shell, this motion is small, and the only relative motion can be assumed to consist of the burned gases being expelled through the nozzle.

Assuming that the actual forces and the data pertaining to the fuel burning and the motion relative to the vehicle are known, the force and moment equations (12.91) and (12.97) should enable us to solve for the translational and rotational motions of the rocket. Equations (12.91) and (12.97), however, are coupled, and no closed-form solution of the complete equations is possible. The translational and rotational motions can be uncoupled by choosing the origin O to coincide with the center of mass of the vehicle at all times. But in general the center of mass moves continuously relative to the vehicle, because of the mass variation, so that we are faced with the problem of determining the position of the center of mass with respect to the rocket at every instant.

Fortunately, in the case of a rocket a number of simplifications can be made by virtue of the fact that the Coriolis forces and the forces due to the unsteadiness of the relative motion are sufficiently small to be neglected. Moreover, it can be assumed that the center of mass does not shift appreciably relative to the vehicle (see Ref. 6, p. 15), so that, taking the origin of the body axes x, y, z to coincide with the center of mass, Eqs. (12.91) and (12.97) reduce to

$$\mathbf{F}_S + \mathbf{F}_B + \mathbf{F}_R = m\mathbf{a}_0 \tag{12.100}$$

and

$$\mathbf{N}_S + \mathbf{N}_B + \mathbf{N}_R = \int_m \mathbf{r} \times [\dot{\boldsymbol{\omega}} \times \mathbf{r} + \boldsymbol{\omega} \times (\boldsymbol{\omega} \times \mathbf{r})]\, dm, \tag{12.101}$$

and we note that translational and rotational motions become uncoupled.

The surface forces are primarily due to aerodynamic effects. Denoting by A_e the exit area of the engine and by A_w the wetted area of the vehicle, shown in Fig. 12.8 as the surfaces AB and ACB, respectively, the surface force and torque vectors can be written

$$\mathbf{F}_S = \int_{A_e + A_w} \boldsymbol{\pi}\, dA, \qquad \mathbf{N}_S = \int_{A_e + A_w} \mathbf{r} \times \boldsymbol{\pi}\, dA, \tag{12.102}$$

where

$$\boldsymbol{\pi} = (p - p_0)\mathbf{u}_n + f\mathbf{u}_t \tag{12.103}$$

FIGURE 12.8

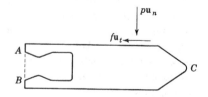

Problems in Spacecraft Dynamics

is called the *stress vector*, in which p is the local static pressure, p_0 is the free-stream static pressure, and f is the local tangential stress exerted by the surrounding medium on the entire vehicle surface. On the other hand, the body force and torque are simply due to gravity

$$\mathbf{F}_B = \int_m \mathbf{g}\, dm, \qquad \mathbf{N}_B = \int_m \mathbf{r} \times \mathbf{g}\, dm, \qquad (12.104)$$

where $\mathbf{g}$ is the gravitational force vector per unit mass.

Using the continuity equation (12.99) and recognizing that the only area where there is a mass efflux is at the exit area A_e, we can rewrite the reactive force and torque in the form

$$\mathbf{F}_R = -\int_{A_e} \mu \mathbf{v}_e\, dA, \qquad N_R = -\int_{A_e} \mathbf{r} \times \mu \mathbf{v}_e\, dA, \qquad (12.105)$$

in which
$$\mu = \rho_e \mathbf{v}_e \cdot \mathbf{u}_n \qquad (12.106)$$

is the mass flow across an exit unit area, where ρ_e and $\mathbf{v}_e$ are the exit mass density and velocity vector, respectively.

Introducing the thrust vector and the associated torque vector

$$\mathbf{F}_T = \int_{A_e} (\boldsymbol{\pi} - \mu \mathbf{v}_e)\, dA, \qquad \mathbf{N}_T = \int_{A_e} \mathbf{r} \times (\boldsymbol{\pi} - \mu \mathbf{v}_e)\, dA, \qquad (12.107)$$

as well as the aerodynamic force and torque vectors

$$\mathbf{F}_A = \int_{A_w} \boldsymbol{\pi}\, dA, \qquad \mathbf{N}_A = \int_{A_w} \mathbf{r} \times \boldsymbol{\pi}\, dA, \qquad (12.108)$$

we can rewrite Eqs. (12.100) and (12.101) in the form

$$\mathbf{F}_T + \mathbf{F}_A + \mathbf{F}_B = m\mathbf{a}_0 \qquad (12.109)$$

and

$$\mathbf{N}_T + \mathbf{N}_A + \mathbf{N}_B = \int_m \mathbf{r} \times [\dot{\boldsymbol{\omega}} \times \mathbf{r} + \boldsymbol{\omega} \times (\boldsymbol{\omega} \times \mathbf{r})]\, dm. \qquad (12.110)$$

Equations (12.109) and (12.110) can be written in the familiar matrix form

$$\{F_T\} + \{F_A\} + \{F_B\} = \{ma_0\} \qquad (12.111)$$

and
$$\{N_T\} + \{N_A\} + \{N_B\} = [I]\{\dot{\omega}\} + [\omega][I]\{\omega\}, \qquad (12.112)$$

where m is the instantaneous mass of the system and $[I]$ is the corresponding moment of inertia matrix about the body axes. It is customary to take the body axes to coincide with the rocket's principal axes, with the implicit assumption that the directions of the principal axes do not change with respect to the rocket.

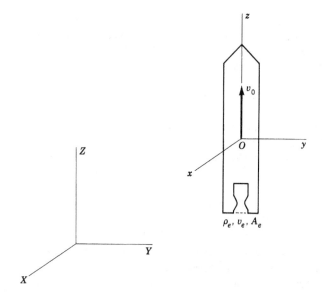

FIGURE 12.9

As a simple illustration, let us consider the case of a rocket traveling vertically upward in vacuum, as shown in Fig. 12.9. Assuming no rotation, Eq. (12.112) can be ignored, whereas Eq. (12.111) has only one component, namely, along the rocket's longitudinal axis. Moreover, the aerodynamic forces are zero.

If the initial mass of the rocket is m_0, the fuel burning rate is constant, and the gas flow is uniform across the exit area, then the mass of the rocket at any time is

$$m = m_0 - \beta t, \tag{12.113}$$

where
$$\beta = \mu A_e, \tag{12.114}$$

in which β is the burning rate. The exit velocity v_e is assumed constant, and since it is opposite in direction to the z axis, the thrust takes the form

$$F_T = p_e A_e + \beta v_e = \text{const}, \tag{12.115}$$

where p_e is the exit pressure. The gravity is also acting in the negative z direction, and, assuming that its magnitude is constant, we have

$$F_B = -mg. \tag{12.116}$$

In view of Eqs. (12.113), (12.115), and (12.116), the z component of Eq. (12.111) yields

$$p_e A_e + \beta v_e - (m_0 - \beta t)g = (m_0 - \beta t)\frac{dv_0}{dt}, \quad (12.117)$$

which has the solution

$$v_0 = v_0(0) + \left(\frac{p_e A_e}{\beta} + v_e\right) \ln \frac{m_0}{m_0 - \beta t}, \quad (12.118)$$

where $v_0(0)$ is the rocket's initial velocity.

For an analysis including both the translational and rotational motions as well as the elastic deformations of a flexible rocket see Ref. 11.

PROBLEMS

12.1 An elliptic transfer orbit of the type shown in Fig. 12.10, where all orbits are coaxial and planar, is generally known as a *Hohmann orbit*. Moreover, the sum of the velocity increments necessary to perform the transfer from the initial to the target orbit is called the *characteristic velocity*. In view of this definition, calculate the characteristic velocity (*a*) when the initial orbit is circular and the target orbit is elliptic and (*b*) when both the initial and the target orbits are elliptic.

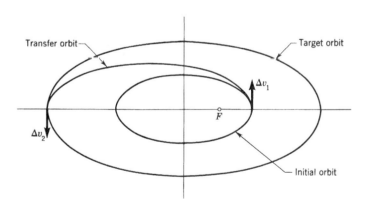

FIGURE 12.10

12.2 To an impulse having two components in the orbit plane and one component perpendicular to the orbit plane there correspond six changes in the orbital elements (Sec. 12.1), which indicates that a single impulse is not sufficient to bring about any desired change in orbit. Make a table showing which orbital changes are possible if only one impulse is to be used.

12.3 Equation (12.19) represents a change in Ω over one period due to the perturbative effect of the earth's oblateness, as described by Eq. (12.11). Derive an analogous expression for $\Delta\omega_{\text{per}}$.

12.4 Check the stability of the equilibrium points E_2 and E_3 of Fig. 12.4 and explain the results.

12.5 Consider the system of Sec. 12.6 and derive the variational equations corresponding to the three equilibrium points E_i, Eqs. (12.70) or (12.71). Test the stability of E_1 by an infinitesimal analysis.

SUGGESTED REFERENCES

1. Auelmann, R.: "Regions of Librations for a Symmetrical Satellite," *AIAA J.*, vol. 1, no. 6, pp. 1445–1447, 1963.
2. Breakwell, J. V., and R. Pringle, Jr.: "Nonlinear Resonance Affecting Gravity-Gradient Stability," *Astrodynamics, Proc. 16th Intern. Astronaut. Congr.*, Gauthier-Villars, Paris, pp. 305–325, 1966.
3. DeBra, D. B., and R. H. Delp: "Rigid Body Attitude Stability and Natural Frequencies in a Circular Orbit," *J. Astronaut. Sci.*, vol. 8, pp. 14–17, 1961.
4. Deutsch, R.: *Orbital Dynamics of Space Vehicles*, Prentice-Hall, Inc., Englewood Cliffs, N.J., 1963.
5. Ehricke, K. A.: *Space Flight*, vol. II, *Dynamics*, D. Van Nostrand Company, Inc., Princeton, N.J., 1962.
6. Gantmakher, F. R., and L. M. Levin: *The Flight of Uncontrolled Rockets*, The Macmillan Company, New York, 1964.
7. Kane, T. R., and P. M. Barba: "Attitude Stability of a Spinning Satellite in an Elliptic Orbit," *J. Appl. Mech.*, vol. 33, pp. 402–405, 1966.
8. Kane, T. R., and D. J. Shippy: "Attitude Stability of a Spinning Unsymmetrical Satellite in a Circular Orbit," *J. Astronaut. Sci.*, vol. 10, no. 4, pp. 114–119, 1963.
9. Kozai, Y.: "The Motion of a Close Earth Satellite," *Astronom. J.*, vol. 64, no. 9, pp. 367–377, 1959.
10. Likins, P. W.: "Stability of a Symmetrical Satellite in Attitudes Fixed in an Orbiting Frame," *J. Astronaut. Sci.*, vol. 12, no. 1, pp. 18–24, 1964.
11. Meirovitch, L.: "General Motion of a Variable-Mass Flexible Rocket with Internal Flow," *J. Spacecraft & Rockets*, vol. 7, no 2. pp. 186–195, 1970.

12. Meirovitch, L., and F. B. Wallace, Jr.: "Attitude Instability Regions of a Spinning Unsymmetrical Satellite in a Circular Orbit," *J. Astronaut. Sci.*, vol. 14, no. 3, pp. 123–133, 1967.
13. Meirovitch, L., and F. B. Wallace, Jr.: "On the Effect of Aerodynamic and Gravitational Torques on the Attitude Stability of Satellites," *AIAA J.*, vol. 4, no. 12, pp. 2196–2202, 1966.
14. Pringle, R., Jr.: "Bounds on the Librations of a Symmetrical Satellite," *AIAA J.*, vol. 2, no. 5, pp. 908–912, 1964.
15. Roberson, R. E.: "Attitude Control of Satellite Vehicles: An Outline of the Problems," *Proc. 8th Intern. Astronaut. Fed. Congr., Barcelona*, 1957, pp. 317–339.
16. Tisserand, F.: *Traité de Mécanique Céleste*, vol. 1, Gauthier-Villars, Paris, 1889.
17. Wallace, F. B., Jr., and L. Meirovitch: "Attitude Instability Regions of a Spinning Symmetrical Satellite in an Elliptic Orbit," *AIAA J.*, vol. 5, no. 9, pp. 1642–1650, 1967.

appendix A

Dyadics

A *dyadic* is a vector operator defined as the linear polynomial

$$\Phi = A_1B_1 + A_2B_2 + \cdots + A_nB_n, \qquad (a)$$

where $A_1, A_2, \ldots, A_n, B_1, B_2, \ldots, B_n$ are vectors. In general the position of the vectors in the pairs A_iB_i ($i = 1, 2, \ldots, n$) is an inclusive part of the dyadic definition and, indeed, the vectors $A_1, A_2, \ldots, A_n$ are called the *antecedents* and the vectors $B_1, B_2, \ldots, B_n$ the *consequents*. The dyadic obtained by interchanging the order of the antecedents and consequents, namely,

$$\Phi_c = B_1A_1 + B_2A_2 + \cdots + B_nA_n, \qquad (b)$$

is known as the *conjugate* of Φ. A dyadic which is equal to its conjugate is said to be *self-conjugate* or *symmetric*. We shall have considerable interest in symmetric dyadics.

The dyadic can be used to form the scalar products

$$C \cdot \Phi = (C \cdot A_1)B_1 + (C \cdot A_2)B_2 + \cdots + (C \cdot A_n)B_n, \qquad (c)$$

and

$$\Phi \cdot C = A_1(B_1 \cdot C) + A_2(B_2 \cdot C) + \cdots + A_n(B_n \cdot C). \qquad (d)$$

The vector C in Eq. (c) is called the *prefactor*, whereas in Eq. (d) it is referred

Dyadics

to as the *postfactor*. Both scalar products of a vector and a dyadic, Eqs. (*c*) and (*d*), yield vectors which are in general different from the vector **C**.

A dyadic consisting of a single term,

$$\mathbf{\Phi} = \mathbf{AB}, \tag{e}$$

is called a *dyad*. Any dyad, however, can be written in the form of a dyadic by expressing any of the vectors **A** and **B**, or both, in terms of their components along the cartesian axes x, y, z multiplied by the corresponding unit vectors **i, j, k**. Indeed we can write

$$\begin{aligned}
\mathbf{\Phi} = \mathbf{AB} &= (A_x\mathbf{i} + A_y\mathbf{j} + A_z\mathbf{k})(B_x\mathbf{i} + B_y\mathbf{j} + B_z\mathbf{k}) \\
&= A_xB_x\mathbf{ii} + A_xB_y\mathbf{ij} + A_xB_z\mathbf{ik} + A_yB_x\mathbf{ji} + A_yB_y\mathbf{jj} + A_yB_z\mathbf{jk} \\
&\quad + A_zB_x\mathbf{ki} + A_zB_y\mathbf{kj} + A_zB_z\mathbf{kk} \\
&= \Phi_{xx}\mathbf{ii} + \Phi_{xy}\mathbf{ij} + \Phi_{xz}\mathbf{ik} + \Phi_{yx}\mathbf{ji} + \Phi_{yy}\mathbf{jj} + \Phi_{yz}\mathbf{jk} \\
&\quad + \Phi_{zx}\mathbf{ki} + \Phi_{zy}\mathbf{kj} + \Phi_{zz}\mathbf{kk}.
\end{aligned} \tag{f}$$

From Eq. (*f*) we conclude that any dyadic $\mathbf{\Phi}$ can be expressed in terms of nine scalar elements $\Phi_{xx}, \Phi_{xy}, \ldots, \Phi_{zz}$. It is not difficult to see that these coefficients are homogeneous quadratic functions of the vector components and, as such, must transform in the same manner as the nine components of a second-rank cartesian tensor. Hence we may regard a dyadic as being formally identical to a second-order cartesian tensor. It may be interesting to point out that in Sec. 4.2 the inertia tensor, which is a second-order tensor, is exhibited in the form of a 3 × 3 symmetric matrix. In fact the nine elements of the dyadic can also be arranged in matrix form, precisely like the nine components of a second-order tensor. Moreover, the scalar product of a dyadic and a vector produces a new vector in the same way as the product of a square matrix and a vector.

The analogy of the dyadic and matrix representations of the inertia tensor can be brought into sharper focus by introducing the *unit*, or *identity*, *dyadic* defined by

$$\mathbf{1} = \mathbf{ii} + \mathbf{jj} + \mathbf{kk}. \tag{g}$$

The unit dyadic **1** is the equivalent of the 3 × 3 unit, or identity, matrix [1]. Indeed the prefactor or postfactor multiplication of the unit dyadic by any vector **A** leaves the vector unaffected

$$\mathbf{A} \cdot \mathbf{1} = \mathbf{1} \cdot \mathbf{A} = \mathbf{A}. \tag{h}$$

Recalling that $\mathbf{r}_i$ denotes the radius vector from the origin of a system of axes to any particle of mass m_i, the inertia tensor can be expressed in the form of the symmetric dyadic

$$\mathbf{I} = \sum_{i=1}^{n} [(\mathbf{r}_i \cdot \mathbf{r}_i)\mathbf{1} - \mathbf{r}_i\mathbf{r}_i]m_i \tag{i}$$

corresponding to the system of particles. As a rigid body can be regarded as a system of particles, we can use the limiting process employed in Sec. 4.2 and write the dyadic corresponding to a rigid body as

$$\mathbf{I} = \lim_{n \to \infty} \sum_{i=1}^{n} [(\mathbf{r}_i \cdot \mathbf{r}_i)\mathbf{1} - \mathbf{r}_i\mathbf{r}_i]m_i = \int [(\mathbf{r} \cdot \mathbf{r})\mathbf{1} - \mathbf{r}\mathbf{r}] \, dm \qquad (j)$$

which we call the *inertia dyadic*.

As a final item of interest, we wish to introduce the *double dot product* of two dyadics defined by the scalar product

$$\mathbf{AB} : \mathbf{CD} = (\mathbf{A} \cdot \mathbf{C})(\mathbf{B} \cdot \mathbf{D}) = \mathbf{C} \cdot \mathbf{AB} \cdot \mathbf{D} = \mathbf{C} \cdot \mathbf{\Phi} \cdot \mathbf{D}. \qquad (k)$$

The result is a scalar, and we note that the double dot product $\mathbf{C} \cdot \mathbf{\Phi} \cdot \mathbf{D}$ is different from $\mathbf{D} \cdot \mathbf{\Phi} \cdot \mathbf{C}$ unless the dyadic $\mathbf{\Phi}$ is symmetric. The form (k) is used in Sec. 4.4 to express the kinetic energy of rotation in terms of the inertia dyadic.

appendix B

Elements of Topology and Modern Analysis

The concept of continuity plays an important role in the theory of differential equations. The classical definition of continuity can be stated as follows: A function f of the real variable x is said to be continuous at $x = x_0$ if, given any real positive number ϵ, there is a positive number δ such that $|f(x) - f(x_0)| < \epsilon$ for all values of x satisfying $|x - x_0| < \delta$. The theory of continuity can be generalized by presenting it in terms of notions from the field of topology, which can be regarded as a branch of general set theory. Continuous transformations from one space to another are referred to as *mappings*. The most important topological spaces are the metric spaces, among which the Euclidean spaces play a special role. We shall briefly review certain concepts and definitions of point set topology with the sole purpose of familiarizing ourselves with the terminology used in the geometric theory of differential equations.

A basic concept in Euclidean geometry is congruence, which represents a type of equivalence expressing the fact that two geometrical figures are identical except for their position in space. A different form of equivalence is defined by affine transformations, in which the shape and size of the geometrical figures need not be the same but straight parallel lines in one figure correspond to straight parallel lines in the equivalent figure.

Topological equivalence is referred to as *homeomorphism*. Two spaces are called topologically equivalent or homeomorphic if one can be transformed into the other by means of a one-to-one continuous transformation which has a continuous inverse transformation. For example, the surfaces of a sphere and a cube are homeomorphic, but the surfaces of a sphere and a torus are not.

If two points in a set can be joined by a continuous curve, the set is said to be *arcwise connected*. Most surfaces encountered in elementary geometry are arcwise connected. An example of a disconnected surface is the hyperboloid of two sheets. If every ordinary closed curve on a connected surface can be contracted continuously into a point without leaving the surface, the surface is said to be *simply connected*. Whereas the surface of a sphere is simply connected, the surface of a torus is not.

Any curve which is topologically equivalent to a circle, i.e., homeomorphic to a circle, is called a *Jordan curve*. The Jordan curve concept enables us to state a very important theorem in plane topology, namely, the Jordan curve theorem (see Sec. B.3).

Attempts have been made to characterize topological spaces by associating with them given entities, such as numbers and groups, with the purpose of finding characterizations which are necessary and sufficient for the spaces to be homeomorphic. For the most part it is easiest to choose the entities to be the same for homeomorphic spaces, for which reason they are called *topological invariants*. However, the characterizations typically are not sufficient for topological equivalence, so that the characterization problem is still with us. A complete solution exists only for two-dimensional manifolds, namely, ordinary closed two-dimensional surfaces, such as a sphere.

B.1 SETS AND FUNCTIONS

We refer to a collection of objects or a family of objects having common and distinguishing properties as a *set*. As an example, the points on a curve form a set of points. The objects of a set are called its *elements* or *members*.

Sets are generally denoted by uppercase letters $A, B, \ldots$ and their elements by lowercase letters $a, b, \ldots$. The statement that a is an element of, or belongs to, A is expressed symbolically as $a \in A$. The negation of $a \in A$ is written $a \notin A$. If all the elements of set A are also contained in set B, then A is said to be a *subset* of B. This statement is expressed $A \subseteq B$, where the case $A = B$ is not excluded. In the event $A \neq B$, the statement that every element of A is contained in B is written $A \subset B$. In this case A is said to be a *proper subset* of B. For example, the set of all real numbers can be thought of as a subset of the set of all complex numbers.

If X is a set with elements x and P is a certain property which the elements of X may have, then the notation $\{x: P(x)\}$ means the set of all $x \in X$ for which the property P is true.

Elements of Topology and Modern Analysis

A set containing no elements is the *empty* set, denoted by ∅ and also called the *void* or *null* set. Every set contains ∅ as a subset.

If A and B are two arbitrary sets, the set consisting of all the elements belonging to at least one of the sets A and B is called the *union* or *sum* of the two sets A and B and denoted by the symbol $A \cup B = \{x: x \in A \text{ or } x \in B\}$. The union of A and B is illustrated in Fig. B.1a. The *intersection* of the sets A and B is the set whose elements belong to both A and B. The intersection is denoted by $A \cap B = \{x: x \in A \text{ and } x \in B\}$ and is shown in Fig. B.1b.

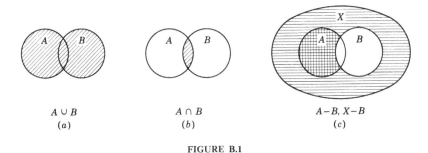

$A \cup B$ $A \cap B$ $A - B, X - B$
(a)   (b)   (c)

FIGURE B.1

The definitions of the union and intersection can be extended to any arbitrary number of sets A_i. If $A \cap B = \emptyset$, the sets A and B are said to be *disjoint*.

If A and B are subsets of the set X, the *difference* is defined as the set of all $x \in X$ which are members of A but not of B and denoted by $A - B = \{x: x \in A \text{ and } x \notin B\}$. The *complement* of B (relative to X) is the set defined by all $x \in X$ such that $x \notin B$, and denoted by $X - B = \{x: x \in X \text{ and } x \notin B\}$. The complement is also denoted by B'. These concepts are illustrated in Fig. B.1c, where the difference is the region shaded vertically and the complement the one shaded horizontally.

The term *function*, or *transformation*, implies a certain relation between the elements of two sets. Let us consider two arbitrary nonempty sets X and Y and assume that there exists a rule f whereby to each element $x \in X$ corresponds a uniquely determined element $y \in Y$. A function consists of three objects: the two sets X and Y and the rule f. The symbol $f: X \to Y$ means that f is a single-valued function whose *domain* is X and whose *range* is contained in Y. We note that the set of y's may or may not comprise all the elements in Y.

The implication of the preceding definition is that a function with domain X and range in Y is a subset of all the ordered pairs (x,y) with $x \in X$ and $y \in Y$ in which each x occurs exactly once. The set containing all the ordered pairs whose first element is a member of X and whose second element is a member of Y is called the *cartesian* or *direct product* of X and Y and denoted by $X \times Y$. Since the pairs are ordered, $X \times Y$ is not identical to $Y \times X$ unless $X = Y$.

Instead of using the terminology that for every $x \in X$ the function assigns a uniquely determined element y in Y, it is common to write simply $y = f(x)$ with the implication that to the element x of the domain there corresponds the element $f(x)$ of the range. The element $f(x)$ is said to be the *image* of the element x under the transformation f. Figure B.2a shows a geometrical interpretation of the concept.

Next let us consider the function f with the domain D and the range R, where $D \subset X$ and $R \subset Y$. If a function can be defined so that the elements (y,x) constitute a subset of $Y \times X$ such that each y occurs exactly once, this subset is a function whose domain is in Y and whose range is in X. This function, denoted by f^{-1}, is called the *inverse* of f and has the domain R and the range D. In the commonly used terminology, we say that $x = f^{-1}(y)$ is the inverse of $y = f(x)$, and the element $f^{-1}(y)$ is called the *inverse image* of the element y (see Fig. B.2b).

A continuous transformation is referred to as a *mapping*. The function f is said to map the set X *onto* the set Y if $f(X) = Y$ and *into* the set Y if $f(X) \subseteq Y$.

Let P be a nonempty set of elements $x, y, \ldots$. The set P is said to be *partially ordered* if for some pairs of elements x, y there exists an ordering relation, denoted by $x < y$, with the properties:

1. If $x \leq y$ and $y \leq z$, then $x \leq z$ (transitivity).
2. For all $x \in P$, $x \leq x$ (reflexivity).
3. If $x \leq y$ and $y \leq x$, then $x = y$ (antisymmetry).

If P is a partially ordered set, an element $x \in P$ is said to be *maximal* if $y \geq x$ implies that $y = x$. Hence, a maximal element in P is an element of P which is not less than or equal to any other element of P.

If S is a subset of a partially ordered set P, an element $x \in P$ is said to be

FIGURE B.2

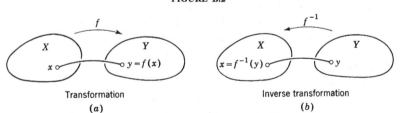

Transformation
(a)

Inverse transformation
(b)

an *upper bound* of S if $a < x$ for every $a \in S$. An element $x \in P$ is called a *lower bound* of S if $x < a$ for every $a \in S$. An element x is said to be a *least upper bound*, or *supremum*, of S, denoted by sup S, if and only if it is an upper bound of S which is less than or equal to every other upper bound of S. Hence, a least upper bound is an upper bound which is a lower bound for the set of all upper bounds. In a similar fashion, a *greatest lower bound*, or *infimum*, of S, denoted by inf S, represents an element which is an upper bound for the set of all lower bounds of S.

B.2 METRIC SPACES

A set possessing some kind of geometric structure is referred to as a *space*. For example, a set of n numbers $x_1, x_2, \ldots, x_n$ defines a *point* x in a Euclidean n-space. But the same numbers can be regarded as defining an n-vector in the same space, so that the concepts of point and vector can be used interchangeably. Since we have considerable interest in spaces for which many of the set properties, such as convergence and continuity, can be related by the notion of *distance*, we are led naturally to the concept of metric spaces.

A nonempty set X is a *metric space* if there is defined a function d with domain $X \times X$ and range in the real number field, such that for all points x, y, and z in X it satisfies the following conditions:

1. $d(x,y) \geq 0$ and $d(x, y) = 0$ if and only if $x = y$.
2. $d(x,y) = d(y,x)$ (axiom of symmetry).
3. $d(x,z) \leq d(x,y) + d(y,z)$ (triangle axiom).

If the above conditions are satisfied, the space X is said to be metrizable and the function d is called the *metric* or *distance function* on X.

Examples of metric spaces are the Euclidean spaces. If x and y are two points on the real line, the distance $d(x,y) = |x - y|$ forms the metric space E^1, namely, the *Euclidean one-space*. If x and y are points consisting of the real number $x_1, x_2, \ldots, x_n$ and $y_1, y_2, \ldots, y_n$ in an n-space, then the metric space defined by the distance $d(x,y) = \left[\sum_{i=1}^{n} (x_i - y_i)^2\right]^{\frac{1}{2}}$ is the *Euclidean n-space* E^n, where $d(x,y)$ is referred to as the *Euclidean distance*. Addition and scalar multiplication in E^n are defined as follows. If $x = (x_1, x_2, \ldots, x_n)$ and $y = (y_1, y_2, \ldots, y_n)$, then the sum of the two points is

$$x + y = (x_1 + y_1, x_2 + y_2, \ldots, x_n + y_n),$$

and if α is a real number, then

$$\alpha x = (\alpha x_1, \alpha x_2, \ldots, \alpha x_n).$$

The point x can be represented geometrically as a vector with the tail at the origin of the space and the head at the point in question. The length of the vector x is defined as the distance from the origin to the point x and denoted by $\|x\| = \left(\sum_{i=1}^{n} x_i^2\right)^{1/2}$, where $\|x\|$ is called the *Euclidean norm*.

The real line can be regarded as a metric space, namely, the space R^1. If a and b are two points in R^1, then the *open interval* from a to b is the subset of R^1 defined by

$$(a,b) = \{x: a < x < b\},$$

whereas the *closed interval* from a to b is

$$[a,b] = \{x: a \leq x \leq b\}.$$

In a similar fashion, we can define the *open-closed interval* $(a,b]$ and the *closed-open interval* $[a,b)$.

Let A be an arbitrary set in the metric space X and x a point in that space. The distance from the point x to the set A is defined by

$$d(x,A) = \inf \{d(x,a): a \in A\},$$

where the symbol inf, denoting the greatest lower bound, was introduced in Sec. B.1. The distance between two sets A and B contained in X is given by

$$d(A,B) = \inf \{d(a,b): a \in A \text{ and } b \in B\}.$$

Hence, if $A \cap B \neq \emptyset$, then $d(A,B) = 0$, or the distance between two sets is zero provided the sets are not disjoint.

The diameter of the set $A \subset X$ is defined by

$$d(A) = \sup \{d(a_1,a_2): a_1 \text{ and } a_2 \in A\}.$$

Hence, the diameter is the least upper bound of the distance between pairs of points in A. If the diameter of a set A is finite, the set A is said to be *bounded*.

An *open sphere* $S(x_0,r)$ in the metric space X is the set of all points $x \in X$ satisfying the condition $d(x_0,x) < r$, where x_0 and r are referred to as the *center* and the *radius* of the sphere, respectively. The open sphere can be written in the form

$$S(x_0,r) = \{d(x_0,x) < r: x \in X\}.$$

In view of the definition of the diameter of a set, a set is bounded if it is contained in some open sphere. The open sphere is also called a *spheroid*. In Euclidean geometry spheroids are at times referred to as *spherical regions*. The set of all points $x \in X$ satisfying the relation $d(x_0,x) \leq r$ is said to be the *closed sphere* $S[x_0,r]$, denoted by

$$S[x_0,r] = \{d(x_0,x) \leq r: x \in X\}.$$

We note that if the space X is the real line R^1, the open sphere reduces to the open interval $(x_0 - r, x_0 + r)$, whereas the closed sphere reduces to the closed interval $[x_0 - r, x_0 + r]$.

If A is a subset of a metric space X, and if $x \in A$, then x is said to be an *interior point* of A if it is the center of some open sphere $S(x,r)$ contained in A. The set of all interior points of A is called the *interior* of A, and denoted by Int (A), or symbolically

$$\text{Int } (A) = \{x : x \in A \text{ and } S(x,r) \subseteq A \text{ for some } r\}.$$

A set $A \subset X$ such that all its points are interior points is referred to as an *open set*. The interval $a < x < b$ of the real line R^1 is an open set since the neighborhood $S(x,\epsilon)$, where $\epsilon = \min(x - a, b - x)$, is contained entirely in that interval. Similarly, the open sphere $S(a,r)$ in an arbitrary metric space R is an open set. If $x \in S(a,r)$, with the implication that $d(a,x) < r$, then we can let $\epsilon = r - d(a,x)$, from which it follows that $S(x,\epsilon) \subseteq S(a,r)$.

For any metric space X and some $x \in X$, any open set which contains x is said to be a *neighborhood* of x. As an example, the open sphere $S(x,r)$ is a spherical neighborhood of x with radius r. Let A be a subset of X. Then a point $x \in X$ is said to be a *limit point* or *accumulation point* of the set A if each spherical neighborhood of x contains at least one point of A different from x. A subset F of the metric space X is referred to as a *closed set* if it contains all its limit points. The subset F is closed if its complement F' with respect to X is open. Exceptions to this are the empty set and the whole space X, which are both open and closed. Examples of closed sets are the closed interval $[a,b]$ and the closed sphere $S[x,r]$.

Let A be a subset of an arbitrary metric space X. The *closure* of A, denoted by $\bar{A}$, is the union of A and all its limit points. If the set A is closed, it must coincide with its closure, $A = \bar{A}$.

Considering again the set $A \subset X$, we say that a point $x \in X$ is a *boundary point* of A if each open sphere centered at x intersects both A and its complement A'. The boundary of A, denoted by ∂A, is the set of all its boundary points. We note the following properties concerning the boundary: (1) the boundary of a set A is equal to the intersection of its closure and the closure of its complement, $\partial A = \bar{A} \cap \bar{A}'$, (2) the boundary of A is a closed set, and (3) the set A is closed if it contains its boundary.

Let $x_1, x_2, \ldots$ be a sequence of points in the metric space X. The sequence x_n ($n = 1, 2, \ldots$) is said to be *convergent to the point* x if for any $\epsilon > 0$ there is a natural number $N(\epsilon)$ such that if $n > N(\epsilon)$, then all x_n are contained in the spherical neighborhood $S(x,\epsilon)$. If the sequence x_n ($n = 1, 2, \ldots$) converges to x, we then write $\lim_{n \to \infty} d(x,x_n) = 0$ and call x the *limit* of the sequence.

B.3 TOPOLOGICAL SPACES

Metric spaces depend on the concept of distance, which is not a topological invariant. Indeed a topological space is not necessarily metrizable, and even if it is metrizable, the metric is not unique. Hence, although the spaces discussed are of a general type, they are defined in terms of distance, a nontopological concept. It turns out that many important definitions and statements concerning metric spaces do not involve the concept of metric itself but only the concept of open (or closed) sets. In metric spaces, and indeed in all topological spaces, the open sets are topological invariants. Hence a topological space can be regarded as a generalization of the metric space concept.

A *topological space* is a set with a family of subsets designated as *open sets*. These open sets must satisfy certain axioms. A set X with a collection of subsets T is said to be a *topology* for X if:

1. The empty set $\varnothing$ and the whole space X belong to T.
2. The union of any number of members of T is a member of T.
3. The intersection of any finite number of members of T is a member of T.

The members of T are called *open sets* of X in this topology, and the elements of X are called its *points*. The open set F containing a point or a set is said to be a *neighborhood* of the point or the set. If X is the whole space, the *closed set* F' is defined as the complement $X - F$ of an open set F. For convenience, the empty set $\varnothing$ and the whole space X are regarded as being the only sets which are simultaneously open and closed, sometimes referred to as *clopen*. But the union of any number and the intersection of a finite number of open sets is open. Hence, by virtue of the complementation rules, it follows that any intersection and any finite union of closed sets is closed.

If $A \subset X$, the intersection of all closed subsets of X which contain A is called the *closure* of A, denoted by $\overline{A}$. The closure $\overline{A}$ is the least closed set containing A. The set A, $A \subset X$, is said to be *dense* (or *everywhere dense*) if $\overline{A} = X$.

The interior of a set A, denoted by Int (A), is the union of all open sets contained in A. Since Int (A) is open and Int $(A) \subset A$, it follows that A is open if and only if $A = $ Int (A).

The *boundary* ∂A of a set A, defined as the intersection of the closure of A and the closure of its complement A', is a closed set. The closure $\overline{A}$ is the set of all points of zero distance from A, whereas the boundary ∂A is the set of all points at zero distance from both A and its complement A'.

A point x is said to be a *limit point* or an *accumulation point* of a set A if every neighborhood of x contains a point of $A - x$. A point x of the set A

is called an *isolated point* of that set if it has a neighborhood which does not contain any points of A different from x.

Next we shall define continuity in terms of topological spaces. Let X and Y be topological spaces and let f be a function with domain X and range Y. The function f is said to be *continuous* at the point $x_0 \in X$ if to each neighborhood V of $f(x_0)$ in Y there corresponds a neighborhood U of x_0 in X such that $f(U) \subset V$. This is equivalent to the statement that f is continuous at x_0 if, for each neighborhood V of $f(x_0)$, the set contains a neighborhood of x_0. The function f is said to be continuous on X if it is continuous at each point of X. A continuous function is called a *mapping*. Hence f is continuous on X if and only if $f^{-1}(V)$ is an open set in X whenever V is an open set in Y.

Consider a function f with domain X and range Y such that the inverse function f^{-1} exists, where X and Y are topological spaces. If both f and f^{-1} are continuous on their domains, f is said to be a *homeomorphism* of X onto Y and the topological spaces, referred to as *homeomorphic*, are said to be equivalent.

The topological image of the set of points of an n-space defined by $\sum_i x_i^2 < 1$ is called an *open n-cell*, or simply an *n-cell*. The Euclidean n-space is an n-cell. The subset consisting of the boundary points of the n-cell, namely, the points satisfying $\sum_i x_i^2 = 1$, form the topological image called the $(n-1)$-*sphere*. The topological image of the closed sphere $\sum_i x_i^2 \leq 1$ is sometimes referred to as a *closed n-cell*, or as a *solid n-sphere*. The open interval $a < x < b$, often called simply the interval, has as its topological image the open one-cell, referred to as an *open arc*. The closed interval $a \leq x \leq b$, commonly referred to as a *segment*, is called a *closed arc*. The one-sphere, representing the topological image of a circle, is generally known as a *Jordan curve*.

Let X be a set, A a subset of X, and F a family of subsets of X such that each point of A belongs to at least one member of F. In this case F is called a *covering* of A. If X is a topological space, and if all the sets in F are open, then F is said to be an *open covering* of A.

If every open covering of A includes a finite subfamily which covers A, the subset A of the topological space X is said to be *compact*. Some properties of compact topological spaces are as follows:

1. A compact subset of a space X is closed.
2. A closed subset F of a compact topological space X is also compact.
3. If f is a mapping whose domain is compact, its range is also compact.

Among the examples of compact sets, we must mention the closed cell and sphere, as they are both topological images of bounded and closed Euclidean sets. In particular, a segment and a Jordan curve, which represent a closed one-cell and one-sphere, respectively, are compact.

A topological space X is said to be *connected* if it is not the union of two sets A and B, where A and B are two disjoint nonempty subsets of X. The space X is connected if it is not the union of A and B such that $A \cap \bar{B} = \varnothing$ and $\bar{A} \cap B = \varnothing$. A connected open nonempty set is often referred to as a *domain*. It is also referred to as a *region*. Such a set plus its boundary points is called a *closed domain*. Examples of connected sets are the open n-cell, the closed n-cell, and the n-sphere, where in the latter $n > 0$.

A domain D is said to be *convex* if the segment joining any two points of D lies entirely in D.

As a final item of interest, we state the *Jordan curve theorem: If J is a Jordan curve in the $x_1 x_2$ plane π, then the complement of J, namely, $\pi - J$, is the union of two disjoint open sets S_i and S_e, each of which has J as boundary.* The set S_i, called the interior of J, is bounded, and the set S_e, referred to as the exterior of J, is unbounded. Notice that the union of the set S_i and the curve J defines a closed domain.

SUGGESTED REFERENCES

1. Hille, E., and Phillips, R. S.: *Functional Analysis and Semi-groups*, American Mathematical Society Colloquium Publications, vol. 31, Providence, R.I., 1957.
2. Kolmogorov, A. N., and S. V. Fomin: *Elements of the Theory of Functions and Functional Analysis*, vol. 1, Graylock Press, Rochester, N.Y., 1957.
3. Lefschetz, S.: *Introduction to Topology*, Princeton University Press, Princeton, N.J., 1949.
4. McCarty, G.: *Topology*, McGraw-Hill Book Company, New York, 1967.
5. Patterson, E. M.: *Topology*, Interscience Publishers, New York, 1956.
6. Simmons, G. F.: *Introduction to Topology and Modern Analysis*, McGraw-Hill Book Company, New York, 1963.
7. Taylor, A. E.: *Introduction to Functional Analysis*, John Wiley & Sons, Inc., New York, 1958.
8. Yosida, K.: *Functional Analysis*, Springer-Verlag OHG, Berlin, 1965.

Name Index

Alfriend, K. T., 425, 451
Aristotle, 1, 2
Arnold, R. N., 407
Auelmann, R., 482, 492

Barba, P. M., 483, 492
Barbasin, E. A., 237
Barnett, S., 259, 262
Bellman, R., 262, 292, 328
Bendixson, I., 199, 205
Bernoulli, John, 55
Bertram, J. E., 262
Bogoliubov, N. N., 302, 306, 322, 328
Born, M., 372, 376, 380
Breakwell, J. V., 428, 451, 474, 492
Brouwer, D., 451
Bruns, H., 416
Byerly, W. E., 451

Cauchy, A. L., 174
Cesari, L., 208, 262, 291, 292, 328
Chetayev, N. G., 236, 251, 262

Clemence, G. M., 451
Coddington, E. A., 199, 208, 262, 292
Corben, H. C., 44, 380
Cunningham, W. J., 328

D'Alembert, J. le Rond, 65
Danby, J. M. A., 451
DeBra, D. B., 474, 492
Deimel, R. F., 407
Delp, R. H., 474, 492
Deutsch, R., 451, 455, 492
Duffing, G., 313

Ehricke, K. A., 455, 492
Einstein, A., 2, 6–8
Euclid, 2, 172, 175, 501, 502
Euler, L., 57, 84, 126, 138, 140, 333

Floquet, G., 264
Fomin, S. V., 506
Foucault, J. B. L., 101, 116, 381

507

Galilei, G., 3, 9
Gantmakher, F. R., 492
Gelfand, I. M., 292
Goldstein, H., 44, 101, 121, 169, 380
Goodstein, R., 389, 407
Griffith, B. A., 121, 169

Hahn, W., 262, 291, 292
Hale, J. K., 292, 328
Hamilton, Sir W. R., 66, 91, 94, 334, 355, 360
Hayashi, C., 328
Hill, G. W., 278, 280, 309
Hille, E., 506
Hurewicz, W., 208, 262
Hurwitz, A., 222

Jacobi, C. G. J., 83, 180, 214, 355
Jordan, C., 181, 195, 506

Kalman, R. E., 262
Kane, T. R., 474, 483, 492
Kelvin, Lord (W. Thomson), 255
Kepler, J., 4, 9, 11, 28, 36, 409
Kolmogorov, A. N., 506
Kozai, Y., 451, 460, 463, 492
Krasovskii, N. N., 237, 262, 292
Krylov, N., 302

Lagrange, J. L., 45, 54, 68, 72, 178, 193, 244, 333, 349, 420, 424, 446
Lanczos, G., 44, 100, 380
Laplace, P. S. de, 178, 415
LaSalle, J., 262
Leech, J. W., 44, 100
Lefschetz, S., 208, 262, 506
Levin, L. M., 492
Levinson, N., 199, 208, 262, 292
Liapunov, A., 176, 193, 209, 227, 231, 245, 270, 288
Lidskii, V. B., 292
Likins, P. W., 482, 492
Lindstedt, A., 299
Liouville, J., 214, 338, 343, 349
Lipschitz, R., 174
Lorentz, H. A., 5–6

McCarty, G., 506
McCuskey, S. W., 451
McLachlan, N. W., 292
MacMillan, W. D., 433, 451
Malkin, I. G., 262, 291, 292
Massera, J. L., 291
Mathieu, E., 278, 282, 309
Maunder, L., 407
Maupertuis, P. L. N. de, 333
Maxwell, J. C., 5–6
Meirovitch, L., 100, 169, 407, 475, 482, 483, 492, 493
Minorsky, N., 208, 328
Mitropolsky, Y. A., 302, 306, 322, 328
Moulton, F. R., 451

Nemytskii, V. V., 208, 262, 292
Newton, Sir I., 3–4, 9

O'Donnell, C. F., 407

Pars, L. A., 100, 380
Patterson, E. M., 506
Phillips, R. S., 506
Plummer, H. C., 451
Plymale, B. T., 389, 407
Poincaré, H., 6, 177, 195, 198, 199, 226, 348, 416
Poinsot, L., 147
Poisson, S. D., 349
Pringle, R., Jr., 428, 451, 474, 482, 492, 493

Rand, R. H., 425, 451
Rayleigh, Lord, 88
Riemann, G. F., 2
Roberson, R. E., 493
Robertson, H. P., 44
Routh, E. J., 83, 86, 222, 250
Rund, H., 380

Schechter, H. B., 428, 451
Schuler, M., 403
Shippy, D. J., 474, 492
Simmons, G. F., 506
Smart, W. M., 451

Name Index

Stehle, P., 44, 380
Stepanov, V. V., 208, 262, 292
Stoker, J. J., 328
Storey, C., 259, 262
Struble, R. A., 208, 328
Sundmann, K. F., 416, 451
Sylvester, J. J., 232
Synge, J. L., 121, 169

Taylor, A. E., 506
Thomson, W. T., 407
Tisserand, F., 461, 493

Van der Pol, B., 205

Wallace, F. B., Jr., 475, 482, 483, 493
Watson, G. N., 292
Webster, A. G., 169
Whittaker, E. T., 100, 169, 292, 333, 380, 451
Wintner, A., 451

Yosida, K., 506

Subject Index

Absolute integral invariant, 347
Acatastatic system, 51
Acceleration:
  absolute, 11
  centripetal, 112
  Coriolis, 112
  relative, 112
Action and reaction, 10, 22
Action integral, 3
Action variables, 365–373
Adjoint equation, 215
Affine transformation, 181, 409
Angle variables, 365–373
Angular momentum, 12, 22
  apparent, 24
  conservation of, 13, 24, 27, 415
  matrix, 129
  of rigid body, 126–130
  of system of particles, 23
  translation theorem for, 130–132
Angular velocity, 43, 108
  of the earth, 113
  matrix, 108, 129

Angular velocity:
  in terms of body axes, 142
  in terms of node-axis system, 142
  vector, 108
Anomaly:
  eccentric, 43, 409
  mean, 43, 409
  true, 409
Antecedents, 494
Aphelion, 33
Apocentron, 33
Apogee, 33
Apsis, 33
Arc:
  closed, 505
  open, 505
Ascending node, 412
  longitude of, 412
Asymptotic stability, 176
  of linear autonomous system, 222
  orbital, 177
  uniform, 176

Atmospheric drag:
  effect on satellite orbits, 463–466
  effect on stability of artillery shells, 255–258
Attitude motion of satellites, 466–483
  gravity-gradient stabilized, 467, 470–475
  spin stabilized, 475–483
Autonomous systems, 174, 209, 217–224
  linear, 217–224
    asymptotically stable, 222
    stable, 222
    unstable, 222
Averaging:
  method of, 322–326
  operator, 323
  principle of, 324
Axial vector, 110

Barbasin and Krasovskii's theorem on asymptotic stability, 237, 241
Barycenter, 414
Bendixson's criterion, 205
Body axes, 124
Body cone, 144
Boundary of set, 503, 504
Bounds of libration of satellites, 481
Brachistochrone problem, 55

Canonical systems, 94, 172
  stability of, 243–252
  variational equations from, 229–231
Canonical transformations (*see* Contact transformations)
Catastatic system, 51
Catenary, 59, 64
Cauchy-Lipschitz theorem, 174, 176, 179
Center, as singular point, 184, 186, 191
Center of gravity, 21
Center of mass:
  of rigid body, 127
  of system of particles, 21
Central force, 25–40
Centrifugal force, 34, 77
Centrifugal potential, 34
Centripetal acceleration, 112
Characteristic in motion space, 174
Characteristic determinant, 180, 218, 222

Characteristic equation, 180, 222
  eigenvalues of, or roots of, 181, 222
  (*See also* Characteristic polynomial)
Characteristic exponents, 266
  for canonical systems with periodic coefficients, 275
Characteristic multipliers, 266
  simple, 267
Characteristic polynomial, 218
  elementary divisor of, 218
    simple, 218
  invariant coefficients of, 218
  (*See also* Characteristic equation)
Characteristic velocity, 491
Chasles' theorem, 126
Chetayev's instability theorem, 236, 241, 291
Class of a function, 11, 174
Closed domain, 506
Closed interval, 502
Closed set, 503
Closed sphere, 502
Combination tones, 321
Complementary Kronecker delta, 22, 414
Complete damping, 247
Complete degeneracy, 370, 372
Complete equations, 180
  stability of, 186
Conditionally periodic motion, 369
  (*See also* Multiply periodic motion)
Configuration space, 46, 172
Congruence, 1, 2, 497
Consequents, 494
Conservation:
  of angular momentum, 13, 24, 27, 415
  of energy, 16, 66, 84, 416
  of generalized momentum, 82
  of linear momentum, 12, 23
Conservative force field (*see* Conservative system)
Conservative system, 15, 16
  canonical, 243–247
  with ignorable coordinates, 249–250
  motion in the large, 189–194
Constraint:
  force of, 48, 62
    work performed by, 49–50
  holonomic, 51, 55
  inequality, 51

Subject Index

Constraint:
  kinematical, 47, 48
  nonholonomic, 51, 54
  scleronomic, 51
  system with, 48–52
Contact transformations, 334–346, 373
  homogeneous, 338
  infinitesimal, 352–354
Continuity, 497, 500, 505
Contraction hypothesis, 6
Control surface, 483
Control volume, 483
Coordinate system (*see* Reference system or frame)
Coordinate transformation, 46, 102–104
Coordinates:
  cyclic, 82, 97, 149
    (*See also* ignorable *below*)
  generalized, 47
  ignorable, 82, 85–88, 249
    (*See also* cyclic *above*)
Coriolis acceleration, 112
Coriolis effect, 115
Coriolis force, 4, 8, 77
  in variable-mass systems, 486, 488
Coulomb force field, 37
  scattering in, 37–40
Critical behavior, 187, 222
  of linear autonomous system, 222
  of variational equations with periodic coefficients, 272
Critical point, 179
Curves of zero relative velocity, 477
Cuspidal motion, 153
Cyclic characteristic (*see* Limit cycle)
Cyclic coordinate, 82, 97, 149
  (*See also* Ignorable coordinate)
Cyclic trajectory, 195, 198
  adjacent, 204
  (*See also* Limit cycle)
Cyclone, 115
Cylindrical region, 175

D'Alembert's principle, 65–66
  generalized, 65
Damping:
  coefficient of, 90, 188
  complete, 247

Damping:
  pervasive, 248
Decrescent function, 290
Degeneracy:
  complete, 370, 372
  m-fold, 370
Degrees of freedom, 46–48
  of a rigid body, 123
Descending node, 412
Determinant:
  characteristic, 180, 218, 222
  Jacobian, 180, 336, 348
  of a matrix, 104, 185, 219
Dirac delta function, 385, 387
Direct method of Liapunov (*see* Liapunov's direct or second method)
Direct precession, 146
Displacement:
  actual, 48
  possible, 48
  virtual, 50, 53
  reversible, 60
Dissipation function (*see* Rayleigh's dissipation function)
Dissipative forces:
  of Rayleigh type, 88–91
  stability in the presence of, 252–258
Distance:
  between a point and a set, 177, 502
  between points, 501
  between sets, 502
Disturbing function, 428–430
  due to gravitational field of oblate planet, 460–461
  resolution of, 447–450
Domain:
  closed, 506
  convex, 506
  open, 506
Domain of attraction, 176, 236
Double pendulum, 78–79
Duffing's equation, 313–322
Dyad, 495
Dyadics, 494–496
  conjugate, 494
  double dot product of, 496
  inertia, 133, 137, 496
  scalar product of vector and, 494
  symmetric, or self-conjugate, 494, 496
  unit, or identity, 495

Eccentric anomaly, 43, 409
Eccentricity, 33, 411
Ecliptic plane, 412
Effective force, 65
Eigenvalue problem, 124
Eigenvalues, 124
   multiplicity of, 218
   nullity of, 218
Eigenvectors, 124
Einstein:
   general relativity theory or gravitational theory of, 2, 8, 11
   special relativity theory of, 2, 7
Element of a set, 498
   greatest lower bound, 501
     (*See also* Infimum)
   image of, 500
   inverse image of, 500
   least upper bound, 501
     (*See also* Supremum)
   lower bound, 501
   maximal, 500
   upper bound, 501
Ellipsoid of inertia, 136
Elliptical three-body problem, 425
Energy, 14–17
   conservation of, 16, 66, 84, 85, 416
     (*See also* Energy integral)
   total, 16
Energy diagram, 17–21
Energy integral, 151, 190, 416
Epsilon symbol, 109
Equations of motion:
   of Euler, 138–140, 143, 147
   of Hamilton, 91–97
   of Lagrange, 72–81, 90
     for quasi-coordinates, 157–160
   of Newton, 10–11, 13
   for relative motion, 428–430
   in terms of arbitrary axes, 160–162
Equilateral-triangle solution:
   in general three-body problem, 419
   in restricted three-body problem, 424
Equilibrium point, 174
   neighborhood of, 175, 179
   at the origin, 175, 227
   stability of, 176
   (*See also* Singular point)
Equilibrium position, static, 59, 62
   (*See also* Equilibrium point)

Equivalent force in variable-mass systems:
   Coriolis, 486, 488
   reactive, 486, 489
   due to unsteadiness of relative motion, 486, 488
Equivalent potential energy (*see* Modified potential energy)
Escape velocity, 35
Ether, luminiferous, 5, 6
Euclidean distance, 501
Euclidean geometry, 1, 2
Euclidean norm or length, 175, 502
Euclidean space, 2, 172, 175, 211, 501
Euler-Lagrange differential equation, 57
Euler's angles, 140–143
Euler's equations of motion, 138–140, 143, 147
   modified, 140
Euler's theorem:
   on homogeneous functions, 84
   on motion of rigid bodies, 126
Extended Hamilton's principle, 68
Extended Liouville's theorem, 343
Extremum, 53

First-approximation equations, 180
   stability of, 187
   Liapunov's theorem on, 226–229
First integrals of motion, 82, 249
First point of Aries, 412, 438
Floquet's theory, 264–270
   applied to second-order systems, 278, 280
Focus (*see* Spiral point)
Force, 1, 2
   aerodynamic, 255, 489
   applied, or impressed, 60
   body, 485
   central, 25
   centrifugal, 4, 8, 34, 77
   conservative, 15
   of constraint, 48, 62
   Coriolis, 4, 8, 77, 486
   Coulomb, 37
   dissipative, 88
   effective, 65
   external, 21
   friction, 15
   generalized, 61

Subject Index

Force:
  gravitational, 8
  gyroscopic, 77
  impulsive, 79–81
  inertia, 65
  internal, 21
  nonconservative, 17, 83
  nonpotential, 17
  potential, 17
  reactive, 486
  restoring, 18
  surface, 485
  due to unsteadiness of relative motion, 486
Force field:
  conservative, 15
  lamellar, or irrotational, 17
Foucault's pendulum, 101, 116–119
Free variation, 54
Frequencies:
  of gyrocompass, 396
    damped, 397
  multi-degree-of-freedom system, 368, 370, 373
    commensurable, 369
  nonlinear system, 299, 300
    of combination tones, 321
    of higher harmonics, 321
    of subharmonic oscillation, 320
Frequency:
  critical, 393
  natural, 19, 298, 302
    of symmetric gyroscope, 384, 388, 392–393
Function, 497–501, 505
  bounded, 175, 289
  class of, 11, 174
  continuous, 497
  decrescent, 290
  domain of, 499
  homogeneous, 232
    quadratic, 74, 137, 232
  indefinite, 232
  Lipschitz, 174, 289
  negative definite, 232, 289
  negative semidefinite, 232, 289
  positive definite, 232, 289
  positive semidefinite, 232, 289
  range of, 499

Function:
  sign-constant or constant with respect to sign, 232
  sign-variable or variable with respect to sign, 232
  single-valued, 499
Fundamental matrix, 212, 217, 264, 266
Fundamental set of solutions, 212

Galilean reference frame (*see* Reference system or frame)
Galilean transformations, 3–6
Galileo's law of inertia of, 3, 10
Galileo's principle of relativity, 3
Gegenschein, as related to the three-body problem, 424
Generalized coordinates, 46–48, 60, 72, 171
Generalized forces, 61, 75
Generalized impulse, 80
Generalized momentum, 80
  conjugate, 82
  conservation of, 82
Generating function, 342, 356, 373
Generating solution, 295
Geometry, 1
  of Euclid, 1
  of Riemann, 2
Gimbal of gyroscope, 382
  inertia effect, 386–389
  walk, 386
Gravitation, law of, 11
  (*See also* Inverse square law)
Gravitational potential energy, 32, 416, 430–438, 460
Gravitational torques, 430–438
Gravity-gradient stabilized satellites, 467, 470–475
Group property, 339, 343
Gyrocompass, 393–398
Gyropendulum, 398–403
Gyroscope, 381–406
  integrating, 403–406
  oscillation of, 382–386
    effect of gimbal inertia on, 386–389
    effect of rotor shaft flexibility on, 389–393
  rate, 403–406

Gyroscopic forces, 77
  stability in the presence of, 252–287
Gyroscopic motion, 149, 381–406
Gyroscopic systems, 87, 149
Gyroscopic terms, 77, 87

Half-trajectory:
  negative, 174, 199
  positive, 174, 199
Hamilton-Jacobi equation, 355–361
  modified, 357
Hamiltonian, 94
Hamilton's canonical equations, 91–97
  for holonomic systems, 94, 172, 243, 273
  stability of, 243–252
  for systems: with nonconservative forces, 95
    with nonholonomic constraints, 95
    with nonpotential forces, 95
Hamilton's characteristic function, 360
Hamilton's principal function, 334, 356
Hamilton's principle, 66–72
  application to continuous systems, 70–72
  extended, 68
Heliocentric reference frame, 412
Herpolhode, 149
Hessian matrix, 93, 230, 274
Hill's equation, 278–282
Hill's infinite determinant, 280–282, 285–286
  perturbation technique based on, 309–313
Hohmann orbit, 491
Holonomic constraint, 51, 55, 61
Holonomic system, 51
Homeomorphism, 497, 505

Identity transformation, 343
Ignorable coordinate, 34, 82, 85–88, 249
  (*See also* Cyclic coordinate)
Impact parameter, 38
Impulse:
  angular, 13
  generalized, 80
  linear, 12, 80
Impulsive forces, 79–81

Impulsive moments, 81
Inclination of orbit, 412
Index of Poincaré, 195–197
Inertia, 3, 8
  for rigid body: dyadic, 130
    ellipsoid, 136
    matrix, 129
    moments of, 128
      principal, 136
    products of, 129
    tensor, 129, 135
Inertial integrals, 434
Inertial space or system, 3, 9, 101
Infimum, 177, 234, 243, 501, 502, 504
Infinitesimal contact transformation, 352–354
Infinitesimal stability, 178, 226, 227
Infinitesimal upper bound, 289
Instability, 176
  of linear autonomous systems, 222
Integral curve, 174
  bounded, 175
Integral invariants, 346–349
  absolute, 347
  relative, 347
Integrals:
  of area, 415
  of motion, 86
    (*See also* First integrals of motion)
Integrating operator in method of averaging, 323
Interval:
  closed, 502, 505
  closed-open, 502
  open, 502, 505
Invariable plane of Laplace, 415
Invariable plane of Poinsot, 148
Inverse square law:
  of Coulomb, 37–40
  gravitational, 11, 30–37
Isolated singular point, 175

Jacobi integral, 83, 96, 244, 418
Jacobi-Liouville formula, 214
Jacobian, 180, 336, 348
Jordan canonical form, 181, 219, 266
Jordan curve, 195, 498
Jordan curve theorem, 195, 506
Jump phenomenon, 317–318

Subject Index

Keplerian orbit, 409, 442, 453, 457
Kepler's equation, 409–413
Kepler's planetary laws, 4, 11
  second law, 28
  third law, 36
Kinematics, 1
  Einsteinian, 7
  Lorentzian, 6
  of rigid body, 123–126
Kinetic energy, 14
  of rigid body, 132–134, 137
    rotational, 133, 137
    translational, 133
  of single particle, 14
  of system of particles, 24
Krasovskii's theorem:
  on asymptotic stability, 237, 241
    (*See also* Barbasin and Krasovskii's theorem on asymptotic stability)
  on instability, 237, 241
Kronecker delta:
  complementary, 22, 414
  ordinary, 22, 103, 414

Lagrange brackets, 349–352
Lagrange multipliers, 54, 61, 76, 79
Lagrange's equations, 72–81
  for holonomic systems, 74, 77, 171
  for impulsive forces, 79–81
  for quasi-coordinates, 157–160, 400, 467
  for systems with nonconservative forces, 76, 88–91
    with nonholonomic constraints, 77
    with nonpotential forces, 76
Lagrange's libration points (*see* Libration points of Lagrange)
Lagrange's planetary equations, 446
Lagrange's theorem, 193, 243–244
Lagrangian, 68
Lagrangian configuration space, 172
Lamellar force field, 17
Least action, principle of, 69, 330–334
Legendre transformation, 93, 342
Level curves, 190
Liapunov function, 210, 231
  for linear autonomous systems, 258–259

Liapunov's direct or second method, 209, 226
  for autonomous systems, 231–243, 258–259, 471–474, 477–482
    geometric interpretation of, 239–243
  for nonautonomous systems, 288–292
Liapunov's theorem:
  first instability theorem, 236, 239–241, 290
  first stability theorem, 234, 239–240, 290
  on instability of conservative systems, 193, 245–246
  on reducible systems, 270
  second instability theorem, 236
  second stability theorem, 235, 239–240, 290
  on stability of first-approximation equations, 226–229
Libration, 365
  bounds of, 481
Libration points of Lagrange, 424
  stability of, 425–428
Limit cycle, 198–206
  semistable, 199
  stable, 198, 199
  unstable, 198, 199
Limit point, 200, 503, 504
Limiting set, 200
Lindstedt's method, 299–302
Linear basis, 212
Linear manifold, 211
Linear momentum, 9, 10, 12
  conservation of, 12, 23
  of rigid body, 126–130
  of system of particles, 22
Linear systems, 210–214
  autonomous, 217–224
    stability of, 222–224, 258–259
  fundamental matrix for, 212
  fundamental set for, 212
  general, 210–217
  homogeneous, 210
  linear basis for, 212
  nonhomogeneous, 216, 221
  with periodic coefficients, 264–270
    reducible, 270
    stability of, 268–269

Linear transformation of coordinates, 102–104
Linearized equations, 180
Liouville's theorem, 338, 349
  extended, 343
Lipschitz condition, 174, 178
Lipschitz constant, 174
Lipschitz function, 174
Long-period perturbation of satellite orbit, 461, 462
Lorentz transformation, 5–6, 8
Lunisolar precession (*see* Precession of the equinoxes)

Many-body problem, 413–416
Mappings, 497, 500
Mass, 1–3, 10
  relativistic, 8
Mathieu's equation, 278, 282–288, 309–313
  principal instability region, 313
Matrix:
  block diagonal, 219
  companion, 219
  determinant of, 104, 185, 219
  diagonal, 135, 219
  of direction cosines, 103
  exponential of, 216
  fundamental, 212, 264
  Hessian, 230
  identity, or unit, 103
  inertia, 129
  inverse or reciprocal of, 103
  monodromy, 265, 274
  nonsingular, 180, 213, 265
  periodic, 264, 270
  positive definite, 233
  principal, 213
  rotation, 105, 141
  similar, 218
  skew-symmetric or antisymmetric, 108
  stability, 259
  symmetric, 129
  symplectic, 275
  trace of, 185, 219
  transpose of, 103
  triangular, 220
Mean anomaly, 43, 409
Mean motion, 43, 409

Mechanics:
  analytical, 45
  celestial, 11, 408
  classical, 1, 9
  Hamiltonian, 329, 339
  Lagrangian, 46, 329, 339
  Newtonian, 1–2, 9
  quantum, 9
  relativistic, or Einsteinian, 7
  variational approach to, 46
  vectorial, 45
Member of a set, 498
  (*See also* Element of a set)
Mercury, motion of the perihelion, 8, 11–12
Metric spaces, 501–503
Metrical properties, 172
Michelson-Morley experiment, 5
Minkowski space, 7, 8
Modified potential energy, 32, 88, 247, 418, 422
Moment of a force (*see* Torque)
Moment of inertia, 128
  principal, 136
Moment of momentum (*see* Angular momentum)
Moment-free rigid body:
  general case, 147–149
    stability of, 228–229, 238–239
  inertially symmetric, 143–147
Momentum:
  angular, or moment of, 12, 22
    (*See also* Angular momentum)
  generalized, 80
  integral, 82, 85–87, 151
  linear, 9, 10, 12, 22
    (*See also* Linear momentum)
Monodromy matrix, 265, 274
Motion space, 173
Multipliers of Lagrange (*see* Lagrange multipliers)
Multiply periodic motion, 369, 373

Natural frequency, 19, 298, 302
Natural system, 77
Negative definite function, 232, 289
Negative semidefinite function, 232, 289
Neighborhood of a point, 503, 504
  spherical, 503

Subject Index 519

Newton, laws of, 4, 9
 first law, 4, 9
 gravitation law, 4, 11, 12
  (*See also* Inverse square law)
 second law, 9, 11
 third law or law of action and
  reaction, 10, 22
Nodal axis, 140, 412
Node in orbital motion:
 ascending, 412, 438
 longitude of, 412
 descending, 412
Node, as singular point:
 degenerate, 183
 stable, 182, 183, 185
 unstable, 182, 185
Nonautonomous system, 174, 263
 linear with periodic coefficients, 264–270, 309–313
  asymptotically stable, 268
  perturbation solution of, 309–313
  reducible, 270
  stable, 269
  unstable, 269
 nonlinear, 313–327
 weakly, 294
Nonconservative systems, 17, 80
 Hamilton's equations for, 95
 stability of, 247–249
 Lagrange's equations for, 76, 88–91
Nonholonomic constraint, 51, 54
Nonholonomic system, 51, 162–164
Nonhomogeneous system, 216, 221
Noninertial system of reference, 101, 104, 112
Nonlinear:
 spring: hard, 207
  soft, 207, 260
 systems or equations, 170–171, 190, 198, 209, 293, 300
  behavior of, 187, 227
  nonautonomous, 313–327
  perturbation solution of, 294–297, 299–309, 313–327
 weakly, 294
Nonnatural system, 77, 87
 stability theorem for, 247
Nonpotential forces, 17
Norm, Euclidean, 175, 502

Null solution, 176
 (*See also* Trivial solution)
Nutation, 142, 152
 of earth's polar axis, 438–442

Oblate planet, effect of gravitational field of, 457–463
Open domain, 175
Open interval, 502
Open set, 503
 connected, 176
Open sphere, 502
Orbit:
 circular, 33, 36
 determination of, 409–413
 eccentricity of, 33
 elliptic, 33, 35, 409
 equation of, 32
 hyperbolic, 33, 34, 38
 inclination of, 412
 Keplerian, 409, 442, 453, 457
  modification of, 453
 open, 34
 osculating, 444
 parabolic, 33, 34
 of planet, 30–37
 of satellite, 30–37
  perturbation of, 457–466
 transfer, 453–457
 rendezvous, 455
Orbital elements, 41, 442–446, 454
 change of, 453–457
 osculating, 444
 variation of, 442–446
Orbital reference frame, 467
Orbital stability, 177, 272–273
 asymptotic, 273
Ordinary point, 174
 (*See also* Regular point)
Orthogonal axes, 104
Orthogonal transformation, 124
 (*See also* Orthonormal transformation)
Orthonormal transformation, 103, 109, 134, 344
Oscillator, harmonic, 17, 360–361, 370–371
 anisotropic, 379
 natural frequency of, 19
 period of, 19

Osculating elements, 444
Osculating orbit, 444

Path in the configuration space, 67
Pendulous weight, 394
Pendulum:
  double, 78–79
  of Foucault, 101, 116–119
  with moving support, 287–288
    damped, 325–327
  simple, 187–189, 365
  spherical, 84–85, 87–88, 96–97
Pericentron, 33
Perigee, 33
  height, 35
Perihelion, 33
Periodic matrix, 264, 270
Periodic solution, 269, 365
  of Duffing's equation, 313–322
  of Hamilton's equations, 273–277
  of Hill's equation, 279
  of Mathieu's equation, 282–287
  stability of, 271–273
Periodicity conditions, 301, 314
Perturbation, 175, 225, 293, 294
Perturbation method, 293
  averaging method, 322–326
  based on Hamilton-Jacobi theory, 372–377
  based on Hill's determinants, 309–313
  fundamental technique, 294–297
  Krylov-Bogoliubov-Mitropolsky, 302–309
  of Lindstedt, 299–302
Perturbative function (*see* Disturbing function)
Perturbed motion, differential equations of, 175, 225
Pervasive damping, 248
Pfaffian form, 48, 51
Phase plane, 178
Phase portrait, 173
Phase space, 173
Picard-Lindelöf theorem, 216
Poincaré-Bendixson theorem, 199–204
Poincaré stability, 177
  (*See also* Orbital stability)
Poincaré's relative integral invariant, 348
Poinsot's geometric description, 147–149

Point of a set, 501
  boundary, 503, 504
  interior, 503
  isolated, 505
  limit or accumulation, 503, 504
  neighborhood of, 503
Poisson brackets, 349–352
Polhode, 149
Positive definite:
  function, 232, 289
  matrix, 232
Positive semidefinite function, 232, 289
Postfactor, 495
Potential energy, 15, 32, 61
  augmented, 61
  gravitational, 32, 416, 430–438, 460
  modified or equivalent, 32, 88, 247, 418, 422
Potential function, 17
Precession, 142, 152
  direct, 146
  of earth's polar axis, 438–442
  of the equinoxes, 438, 441
  fast, 156
  retrograde, 146
  slow, 156
  steady, 155–156
Prefactor, 494
Principal axes, 134–137
Principal matrix, 213
Principal minor determinants, 233
Principal moments of inertia, 136
Principle:
  of d'Alembert, 65–66
    generalized, 65
  of Hamilton, 66–72
    extended, 68
  of least action, 69, 330–334
  of virtual work, 59–64
Products of inertia, 129
Pseudovector, 110

Quadratic form, homogeneous, 74, 137, 232
Quadric surface, 136
Quasi-coordinates, 139, 157
  Lagrange's equations for, 157–160
Quasi-harmonic system, 298, 301
Quasi-linear system, 294

Subject Index

Rayleigh's dissipation function, 88–91, 95, 248
Reducible system, 270
Reference system or frame, 2, 3
   heliocentric, 412
   inertial or Galilean, 3, 9
   noninertial, 112
   orbital, 467
   rotating, 104–119
      motion relative to, 110–119
Region, 175, 506
   cylindrical, 175
   spherical, 175, 502
Regular point, 174, 179
Relative motion, equations of, 428–430
Relativistic mass, 8
Relativity principle of Galileo, 3
Relativity theory of Einstein:
   general or gravitational, 2, 8, 11
   special, 2, 7
Representative point, 46, 172
Repulsive central force (see Scattering in inverse square field)
Retrograde precession, 146
Rheonomic system, 51
Riemannian geometry, 2
Riemannian space, 8
Rigid body, 122
   angular momentum of, 126–130
   center of mass of, 127
   equations of motion for, 137–138
      arbitrary axes, 160–162
   kinematics of, 123–126
   kinetic energy of, 132–134
      rotational, 133
      translational, 133
   linear momentum of, 123–126
Rocket dynamics, 487–491
Rolling of a coin, 162–164
Rotating reference system, 104–110
Rotation:
   finite, 106
   improper, 110
   infinitesimal, 107
   matrix, 105, 141
   proper, 110
Routh-Hurwitz criterion, 222–224
Routhian, 86, 151, 249
Routh's method for ignoration of coordinates, 83, 85–88

Routh's theorem, 250
Rutherford's scattering formula, 39

Saddle point, 182, 186, 191
Satellite orbit, 30–37
   perturbation of, 457–466
      by atmospheric drag, 463–466
      by gravitational field of oblate planet, 457–463
      long-period, 461, 462
      secular, 461, 462
      short-period, 461, 462
   regression of the nodes, 459
Satellite rotational motion (see Attitude motion of satellites)
Scattering in inverse square field, 37–40
   angle, 38
   fixed center of force, 37
   moving center of force, 40
   Rutherford formula for, 39
   total cross section of, 40
Schuler period, 403
Schuler tuning, 398–403
Scleronomic system, 51
Second method of Liapunov (see Liapunov's direct or second method)
Secular perturbation of satellite orbit, 461, 462
Secular terms, 294, 297–299
Segment (see Interval, closed)
Segment without contact, 200
Self-adjoint system, 215
Self-excited oscillation, 199
Self-sustained oscillation, 199
Semimajor axis of ellipse, 36
Semiminor axis of ellipse, 410
Separable systems, 361–365
Separation of variables, 361, 372
Separatrix, 193
Set(s):
   boundary of, 503, 504
   bounded, 502
   cartesian product or direct product of, 500
   closed, 503, 504
   closure of, 504
   compact, 505
   complement of, 499
   connected, 175, 506

Set(s):
  connected: arcwise, 498
    simply, 498
  dense, 504
  diameter of, 502
  difference of, 499
  disjoint, 499
  distance between, 502
  empty, null, or void, 499
  interior of, 503
  intersection of, 499
  isolated point of, 505
  limit point of, or accumulation point of, 503, 504
  open, 503, 504
  partially ordered, 500
  union of, or sum of, 499
Short-period perturbation of satellite orbit, 461, 462
Significant behavior, 222
  of linear autonomous systems, 222
  of variational equations with periodic coefficients, 272
Similarity transformation, 125, 181, 218, 266
Simple pendulum, 187–189, 365
Simply periodic motion, 366
Simultaneity, 4
  absolute, 7
  relative, 8
Single-axis gyroscope, 403
Singular point, 174
  center, or vortex, 184
  isolated, 175
  node, 182
    degenerate, 183
  nonelementary, 192
  at the origin, 175
  saddle point, 182
  spiral point, or focus, 184
  (See also Equilibrium point)
Sleeping top, 156
  stability analysis by Liapunov's direct method, 251–252
Solidification, principle of, 487
Space, 1, 8
  absolute, 2, 4
  configuration, 46, 172
  Euclidean, 2, 175, 211, 501
  inertial, 3

Space:
  Minkowski, or world, 7, 8
  phase, 173
  Riemannian, 8
Space cone, 144
Sphere:
  center of, 175, 502
  closed, 502
  open, 175, 502
  radius of, 175, 502
  solid, 505
Spherical domain, 232
Spherical pendulum, 84–85, 87–88, 96–97
Spherical region, 175
Spin, 142
Spin-stabilized artillery shell, 255–258
Spin-stabilized satellite, 467, 475–483
Spiral point:
  stable, 184, 186
  unstable, 184, 186
Stability:
  asymptotical, 176
  asymptotically orbital, 177, 273
  global, 176
  infinitesimal, 178, 226, 227
  Lagrange, 178
  Laplace, 178
  in the large, 226
  Liapunov, 176
  of linear autonomous systems, 222–224
  of linear systems with periodic coefficients, 268–269
  orbital, 177, 272–273
  of periodic solutions, 271–277
    general variational equations, 271–272
    variational equations from canonical systems, 273–277
  Poincaré, 177
  temporal, 255
  uniform, 176
Stability matrix, 259
Stationary solutions in three-body problem, 420
Stationary value:
  of a function, 53–55
  of an integral, 55–59
Straight-line solution:
  general three-body problem, 419
  restricted three-body problem, 424

Subject Index 523

Strutt diagram, 286
Subharmonic oscillation, 320
Subset, 498
   proper, 498
Subspace, 47
Successive approximations, method of, 215
Supremum, 234, 243, 501, 502, 504
Surfaces of zero relative velocity:
   rotating satellites, 471
   three-body problem, 423
Sylvester's theorem or criterion, 232–233
Symmetric top (*see* Top)
Symplectic matrix, 275
System:
   acatastatic, 51
   autonomous, 174
   catastatic, 51
   conservative, 15, 16
   holonomic, 51
   natural, 77
   nonautonomous, 174
   nonconservative, 17
   nonholonomic, 51
   nonnatural, 77
   rheonomic, 51
   scleronomic, 51
System of coordinates (*see* Reference system or frame)
System of particles:
   angular momentum of, 23
      apparent, 24
      conservation of, 24
   center of gravity of, 21
   center of mass of, 21
   kinetic energy of, 24
   linear momentum of, 22
      conservation of, 23
System of reference (*see* Reference system or frame)

Temporal stability, 255
Tensor of first rank, 134
Tensor of second rank, cartesian, 135
Three-body problem, 416–428
   general, 416–420
   restricted, 420–425
      elliptical, 425

Three-body problem:
   restricted: stability of equilibrium points, 425–428
Time, 1, 8
   absolute, 2
   as conjugate canonical variable, 357, 360
   local, 2, 6
   of pericentron passage, 409
Top:
   motion of, 149–157
   sleeping (*see* Sleeping top)
Topological invariants, 498
Topological properties, 172
Topological spaces, 504–506
   homeomorphic, 505
Torque:
   due to single force, 13
   due to system of forces, 23
Trajectory in phase space, 174
   closed, 177, 195
   cyclic, 195
Transfer orbit, 453–457
   rendezvous, 455
Transformation:
   affine, 181, 409
   contact or canonical, 334–346
      homogeneous, 338
      infinitesimal, 352–354
   Galilean, 3, 8
   general coordinate, 46
   identity, 343
   Jacobian of, 180, 336, 348
   Legendre, 93, 342
   linear coordinate, 102–104
   Lorentzian, 5–6, 8
   orthogonal, 124, 344
   orthonormal, 103, 109, 134, 344
   point, 340
   set, 499
   similarity, 125, 181
   theory, 334–377
Translation theorem for angular momentum, 130–132
Transversal, 200
Trivial solution, 175, 211
   (*See also* Null solution)
Trojans, as related to three-body problem, 424
True anomaly, 409

Two-body problem, 25–37
  inverse square law, 30–37

Uniform stability, 176
Unperturbed motion, 175, 225

Van der Pol's oscillator, 205–206, 297
Variable-mass systems, 483–493
Variation, 53, 331
Variation in latitude, 146
Variation of orbital elements, 442–446
Variational equations, 209, 225–226
  with constant coefficients, 226–231
    from canonical systems, 273–277
    for three-body problem, 425–427
  general, 225–226
  with periodic coefficients, 271–277
    from canonical systems, 273–277
Variational principle, 59

Vector(s):
  linearly dependent, 211
  linearly independent, 211
  space, 211
  unit, 211
Velocity, 2, 9, 10
  absolute, 10
  of escape, 35
  of light, 4
Vernal equinox, 412
  regression of, 441
Virtual displacements, 53, 56
  reversible, 60
Virtual work, 50, 60
  principle of, 59–64
Vortex (see Center, as singular point)

Work, 14–17
  function, 17
  virtual (see Virtual work)
World space (see Minkowski space)